Engineering Materials

This series provides topical information on innovative, structural and functional materials and composites with applications in optical, electrical, mechanical, civil, aeronautical, medical, bio- and nano-engineering. The individual volumes are complete, comprehensive monographs covering the structure, properties, manufacturing process and applications of these materials. This multidisciplinary series is devoted to professionals, students and all those interested in the latest developments in the Materials Science field, that look for a carefully selected collection of high quality review articles on their respective field of expertise.

Indexed at Compendex (2021) and Scopus (2022)

Teboho Clement Mokhena ·
Mokgaotsa Jonas Mochane ·
Emmanuel Rotimi Sadiku · Suprakas Sinha Ray
Editors

Biomaterials as Green Flame Retardants

Springer

Editors
Teboho Clement Mokhena
Nanotechnology Innovation Centre (NIC)
Advanced Materials Division
Mintek
Randburg, South Africa

Mokgaotsa Jonas Mochane
Department of Chemistry
College of Science, Engineering,
and Technology, Florida Science Campus
University of South Africa
Roodepoort, Johannesburg, South Africa

Emmanuel Rotimi Sadiku
Department of Chemical Metallurgical
and Materials Engineering (Polymer
Technology Division)
Institute of Nano Engineering Research
(INER) Tshwane University of Technology
Pretoria, South Africa

Suprakas Sinha Ray
Centre for Nanostructures and Advanced
Materials
DSI-CSIR Nanotechnology Innovation
Centre, Council for Scientific and Industrial
Research
Pretoria, South Africa

ISSN 1612-1317 ISSN 1868-1212 (electronic)
Engineering Materials
ISBN 978-981-97-6870-7 ISBN 978-981-97-6871-4 (eBook)
https://doi.org/10.1007/978-981-97-6871-4

This Springer imprint is published by the registered company Springer Nature Singapore Pte Ltd.
The registered company address is: 152 Beach Road, #21-01/04 Gateway East, Singapore 189721, Singapore

Preface

This book discusses the emergence of biomaterials as viable and efficient flame retardants (FRs), poised to replace a majority of traditional FRs. Their exploration as flame retardants represents a significant shift toward sustainability in materials science. These biomaterials, derived from abundant and renewable resources, offer an eco-friendly alternative to traditional, often hazardous flame retardants. The research into the application of biomaterials in this context is not only promising but also necessary, as it aligns with the global push for environmentally responsible manufacturing practices. The use of biomaterials such as bio-based polysaccharides, amino acids, phosphates, and phenolic compounds has been extensively studied, revealing their potential to act as effective flame retardants. These materials, often sourced from agricultural by-products, waste streams, or directly from nature, present a dual benefit: reducing reliance on non-renewable resources and addressing waste management issues. The development of biomaterials for flame retardancy involves understanding their decomposition process, flame-retardant mechanisms, and the interaction with the materials they are intended to protect. This knowledge is crucial for optimizing their performance and ensuring safety standards are met. The modification methods for these biomaterials are diverse, including chemical treatments and physical alterations, to enhance their flame-retardant properties. The integration of biomaterials into various products, from packaging materials to construction components, could significantly reduce the fire hazards associated with polymers and other flammable materials. This integration not only improves the safety profile of the resultant products but also contributes to a circular economy, where materials are reused and recycled, minimizing waste and environmental impact. The challenges in this field, however, are not insignificant. Ensuring that biomaterial-based flame retardants meet the stringent performance requirements for different applications is a complex task. Additionally, the scalability of production and the economic viability of these materials must be addressed to facilitate widespread adoption. The book delves into the properties of biomaterials and their applications as FRs, shedding light on the significant strides made in leveraging biomaterials as flame retardants across diverse sectors. It aims to highlight the prospects and impediments associated with the transition to biomaterial-based FRs from conventional ones, with the ultimate objective of fostering a sustainable society.

Randburg, South Africa — Teboho Clement Mokhena
Johannesburg, South Africa — Mokgaotsa Jonas Mochane
Pretoria, South Africa — Emmanuel Rotimi Sadiku
Pretoria, South Africa — Suprakas Sinha Ray

Contents

Biomaterials as Green Flame Retardants

W. Moloto, M. P. Motaung, B. Ntsendwana, T. C. Mokhena, and M. J. Mochane

Abstract Flame-retardants (FR) are essential for improving the combustibility of different materials. However, conventional flame retardants commonly contain toxic and environmentally hazardous substances environment making the development of sustainable and renewable alternatives critical. Lignin, chitosan, and starch are natural polymers derived from plant and seafood sources with significant potential in the development of green flame-retardants. Various mechanisms which includes char formation, gas-phase inhibition, and heat absorption, enable these materials to be used as flame retardants. This chapter focuses on the chemical structures, sources and modification of these bio-based materials, as well as the impact on flame retardancy. The modification of these polymers are being investigated, which demonstrate how incorporating with additives can improve their flame retardancy while eliminating fire dangers, in addition to overcome environmental concerns showed by conventional flame retardants. When modified, they exhibit promising flame-retardant properties, making them ideal candidates for green flame retardant formulations. They are promising candidates for green flame retardant compounds as their composition can be modified to exhibit promising flame-retardant qualities. The chapter further addresses current developments, problems, and future prospective future, underlining the significance of bio-based materials in promoting green flame retardant technology.

Keywords Fire retardant · Bio-based materials · Bio-composite · Lignin · Starch · Thermal stability · Flammability

---

W. Moloto · M. P. Motaung · B. Ntsendwana (✉) · T. C. Mokhena
DSI/National Innovation Centre, Advanced Materials Division, 200 Malibongwe Drive, Private Bag X 3015, MintekJohannesburg, Gauteng, South Africa
e-mail: bulelwan@mintek.co.za

M. J. Mochane
Department of Life Sciences, Central University of Technology, Free State, Bloemfontein, South Africa

Department of Chemistry, College of Science, University of South Africa, Johannesburg, South Africa

T. C. Mokhena et al. (eds.), *Biomaterials as Green Flame Retardants*, Engineering Materials, https://doi.org/10.1007/978-981-97-6871-4_1

1 Introduction

Polymers such as plastics, rubbers and fibers are widely used in the fields of construction, chemical engineering and military. However, most of these polymer materials have the problem of easy combustion. Thus, this lead to the number of major fires is increasing. The application of flame retardant technology in these fields has drawn a lot of interest in recent years [1–3]. A lot researchers are developing and deploying so-called flame-retardants to improve flame resistance of the resultant. Flame-retardants are regarded as substances that are capable of inhibiting the spread of fire by suppressing the chemical reactions that is responsible for burning process or by the formation of a protective layer to prevent fire-related damage [4, 5].

There are certain properties FR materials must possess to be effective in stopping fires, (i) the material must have high thermal stability, (ii) the materials should have hydroxyls, amines, carboxylic acids, that may go under chemical metamorphism, (iii) the material must have a higher charring ability and (iv) elements that have ability to enhance flame retardancy, such as P, N, Si, must be part of the materials [6–8]. The common flame-retardants are the halogenated and non-halogenated. Halogenated FR have gained more attention due to their affordability and effective. However, they contain bromine or sometimes chlorine linked to carbon, bioaccumulation and have some human safety concerns. Although, they have been commonly used, the environmental and toxicity concerns associated with halogenated-based compounds which have resulted with some being discontinued. Flame retardant fabrication is generally based on non-renewable chemical ingredients, such as ammonium polyphosphate, pentaerythritol, melamine etc. [9, 10]. The development of biomass FR has received significant attention due to the excessive consumption of non-renewable resources [11–13].

Green flame retardant materials based on biomass have received a significant of interest due to their potential of environmentally friendly, green, and sustainable development [11, 13, 14]. Polysaccharides derived biomass, such as starch, chitosan and lignocellulose can be used for carbon source due to their rich hydroxyl structure and dehydration catalyzed by acid source. In addition, some proteins and amino acids found in biomass can operate as gas sources during heat decomposition by releasing non-flammable gases, such as ammonia (NH_3) [7, 15, 16]. Carbon sources are becoming more expensive and non-renewable, deploying clean, affordable, and ecological carbon sources have emerged an innovative issue and research focus in the field of flame-retardants. However, nature has presented mankind with a novel solution to this challenges [14, 17].

In combustion of FR, processes such as formation of char on a polymer surface, cooling, dilution, intumescence, and gas phase reaction occur simultaneously. Char is formed from the accumulation of stable carbonaceous residue on the surface. The FR act as a barrier to retard any gaseous products from starting flames and this subsequently act as a protective layer on polymer surface from any heat or air [18, 19]. The process of cooling is referred as a heat sink mechanism which involves the endothermic decomposition of FR to cool the pyrolysis zones on the combustion surfaces. Dilution occurs in the condensed gas phase. When condensed, FR dilutes the polymer, lowering the concentration of decomposition-derived combustible gases. When activated in gas phase, the combustible gases are diluted by breakdown into inert gas phase. Intumescence occurs when FR swells with degrading material, resulting in the formation of porous foamed structures that function as a barrier to pyrolysis product, air, and heat. The gas phase reaction occurs when FR in the vapor phase interacts with free radicals in the flame spread [20, 21].

It is therefore, the development of environmentally friendly flame retardants with novel, efficient, and low toxicity have become vital area for current research. Bio-based material have attracted extensive attention due to their low price, wide source and easy access. They contain phenolics such as lignin and tannin, lipids such as cardanol and vegetable oils, and carbohydrates such as cellulose, starch, chitosan, and alginate [13, 22]. They can be used as carbon sources for intumescent flame retardants because they are cross-linked to carbon through esterification during combustion. Phytic acid and dopamine can also chelate with transition metals, thus showing excellent flame retardancy and smoke suppression. In addition, DNA is a natural three-source integrated intumescent flame retardant (IFR) [23, 24]. Because the pentose, phosphoric acid, and nitrogen base in its structure can be used as carbon source, acid source, and gas source, respectively. Many bio-based materials can be used as flame retardants in a wide range of biological, construction and other applications. Therefore, any development of FR must be accompanied with rigorous evaluation in terms of human health and safety, and also compounds assessed for their environmental friendliness. It is common course that regulations are more enforced in other sectors than others. For examples, aerospace industry has more stringent regulation than in automobile industry. So, it is also important to employ rigorous method to examine flammability quality and fire retardants function of the biomaterials for them to have a wide range of applicability in many sectors [21, 25, 26].

2 Biomaterials

2.1 Lignin

Lignin is one of the most readily available biopolymers and very important component in wood (see Fig. 1). The structure of lignin is one of the most complex with a three-dimensional amorphous polymer made of phenlypropane units based on the tri-monolignols (i.e. p-coumaryl alcohol, sinapyl and coniferyl) [27]. Uses of lignin range from being a source for other compound (i.e. depolymerization of lignin to generate novel chemicals, e.g., acetophenones and phenolic compounds), blending of polymers, as a reactive material in other polymers and resins. Lignin has also been used as functionalization of fire safety of acrylonitrile butadiene styrene (ABS), epoxy resin (EP), polypropylene (PP), polybutylene succinate (PBS), phytic acid

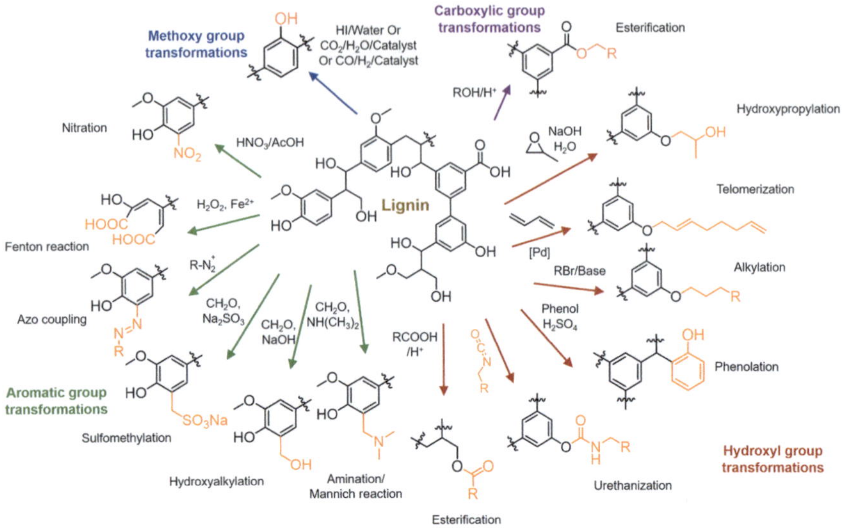

Fig. 1 A summary of chemical transformations to diversify chemical functionality on lignin [29]

(PA) and many others [28]. Three cinnamyl alcohol monomers; coniferyl alcohol, sinapyl alcohol and paracoumaryl make up lignin composition**.**

The attraction of lignin has been brought about high thermal resistivity, mechanical stability and are versatile that they can be used as carbon source for design of intumescent systems in combination with other FR additives. Lignin are important as an additives for polymers. The source of lignin is important as it determined the thermal decomposition. Studies have also shown that lignin can be used as FR and toughening agent when added to polypropylene matrix. In addition, the char residue act as a combustion inhibitor as it act as a protective layer that slows down combustion.in this section lignin different applications in FR will be discussed.

2.2 *Surface Coating Technique*

There are several techniques used in improving bio-flame retardants and lignin based FR are no exception. One of the method is known coating and an example of such as method is layer by layer self-assembly technology (Fig. 2). This technique relies primarily on electrostatic interactions between charge nanoparticles and/or polyelectrolytes [30]. The technique is also allows for variety of donor/acceptor interactions. The sequential process charge substrate alternating either through chemical treatment or via plasma in aqueous environment containing flame retardancy improving components. Typically, the method include three layers or in some cases four layers of flame retardant materials. The coating thickness of FRs is usually ranges between 100 nm (10^{-7}) and 1 μm (10^{-6} m). There are endless chemistries that exist between

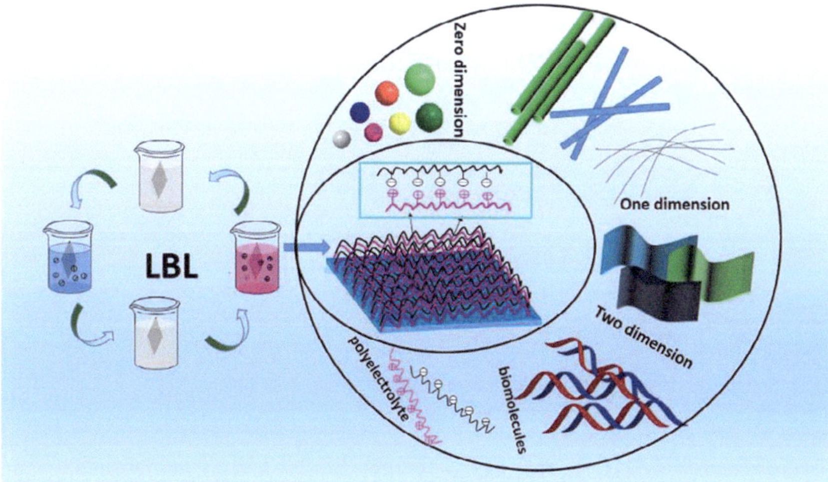

Fig. 2 Schematic of the layer by layer (LBL) process with a wide variety of assembly materials. Reprinted with permission from [32]. Copyright Elsevier

layers but for those used in multilayer fire protective coatings can be categorized by distinctive flame retardant mechanisms employed, for example, passive barriers, or intumescent [31]. Some of the work which employed coating technique are discussed in proceeding section.

Hu et al. [33] fabricated a lignin based FR polyurethane to replace petroleum based polyol with aromatic polyol. In other words, lignin was used as a precursor for synthesis of polyurethane films. Due to high char yield of lignin, the combustion of the polymer was reduced significantly. The lignin based FR was synthesized by reacting lignin with 9, 10-dihydro-9-oxa-10-phos-phaphenanthrene-10-oxide (DOPO) and diphenylmethane diisocyanate (MDI). The FR was fabricated by modifying prepolymer lignin-based polyurethane with lignin-based flame-retardant (LBFR). One of the key finding was that, loading of LBFR onto flame-retardant polyurethane (FLPU) increases the LOI value 29.9% with excellent flame retardant properties with 20% LBFR loading [33]. Furthermore, the LMD loading smoothen the carbon layer surface of FLPU and even more continuous with more intact structure in which the pores become smaller. The high loading of lignin also helps in the formation of denser carbon layer that is compact which as a barrier between gas and phases as a means to block heat and oxygen which subsequently reduce the release of combusting gases. The FR mechanism of FLPU is shown in Fig. 3.

In another study, a fabricated bio-based FR based on industrial lignin chemically grafted with phosphorus (P), copper (Cu) and nitrogen (N) elements as additives of wood-plastic composites (WPC) showed enhanced improved thermal stability and improved flame retardancy of the resulting WPC [34]. WPC is used widely in construction, housewares, etc. The materials used in wood fibre as a reinforcement material and matrix material is comprise of thermoplastics, such as polyethylene,

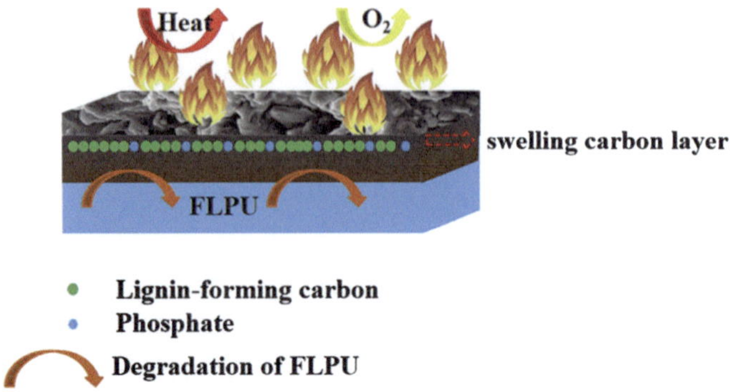

Fig. 3 Schematic flame retardant mechanism of FLPU. Reproduced from [33]

polypropylene, polystyrene, etc., with polyvinyl chloride and polystyrene being most commonly used. Because the reinforcement material have low thermal stability, it is important to find alternative materials to substitute which have higher thermal stability.

The functionalized-lignin (F-lignin) because of presence of Cu^{2+} catalyst on the char formation, and the presence of flame retardancy elements (i.e. P and N) [34]. Unlike the unmodified liginin (O-lignin), the F-lignin reduced heat rate total heat release, and combustion and its smoke during combustion process. At elevated temperatures, lignin gives off high char yield. Studies have used lignin as an additive in polymer-based FR. For example, lignin intercalated montmorillonite (LM) was used as a intumescent FR as a supplement to polybutylene succinate [35]. The LOI value of 36% was attained due to the presence of LM. This results from the synergy between lignin and montmorillonite. The presence of LM did not affect the mechanical properties of polybutylene succinate. The coexistence of montmorillonite and lignin induce an expanded silicate carbon layer in the presence of ammonium polyphosphate (APP). Most importantly, the presences of lignin allows the generation of large amount of inert gas in the combustion stage and the coexistence of intumescent FR and LM improves the efficiency and performance of flame retardant.

Further importance of lignin based FR was demonstrated by Wu et al. [36], in which co-precipitation was used to fabricate lignin-based surfactant sodium lignosulfonate (SLS) modified layered double hydroxide (LDH) FR and they were incorporated into polypropylene (PP) and LDH-LS composites (PP/LDH-LS) using melt bending technique (see Fig. 4). When compared to neat PP, it was found that peak heat release rate (pHRR), total heat release (THR), and total smoke release (TSR) decreased by approximately 63%, 25%, and 43%, respectively. The limiting oxygen index (LOI) value increased from ~18 to 29% with the addition of 20% of LDH-LS. The resultant composites also displayed UL-94 V-0 rating, whereas PP showed no rating. The synergy between LDH and LS, which was able to produce a protective char layer, was attributed with the improved fire resistance performance. During combustion, a thick layer of high-quality carbon successfully prevented the transfer of heat, oxygen, and radicals.

In another study, a novel lignin-based FR for epoxy resin was prepared by a one-step reaction between APP, melamine (MEL) and Al-lignin [37]. TGA analysis showed that the decomposition rate of F-lignin@APP20/EP was 11.8% min^{-1} which was lower than EP and Al-lignin by almost 8% min^{-1} and 4.7% min^{-1}, respectively. It was also found that the char residue of F-lignin@APP20/EP was 32.5wt.%, which is twice that of EP. The LOI value increased with an increase in both Al-lignin and Fe-lignin@APP content. The presence of Fe-lignin@APP resulted in LOI value exceeding 30% and with excellent UL-98 rating (Table 1).

In order to measure the true effect of Al-lignin and F-lignin@APP on FR, the cone calorimetry test were carried out. In which parameters such as HRR, THR and smoke production rate (SPR) were measured (Fig. 5). In Fig. 5 the PHRR of Al-lignin20/EP and F-lignin@APP20/EP is reduced to 571 kW m^{-2} and 385.4 kW m^{-2} respectively,

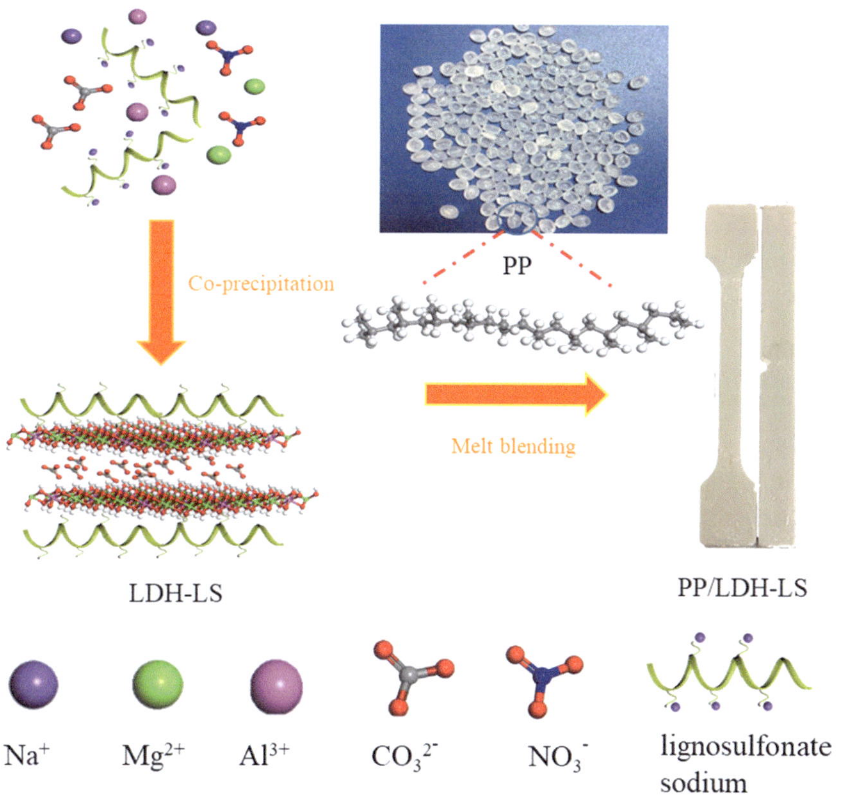

Fig. 4 Polypropylene (PP) and lignin-based surfactant sodiumlignosulfonate (SLS) modified layered double hydroxide (LDH) flame retardant (LDH-LS) composites (PP/LDH-LS) were prepared obtained by melt blending method. Reprinted with permission from [36]. Copyright Elsevier

Table 1 Flame retardant of EP, Al-lignin 20/EP and F-lignin@APP 20/EP. Reproduced with permission from [37]

Sample	EP (wt.%)	DDM (wt.%)	Al-lignin (wt.%)	F-lignin@APP (wt%)	LOI (%)	UL-94 rating
EP	83.3	16.7		–	26.7	–
Al-lignin 15/EP	70.8	14.2	15	–	27.9	–
Al-lignin 20/EP	67.6	13.3	20	–	28.5	–
F-lignin@APP 15/EP	70.8	14.2	–	15	34.2	–
F-lignin@APP 18/EP	68.3	13.7	–	18	34.7	V1
F-lignin@APP 20/EP	67.7	13.3	–	20	36.1	V0

compared to the 703.6 kW m^{-2} of pure EP. This is an indication of higher ability to retard heat hazard when burned. The THR result were consistent with that of PHRR, with F-lignin@APP20/EP was 53.1 kW m^{-2} which is half of 108.9 kW m^{-2} EP. In a nutshell, it is evident that heat release rate of EP occurring during combustion cane be reduced by the presence of F-lignin@APP. Noticeably, the TTI of EP is much higher than that of F-lignin@APP20/EP due to the phosphoric and pyro-phosphoric acids degradation in F-lignin@APP, which can further aid decomposition and induce dehydration of EP [38]. The highest char residue are observed in both Al-lignin and F-lignin compared to EP which in effect suggest potential to reduce exothermic rate.

2.3 *Chemical Functionalization*

Solihat et al. [39] indicated that lignin can be easily be chemically functionalized due to the presence of available chemical reactive sites such as carboxyl, hydroxyl, carbonyl, methoxyl group and others. There are four major chemically functionalized lignin used in FR namely, (i) phosphorus modified, (ii) nitrogen modified lignin, (iii) phosphorus-nitrogen modified lignin, (iv) phosphorus-nitrogen modified lignin-containing metal ions and (iv) combination of lignin with silica-containing flame retardants as shown in Fig. 6 [39]. Studies suggest that the phosphorus and nitrogen addition can improve flame retardancy of lignin by forming char and inert gases. Nitrogen in FR can influence the formation of inert gases (ammonia and or nitrogen) in the flame retarded system. On the other hand, phosphate can facilitate formation of char which can slow down heat transfer and therefore prevent oxygen entry and polymer degradation [34, 40].

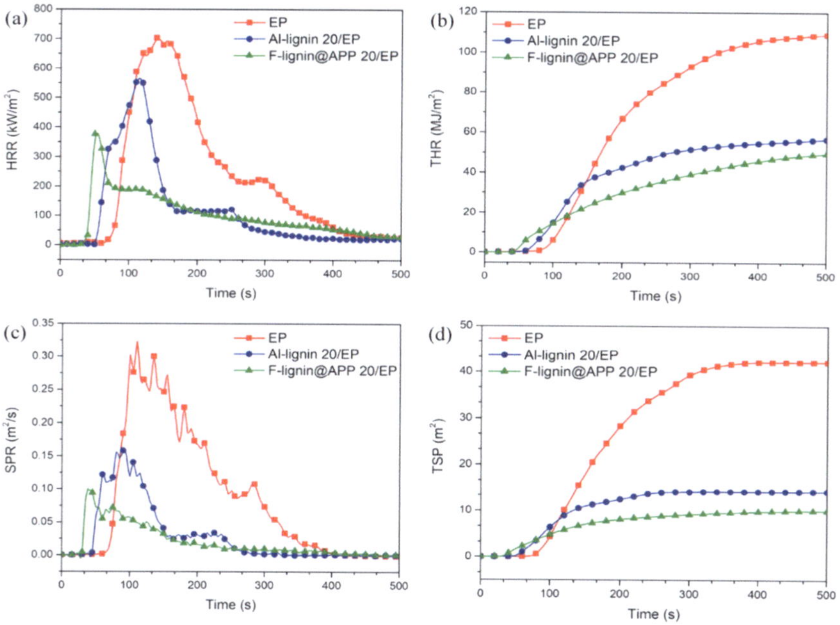

Fig. 5 HRR (**a**), THR (**b**), SPR (**c**) and TSP (**d**) curves of EP, Al-lignin 20/EP and F-lignin@APP 20/EP. Reprinted with permission from [37]. Elsevier copyright

2.3.1 Phosphorus-Nitrogen Modified Lignin

One of the effective way of investigating thermal stability of materials is through TGA. Phosphorus-nitrogen (PN) lignin was used to modify unsaturated polymer (UP). The TGA results confirmed that PN lignin had a lower initial decomposition temperature (170 °C) compared to pure lignin (277 °C) (see Table 2). Lower initial decomposition temperature is important as it can effectively protect the material from heat and flame because of the char residue formed. From Table 2, it can be observed that the burning rate of UP decreased upon the addition of neat lignin by ~29% and functionalized lignin by ~32%. The addition of PN to lignin allowed the flame retardancy to be higher than the neat lignin which suggest the importance of PN-lignin to improve flame retardancy. Similarly, polypropylene flame retardant property was improved with the addition of alkali lignin and addition of P-N. The char residue has improved by almost 50% that is responsible for improved flame retardancy. PP/30PN-lignin composite had lower pHRR and THR when compared to neat PP demonstrating that the addition of PN lignin capable of enhancing flame retardant property. The synergistic effect of melamine phosphate modified lignin (MAP lignin) and rice husk ash (RHA) on mechanical and thermal stability of poly(3-hydroxybutyrate-co-4-hydroxybutyrate) (P34HB) (Table 2). The significant reduction of PHRR and THR of P34HB noticeably by ~43 and ~24% was recorded, respectively (Table 2). MAP

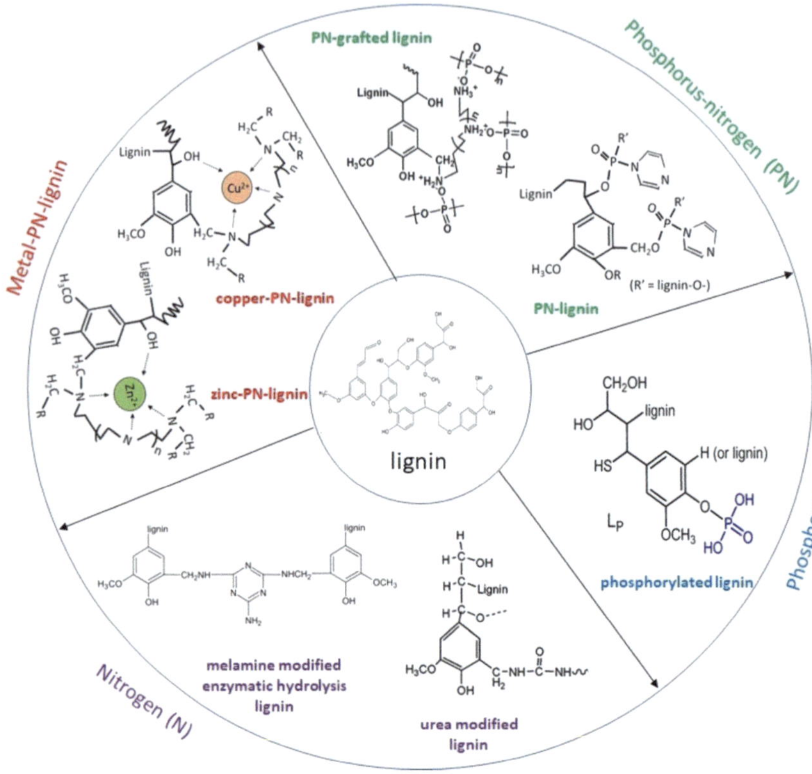

Fig. 6 Different chemical modified lignin [39]

lignin also suppress CO and CO_2 production rate. Barrier effect caused by compact char residue with intact structure enhanced flame retardancy.

2.3.2 Nitrogen Modified Lignin

Phosphorus-Nitrogen Modified Lignin-Containing Metal Ions

There are several ways in which lignin can be functionalized, that include modification of its aromatic rings, or through hydroxyl groups or even via modification of methoxy groups [28]. Unmodified lignin are variable and have low chemical reactivity and when unmodified can also produce brittle foams with soft and hard segments, which weakens the mechanical properties of the material being modified (e.g. weakened polyurethanes) [28].

Sodium lignosulfate (SLS) has been used as lignin flame retardant material where ammonium polyphosphate was decomposed during combustion process to produce

Table 2 Thermal stability and cone calorimeter measurements of composites added with phosphorus-nitrogen modified lignin (PN-lignin)

Composite	T_i (°C)	T_{max} (°C)	Char residue	T_{ign}	PHRR (kW m^{-2})	THR (MJ/m^2)	Burning rate (mm^3/second)	References
PP	324	418	0	49	1350	87.3	–	[40]
PP/lignin	316	472	12.5	34	382/333	76	–	[40]
PP/20PN-lignin	325	479	12.4	38	380	74.2	–	[40]
PP/30PN-lignin	361	483	21	39	360	69.7	–	[40]
P34HB	–	301/353	3.5	55	526	64.5	–	[41]
P34HB-10A-lignin	–	250/320	15.2	35	432	54.6	–	
P34HB-20A-lignin	–	300/348	14.9	54	348	57.8	–	[41]
P34HB-30A-lignin	–	300/342	20.1	68	328	53.4	–	[41]
P34HB/30MAP-lignin	–	299/340	26.3	83	301	45.8	–	[41]
UP	277	411	–	–	–	–	64.18	[42]
UP/neat lignin	175	430	–	–	–	–	49.50	[42]
UP/PN lignin	170	445	–	–	–	–	43.66	[42]

non-combusting gas, in which oxygen content in the system is diluted and combustion is slowed down [36]. SLS is important as it can produce large amount of non-combusting gas and prevent combustion and it can also promote dehydration and decarboxylation for large carbon residue due to presence of sodium [36].

2.4 *Lignin Composites*

One of the effective way of improving the performance of lignin-based flame retardants is through the formation of composites with nanomaterials. One of the important nanomaterials are carbon nanotubes (CNTs) with higher thermal stability and higher thermal conductivity. Mass loss rate and peak heat release rate of FR can be decreased with the addition of CNTs. The use of polymer enriched CNTs for FR have been explored. For example, carbon nanotubes have been used to form composite with lignin and potassium carbonate coated on roving [43]. It is widely know at this point that lignin it is itself a natural flame retardant agent with its high char yield. However, the use of nanomaterials such as carbon nanotubes (CNTs) not only strengths the structure of FR but also act as a catalyst for the formation of char.

The addition of CNTs also promotes uniform distribution of lignin in the cellulose matrix [43]. The study in fact revealed that the presence of CNTs increased the thermal stability according to their observed thermogravimetric analysis (Fig. 7).

The DTG curve upon addition of high content of CNTs there was no decomposition at 360 °C, which is a curve in which most of the flammable gases get released. Furthermore, only 29% of L/CNTs decomposes at 600 °C. One other observation that improved the decomposition of the composite, it is the uniform distribution of CNTs on to the lignin matrix as showed by TEM images (Fig. 7). The increase in rigidity of lignin is effected after mixing with CNTs. This is due to the aqueous suspension of CNTs wrapped by lignin oligomers and interaction with carbon rings through pi-pi stacking [44].

The LOI results on the other hand, suggested that pure lignin coating can increase the parameter by 26.8% which was far more than raw cotton (at ~17%) due to better char formation and ability to form a inhibiting barrier between volatile gases released during combustion and oxygen (Table 3). Noticeably, the higher content loading of CNTs when used during the dispersion increases the fire retardant resulting in LOI reaching ~32% for C/L/CNT. The improvement in LOI was due to the enhancement of lignin-char gas barrier formation effected by formation of mechanical network formed by CNTs and also the acceleration of char form of lignin induced by higher thermal conductivity [43].

Research has also should that composites of lignin allow for synergistic effect between materials when deployed in flame retardant application. For instance, the combination of polyphosphate and lignin allows polyphosphate to act as a dual-gas

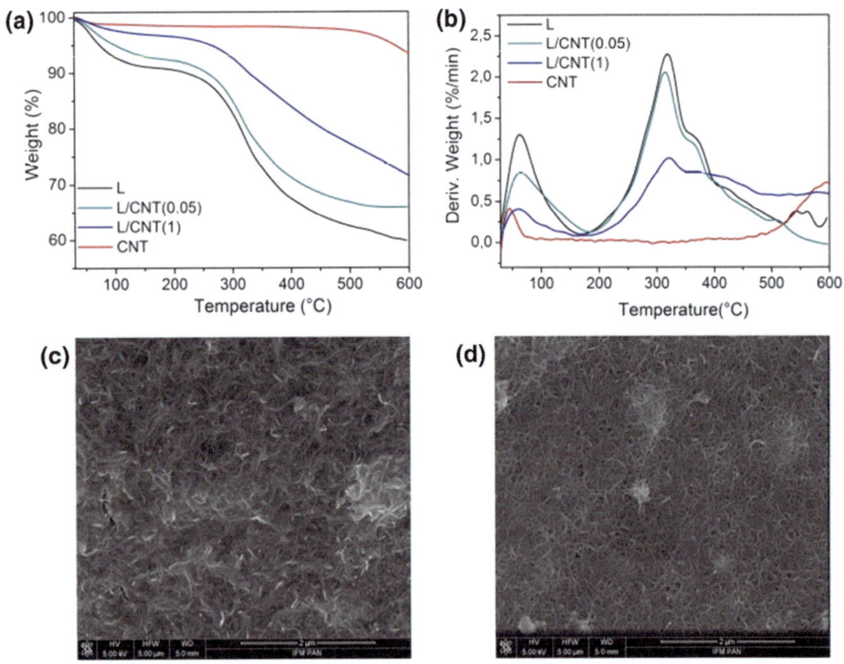

Fig. 7 TG (**a**) and DTG (**b**) curves of CNTs, L and L/CNT composites; SEM images of: **c** L/CNT (0.05) and **d** L/CNT [44]

Table 3 Limiting oxygen index of investigated samples [43]

Sample	LOI
C	17.1
C/L	26.8
C/L/CNT(0.05)	25.0
C/L/CNT(0.3)	27.5
C/L/CNT(1)	31.7
C/L/K_2CO_3	32.5
C/L/CNT(1)/K_2CO_3	38.5

source and acid source. This is important because when lignin operate as carbon source, inorganic acid from polyphosphate allows crosslinking, expansion and foam of the system. Specifically, the lignin causes system to expand and foam due to the dehydration and its carbonization and the non-combusting gas released by polyphosphate also responsible for the forming and expansion of the system [39]. In the process, intumescent is formed by combination of two chemicals with synergistic effect. In a similar example, physical blending of kraft lignin and APP showed a reduction of pHRR of polyamide-11 by 66% relative to pure polyamide-11.

The LOI reached 26.3% due to the synergistic effect of lignin and APP in the composite [45]. Composites of lignin are therefore important as demonstrated by many studies and thus important in improving FR efficient and effectiveness in averting fires. The formation of composite between PP, TiO_2 and lignin (i.e. PP-TiO_2/lignin) combined by melt mixing extrusion have supported this claim. In which cone calorimetry tests showed that the pHRR and the THR reduced by ~34% and ~36%, respectively, induced by the addition of 25 wt.% TiO_2-lignin to PP [46]. Many factors are responsible for this improvement, but the SEM images (Fig. 8) suggested that improved dispersion due to the interaction of TiO_2 and lignin in the PP matrix during extrusion process, may have also had a huge impact [46].

3 Chitosan as Flame Retardant

Before considering chitosan in FR application, one must first consider its thermal behavior. Accordingly, thermal gravimetric analysis has revealed two important decomposition temperatures, i.e. one occurring between 30 and 110 °C due to evaporation of moisture. The second decomposition temperature occurs at 200 and 340 °C attributed to decomposition of polymer chains and the overall weight loss is not more than 50%. It is for this reason that chitosan is even considered for designing FR. In many of its application in FR, chitosan is mainly used as a chemical modification and not as a substrate. Chitosan and its derivatives are shown in Fig. 9.

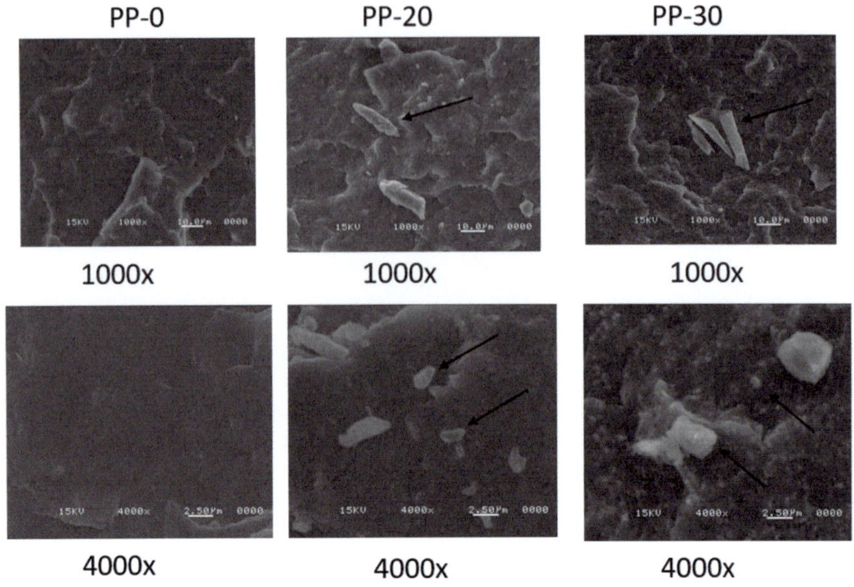

Fig. 8 Scanning electron micrographs of PP-0, PP-20, and PP-30 at 1000 and 4000 [46]

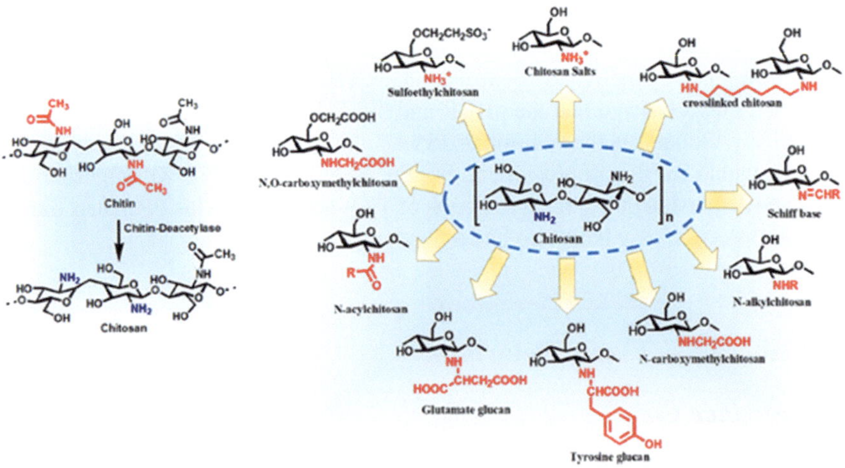

Fig. 9 Structures of functionalized chitosan derivatives and its synthesis from chitin [47]

The combination of eco-friendly biocomposites combining thermoplastic starch, flax fabrics and chitosan have been used for FR application. The thermoplastic starch-flax fabric composite were fabricated using compression technique and the detailed method was reported in the work by Kim and colleagues [48]. In the study, the loading of chitosan effect on the biocomposites was investigated. In particular, the 6

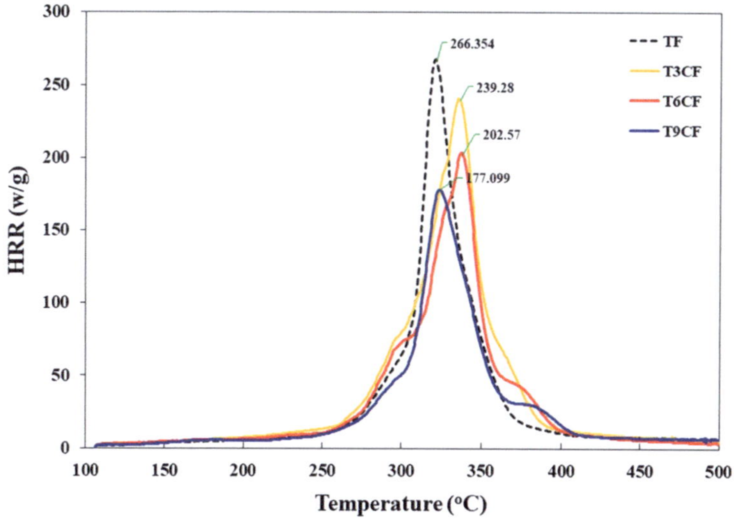

Fig. 10 Heat Release Rate (HRR) versus time curves for thermoplastic starch-flax fabric (TF) and thermoplastic starch-chitosan-flax fabric (TXCF, where X indicates the weight % of chitosan). Reprinted with permission from [49]

wt. % of chitosan reached a self-extinction in horizontal flame tests, UL94-V-0 rating and limiting oxygen index above 45% versus 21% of thermoplastic starch (Fig. 10). Interestingly, the HRR was reduced with chitosan loading by 10%, 24%, and 34% in the composites containing 3, 6 and 9 wt. % biomacromolecule [49]. There was also a higher charring effect due to the presence of chitosan. Chitosan did not only improve the FR capability but also improved the mechanical behavior of the composite, the modulus and also the tensile strength [49, 50]. In fact in Table 4, the study revealed that the effect of chitosan incorporation into the biocomposite matrix enhanced the tensile strength due to the interaction between chitosan and thermoplastic starch see Table 4 [49].

3.1 Surface Coating Technique

A layer by layer self-assemble (LBL) technology has been used to modify flame retardant materials. For example, a quad-layer coating of chitosan sepiolite-chitosan-APP was by LBL on fabrics. The presence of the layer induced improvement of flame retardant of the fabrics. This gave rise to ~25% LOI value of fabric and a subsequent decrease in pHRR value by over 32%. The LBL increased carbonization of fiber in the combustion process, giving rise to stable carbon layer. The fiber flame retardancy improved significantly as a results of quad-layer coating [51]. Chitosan due to its excellent cationic properties has been best suited as surface functional additive for

Table 4 Tensile test results of TPS composites

Composite	Tensile strength (MPa)	Tensile modulus (GPa)	Elongation break (%)
Thermoplastic-starch (TPS)	2.92	0.16	11.41
Thermoplastic flax fabric (TF)	16.45	0.60	5.59
Thermoplastic starch-chitosan-flax fabric (3 wt.%) (T3CF)	17.64	0.63	8.035
Thermoplastic starch-chitosan-flax fabric (6 wt.%) (T6CF)	24.03	0.64	16.228
Thermoplastic starch-chitosan-flax fabric (9 wt.%) (T9CF)	20.78	0.66	7.776

Reproduced with permission from [49]

FR and the LBL method is best suited for coating it to polymer surface. Some of the best example include the LBL assembly coating of chitosan and phosphorylated cellulose nanofibers on polyurethane (PU) form. The study showed a decrease in pHRR by over 30% compared to unmodified PU.

The presence of chitosan improved the flame retardance of PU composite and any droplet phenomenon [52]. Carbon can also be catalyzed by chitosan/phytic acid coating using LBL technology in the condensed phase which allows the protection of the matrix [27]. In fact Fang et al. have demonstrated this by using LBL method for chitosan/phytic acid (chitosan/PA) bio-based intumescent flame retardants material coatings on polyester/cotton blended fabrichisan [25]. The study found that the bio-based FR allows polyester cotton blended fabric to have increased flame retardant properties and therefore retarding any dripping that may occur. In another study, polyethylene glycol terephthalate (PGET) fabric improved its flame retardancy when it was coated with chitosan and ammonium polyphosphate by LBL assembly method [53]. Flame retardancy of PGET increased due to the LOI increase from 21.1 to 26.2%, which illustrated the effect of chitosan and ammonium polyphosphate [53].

Thermoplastic polyurethane was incorporated with cross-linked chitosan with bis-4-formylphenyl-phenyl-phosphonate together combined with ammonium polyphosphate and organo-modified montmorillonite (i.e. cloisite 30B) (Fig. 11). The composite showed self-extinction in vertical flame spread tests at 10 wt. % FR loading. Similarly, there was high increase in limiting oxygen index (29.0% versus 20.8 for filled and unfilled thermoplastic polyurethane, respectively). This materials also induced a reduced THR by ~34%, whilst the PHRR reduced by about 74%, and TSR by ~88%. The concurrent catalytic carbonization of nanoclay and dehydration reaction reactions promoted by phosphonic groups arising from phosphonate conversion in the cross-linked chitosan [54].

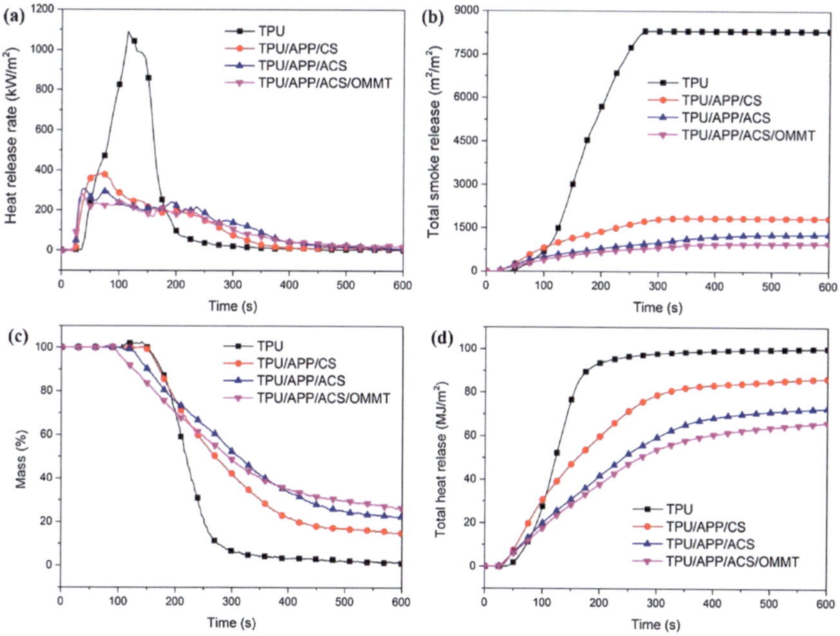

Fig. 11 Heat release rate (HRR) (**a**), Total heat release (THR) (**b**), Mass loss (**c**), and Total smoke release (TSR) (**d**) curves of thermoplastic polyphosphate (TPU); thermoplastic polyurethane containing 5 wt.% of ammonium polyphosphate and 5 wt.% of chitosan (TPU/APP/CS); thermoplastic polyurethane containing 5 wt.% of ammonium polyphosphate and 5 wt.% of chitosan cross-linked with bis-(4-formylphenyl)-phenyl-phosphonate (TPU/APP/ACS); thermoplastic polyurethane containing 4.5 wt.% of ammoniumpolyphosphate, 4.5 wt.% of chitosan crosslinked with bis-(4-formylphenyl)-phenyl-phosphonate and 1 wt.% of Cloisite 30B, an organo-modified montmorillonite (TPU/APP/ACS/OMMT). Reprinted with permission from [54]. Copyright Elsevier

LBL coating technique was also used in phytic acid (PA), chitosan (CHI), and biochar (BC) for its flame retardancy to cotton fabric (see Table 4). The TGA analysis have showed the effect of treating fabric with CHI, PA, and BC. There was a shift of T_{onset} and T_{max} to lower temperature with the addition of biomolecules. The decrease of T_{max} was by 90 °C, 81 °C, 55 °C, 92 °C, 111 °C, and 92 °C for the PA-COT, PA/CH–COT, PA/BC–COT, PA/CH/BC (5%)-COT, PA/CH/BC (7.5%)-COT, PA/CH/BC (10%)-COT, respectively, compared to untreated cotton fabric. Furthermore, the char residue increased by 23.8 wt.%, 27.6 wt.%, 12.1 wt.%, 39.3 wt.%, 40.5 wt.%, and 35.2 wt.%, respectively. PA/CH/BC (7.5%)-COT exhibited lower gas release temperature with minimal hydrocarbon gas release. CHI and BC synergistic effect improves the flame retardancy of PA flame retardants. More so, it decomposes at lower temperatures and produce large amount of carbon and therefore improves the thermal stability of the cotton fabric. The LOI of the composites was found to range between 26.0 and 28.0% [55] and that of untreated cotton fabric was far below the standard at 18.6% the noticeable increase above the standard was seen

Table 5 Formulations and UL-94 ratings and corresponding composites [59]

Sample	LLPDE%	MCHP%	OMMT%	UL-94 rating
LLPDE	100			No rating
PE-1	75	25		V-2
PE-2	74	25	1	V-2
PE-3	70	30		V-1
PE-4	60	30	1	V-0

after treatment with BC, CHI, and PA, with improvement of LOI in PA/CHI–COT and PA/BC–COT. The highest increase was observed when cotton fabric was treated with 10%BC reaching LOI value of 66.8%.

In a similar study, melamine phosphate was added to nano chitosan (nCH) and polyvinyl alcohol (PVA) solution for the fabrication of FR for coatings for cotton fabric. Furthermore, melamine salt of chitosan phosphate (MCHP) was prepared with mixed PVA and also used to coat cotton fabric [56]. The LOI values were enhanced after the addition of PVA/MCHP, PVA/nCH/MP, MCHP30/alone and nCH30/alone coatings to cotton fabrics (see Table 5). The increase in loading of chitosan increased LOI of the control by 9.2%, 16.6% and 40.7% in nCH10, nCH20 and nCH30. Furthermore, MCHP30/alone and nCH30/ alone samples increased LOI value of cotton by 24.3% and 21.2%, respectively.

3.2 Flame Retardant Mechanism of MCHP and nCH/MP

In terms of the mechanism, the control sample based on Fig. 12 below, shows that the flame propagate from lower end to the upper end and there is no char residue. On the other hand, Fig. 12b–g showed the PVA/MCHP, PVA/nCH/MP and PVA/MP coatings were successful in retarding flame and thereby saving cotton fabric. nCH/MP showed discontinuous char layer on the fabric surface and this behavior is also observed in MP30 sample. In contrast, MCHP form an intact and coherent char layer on cotton fabric surface [56].

Another study used padding drying method to coat crayfish-based chitosan on cotton fabric [57]. The study also investigated the combination of nano-TiO_2 and chitosan as expected, the TGA test showed that the thermal resistance and flame retardancy performance of denim fabrics. The nano-TiO_2 was responsible for increase in thermal resistance of blue dyed denim and also reduce mass loss. The coating of fabric with chitosan and nano-TiO_2 improved the post flame burning performance of the fabrics and also decrease the burning time. Unlike the uncoated fabrics which burn rapidly, the coated fabrics slows down the flame propagation [57].

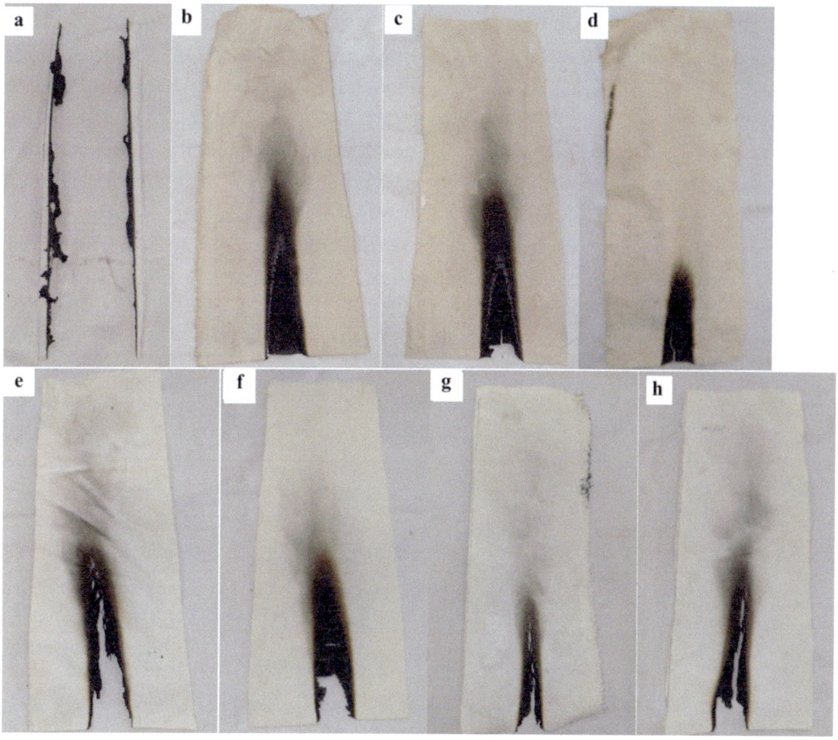

Fig. 12 Digital photographs for **a** control, **b** MCHP10, **c** MCHP20, **d** MCHP30, **e** nCH10, **f** nCH20, **g** nCH30 and **h** MP30 samples after vertical flame spread test. Reprinted with permission from [56]. Copyright Elsevier

3.3 Chemical Modification of Chitosan

As abovementioned, chemical modification of biopolymers can enhance the properties of flame retardants. Commonly, nitrogen and phosphorus are used together and this is because of their synergistic effect in which when nitrogen promote the production of inert gases whereas phosphorus improves the char density formation. Melamine is a good source of nitrogen and sodium pyrophosphate is a good source for phosphorus. The effect of P-N on chitosan were demonstrated by when chitosan was used as a multifunctional chemical agents and glyoxal was used as crosslinking agent.

The LOI value is directly related to flammability of the product and therefore is must be greater than 26%. For instance, the LOI value of cotton fabric is ~18% and thus it is flammable material. Patankar et al. showed that treating the fabric with chitosan can improve its thermal resistance performance [58]. Noticeably, the thermal stability and thereby increase in LOI was attributed by the P-N synergism. In Fig. 13a, the LOI of treated fabric with chitosan is directly proportional to the concentration.

When the concentration of chitosan is 200 g L^{-1} the cotton fabric has the highest LOI value of 32%, and thus signaling a good flame retardancy. Vertical flammability test are important to investigate the combustion mechanism of the materials. In Fig. 13b, the untreated cotton fabric shows a total degradation without any char residue and flame rate of 13 s. It can been seen that the increase in concentration of modified chitosan resulted in char length of 3.9 cm after a flame time of 6.7 s and after a glow time of 30.1 s at 200 g L^{-1} of modified chitosan.

The impact of P-N synergism is confirmed by the improved thermal properties of cotton fabric. Hassan and others, did similar work, to demonstrate the synergistic

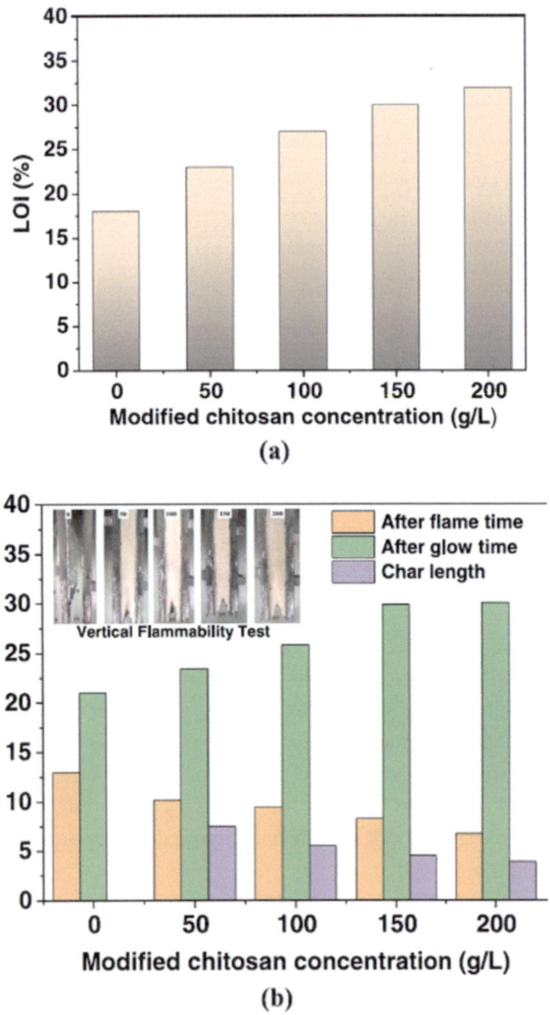

Fig. 13 Effect of concentration of modified chitosan on **a** limiting oxygen index and **b** vertical flammability (VF) test of cotton fabric. Reprinted with permission from [58]. Copyright Elsevier

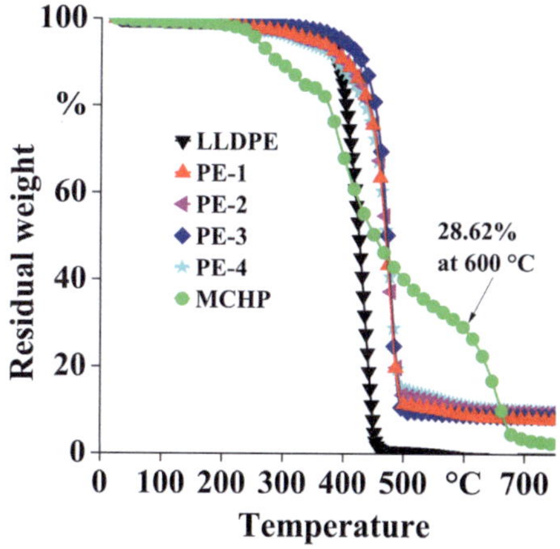

Fig. 14 TGA curves of MCHP, LLDPE and its composites. Reprinted with permission from [59]

effect of melamine salt of chitosan phosphate (MCHP) and organic modified montmorillonite (OMMT) on low linear polyethylene (LLDPE) [59]. The role of phosphorus in this study, was to activate chain stripping, crosslinking, and char residue formation on the LLPDE surface. The introduction of MCHP and MCHP/OMMT raised the T_{max} of LLPDE by 30–35 °C. The addition of the MCHP and MCHP/OMMT also increase the char density at 750 °C (see Fig. 14). UL-94 tests have showed LLDPE is flammable and thus had no rating in the vertical burning test. V-2 rating was attained for PE-1 and PE-2, however, the addition of MCHP (30 wt.%) (PE-3) and 1 wt.% allowed LLPDE to reach V-1 rating (Table 5). Furthermore, the mixture of 30 wt% of MCHP and 1 wt% OMMT (PE4) allows LLPDE to even reach V-0 rating. This indicate the importance of chemical modification as it allows for greener and more effective flame retardants.

4 Starch as Flame Retardant

Starch is a natural polymer produced by plants; similarly to other bio-materials (e.g., lignocellulose and chitosan), it is abundance, and biodegradable [60]. Starch is a polysaccharide composed of D-glucose units with $\alpha(1 \rightarrow 4)$ glyosidic linkages in linear amylose and $\alpha(1 \rightarrow 4)$ linkages and circa 5% $\alpha(1 \rightarrow 6)$ glyosidic branch linkages in amylopectin [61, 62]. Starch can be utilized as a flame retardant due to its ability to form a protective layer when exposed to heat and flames. Starch-based flame retardants are commonly used in textiles, paper products, and some polymers [63, 64]. Starch can be use in various applications in food productions, pharmaceuticals, water treatment and biotechnology [12, 63, 65, 66]. Starch is a valuable compound that has flame retardant characteristics. Starch consists of essential functional groups that it afford its application as FR for a number of substances to make them less susceptible to flame and to limit the spread of flames in the event of a fire [3, 46,

48]. Starch produced from plants such as corn, wheat, or potatoes has the ability to act as a FR due to its distinctive character [50–53]. One of the key properties of starch that makes it an excellent flame retardant when exposed to heat is its ability to form a protective layer [60, 65]. Starch can gelatinizes, absorbs water and expands when subjected to heat. This process forms a thick, burnt coating on the surface of the protected material, acting as a barrier between the material and the environment [67, 68].

Pigolwska et al. [69] demonstrate the properties of corn starch flame retardant that was used as an additive to the conventional electrolytes in lithium-ion batteries. They tested the maize starch flame retardant property and discovered that 5% of starch was the optimal level to achieve desired properties. It was reported that the presence of starch the as-prepared electrolyte exhibited visible flame for 30 s and ceased after 50 s [69]. In addition Shao et al. [60] used starch and other bio-materials (such as distarch phosphate (DSP) and hydroxypropyl distarch phosphate (HPDSP)) as "green" carbon sources with ammonium polyphosphate (APP) to prepare flame retardant reconstituted tobacco sheet (RTS) by paper-making process. The effect of these starch-based flame retardants on their thermal degradation and combustion properties was preliminarily investigated. Micro-scale combustion calorimetry (MCC) results showed the value of the heat release of modified RTS was lower than that of pure RTS, demonstrating the restrained combustion behavior of modified RTS. Their work could highly expand the application of starch and its derivatives in flame retardant and tobacco industry [60]. Because starch has a high water content, it works effectively as a flame retardant. It has the ability to both absorb flame heat and create an inhibitor to stop further combustion [70].

Starch also creates a carbonaceous char that acts as an insulator and keeps the flames from expanding [25, 53, 67]. The mechanism of starch flame retardant occurs in three stages as shown in Fig. 15. When starch is heated to high temperatures, it undergoes a chemical transition that results in a charred, carbonaceous residue.

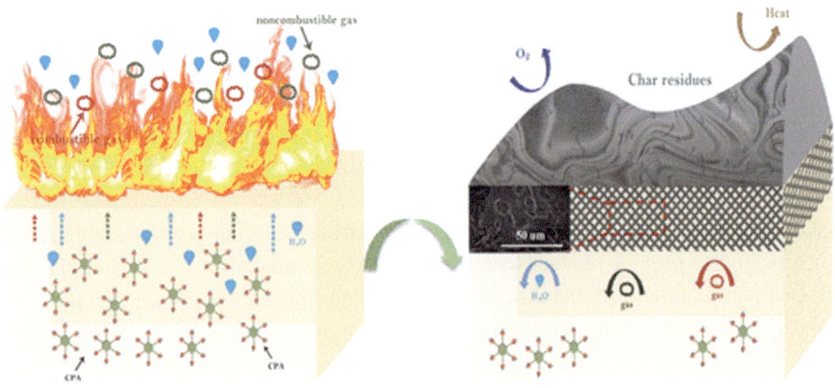

Fig. 15 Schematic illustrations of the flame retardant mechanism of starch composites [71]

(i) This char coating works as a protective barrier, shielding the underlying material from heat and flames and therefore delaying combustion.
(ii) Heat transfer reduction: by absorbing some of the heat energy, starch-based flame retardants can limit the passage of heat from the flame to the material. This contributes to the material not reaching its ignition temperature.
(iii) Release of fire-suppressing gases: when exposed to high temperatures, several starch-based compounds emit fire-suppressing gases. These gases have the potential to dilute the combustible gases created during burning, so inhibiting the combustion process.

While starch can be useful as a flame retardant in some applications, it may not be appropriate for other materials or fire conditions. The efficiency of starch-based flame retardants is affected by a variety of factors, including the type of material being treated, starch formulation employed. Furthermore, there are concerns about the environmental and health effects of certain chemical flame retardants, researchers are always attempting to find safer and more sustainable alternatives, which may include bio-based materials such as starch [72, 73].

4.1 Modification of Starch

Flame retardants are chemicals or materials that are added to other materials to reduce their flammability and slow down the spread of fires. In general, natural starch can be transformed into thermoplastic starch in the presence of a plasticizer (such as glycerol). This distinctive feature makes it a viable alternative in the food, textile, paper, and other industries. Starch-based flame retardants are typically used in textiles, paper products, and some plastics. Starch can be chemically modified to make it more fire resistant. This usually entails adding flame retardant chemicals or additions to the starch [13, 74, 75]. Starch has been modified using a variety of physical and chemical processes to make it an efficient fire retardant. Chemical modification procedures for starch include cross-linking, grafting polymerization, oxidation, etherification, and esterification [60, 76].

4.1.1 Cross-Linking Modification

Cross-linking or cross-ligation is the joining of linear or branched polymer chains. In the cross-linking procedure of the starch molecules, the chemicals react with the hydroxyl groups, forming ether or ester linkages [77]. Cross-linking has previously been used to improve the thermal stability and mechanical strength of starch films. Cross-linking starch films with various chemical agents has greatly reduced swelling force and increased water resistance. Additionally, cross-linking improves starch film stability during freezing and melting [71, 76]. Furthermore, cross-linked starch generates a char coating when exposed to heat, which can act as a barrier against further combustion [60, 65]. Starch can be cross-linked with phosphate to form chemical modified distarch phosphate (DSP), hydroxypropyl distarch phosphate (HPDSP) to improve its solubility and swelling capacity. Moreover, starch and its derivatives (e.g., DSP, HPDSP etc.) have the potential been used as carbonization agent due to their structure of multi-hydroxyl group. On the other hand, ammonium polyphosphate (APP) has been widely used as acid source to yield polyphosphoric acid on heating and cooperate with carbon source to form an intumescent flame retardant system promoting char formation. Therefore, APP, ST and its derivatives have the potential to work as IFR synergistically in application [60].

Recently, cross-linked esterified starch was prepared using a one-step method, where maize starch was selected as the raw material, sodium trimetaphosphate as the cross-linking agent, and acetic anhydride as the esterifying agent, respectively. In addition, CES displayed increased shear and acid tolerance compared to the original waxy maize, and their lowest differences between waxy maize and CES were only 0.62% and 0.59%, respectively. In summary, a novel method for starch modification was provided [78].

4.1.2 Grafting Polymerization

Grafting polymerization is a method frequently used to modify the characteristics of polymers, such as starch, to add new characteristics or improve those that are currently present. Grafting polymerization is an effective method to produce starch fire-resistant. There are many industries where fire-retardant starch can be used, including textiles, paper, and packaging materials. It is significant to note that the particular grafting polymerization environments and monomers may differ based on the desired fire-retardant characteristics and the expected use for the modified starch. Effective monomers (phosphorus-containing substances, nitrogen-containing substances, and halogen-containing substances) that can be grafted onto the starch structure to provide flame resistance [1, 13, 74]. Xia et al. [79] reported modification of starch-based flame retardants through surface hydrophobicity to improve the interfacial compatibility with polypropylene (PP) (as shown in Fig. 16). PP composites exhibit total flame retardant concentration of 30 phr with limiting oxygen index (LOI) values as high as 33.5% with a UL-94 V-0. The synergistic effect of starch-based flame retardants and expandable graphite (EG) shows PP with improved flame retardancy and anti-dripping ability, which can be ascribed to the combined chemical-physical intumescence action in increasing charring layer quality. Furthermore, char residue analysis indicated the synergistic flame retardant mechanism [79].

The preparation of flame retardants based on starch has been explored. Although starch powder is flammable and starch gel displays poor flame retardancy, starch has an abundance of hydroxyl groups with high reactivity, and thus starch-based flame-retardant can easily be grafted onto the polymeric structure. For example, starch can be esterified using a large amount of phosphoric acid without formaldehyde to attach multiple phosphoric acid groups and then converted to ammonium salts to realize flame retardancy. Further, can be modified by the presence of metal ions also

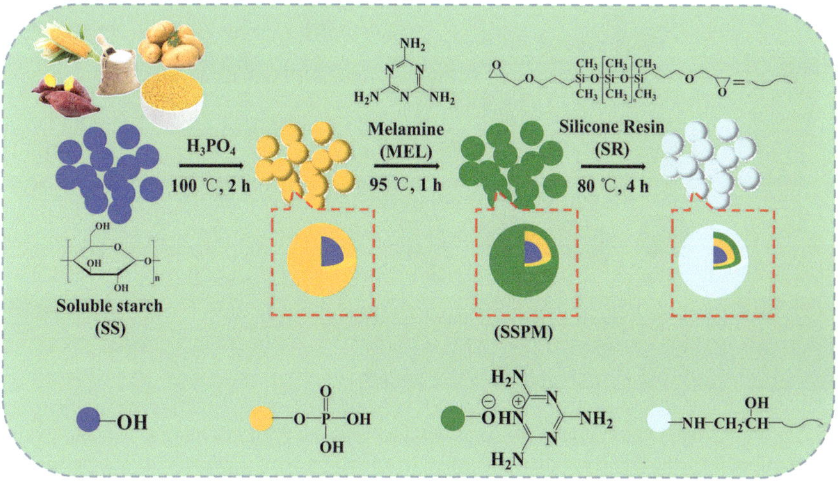

Fig. 16 The illustration for preparation and modification process of MSSPM [79]

exhibit good flame retardancy [61]. A novel polysaccharide flame retardant 2, 6-dimethoxy polysaccharide ammonium phosphate (DOPSAP), was synthesized. The results DOPSAP was successfully grafted onto the cotton fabric via P-O-C covalent bonds. In cotton fabrics treated with 30 wt.% DOPSAP, the char length and limiting oxygen index (LOI) were 42 mm and 49.4%, respectively, and remained at 52 mm and 37.5% after 50 LCs in the abrasive washing conditions (AATCC 61–2013 3A washing standard). These findings demonstrated the superior durability and flame retardancy of cotton fabrics treated with DOPSAP [80].

4.1.3 Surface Coating Technique

Starch-based flame retardant coatings can be applied to a variety of materials to improve their fire resistance. These coatings are frequently utilized in fields including textiles, paper, and wood items where fire safety is an issue [1, 60, 61, 68]. Different flame retardant compounds are often incorporated into the formulation in order to enhance the starch-based coating performance to reduce flammability. These additives may include substances based on phosphorus, substances containing nitrogen, or other substances that hinder burning. Pigolwska et al. demonstrate the properties of corn starch flame retardant that used as an additive to the conventional electrolytes in lithium-ion batteries. They tested the maize starch flame retardant for viscosity, flash point, SET (self-extinguishing time), and conductivity, and discovered that 5% of additional starch was the optimal level. This was confirmed by the estimated kinetic and thermodynamic properties, as well as the observation that the system did not gel while subjected to heat [69]. Choi et al. developed an eco-friendly, high-performance flame-retardant coating agent was by layering polyurethane foam

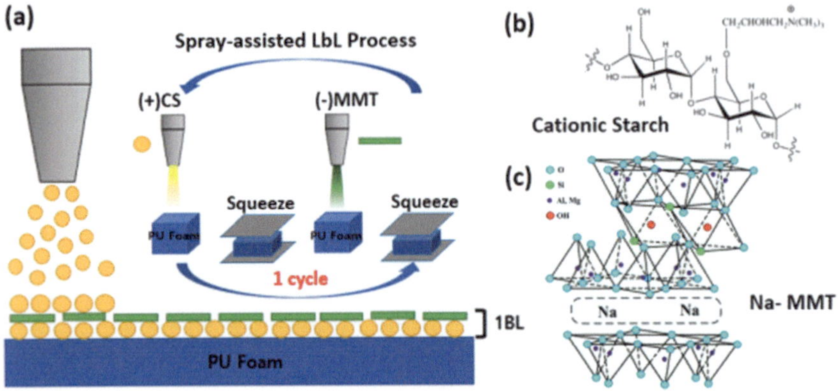

Fig. 17 **a** Schematic illustration of the spray-assisted layer-by-layer assembly of cationic starch–montmorillonite [CS/MMT]; **b, c** chemical structures of cationic starch and sodium montmorillonite platelet shown here is reprinted with permission from Ref. [65]

with cationic starch (CS) and montmorillonite (MMT) nano-clay using a sprayer via spray-assisted layer-by-layer (LbL) method as shown in Fig. 17. A flame test revealed better fire retardancy based on the measurement of combustion speed, flame size, and residues of the LbL-coated foam, and a cone calorimetry test further verified that the heat release rate and total heat release of the coated foam dropped by approximately 50% [65].

Younis et al. reported a synthesis of novel coating materials to enhance thermal and mechanical properties of bagasse paper. Paper sheets were coated by immersing bagasse paper in a coating mixture (with 0.5% cychlodiphosph(V)azane/$CaCO_3$ solution, 1.5%, $NaHCO_3$ solution, and cellulose nanocrystals (CNCs) mixed with 5.0% of starch). The burning rate (BR) was investigated. Therefore, the bagasse paper treatments can be utilized as packaging materials. In addition, compared to the untreated specimen, the resist fire increased by up to 27.5% with CNCs addition, without burning at room temperature. According to their results some additives may generate a layer on the top of the sheet after coating [81].

5 Starch Nanocomposite

The nanotechnological advance of the modern world together with the need to get materials that have good properties for the most varied applications has been increasing. Thus, the search for materials with functionality, nanotechnological properties, low cost and sustainable makes the bionanocomposites turn into materials of great interest [66, 74]. Palygorskite was successfully incorporated in both polymer matrices incorporated from the freeze-drying method, bionanocomposites foam based on pectin or rice starch was obtained. EDS-SEM confirmed that palygorskite was well dispersed as shown in Fig. 18. Additionally, the infrared spectrum was used to confirm the interaction between the species ascribed to silanol groups from palygorskite and –C–H groups from rice starch changed their stretching vibration band of –OH. Thermogravimetric measurement with a rise in breakdown temperature further confirmed the thermal stability of bionanocomposite foams [62].

Bionanocomposites are hybrid materials designed from a continuous phase, usually organic, and dispersed phase (inorganic). Both phases come from nature, where at least one has a nanoscale structure. The use of bionanocomposites has stood out against the conventional nanocomposites, due to the properties that this type of material can offer, like as biodegradability and biocompatibility [62, 74]. Due to the advantages that bionanocomposites have over traditional nanocomposites, including biodegradability and biocompatibility, they have gained popularity. For the organic phase, starch biopolymers and pectin have been used for the preparation of bionanocomposites. Pectin is a polysaccharide extracted from citrus fruits, where its structure has carboxylic groups conferring a negative character, which consists of

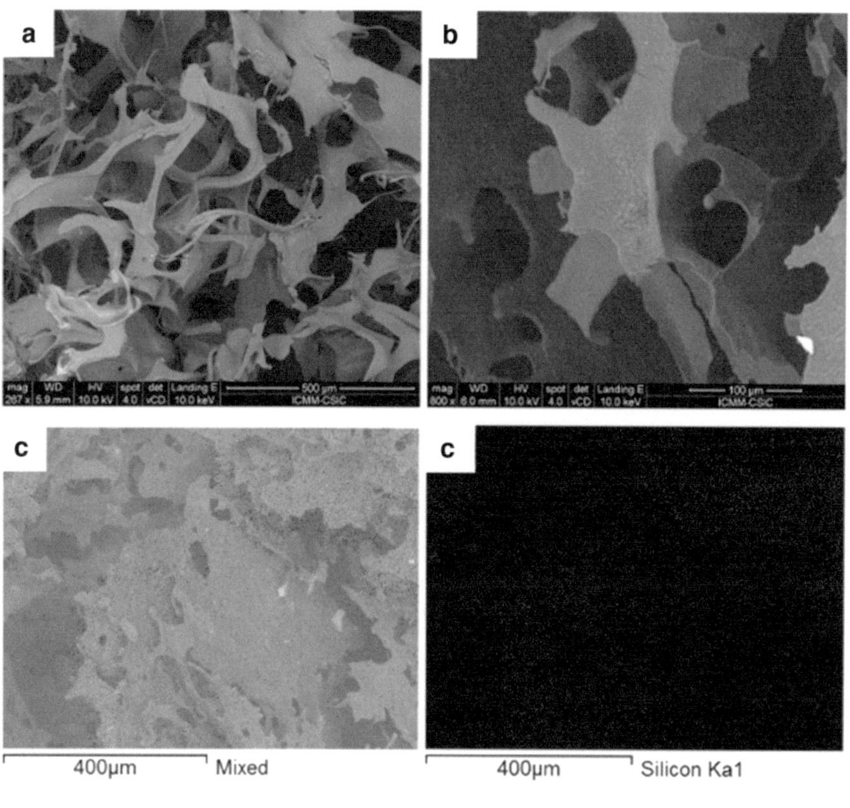

Fig. 18 SEM images from raw rice starch (**a**), solid foam with 25% of palygorskite (**b**), electron image for EDS analysis (**c**) and mapping of silicon (**d**)

residues of uronic acid [61, 74, 82]. Pectin has been widely used in the food industry and in controlled drug release studies because it has resistance to enzymes, such as amylase. However, also because it is not expensive, non-toxic, biodegradable and sustainable. The pectin is still capable of crosslinking under the presence of bivalent cations, but also of binding to other species by carboxylic groups giving specificity in the development of a bionanocomposite [62]. The fire resistance of materials made from starch can be improved through adding nanoparticles like carbon nanotubes or nanoclay. The thermal stability and fire-retardant characteristics of starch can be enhanced by these nanoparticles.

6 Limitation and Future Perspectives

Biomaterials have low intrinsic flame retardant ability in their natural state. These materials may provide some fire resistance, however they are ineffective when compared to synthetic flame retardants. Due to this drawback, they may be unsuitable for applications requiring severe fire safety regulations. For these biomaterials to comply with safety requirements and efficiently eliminate combustion, they frequently require modification or combination with other flame retardant chemical compounds. However, due to their intrinsic volatility and sensitivity to environmental conditions, these natural polymers can be difficult to modify into nanomaterials or nanocomposites. Even though these materials are often thought of as ecologically beneficial, one of the primary problems with their use is still their long-term stability. In addition, the modification of biopolymers may have possible unfavorable environmental effect, in particular if they are modified or coupled with other chemicals.

Furthermore, incorporating additives to these bio-based materials can sometimes have a negative impact on their mechanical properties (such as strength and flexibility) of the final composite materials. Moreover, it can be challenging to achieve uniform distribution and stability with the additive substance. Although these biomaterials are renewable, their extraction and processing can be more expensive than standard flame retardants, which may limit their commercialization.

Lignin, chitosan, and starch are natural polymers with different solvent solubility properties, modifying them might be difficult due to their solubility. The solubility of lignin in water and many organic solvents can be a challenge in modification. While chitosan and starch are hydrophilic and can absorb moisture which could hinder its flame retardant abilities.. Biopolymers can deteriorate over time when exposed to components like moisture and sunlight. The long-term flame-retardant abilities of materials coated or modified with these polymers may be degraded. In addition, lignin can decompose at high temperature emitting potentially flammable volatile compounds. These biomaterials may not be biocompatible with all the material, limiting their use to limited applications. Although these natural polymers are generally considered harmless, it is important to carefully consider any toxicity or allergy risks.

One approach to overcome these limits is to create synergistic mixture of lignin, chitosan, and starch with additional flame retardants. This can retain their sustainability qualities whereas enhancing their flame retardancy. These materials can be chemically modified to increase their flame retardancy ability by esterifying, phosphorylating, or grafting flame-retardant groups onto them. In addition, chemical modifications of these bio-materials can be performed to enhance the solubility and thermal stability of the fire retardant materials. The mobility and flame retardant efficiency of these bio-materials can be significantly enhanced via nanotechnology to make nanoparticles or nanocomposites. Green flame retardant can be produced by integrating bio-based materials with biocomposites (such as natural fibers, biodegradable polymers).

7 Conclusion

In summary, this chapter aimed at showcasing some of the important of developing "green" flame retardants from lignin, chitosan, and starch is a promising strategy to improve fire safety while reducing environmental and health risks. These environmentally friendly flame retardants will very probably continue to be improved upon and used in a range of industries as a result of ongoing research and innovation in this area, helping to create a more sustainable future. Biopolymers such as lignin, chitosan, and starch have been studied as potential environmentally beneficial alternatives to traditional flame retardants due to their biodegradability and non-toxic nature make them more environmentally friendly and safer for human health. The volatile combustible gases during degradation are prevented through inducing char formation to reduce the flammability of materials. Furthermore, it may produce an insulation char layer on the surface of the material that prevents further degradation. Biomaterials are promising alternative for development of green flame retardants due to their abundant in nature and possesses a complex chemical structure. These biomaterials can be modified into nanobiomaterials, which offers novel and long-term options for flame retardancy. The advantages of implementing these natural materials are enormous, however there are still challenges to be solved in terms of performance optimization and processing techniques. The goal is to reduce carbon print and pollution in the development of new flame retardants, it is important to ensure the efficiency and the integrity of the new flame retardant remain competitive to the halogenated flame retardants. In eliminating the use of conventional flame retardants, offers the way to a safer and more environmentally friendly future by removing the use of conventional flame retardants, decreasing our environmental footprint and ensuring safety without harming human health or the ecology.

References

1. Kundu, C.K., Wang, X., Song, L., Hu, Y.: Chitosan-based flame retardant coatings for polyamide 66 textiles: one-pot deposition versus layer-by-layer assembly. Int. J. Biol. Macromol. **143**, 1–10 (2020)
2. Rad, E.R., Vahabi, H., Thomas, S., De, A.A.R., Saeb, M.R.: Bio-epoxy resins with inherent flame retardancy. Prog Org Coatings. **135**, 608–612 (2019)
3. Hussin, M.H., Nelson, J., Eng, N., Hanis, N., Latif, A., Brosse, N., et al.: A recent advancement on preparation, characterization and application of nanolignin. Int. J. Biol. Macromol. **2022**(200), 303–326 (2021)
4. Sun, Y., Yu, B., Liu, Y., Bai, F., Yan, J., Wang, J., et al.: Design of 2D charring-foaming agent for highly efficient intumescent flame retardant polylactic acid composites. Compos Commun. 101720 (2023)
5. Wu, T., Qiu, J.D., Xu, W.H., Du, Y., Zhou, W.L., Xie, H., et al.: Efficient fabrication of flame-retarding silicone rubber/hydroxylated boron nitride nanocomposites based on volumetric extensional rheology. **435** (2022)
6. Sharmin, E., Zafar, F.: Introductory chapter: flame retardants. Flame Retard. 3–16 (2019)
7. Nurul, M., Mohammad, A., Antov, P., Savov, V., Fatriasari, W., Seng, L, et al., Current progress of biopolymer-based flame retardant. Osvaldov a. **205** (2022)
8. Xiao, D., Wang, Z.B., Gohs, U., Harre, K., Wang, D.Y.: A novel highly-efficient bio-based fire retardant for poly (lactic acid): synthesis, preparation, property and mechanism. Chem. Eng. J. **446**(P1), 137092 (2022)
9. Kervran, M., Shabanian, M., Vagner, C., Ponçot, M., Meier-haack, J., Laoutid, F., et al.: Flame retardancy of sustainable polylactic acid and polyhydroxybutyrate (PLA/PHB) blends. Int. J. Biol. Macromol. **251** (2023)
10. Vahabi, H., Laoutid, F., Mehrpouya, M., Reza, M., Dubois, P.: Flame retardant polymer materials: an update and the future for 3D printing developments. Mater. Sci. Eng. R **2021**(144), 100604 (2020)
11. Yu, X., Su, X., Liu, Y., Yu, D., Ren, Y., Liu, X.: Biomass intumescent flame retardant polyacrylonitrile composite: flame retardancy, smoke suppression and recycling. Compos. Part A: Appl. Sci. Manuf. **173** (2023)
12. Xia, Y., Chai, W., Liu, Y., Su, X., Liao, C., Gao, M., et al.: Facile fabrication of starch-based , synergistic intumescent and halogen-free flame retardant strategy with expandable graphite in enhancing the fire safety of polypropylene. Ind. Crop. Prod. **184**, 115002 (2022)
13. Lokhande, P.E., Chavan, U.S., Nurul, M., Mohammad, A., Antov, P., Savov, V., et al.: Efficient fabrication of flame-retarding silicone rubber/hydroxylated boron nitride nanocomposites based on volumetric extensional rheology. Chem. Eng. J. **118**, 108184 (2022)
14. Zhou, X., Qiu, S., Mu, X., Zhou, M., Cai, W., Song, L., et al.: Polyphosphazenes-based flame retardants: a review. Compos. Part B. **202**, 108397 (2020)
15. Marinovic, A.: Fluorescent carbon dots as sensitizers for nanostructured solar cells
16. Article, R., Lu, Y., He, Q., Fan, G., Cheng, Q., Song, G.: Extraction and modification of hemicellulose from lignocellulosic biomass: a review. 779–804 (2021)
17. Visakh, P.M., Semkin, A.O., Rezaev, I.A., Fateev, A.V.: Review on soft polyurethane flame retardant. Constr. Build. Mater. **227**, 116673 (2019)
18. Guo, Y., Cheng, C., Huo, T., Ren, Y., Liu, X.: Highly effective flame retardant lignin/ polyacrylonitrile composite prepared via solution blending and phosphorylation. Polym. Degrad. Stab. **181**, 109362 (2020)
19. Batistella, M., Roux, J.C., Saout, G., Callone, E., Xenopoulos, C.: Use of fly ash as synergistic and reactive component of flame retardant system in polylactide. Polym. Degrad. Stability **211**, 1–11 (2023)
20. Li, M., Prabhakar, M.N., Song, J.: Synthesis of novel Si–P–N complex coating for the fabrication of flame-retardant and hydrophobic cotton textile potentially suitable for rail seat cover. Prog. Org. Coatings **172**, 107144 (2022)

21. Cheng, X.W., Guan, J.P., Tang, R.C., Liu, K.Q.: Phytic acid as a bio-based phosphorus flame retardant for poly (lactic acid) nonwoven fabric. J. Clean Prod. **124**, 114–119 (2016)
22. Xu, Y.J., Qu, L.Y., Liu, Y., Zhu, P.: An overview of alginates as flame-retardant materials: pyrolysis behaviors, flame retardancy, and applications. Carbohydr. Polym. **260**, 117827 (2021)
23. Ning, Y., Liu, R., Chi, W., Liu, W., Zhu, Q., Xu, S., et al.: Simultaneously flame-retardant and antibacterial films from κ -carrageenan/carboxylated cellulose nanofibril/phytic acid for the preservation of cooked pork. Food Hydrocoll. **143**, 108915 (2023)
24. Dong, Y.Q., Bai, W.B., Zhang, W., Lin, Y.C., Jian, R.K.: Bio-based phytic acid @ polyurushiol—titanium complex coated cotton fabrics with durable flame retardancy for oil-water separation. Int. J. Biol. Macromol. **235**, 123782 (2023)
25. Fang, Y., Sun, W., Li, J., Liu, H., Liu, X.: Eco-friendly flame retardant and dripping-resistant of polyester/cotton blend fabrics through layer-by-layer assembly fully bio-based chitosan/phytic acid coating. Int. J. Biol. Macromol. **175**, 140–146 (2021). https://doi.org/10.1016/j.ijbiomac.2021.02.023
26. Fang, Y., Wu, J., Chen, Y., Wu, L.: Durable flame retardant and anti-dripping of PET fabric using bio-based covalent crosslinking intumescent system of chitosan and phytic acid. Prog. Org. Coatings. **183**, 107785 (2023)
27. Wang, M., Yin, G.Z., Yang, Y., Fu, W., Díaz Palencia, J.L., Zhao, J., et al.: Bio-based flame retardants to polymers: a review. Adv. Ind. Eng. Polym. Res. **6**(2), 132–155 (2023)
28. Vahabi, H., Rastin, H., Movahedifar, E., Antoun, K., Brosse, N., Saeb, M.R.: Flame retardancy of bio-based polyurethanes: opportunities and challenges. Polymers (Basel) **12**(6) (2020)
29. Ganewatta, M.S., Lokupitiya, H.N., Tang, C.: Lignin biopolymers in the age of controlled polymerization. Polymers (2019)
30. Jiang, J., Trundle, P., Ren, J., Cheng, Y.L., Lee, C.Y., Huang, Y.L., et al.: An overview on the use of lignin and its derivatives in fire retardant polymer systems. Intech **34**(8), 57–67 (2010). https://doi.org/10.1007/s12559-021-09926-6
31. Lazar, S.T., Kolibaba, T.J., Grunlan, J.C.: Flame-retardant surface treatments. Nat. Rev. Mater. [Internet]. https://doi.org/10.1038/s41578-019-0164-6
32. Qiu, X., Li, Z., Li, X., Zhang, Z.: Flame retardant coatings prepared using layer by layer assembly: a review. Chem. Eng. J. **334**, 108–122 (2018). https://doi.org/10.1016/j.cej.2017.09.194
33. Wang, Y., Zhang, Y., Liu, B., Zhao, Q., Qi, Y., Wang, Y., et al.: A novel phosphorus-containing lignin-based flame retardant and its application in polyurethane. Compos. Commun. **21**, 100382 (2020). https://doi.org/10.1016/j.coco.2020.100382
34. Liu, L., Qian, M., Song, P., Huang, G., Yu, Y., Fu, S.: Fabrication of green lignin-based flame retardants for enhancing the thermal and fire retardancy properties of polypropylene/wood composites. ACS Sustain. Chem. Eng. **4**(4), 2422–2431 (2016)
35. Yue, X., Li, C., Li, Y.: Using colloidal lignin intercalated montmorillonite nanosheets as synergistic and reinforced agent for flame-retardant poly(butylene succinate) composites. Polym. Adv. Technol. **32**(6), 2552–2565 (2021)
36. Wu, K., Xu, S., Tian, X.Y., Zeng, H.Y., Hu, J., Guo, Y.H., et al.: Renewable lignin-based surfactant modified layered double hydroxide and its application in polypropylene as flame retardant and smoke suppression. Int. J. Biol. Macromol. **178**, 580–590 (2021). https://doi.org/10.1016/j.ijbiomac.2021.02.148
37. Liang, D., Zhu, X., Dai, P., Lu, X., Guo, H., Que, H., et al.: Preparation of a novel lignin-based flame retardant for epoxy resin. Mater. Chem. Phys. **259**, 124101 (2021). https://doi.org/10.1016/j.matchemphys.2020.124101
38. Tan, Y., Shao, Z.B., Yu, L.X., Long, J.W., Qi, M., Chen, L., et al.: Polymer Chemistry mechanical property (2016)
39. Nurfajrin, N., Alif, S., Hidayat, F., Nurul, M., Mohammad, A.: Recent developments in flame-retardant lignin-based biocomposite: manufacturing, and characterization. J. Polym. Environ. 4517–4537 (2022)
40. Yu, Y., Fu, S., Luo, X., Jin, Y., Lu, F., Wu, Q.: Functionalized lignin by grafting phosphorus-nitrogen improves the thermal stability and flame retardancy of polypropylene. Polym. Degrad. Stab. **97**(4), 541–546 (2012). https://doi.org/10.1016/j.polymdegradstab.2012.01.020

41. Wu, W., He, H., Liu, T., Wei, R., Cao, X., Sun, Q., et al.: Synergetic enhancement on flame retardancy by melamine phosphate modified lignin in rice husk ash filled P34HB biocomposites. Compos. Sci. Technol. **168**, 246–254 (2018). https://doi.org/10.1016/j.compscitech.2018.09.024
42. Abu, M.Y., Conference, M.E.: Phosphorus/Nitrogen Grafted Lignin as a Biobased Flame Retardant for Unsaturated Polyester Resin, pp. 429–434. Springer Nature Singapore (2020)
43. Łukawski, D., Grześkowiak, W., Lekawa-Raus, A., Widelicka, M., Lisiecki, F., Dudkowiak, A.: Flame retardant effect of lignin/carbon nanotubes/potassium carbonate composite coatings on cotton roving. Cellulose **27**(12), 7271–7281 (2020)
44. Pérez, E.M., Martín, N.: π-π Interactions in carbon nanostructures. Chem. Soc. Rev. **44**(18), 6425–6433 (2015)
45. Rault, F., Salaün, F.: Influence of ammonium polyphosphate/lignin ratio on thermal and fire behavior of biobased thermoplastic: the case of polyamide. Materials (2019)
46. Andrade-guel, M., Cabello-alvarado, C., Avila-orta, C.A., Marissa, P., Cadenas-pliego, G., Yahaira, P.: Green flame-retardant composites based on PP/TiO_2/Lignin obtained by melt-mixing extrusion. Polymers (Basel). **14**, 1–13 (2022)
47. Fabric, P., Ren, Y., Tian, T., Jiang, L.: Fabrication of chitosan-based intumescent flame retardant coating for improving flame retardancy of polyacrylonitrile. Molecules (2019)
48. Kim, K., Ryu, J.H., Lee, D.Y., Lee, H.: Bio-inspired catechol conjugation converts water-insoluble chitosan into a highly water-soluble, adhesive chitosan derivative for hydrogels and LbL assembly. Biomater. Sci. **1**(7), 783–790 (2013)
49. Prabhakar, M.N., Song, J.: Fabrication and characterisation of starch/chitosan/flax fabric green flame-retardant composites. Int. J. Biol. Macromol. **119**, 1335–1343 (2018). https://doi.org/10.1016/j.ijbiomac.2018.07.006
50. Malucelli, G.: Flame-retardant systems based on chitosan and its derivatives: state of the art and perspectives. Molecules **25**(18) (2020)
51. Battegazzore, D., Frache, A., Carosio, F.: Layer-by-layer nanostructured interphase produces mechanically strong and flame retardant bio-composites. Compos. Part B Eng. **200**, 108310 (2020). https://doi.org/10.1016/j.compositesb.2020.108310
52. Carosio, F., Ghanadpour, M., Alongi, J., Wågberg, L.: Layer-by-layer-assembled chitosan/phosphorylated cellulose nanofibrils as a bio-based and flame protecting nano-exoskeleton on PU foams. Carbohydr. Polym. **202**, 479–487 (2018). https://doi.org/10.1016/j.carbpol.2018.09.005
53. Fang, Y., Liu, X., Tao, X.: Intumescent flame retardant and anti-dripping of PET fabrics through layer-by-layer assembly of chitosan and ammonium polyphosphate. Prog. Org. Coatings **134**, 162–168 (2019). https://doi.org/10.1016/j.porgcoat.2019.05.010
54. Zhang, S., Liu, X., Jin, X., Li, H., Sun, J., Gu, X.: The novel application of chitosan: effects of cross-linked chitosan on the fire performance of thermoplastic polyurethane. Carbohydr. Polym. **189**, 313–321 (2018). https://doi.org/10.1016/j.carbpol.2018.02.034
55. Abou-okeil, A., El-sawy, S.M., Abdel-mohdy, F.A.: Flame retardant cotton fabrics treated with organophosphorus polymer. Carbohydr. Polym. **92**(2), 2293–2298 (2013). https://doi.org/10.1016/j.carbpol.2012.12.008
56. Makhlouf, G., Abdelkhalik, A., Ameen, H.: Preparation of highly efficient chitosan-based flame retardant coatings with good antibacterial properties for cotton fabrics. Prog. Org. Coatings **163**, 106627 (2022). https://doi.org/10.1016/j.porgcoat.2021.106627
57. Erdo, S.: Coating improves the flame retardancy of dyed and undyed denim fabrics by increasing the charring. **51**, 1252–1278 (2022)
58. Patankar, K.C., Biranje, S., Pawar, A., Maiti, S., Shahid, M., More, S., et al.: Fabrication of chitosan-based finishing agent for flame-retardant, UV-protective, and antibacterial cotton fabrics. Mater. Today Commun. **33**, 104637 (2022). https://doi.org/10.1016/j.mtcomm.2022.104637
59. Hassan, M., Nour, M., Abdelmonem, Y., Makhlouf, G.: Synergistic effect of chitosan-based flame retardant and modified clay on the flammability properties of LLDPE. Polym. Degrad. Stab. **133**, 8–15 (2016). https://doi.org/10.1016/j.polymdegradstab.2016.07.011

60. Shao, N., Qu, Y., Hou, L., Hu, Y., Tian, Z., Gao, Y., et al.: Effect of starch-based flame retardant on the thermal degradation and combustion properties of reconstituted tobacco sheet. Cellulose **28**(2), 741–755 (2021). https://doi.org/10.1007/s10570-020-03587-8
61. Liu, P., Ling, J., Mao, T., Liu, F., Zhou, W., Zhang, G., et al.: Adhesive and flame-retardant properties of starch/Ca^{2+} gels with different amylose contents (2023)
62. Almeida, J.A., Oliveira, A.S., Rigoti, E., Neto, J.C.D., De, A.A.C.S.: Design of solid foams for flame retardant based on bionanocomposites systems. Appl. Clay Sci. **180**, 105173 (2019)
63. Muzammil, M., Ejaz, M., Shah, R., Afaq, S.K., Song, J.: A bio-based approach to simultaneously improve flame retardancy, thermal stability and mechanical properties of nano-silica filled jute/thermoplastic starch composite. Mater. Chem. Phys. **289**, 126485 (2022)
64. Lu, Y., Zhao, P., Chen, Y., Huang, T., Liu, Y., Ding, D., et al.: A bio-based macromolecular phosphorus-containing active cotton flame retardant synthesized from starch. Carbohydr. Polym. **298**(2), 120076 (2022)
65. Choi, K.W., Kim, J.W., Kwon, T.S., Kang, S.W., Song, J., Park, Y.T.: Mechanically sustainable starch-based flame-retardant coatings on polyurethane foams (2021)
66. Prabhakar, M.N., Song, J.: Fabrication and characterisation of starch/chitosan/flax fabric green flame-retardant composites. Int. J. Biol. Macromol. **119**, 1335–1343 (2018)
67. Prabhakar, M.N., Shah, R., Song, J.: Improved flame-retardant and tensile properties of thermoplastic starch/flax fabric green composites. Carbohydr. Polym. **168**, 201–211 (2017)
68. Wang, S., Zhang, P., Li, Y., Li, J., Li, X., Yang, J., et al.: Recent advances and future challenges of the starch-based bio-composites for engineering applications. Carbohydr. Polym. **2023**(307), 120627 (2022)
69. Pigłowska, M., Kurc, B., Rymaniak, Ł: Starch as the flame retardant for electrolytes in. Materials (Basel) **15**, 523 (2022)
70. Zhang, J., Liu, B., Zhou, Y., Essawy, H., Zhao, C., Wu, Z., et al.: Gelatinized starch-furanic hybrid as a biodegradable thermosetting resin for fabrication of foams for building materials. Carbohydr. Polym. **298**, 120157 (2022)
71. Wang, D., Wang, Y., Li, T., Zhang, S., Ma, P., Shi, D., et al.: A bio-based flame-retardant starch based on phytic acid. ACS Sustain. Chem. Eng. **8**(27), 10265–10274 (2020)
72. Ji, W., Wang, D., Guo, J., Fei, B., Gu, X., Li, H., et al.: The preparation of starch derivatives reacted with urea-phosphoric acid and effects on fi re performance of expandable polystyrene foams. Carbohydr. Polym. **2020**(233), 115841 (2019)
73. Passauer, L.: Thermal characterization of ammonium starch phosphate carbamates for potential applications as bio-based flame-retardants. Carbohydr. Polym. **211**, 69–74 (2019)
74. Phanthong, P., Reubroycharoen, P., Hao, X., Xu, G., Abudula, A., Guan, G.: Nanocellulose: extraction and application. Carbon Resour. Convers. **1**(1), 32–43 (2018)
75. Nurul, M., Mohammad, A., Sherwyn, T., Garba, Z.N., Trache, D., Uyama, H., et al.: Recent progress in cellulose-based composites towards flame retardancy applications. Polymer (Guildf) **244**, 124677 (2022)
76. Gebresas, G.A., Szabó, T., Marossy, K.: A comparative study of carboxylic acids on the cross-linking potential of corn starch films. J. Mol. Struct. 1277 (2023)
77. Basak, S., Raja, A.S.M., Saxena, S., Patil, P.G.: Tannin based polyphenolic bio-macromolecules: creating a new era towards sustainable flame retardancy of polymers. Polym. Degrad. Stab. **189**, 109603 (2021)
78. Xue, X., Liang, Q., Gao, Q., Luo, Z.: One-step synthesis of cross-linked esterified starch and its properties. Appl. Sci. **12**(8) (2022)
79. Xia, Y., Chai, W., Liu, Y., Su, X., Liao, C., Gao, M., et al.: Facile fabrication of starch-based, synergistic intumescent and halogen-free flame retardant strategy with expandable graphite in enhancing the fire safety of polypropylene. Ind. Crops Prod. **184**, 115002 (2022)
80. Lu, Y., Ding, D., Liu, Y., Lu, Y., Zhang, F., Zhang, G.: A high durable polysaccharide flame retardant based on phosphorus element for cotton fabrics Author links open overlay panel (2023)

81. Younis, A.A., Mohamed, S.A.A., El-sakhawy, M.: Fire resistant and mechanical properties of bagasse packaging paper coated with hexachlorocyclodiphosph (V) azane/starch/$NaHCO_3$/$CaCO_3$/cellulose nanocrystals composite q. Egypt. J. Pet. **31**(3), 55–64 (2022)
82. Khodayari, A., Thielemans, W., Hirn, U., Van, V.A.W., Seveno, D.: Cellulose-hemicellulose interactions—a nanoscale view. Carbohydr. Polym. **270**, 118364 (2021)

Flame Retardant Properties of Different Polymers

K. Selatile, E. R. Sadiku, S. S. Ray, M. J. Mochane, and Teboho C. Mokhena

Abstract Flame retardancy in polymers, is one of the key features to be established before their potential usage in a wide variety of applications is ascertained. The polymers covered in this chapter are either inherently flame retardant, chemically modified, or with flame-retardants inclusions, as additives for polymers. The chapter gives an overview of the flammability and thermal degradation of various polymers. To investigate the flame retardant properties, fire test techniques, such as UL-94, limiting oxygen index (LOI), ignition prevention properties, reduction in the rate of heat release, and smoke generation, are also discussed. The flammability studies of various polymers by using different flame-retardants (FRs), including bio-based alternatives and flame retardant efficiencies, are also discussed.

Keywords Flame retardant · Polymers · Flammability · Combustion · Mechanisms

K. Selatile · E. R. Sadiku
Division of Polymer Technology, Department of Chemical, Metallurgical and Materials Engineering and Institute of Nano Engineering Research, Tshwane University of Technology, Pretoria, South Africa

S. S. Ray
Centre for Nanostructures and Advanced Materials, DSI-CSIR Nanotechnology Innovation Centre, Council for Scientific and Industrial Research, Pretoria 0001, South Africa

M. J. Mochane
Department of Life Sciences, Central University of Technology, Free State, Bloemfontein, South Africa

T. C. Mokhena (✉)
DSI/Mintek-Nanotechnology Innovation Centre, Advanced Materials, Mintek 2125, Randburg, South Africa
e-mail: mokhenateboho@gmail.com

K. Selatile
Department of Chemical and Materials Engineering, College of Science, University of South Africa, Johannesburg, South Africa

M. J. Mochane
Department of Chemistry, College of Science, University of South Africa, Johannesburg, South Africa

T. C. Mokhena et al. (eds.), *Biomaterials as Green Flame Retardants*, Engineering Materials, https://doi.org/10.1007/978-981-97-6871-4_2

1 Introduction

Polymeric materials have gained considerable popularity in a wide variety of applications, such as: automotive, aerospace, computer industries, and many other applications, due to their remarkable properties, e.g., low weight and ease of processing [1]. A majority of polymers, originate from petroleum hydrocarbons and are therefore, highly flammable. Synthetic polymers in their pristine form, usually have undesirable properties for most applications. In addition, un-modified or pristine polymer materials, lack the ideal flame-retardant properties. Therefore, through modification with the use of additives, the properties of polymers are enhanced in order for them to meet the requirements for specific applications. The most commonly used additives, include: lubricants, plasticizers, stabilizers, coupling agents, flame retardants, and reinforcements [2–4]. Due to their high flammability, improvement in the fire retardant property of polymers, is a major challenge in order to extend their usage to most applications [5]. Safety requirements are rising with regards to polymers' reactions to fires and their fire resistance behaviours. Therefore, in the past decades, the use of fire retardants, has become increasingly important.

The burning of organic substances, such as carbon and hydrogen-rich polymers, is a complex process that is affected by various factors, such as: temperature, atmospheric conditions, chemical and physicochemical properties of a burning polymer material. During the burning process, the polymer undergoes different phases, viz.: heating, phase transitions, degradation, decomposition, oxidation, ignition, combustion, and propagation. Therefore, flame retardants offer solutions to prevent the ignition of the polymeric materials, or lower the heat released during their combustion, thereby, disturbing the overall processes involved in polymer combustion by using various physical and/or chemical strategies.

Various FRs, such as: halogen, i.e., phosphorus (P), nitrogen (N), and silicon-containing, i.e., minerals (metal hydroxides, layered double hydroxides, and clays), borate-based, organic silane and other compounds, have been employed, and halogenated compounds are considered old-generation FRs since they have been phased-out due to their production of poisonous products during their use. The newly developed FRs are free from halogen and they make use of the less hazardous non-halogenated P-based compounds [6–8]. The latest development of environmentally-friendly FRs is one of the promising strategies of replacing the halogenated compounds and improving the FR properties of polymeric materials. In addition, their availability and support for increased use of bio-based polymers, conserves sustainability with good environmental impact.

Bio-based polymers are a new class of polymers that are produced from renewable resources, and they are characterized by the presence of the renewable carbon, as opposed to the petroleum-based carbon, found in the synthetic polymers. The effectiveness of flame retardants is affected by various factors, such as: polymer-additive compatibility, additive migration, phase separation, etc. Intumescent flame retardant (IFR) additives have been used extensively since they offer various benefits, such as: reduced smoke generation, less toxicity, minimal flame dripping, etc. The performance of the various FRs, as evaluated by using various test methods, and the LOI test and the vertical burning measurement (UL-94), are regarded as the main polymer flammability evaluation procedures for the determination of the FRs effectiveness, towards the inhibition of the combustion of the materials.

2 General Classification of Polymers

Polymers are generally, categorized as thermoplastics or thermosets (Fig. 1). These two main types of polymer classes, differ by the bonding between the polymer chains (chemical structure). Thermosets have a linear molecular structure, since they have primary bonds between the chains. Thermoplastics, on the other hand, are cross-linked, they have secondary bonds, and can melted or molded. The two names/types are also associated with their thermal and processing properties. Thermoplastics are either amorphous or crystalline and can be semi-crystalline. Crystalline polymers are usually denser than amorphous polymers due to the tight packing of their long-chain molecules [1]. Examples of crystalline thermoplastics are polyethylene (PE), LLDPE (linear low density polyethylene), polypropylene (PP), polyesters, and polytetrafluoroethylene (PTFE). LDPE (low density polyethylene) is semi-crystalline. HDPE (high density polyethylene) is highly crystalline. Amorphous thermoplastics, include polyvinyl chloride (PVC), polystyrene (PS), and polycarbonate (PC).

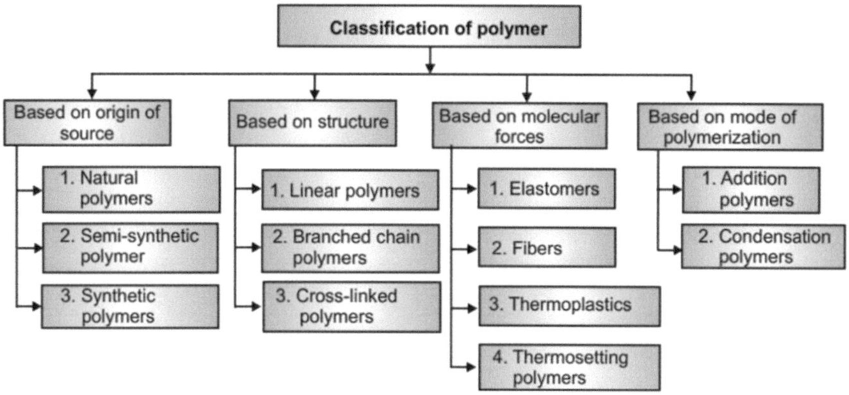

Fig. 1 Classification of polymers based on different categories [9]

3 Polymer Combustion and Flame Retardancy

Polymers are highly combustible materials. This is due to their chemical structures that are mainly, composed of carbon and hydrogen. Their combustion releases smoke, in combination with the production of corrosive and toxic gases. For the combustion process to take place, both oxidizing and reducing agents are required. The oxidizing agent is the oxygen in the air and the reducing agents are the combustibles. Heating is caused by external sources, contributing thermal energy through conduction, convection, and radiation, while chemical reactions, induced within the material (fermentation oxidation), and exothermicity of the combustion process take place. During combustion, the heat source causes the polymer to heat-up until the polymer bond scission (degradation), is induced. As the polymer fragments, the flammable volatiles diffuse into the air and form a combustible gaseous mixture, known as fuel. When the auto-ignition temperature is reached, the fuel ignites, and heat is released. This is the temperature at which the activation energy of the combustion process is reached. On the other hand, with the help of an external heating source, such as: a spark or a flame, a fuel can also ignite at a lower temperature (known as the flash point), or when the ambient temperature exceeds the ignition temperature of the polymer. A polymer further degrades since part of the heat is fed back to the polymer. As the polymer continues to burn, the scattering of the volatiles in a cyclic manner, releases smoke, heat, and toxic gases (pyrolytic constituents), until the polymer is totally burnt out. The combustion cycle is maintained for as long as the new decomposition reactions are induced on the polymer surface and thus, producing more combustible possibilities [5, 10]. The combustion process is depicted in Fig. 2.

4 Polymer Degradation Reactions

The type of polymer determines the amount of energy required to initiate combustion and this depends on the material physical properties. These properties, include: heat storage capacity, enthalpy of fusion and the degree of crystallinity of the polymer material. Semi-crystalline thermoplastics, undergo polymer softening, melting and dripping, during the heating process. Dripping poses a threat as it serves as a second source of fire, thus, igniting other materials and causing flame propagation [12]. On the other hand, with amorphous thermoplastics and thermosets, the heating stage leads straight to the material decomposition, owing to the absence of the polymer melting point. The thermal decomposition of polymers, involves the breaking of the polymer C–C covalent bonds. This process requires energies that are higher than the binding energy of the covalent bonds (200–400 kJ/mol). The mechanism for polymer decomposition depends on a number of factors, such as the: bond strength and presence of oxygen in the gas or solid phases. Thermal decomposition is either an oxidizing or a non-oxidising process. The non-oxidizing mechanism, initiates with chain scission and can occur in either of two methods, viz: through the formation of

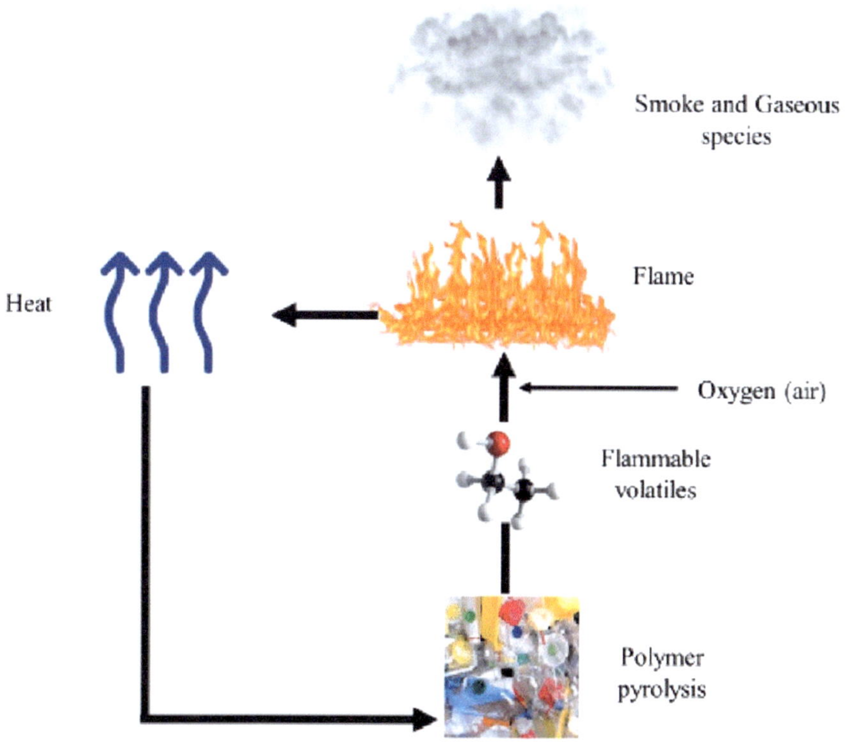

Fig. 2 Plastic/polymer combustion. Reproduced with permission from Ref. [11]. Copyright 2022. Elsevier science Ltd

free radicals (R_1–CH_2–CH_2–R_2 → R_1–CH_2 + CH_2–R_2), and through the migration of the H atoms and the formation of reactive C double bond (R_1–CH_2–CH_2–CH_2–R_2 → R_1–CH = CH_2 + CH_3–R_2). A non-oxidizing thermal decomposition also occurs at temperatures exceeding 300 °C and the degree of decomposition is considerably faster than the rate oxygen diffuses in the solid phase. The polymer degradation gases mix with the oxygen through convection and diffusion, into the layer near the polymer surface, is followed by radicals' formation and eventually, ignition occurs. The ignition can either be activated by an external source (flash-ignition) or self-induced ignition. New combustible gases are generated as the combustion gases increase the polymer temperature. Subsequently, combustion continues, even without an external heat source.

On the other hand, for the oxidizing thermal decomposition, the reaction between oxygen and the polymer, leads to the formation of short chain products, e.g., carboxylic acids, while releasing the reactive species: $H^{\cdot}$ and $OH^{\cdot}$ [5].

5 Polymer Fire Retardancy Performance

The flammability criteria of polymers are characterized by their ignitability, flame-spread rate and heat release rate. These properties are often, measured using various flammability tests, of which the most important include Limited Oxygen Index (LOI), UL 94 V and the Cone calorimeter [5].

The heat release rate is calculated with Eq. 2.1.

$$\Delta Q_c = \Delta W + \Delta Q_h + \Delta U \tag{2.1}$$

where: Q_c (kJ kg^{-1}) is the chemical energy (heat) released, W is the work done or work output (kJ kg^{-1}), Q_h denotes the heat produced or transferred (kJ kg^{-1}) and U is the internal energy (kJ kg^{-1}).

5.1 Laboratory Fire Testing

5.1.1 Limited Oxygen Index (LOI) Testing

This test is conducted to determine the relative flammability of materials [13]. The value of the LOI is defined as the minimal oxygen [O_2] concentration in the oxygen/nitrogen [O_2/N_2] mixture that either maintains the flame combustion of the material for 3 min or consumes a length of 5 cm of the sample when placed vertically upwards (Eq. 2.2) [5].

$$LOI = 100\frac{[O_2]}{[O_2] + [N_2]} \tag{2.2}$$

Air contains ~21% oxygen, therefore, materials with the LOI value less than 21, are considered combustible, whereas, those with the LOI value above 21 are classified as self-extinguishing. Combustible materials readily undergo combustion, whereas with the self-extinguishing materials, combustion cannot continue at ambient temperature, therefore, it requires the help of an external heat source. Higher LOI values translate to better flame retardant property [5]. The LOI test is extensively used to compare the flammability of various polymeric materials (Table 1) [5, 14–17]. It is evident that a polymeric material's flammability is greatly influenced by its structural characteristics. For example, the structure of poly(vinyl chloride) (PVC) contains a significant quantity of chlorine. Chlorine, as a halogen, has strong fire resistant qualities. In this case, chlorine acts as a flammable gas scavenger by facilitating the developments of char and hydrochloric acid (HCl), which slow down the reactions of the combustible compounds in gas phase [14]. The same applies for poly (tetrafluoroethylene), which has the greatest LOI value of ~95% of all polymers. Tetrafluoroethylene, or poly(tetrafluoroethylene), is a fluoropolymer, meaning that

Table 1 Flash-ignition temperature, self-ignition temperature and correlated LOI values for selected representative polymers

Polymer	Self-ignition temperature (°C)	Flash ignition temperature (°C)	LOI (%)	References
Polyethylene	340	350	18	[5, 17]
Polyoxymethylene	–	–	15	[16, 17]
Polyethylene oxide	–	–	15	[16, 17]
Polyethyelene terephthalate	–	–	20	[16, 17]
Polybutadiene	–	–	18	[16]
Polypropylene	320	350	18	[5]
Polycarbonate			26	[16]
Polyphenylene oxide			28	[16]
Styrene acrylonitrile			19	[16]
Poly(dichlorostyrene)			30	[16]
Polystyrene	350	490	18	[5]
Poly(vinylchloride) (PVC)	390	450	42	[14]
Poly(tetrafluoroethylene)	560	580	95	[15]
Acrylonitrile–butadiene–styrene terpolymer	390	480	19	[5, 17]
Poly(methyl methacrylate) (PMMA)	300	430	18	[5]
Polyvinyldene flouride (PVDF)	–	–	42–44	[15, 16]
Ethylene chlorotrifluoroethylene (ECTFE)			>52	[15]
Ethylene tetrafluoroethylene (ETFE)	510–515	470	30–32	[15]
Poly(acrylonitrile) (PAN)	480	560	27	[5, 17]
Polyamide 6	420	450	25	[5, 17]
Polyamide 66	490	530	24	[5, 17]

its main constituents are carbon and fluorine, giving it exceptional flame resistance and thermal stability [15].

5.1.2 Cone Calorimeter

Cone calorimeter is used to measure the fire behaviour of various polymers. It measures the decreasing oxygen concentration of samples that are subjected to combustion. The gas flow, oxygen, CO and CO_2 concentrations and smoke density, are measured in the exhaust duct. The quantity of the heat released per unit of time and the surface area are calculated by using the measurements of heat flow and oxygen concentration. The heat release rate (HRR) is expressed in kW/m^2. Most polymeric materials release a certain amount of heat proportional to the amount of oxygen consumed, while burning, as according to Huggett's statement [18]. Other parameters, e.g., the total heat released (THR), expressed in kJ m^{-2}, time to ignition (TTI), time of combustion or extinction (TOF), the mass loss during combustion, quantities of CO and CO_2, and total smoke released (TSR), are also characterized by the cone calorimeter [5].

The ignition time(s) is/are the time/s taken for the whole surface of the polymer sample to sustain burning, and it is determined visually. Long ignition times translate to low probability for fire/combustion to start [19].

The peak rate of heat release (kW/m^2) indicates the intensity and persistence of the fire and how fast the fire will spread. Low values of the peak rate of heat release are preferred or since they indicate better performance [19–21].

The fire performance index (FPI) ($s.m^2\ kW^{-1}$) is the ratio of the ignition time to the peak rate of heat release. It indicates the time to flashover in a full-blown fire or the time remaining to escape in a full-blown fire occurrence. It gives a complete valuation of the fire performance of polymers in a cone test procedure. Materials with high values of FPI, indicate a better performing materials.

Smoke is usually evaluated with two factors: the average rate of smoke emitted within the first 3 min of burning ($m^2/s.m$), and the total smoke generated during the test ($m^2\ m^{-2}$). A good performing FR lowers both the smoke emission and generation, this effect usually increases with increase in the concentration of the FRs [19].

5.1.3 Underwriters Laboratory-94 (UL-94 V)

The UL-94 test method is a vertical burning test and it is used to evaluate the ignitability and flame spread and it categorizes the FR levels of various polymers [22]. It is based on how long it takes for the flame to spread and the polymer to resist dripping. The highest ranking of this test is assigned to the self-extinguishing polymers without dripping. The same also applies for the slowly burning polymers.

6 Flame Retardancy Mechanisms

Flame retardant polymers inhibit the initial stages of combustion, either by the liberation of non-flammable fumes or by producing char during burning. A flame retardant does not burn, but inhibits or delays the flammability of the material. The fundamental properties of the best performing flame retardants are: increasing the limiting oxygen index (LOI) value, decreasing the HRR peak and suppress smoke. The mechanisms for flame retardants vary, depending on the type of flame retardant, its interaction with the polymer, and the chemical structure. The degree of crosslinking is determined by the functional groups present in the flame retardant., Based on the mechanism, there are four types of flame retardants (FRs), viz: gas phase, endothermic, solid phase, and intumescent flame retardants [10].

The gas phase flame retardants, function by liberating the non-flammable gases into the flame area, thereby, reducing the amount of oxygen and heat intensities, since these are the main causes of flame dissemination or propagation. This also reduces the quantity of airborne components (combustible gaseous mixture or fuel) because the heat released into the flame zone area, is restricted by the flame retardants. Flame retardants, such as mineral fillers (e.g., aluminum trihydroxide) release acids and non-flammable gases, such as CO_2, SO_2 and some H_2O. The oxygen content of the flame is thus, reduced, while the flame zone area cools down, in the process [23, 24]. Halogen acts in the vapor phase through the scavenging reactive free radicals by releasing halogen radicals that prevent combustion [25].

On the other hand, the flame retardant activity of endothermic fire retardants, is presented by the formation of endothermic radicals that absorb the heat and cool down the flame zone area, thereby, reducing the flammability of the material [26, 27]. A typical example is hydrated fillers. These lower the flame temperature to the value less than the polymer degradation temperature, since they decompose endothermically. The temperature is also lowered to the value below the combustion temperature. Polymers comprising metal hydroxides, antimony and phosphorus, fall under this group.

Solid phase flame retardants produce a char layer that inhibits the transmission of heat from the flame to the polymer surface. The char layer also shields the polymer material against oxygen exposure, since both the heat and oxygen, are essential for the propagation or dissemination of the fire flame. Phosphorous (P) and Sulphur (S) derivatives, are typical examples of solid phase flame retardants. During combustion, they form corresponding acids responsible for the dehydration process that leads to the formation of the char layer onto the polymer surface [28, 29].

Intumescent flame retardants (IFRs) produce expandable char layers on the polymer surfaces to shield the surface from oxygen exposure. The expandable char layers may necessitate the need to prepare the polymer surfaces with char-forming agents. These flame retardants have special properties or features, such as anti-dripping property, halogen-free composition, low smoke production, and low amount of toxic gases. IFRs consist of three components in its system, viz: an acid source, carbonization or char forming agent, and a blowing agent [30], and these contribute towards the quality of the char formation. When exposed to fire, charring FR systems form an insulating layer (upon dehydration and carbonization) at the surface of the burning polymer material, decreasing both the thermal and oxygen diffusion, as well as the volatilization of combustible products. Graphene-based and clay-containing metal oxides and most of the *N*- or *P*-based salts, display intumescent flame retardant (IFR) activity [28, 31].

The main purpose of flame retardants (FR) is to delay or prevent the polymer combustion reaction. The flame inhibition process is either physical or chemical. The physical process proceeds by cooling, formation of a protective layer or fuel dilution. The chemical process occurs through a reaction in the gas phase or the condensed phase mechanisms. The condensed phase takes place in polymer pyrolysis. In the gas phase, the FRs inhibit the release of substances that can react with the free radicals in order to terminate the chain reaction. FRs create disturbances in the various processes involved in polymer combustion. These processes, include: heating, pyrolysis, ignition, and propagation or dissemination of thermal degradation [5, 29].

6.1 Physical Mode

Through the FR that consumes heat, there is the reduction of the reaction temperature to below the polymer combustion temperature. For example, magnesium hydroxide releases water vapour, once the temperature reaches the range of between 200 and 300 °C. Moreover, FR produces inert gases, such as H_2O and CO_2 upon decomposition, thus, diluting or reducing the concentration of the combustible gas mixture in the process, which therefore, restricts ignition. Some FR additives function by forming a protective layer thus, preventing contact between the gas phase (where combustion takes place) and the solid phase (where thermal degradation takes place). This means that fuel gas is separated from the oxygen and thus, the combustion process is unsustainable [5].

6.2 FR Chemical Mode of Action

The chemical action for FR can occur either in the gas phase or the condensed phase. The formation or release of $Cl^{\cdot}$ and $Br^{\cdot}$ radicals by the FRs in the gaseous phase, stops the free radical mechanism of the combustion process. These radicals then, react with the highly reactive $H^{\cdot}$ and $OH^{\cdot}$ and produce less reactive molecules to decrease the reaction exothermicity. This leads to a reduction in the temperature as well as a reduction in the production of fuel. On the other hand, in the condensed phase, firstly, the flame is moved away from the flame action zone through dripping, as the FR speeds-up the rapture of the polymer chains. Secondly, a char layer is formed on the polymer surface, thereby, acting as an insulation layer separating the gas and the condensed phases [5]. The inflammable char layer is either rubbery, glassy or foamy residues, protecting the polymer layer from radicals and oxygen exposure and heat transfer. Examples of the FR additives for condensed phase, include char-forming agents, such as derivatives of polyols and polyphenols; *B*-, *Si*- or *P*-based compounds forming the solid protective layer upon decomposition; metal salts catalysing the char formation and fire-resistant fillers that inhibit heat transfer.

7 Flame Retardant Polymers

Polymers can be modified with FR additives, either as fillers or at the molecular level. Flame retardant additives can be part of a polymer (as fillers) or monomer (at the molecular level). The resulting polymer is, therefore, inherently flame retardant. Therefore, the two procedures used to prepare flame retardant polymers are the additive and reactive methods (also known as the two categories for flame retardants) [5].

The additive method involves the physical mixing of flame retardant additives with the polymer, therefore, there is no chemical bonding between the additives and the polymer matrix. The FR additives are often, distributed uniformly all over the polymer surface through physical interactions (hydrogen-bonding, van der Waals forces, and ionic interactions). This method therefore, does not successfully improve the FR properties of polymers due to the absence of covalent bonding between the polymer matrix and the additives. In addition, due to the high additive loadings required, the resulting FR polymers are more likely, to experience leaching, phase separation and inferior mechanical properties. Examples of such FR additives, include mineral fillers, organic compounds, etc., [5, 10].

The reactive method on the other hand, unlike the physical mixing, incorporates the reactive additives into the polymer network in the form of a monomer, chain extender or polymer precursor, during the polymer synthesis process. They can also be introduced to the polymer through a post-reaction process (e.g., via chemical grafting). Therefore, this method introduces covalent bonds between the FR additives and the polymer matrix or chains. This method can also be referred to, as an intrinsically flame retardant technology. Phosphorus, nitrogen, halogen and silicon-based compounds are used as the reactive FRs, while phosphorus-based are more compatible with the thermoplastics and the thermosets [10, 32]. Halogen, however, is restricted for most applications due to its production of toxic gases and large amounts of smoke, upon burning [33]. Although less developed, this method carries some of the advantages, such as: lower additive loading, less corrosiveness, better polymer and FR additive stability [34].

7.1 Poly(Vinyl Chloride) (PVC)

PVC, with the chemical structure represented as $(C_2H_3Cl)_n$, generates chlorinated benzodioxins and benzodifurans during combustion or house fires. It was reported that dioxin generated during house fires is in the range of 0.47 to 23 g Toxicity Equivalent (TEQ) y^{-1}, thus, leading to between 20 000–50 000 g (TEQ) of dioxin that is released annually, into the environment, as estimated by the Environmental Protection Act (EPA) [35].

PVC may possess good fire retardant properties (inherently flame resistant) due to its high chlorine content of about 56.8% (Fig. 3). In addition, the rapid evolution

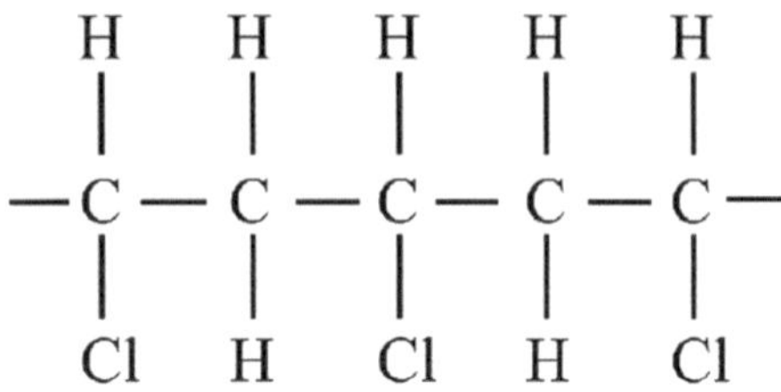

Fig. 3 The molecular formula of polyvinyl chloride. Reproduced with permission from Ref. [37]. Copyright 2018, F1000 Research

of hydrogen chloride (HCl) during the PVC combustion, forms carbonized products responsible for conjugated and cross-linked structures. These structures therefore, offer some degree of protection to the PVC polymer backbone during combustion [36]. The combustion reaction of PVC therefore, terminates upon the removal of the heat source. This is true for PVC materials, such as pipe fittings and windows, since they contain no additives, e.g., plasticizers [19]. Therefore, such rigid PVC formulations meet some of the flame retardancy specifications in the absence of any flame retardant additives. However, the incorporation of some additives, e.g., high levels organic additives, interfere and lowers the flame retardant properties of PVC. Plasticizers present in flexible PVC also compromise the fire resistance of the polymer. Therefore, for more stringent specifications, such as the: LOI value, heat release, smoke evolution, or extent of burning in cable tests, flame-retardants and smoke-suppressants are usually incorporated into plasticized or flexible PVC. Usually, pristine synthetic polymers may not have the desirable flame-retardancy properties in order to meet the desired requirements of a specific application. Therefore, additives or fillers are used to modify the polymer to enhance the properties to meet such requirements. The flame retardant and smoke suppression performances of various polymers (pure or plasticized), can be determined through the oxygen index (OI) and cone calorimetry [19]. The flame retardants, commonly used in PVC, include the followings: antimony(III) trioxide (Sb_2O_3), Halogenated FRs and Inorganic FRs (mineral additives).

When forced to burn, PVC generates black HCl-containing smoke. During combustion, PVC degrades to form hydrogen chloride (HCl), polyene and aromatic compounds, which further condense to produce tar. PVC begins to degrade at temperatures above 100 °C, due to the gradual loss of HCl along the PVC chain (known as un-zipping reaction), thus, leading to the formation of polyene structures. Upon further heating, further reactions, e.g., chain scission, take place and lead to polymer degradation and reduction in mechanical properties.

7.2 *Thermal Decomposition of PVC*

The thermal decomposition mechanism of PVC, gives an insight into the flammability performance and properties of PVC. Due to its thermal instability, it is generally, accepted that the thermal degradation mechanism of PVC, follows a chain mechanism, known as the zipper elimination, as HCl evolves. In addition, slow HCl removal, produces isolated double bonds and then, dehydrochlorination continues rapidly because of the stimulation of allylic chlorine. The individual reaction steps proposed for the decomposition of PVC, are the free radical, ionic or molecular intensive mechanisms [38]. The radical chain mechanism starts by the scission of the C–Cl bond and the dissemination by detachment of the hydrogen (H) from the neighboring carbon atom. At the same time, another Cl atom is formed, while a macro-radical is regenerated. The Cl is eliminated by the macro-radical to form a double bond [39].

The PVC polymer chains dehydrochlorinate via an ionic mechanism. The early stages of the PVC thermal degradation, lead to crosslinking, formation of aromatic volatiles and Diels's Alder reaction and finally, a char formation. Oxidation of polyenes takes place in the presence of air, and the secondary reactions tend to put an end to the polyenes structures, one of which seems to be an inter-molecular Diels–Alder process that forms crosslinking. The formation of these cross-linked structures, is more likely to contribute towards the flame retardancy in PVC. The reason being that the provision of the flame with combustible fragments is inhibited, while the other mechanism is a cyclization process with radical elimination that leads to the formation of benzene. Benzene, however, is said to be relatively combustible, and when exposed to the flame, it forms smoke since it tends to condense into larger polyaromatics and soot in the gas phase. HCl on the contrary, is necessary for extinguishing the flame.

7.3 Polyamides

Polyamides (PAs), are thermoplastic engineering polymers with a repeating amide functionality, i.e., they contain amide bonds in the main chain, and can be either synthetic or naturally occurring. Synthetic PAs depend on the type or structure of their monomers, and they are either aliphatic, partially aromatic, semi-aromatic or aromatic. Aliphatic PAs, also known as nylons, are the linear PA6, PA66, PA12, etc. Apart from PA66, the aliphatic PAs completely burn without forming any residues. Although possessing excellent processing properties, they are highly flammable and require FR additions in order to meet the fire safety requirements in various applications. Examples of the naturally-derived polyamides, include: wool, silk, collagen, β-keratin and polyglutamic acid. Wool contains large amounts of L-amino acids, high moisture absorption of up to 15%, with low H-content of about 6%, and N- and S-contents of ~16 and ~3%, respectively. All these properties render it intrinsically flame retardant, with ignition temperatures of up to 600 °C, low total heat release (THR) (20.5 kJ/g) and LOI values >21% (i.e. 25–28%) without any FR additives [40]. The FR properties are in addition to char forming properties, accompanied with low smoke emissions [34]. The synthetic PAs, are known as aliphatic, partially aromatic, semi-aromatic and aromatic. The aliphatic PAs, unlike the natural PAs are highly flammable due to their high H-content and low water absorption. They have

LOI values of 20–26% and once ignited, they self-extinguish (through shrinking and dripping) due to the low melt viscosity of PAs. The combustion of PA is accompanied by high HRR and THR, with emission of small quantities of toxic gases and smoke. Introducing aromatic units for the replacement of aliphatic monomers is another way of enhancing the fire resistance of PAs, in addition to enhancement in mechanical and thermal properties. In the same manner, semi-aromatic and partially aromatic PAs, incorporating aromatic monomers in the macromolecular backbone, have higher LOI values that the aliphatic counterparts [41]. The aromatic content is about 50% for semi aromatic PAs, and up to 85% for partially aromatic Pas. Fully aromatic PA (aramids), with the aromatic component >85%, exhibit inherent flame retardancy among the PAs. They are barely flammable, with low THR, high T_g and high char formation. During combustion, the homolytic and hydrolytic cleavage of amide bonds release C-containing aromatic molecules. Therefore, the FR properties are affected by the presence of water and substituents, and crystallinity. Usually, aramids may not require further flame retardancy modification and can thus be applied in fire resistant applications, such as clothing, firefighting and the military. Non-reactive additives or fillers are commonly used for PAs. however, the mechanically mixed additives lack compatibility with the polymer and are therefore highly susceptible to aggregation, phase separation and leaching. Alternatively, incorporation of monomers with FR additives utilized during the polymerization process and this reactive additive approach, has proven to be more effective [34].

7.4 PA66 Blends

Polyamide 66 has poor flame retardancy with LOI value of 22.5%, in addition to release of toxic smoke when exposed to flames [42, 43]. Therefore, PA66 needs to be alloyed to improve its overall performance. It has been blended with PPO, in the presence of compatibilizers for interface compatibility. Polyphenylene oxide (PPO), also known as polyphenylene ether (PPE), is an amorphous polymer with excellent heat resistant, low moisture absorption ability and remarkable hydrolysis resistance. Owing to its unique molecular structure and the LOI value of approximately 28%, PPO functions as a good flame redundant. Due to its poor processability, PPO is often blended with polystyrene (PS) or nylon. In addition, PPO exhibits good char forming properties, especially when used in conjunction with polyamide [44], high impact PS [45], and poly(butylene terephthalate [46]. In a study by Zhang, Yang [42], there were no chemical bonds between the PA66/PPO blend and the FRs (as compared to the reactive type), as confirmed by the FTIR results. The LOI value increased from 24% for the PA66/PPO, to 33.2% and 41% for PA66/PPO–P@Si and A66/PPO-RP, respectively. The UL-94 V-2 level of the PA66/PPO was enhanced to UL-94 V-0 level upon addition of the FRs (PA66/PPO–P@Si and A66/PPO–RP).

On the other hand, PPO and polystyrene (PS) are compatible and may require no compatibilizers. The LOI value for PS is only 18%, therefore the highly flammable PS burns with a dense dark smoke accompanied with dripping behaviour. The good char forming ability of PPO promotes the formation of a dense thick char layer in the presence of intumescent FR. The LOI was enhanced from 18% for the unmodified PS to as high as 30.5% for the composite (PS/PPO/intumescent FR) [47].

7.5 Polylactic Acid (PLA)

PLA is one of the promising commercialised biopolymers due to exceptional properties, such as biodegradability, non-toxicity, mechanical strength and easy processing [48, 49]. The application of PLA is, unfortunately, limited due to its high flammability. PLA undergoes constant melt dripping upon ignition. This can be observed by violent flames accompanied with melt dripping behaviour upon PLA ignition, during the vertical combustion and LOI tests. [50]. Most of the reports indicated that PLA cannot be ranked in the UL-94 vertical burning test with LOI values ranging in between 19 and 21% [49, 51]. The introduction of FRs into PLA is essential to improve the overall fire performance [49]. The use of a greener intumescent FR system, whereby starch and/or lignin replace the usually used ammonium polyphosphate (APP)/pentaerythritol (PER) system, was evaluated in a few studies [52, 53]. These result in ~50% reduction in the peak of heat release and a 32% obtainable for the LOI value.

8 Development of Bio-Based FRs

Bio-based, eco-friendly FR are increasingly becoming popular in the past years, as they are continuously replacing the petroleum-based FRs [51, 54, 55]. Natural products, e.g., lignocellulose from certain plants display a defence action against fire attack. A classic example is an inherently fire resistant cork oak tree containing a waxy substance composed of polyaliphatic and polyaromatic domains, interlinked by glycerol. These naturally occurring compounds have the ability to produce a char layer when exposed to fire [56]. A majority of bio-based FRs can perform as both the condensed and gas phases for the flame retardancy in most polymers. These bio-based flame retardants have been commonly applied in various polymers, such as: PP, polyethylene (PE), polylactic acid (PLA), epoxy resin (EP) [56].

8.1 Bio-Based Flame Retarded Systems

Cellulose is one of the most abundant naturally-occurring polymers and formed by the chains of glucose units [57–59]. The degree of polymerization, in the range of between 100 to 40 000 is dependent on the origin and treatment method for extraction. The tri-dimensional chain structure of cellulose is stabilized by hydrogen bonds, van

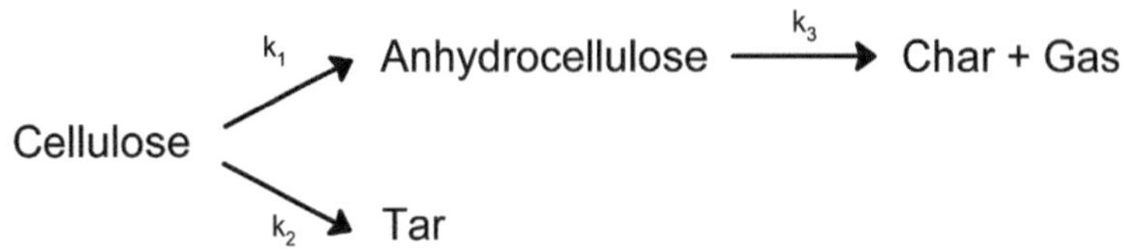

Fig. 4 Kinetic model for cellulose pyrolysis proposed by Broido. Reproduced with permission from Ref. [62]. Copyright 2017. Elsevier Science Ltd

der Waals bonds and hydrophobic interactions. The arrangement of cellulose chains, results in the formation of microfibrils that are assembled together to form cellulose fibers containing both crystalline and amorphous regions [57, 59, 60]. Cellulose presents remarkable char forming properties. The thermal decomposition of cellulose, is presented by two competing reactions, as depicted in Fig. 4. The first step is the formation of anhydrocellulose at temperatures, well below 280 °C. At higher temperatures, an endothermic un-zipping reaction is initiated. This un-zipping reaction leads to the formation of tar (k_2), while anhydrocellulose forms char and gas, through exothermic decomposition (k_3). The thermal decomposition also leads to the complex process of char formation. During cellulose pyrolysis, the thermal degradation conditions, determine the amount of char formed and its thermal stability [61]. The dehydration reaction favoured at lower temperatures, results in char formation.

Various cellulose surface chemical modification procedures, e.g., esterification, have been undertaken for improving the FR properties. Cellulose can be functionalised with ease due to the presence of functional groups, e.g., the hydroxyl functional groups. The soaking of cellulose fiber in phosphoric acid and urea mixture then, neutralized with a base to yield phosphorylated cellulose. Modification through phosphorylation, enhances the char forming ability, thermal degradation of cellulose, and other FR properties, e.g., decreases in the initial and peak degradation temperatures. Phosphorylated microcrystalline cellulose (MCC-P), when incorporated into polylactic acid (PLA), can improve the FR properties by decreasing the pHRR and UL 94 V-0 rating. Although PLA is a bio-based polymer with good thermal and processing properties that have attracted various uses in a variety of applications, the high flammability restricts its usage certain applications. Micro-crystalline and nanocrystalline cellulose, combined with bio-based phosphorus (aluminum phytate), have been used to improve the thermal stability and FR properties of PLA. The cellulose/phosphorus combination is introduced by using, either the chemical grafting or melt blending methods. The charring effect of PLA was enhanced (the high specific surface area of the nano-sized cellulose, favoured the formation of a more effective char layer), while the pHRR was reduced [62, 63].

Starch is a polysaccharide, consisting of amylose, and amylopectin. Amylose is a linear polymer, made of glucose units, while amylopectin is a heavily branched polymer with small glucose segments, linked to the main glucose chain. Starch is sourced from plant-based proteins through some chemical treatment methods [64]. The three stages of thermal degradation of starch, include dehydration due to the loss of water, followed by a chemical dehydration when the temperature reaches 300 °C and forming ether segments and then, water liberation due to thermal condensation between the OH groups. Dehydration of the neighbouring OH groups also takes place and results in the formation of carbon double bond and ring scission. The aromatic benzene and furan structures are produced at elevated temperatures. The final stage is the he formation of large conjugated aromatic structures, combined with the carbonization reactions, as the temperature reaches 500 °C [62]. In an intumescent flame retardant (IFR) systems, starch can function as a carbon source due to the presence of polyhydroxy groups, which are cross-linked to carbon, through esterification reaction when exposed to the fire. This is thus, very useful for the release or formation of the carbonaceous layer to protect the material against the action of heat flux or flame (by restraining oxygen diffusion and heat transfer). Ammonium polyphosphate (APP), together with pentaerythritol (PER), have both been used in an intumescent system as the acid source and carbonization agent, respectively. As the acid source, the APP releases phosphoric acid, and at the same time, releases ammonia as a blowing agent, needed to release some non-combustible gases upon heating, thereby diluting the oxygen content to terminate the combustion process [65]. The released phosphoric acid catalyses the degradation of the thermoplastic polymer and the carbonization agent, as the temperature rises, followed by the formation of some radicals. The degraded products are trapped in the solid phase, by the carbonizing agent [66]. The incorporation of a 12 wt% APP and 28 wt% starch into PLA, led to 32% LOI value, from a value of 24% for pure PLA [52]. A 41% LOI value was obtained for PLA incorporated with 10 wt% starch and 20 wt% intumescent flame retardant (IFR) composed of microencapsulated ammonium polyphosphate (MCAPP) [67]. Wang, Ren [68] evaluated the performance of PLA foams phosphorus-containing flame retardant (FR) and starch as the natural charring agent and the LOI value increased from 18 to 30.6%. Oxidized starch (OS) (by sodium perborate) was blended into PLA in the presence of ammonium polyphosphate (APP), a halogen-free FR used as acid donor in the intumescent system, in the preparation of PLA/OS/APP composites. The authors observed a reduction in the HRR value due to the char layer formed in the condensed phase on the polymer surface. The phosphate layer stabilized the char residues during combustion. In addition, the superior FR properties of the PLA were ascribed to the limiting of the total amount of fuel for the combustion process [69].

Chitosan (CSI) is a naturally occurring, renewable, bio-based linear amino polysaccharide comprising of glucosamine and *N*-acetylglucosamine units. As a polysaccharide, it is one of the most common biomaterial and it is the most plentiful main constituent of some living organisms, and it is a fully and/or partially, deacetylated form of chitin. It is a natural polymer with skeletal amino groups. The hydroxyl groups of the carbohydrate structure can function as char forming agents for the FR systems. Chitosan has also presented itself as an anti-dripping and self-extinguishing agent in polymers, e.g., PLA [70]. Both the modified and unmodified CSI, are employed as natural FRs and char forming agents. CSI was modified by cross-linking it with bis-(4-formylphenyl)-phenyl-phosphonate, followed by its incorporation into thermoplastic urethane (TPU), combined with APP and organic modified montmorillonite (OMMT), resulted in an UL-94 V-0 rating, and a reduction in the flammability since the LOI of 20.8%, increased to 29.0%, with the peak heat release rate decreasing from 1090 to 284 kW/m^2. The reduction in HRR is due to the formation of the char residue, caused by the catalytic carbonization of OMMT and the esterification of polyphosphoric acid [71]. A positively charged CS and a negatively charged phytic acid salt (PA-Na), assembled with ammonium polyphosphate (APP) in the preparation of a core–shell structured APP@CS@PA–Na, as a bio-based FR incorporated into PLA, improved the flame retardancy of PLA. The LOI value of 20.0% for PLA increased to 30.5% with the addition of the developed CS based FR, with UL-94 V-0 rating [48]. Another study by Ma, Wu [50], using a modified CS was prepared with phenylphosphoryl dichloride and tetraethylenepentamine for the preparation for bio-based CS fire retardants, namely, phenylphosphorylate CS (PhPCS) and phenylphosphoramidated CS (PhPNCS), respectively. These bio-based FRs showed exceptional fire retandancy proficiency towards PLA, as depicted in Table 2. Both FRs of the PLA composites (i.e., PLA/PhPCS and PLA/PNCS) or biocomposites, achieved a V-0 rating for the UL-94 vertical combustion test (since they are self-extinguishable after two ignitions) and the LOI values increased from 20.0% to 29.0% and 31.0% for the PLA/3wt%PhPCS and PLA/5wt%PhPCS, respectively. For PNCS, the slight increases of 27.0% and 29.0% were due to the higher CH_2 content than the PhPCS. In addition, the typical melt-dripping behaviour of the PLA polymer was inhibited. Other indications of the improved flame retardancy for the PLA were the reductions in pHRR, THR and TSP. In addition, the decomposition of the bio-based FRs (PhPCS and PhPNCS), resulted in the release inert gaseous NH_3, which is believed to have contributed towards the dilution of the combustion gases and inhibited a further combustion of PLA. The total smoke production of the composites increased with the FRs content, due to the production of molecular volatile gases during the initial phases of combustion. These gases result from the cleavage of glycosidic PhPCS bonds. As the PLA chains broke down at high temperatures, macromolecular chain free radicals were generated. This produced short chain molecular gases in the process. Further breaking down of the PLA chains was inhibited by the decomposing *P*-containing radicals that react with the H· and OH· radicals produced by PLA, as the combustion process proceeded. Therefore, more CO_2 was produced.

The proposed mechanism for flame redardancy and charring for PLA/PNCS biocomposites, is exemplified in Fig. 5. The first stage is the decomposition of the tetraethylene pentimide unit of PhPCNS, resulting in the release of gases, i.e., the small chain hydrocarbon gases and inert gases, such as NH_3 and H_2O. These will then, decrease the concentration of the volatile combustible gases. The second stage is the dehydration of the PLA main chains, as activated by the PhPNCS phosphate groups, thus, releasing the phosphorus-oxygen and benzene radicals in the gas phase. The phosphorous-rich radicals generated, then react with the ·H and ·OH radicals generated from the PLA and thus, preventing the decomposition of PLA through the combustion process. A further increase in temperature, decomposes PhPNCS into short chain products, such as polyols. After the formation of aromatic compounds, crosslinking and carbonization of the PLA/PhPNCS, a char layer containing the

Table 2 Flame-retardant results of the PLA/PhPCS and PLA/PhPNCS biocomposites [50]

Sample	LOI (%)	t_1/t_2 (s/s)	Dripping	Igniting cotton	UL-94 rating	UL-94 rating after immersing in water for 24 h
PLA	20	/	Y	Y	No rating	No rating
PLA/ 3wt%PhPCS	29	1/0	Y	N	V-0	V-0
PLA/ 5wt%PhPCS	31	0/0	Y	N	V-0	V-0
PLA/ 3wt%PhPNCS	27	3/0	Y	N	V-0	V-0
PLA/ 5wt%PhPNCS	29	2/0	Y	N	V-0	V-0

polyphosphate and polycyclic aromatic ring was produced at high temperatures. This layer functioned as a physical barrier that prevented the release of combustible gases, while inhibiting a further combustion of the underlying PLA layer, in the condensed phase (stage 3) [50].

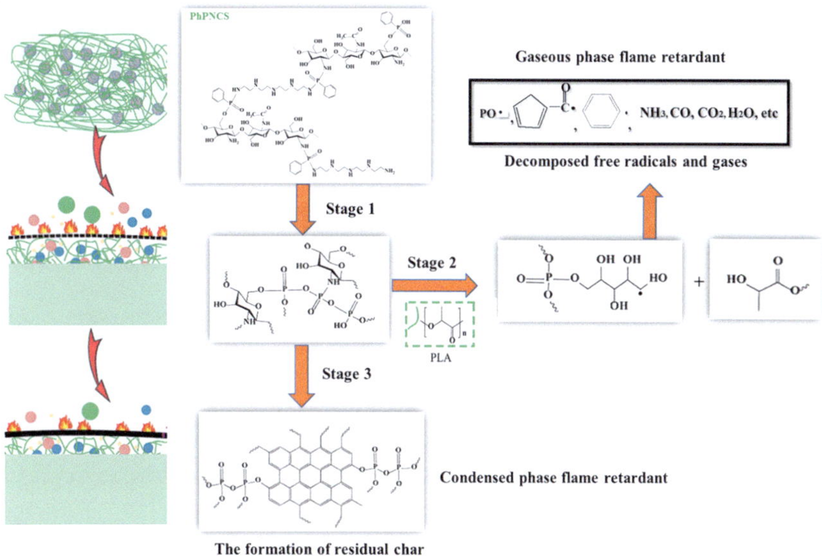

Fig. 5 The probable flame-retardant mechanism of PLA/PhPNCS biocomposites [50]

9 Conclusion and Future Perspective

In this chapter, the flammability properties and behaviours of different polymers, such as: PVC, PLA, PA, were discussed. The polymers can either be intrinsically flame-retardant or lack flame retardant characteristics. Polymers exhibiting intrinsic flame retardancy, exhibited LOI values above 21%, while the highly flammable polymers, displayed low values below 21%. However, since most applications require high performance polymers that also undergo stringent fire tests (LOI, cone calorimetry, UL-94, etc.,) they therefore, require modifications through the inclusion of fire retardant additives. However, other polymer properties, such as mechanical strength may decrease with the addition of the FRs additives. Although various halogen-free flame retardant additives have been utilized, but due to environmental concerns, recent developments however, have led to the use of bio-based alternatives. The bio-based flame-retardants have also shown excellent performance towards improving the flammability performance of the polymers. Combustion of polymers is a very complex process, and therefore, requires in-depth analysis towards the understanding of their combustion behaviors.

References

1. Brinson, H., Brinson, L.: Characteristics, Applications and Properties of Polymers (2008)
2. Almeida, S., et al.: A brief evaluation of antioxidants, antistatics, and plasticizers additives from natural sources for polymers formulation. Polymers **15**(1), 6 (2023)
3. Bodaghi, A.: An overview on the recent developments in reactive plasticizers in polymers. Polym. Adv. Technol. **31**(3), 355–367 (2020)
4. Lu, J.Z., Wu, Q., McNabb, H.S.: Chemical coupling in wood fiber and polymer composites: a review of coupling agents and treatments. Wood Fiber Sci. 88–104 (2000)
5. Laoutid, F., et al.: New prospects in flame retardant polymer materials: From fundamentals to nanocomposites. Mater. Sci. Eng. R. Rep. **63**(3), 100–125 (2009)
6. Velencoso, M.M., et al.: Molecular firefighting—how modern phosphorus chemistry can help solve the challenge of flame retardancy. Angew. Chem. Int. Ed. **57**(33), 10450–10467 (2018)
7. Rabe, S., Chuenban, Y., Schartel, B.: Exploring the modes of action of phosphorus-based flame retardants in polymeric systems. Materials **10**(5), 455 (2017)
8. Savas, L.A., Dogan, M.: Flame retardant effect of zinc borate in polyamide 6 containing aluminum hypophosphite. Polym. Degrad. Stab. **165**, 101–109 (2019)
9. Mistry, S.: Classification of Polymer (2022). Available from: https://solutionpharmacy.in/classification-of-polymer/
10. Sinha Ray, S., Kuruma, M.: Polymer combustion and flame retardancy. In: Sinha Ray, S., Kuruma, M. (eds.) Halogen-Free Flame-Retardant Polymers: Next-generation Fillers for Polymer Nanocomposite Applications, pp. 5–9. Springer International Publishing, Cham (2020)
11. Mensah, R., et al.: A review of sustainable and environment-friendly flame retardants used in plastics. Polym. Testing **108**, 107511 (2022)
12. Zhan, J., et al.: Flame-retardant and anti-dripping properties of intumescent flame-retardant polylactide with different synergists. Polym.-Plast. Technol. Eng. **53**(4), 387–394 (2014)
13. Camino, G., et al.: The oxygen index method in fire retardance studies of polymeric materials. J. Appl. Polym. Sci. **35**(7), 1863–1876 (1988)

14. Hirschler, M.M.: Poly(vinyl chloride) and its fire properties. Fire Mater. **41**(8), 993–1006 (2017)
15. Ebnesajjad, S.: 18—properties of tetrafluoroethylene homopolymers. In: Ebnesajjad, S. (ed.) Fluoroplastics (Second Edition), pp. 396–440. William Andrew Publishing, Oxford (2015)
16. Imhof, L.G., Stueben, K.C.: Evaluation of the smoke and flammability characteristics of polymer systems. Polym. Eng. Sci. **13**(2), 146–152 (1973)
17. Lyon, R.E., Janssens, M.L.: Polymer flammability. In: Encyclopedia of Polymer Science and Technology, pp. 1–70.
18. Huggett, C.: Estimation of rate of heat release by means of oxygen consumption measurements. Fire Mater. **4**(2), 61–65 (1980)
19. Thomas, N.L.: Zinc compounds as flame retardants and smoke suppressants for rigid PVC. Plast., Rubber Compos. **32**(8–9), 413–419 (2003)
20. Shan, G., et al.: Flame-retardant polymer nanocomposites and their heat-release rates. J. Hazard. Toxic Radioactive Waste **19**(4), 04015006 (2015)
21. Chen, D.-J., et al.: Making polycarbonate flame retardant: flame retardant selection and calorimetric analyses. Polym. Testing **117**, 107876 (2023)
22. Joseph, P., Tretsiakova-McNally, S.: Melt-flow behaviours of thermoplastic materials under fire conditions: recent experimental studies and some theoretical approaches. Materials **8**(12), 8793–8803 (2015)
23. Wazarkar, K., Kathalewar, M., Sabnis, A.: Reactive modification of thermoplastic and thermoset polymers using flame retardants: an overview. Polym.-Plast. Technol. Eng. **55**(1), 71–91 (2016)
24. Chattopadhyay, D.K., Raju, K.: Structural engineering of polyurethane coatings for high performance applications. Prog. Polym. Sci. **32**(3), 352–418 (2007)
25. Morgan, A.B., Gilman, J.W.: An overview of flame retardancy of polymeric materials: application, technology, and future directions. Fire Mater. **37**(4), 259–279 (2013)
26. Gu, J.-W., et al.: Study on preparation and fire-retardant mechanism analysis of intumescent flame-retardant coatings. Surf. Coat. Technol. **201**(18), 7835–7841 (2007)
27. Lewin, M., Weil, E.D.: Mechanisms and modes of action in flame retardancy of polymers. Fire Retardant Mater. **1**, 31–68 (2001)
28. Kandola, B.K., Horrocks, A.R.: Complex char formation in flame-retarded fiber/intumescent combinations: physical and chemical nature of char1. Text. Res. J. **69**(5), 374–381 (1999)
29. Horrocks, A.R., Price, D., Price, D.: Fire Retardant Materials. Woodhead Publishing (2001)
30. Chen, X., Jiao, C.: Flame retardancy and thermal degradation of intumescent flame retardant polypropylene material. Polym. Adv. Technol. **22**(6), 817–821 (2011)
31. Xiao, F., et al.: Combination effect of zinc borate in intumescent polybutylene succinate systems: study of the carbonaceous structures using solid-state NMR. Polym. Degrad. Stab. **206**, 110192 (2022)
32. Lomakin, S.M., Zaikov, G.E.: Ecological Aspects of Polymer Flame Retardancy, vol. 10. Vsp (1999)
33. Sen, A.K., et al.: Preparation and characterization of low-halogen and nonhalgoen fire-resistant low-smoke (FRLS) cable sheathing compound from blends of functionalized polyolefins and PVC. J. Appl. Polym. Sci. **43**(9), 1673–1684 (1991)
34. Drigo, N., Gaan, S.: Intrinsically flame retardant polyamides: research progress in the last 15 years. Adv. Indus. Eng. Polym. Rese. **6**(2), 95–131 (2023)
35. Carroll, W.F., Jr.: Is PVC in house fires the great unknown source of dioxin? Fire Mater. **20**(4), 161–166 (1996)
36. Harper, C.A.: Handbook of Building Materials for Fire Protection, vol. 441. McGraw-Hill New York (2004)
37. Arahman, N., et al.: Morphology and performance of polyvinyl chloride membrane modified with Pluronic F127. F1000Research **7**, 726 (2018)
38. Salovey, R., Bair, H.: Degradation of poly (vinyl chloride). J. Appl. Polym. Sci. **14**(3), 713–721 (1970)
39. Levchik, S.V., Weil, E.D.: Overview of the recent literature on flame retardancy and smoke suppression in PVC. Polym. Adv. Technol. **16**(10), 707–716 (2005)

40. Benisek, L.: Flame retardance of protein fibers. In: Lewin, M., Atlas, S.M., Pearce, E.M. (eds.) Flame-Retardant Polymeric Materials, pp. 137–191. Springer US, Boston, MA (1975)
41. Morgan, P., Kwolek, S.: Polymides from phenylenediamines and aliphatic diacids. Macromolecules **8**(2), 104–111 (1975)
42. Zhang, Z., et al.: Effect of the flame retardants and glass fiber on the polyamide 66/polyphenylene oxide composites. Materials **15**(3), 813 (2022)
43. Li, Y., Liu, K., Xiao, R.: Preparation and characterization of flame-retarded polyamide 66 with melamine cyanurate by in situ polymerization. Macromol. Res. **25**(8), 779–785 (2017)
44. Gao, J., et al.: A facile method to improve thermal stability and flame retardancy of polyamide 6. Compos. Commun. **13**, 143–150 (2019)
45. Chen, Y., Zang, C., Zhu, X.: Toughening and strengthening modification of flame-retardant high-impact polystyrene. In: Journal of Physics: Conference Series. IOP Publishing (2023)
46. Balabanovich, A., Engelmann, J.: Fire retardant and charring effect of poly (sulfonyldiphenylene phenylphosphonate) in poly (butylene terephthalate). Polym. Degrad. Stab. **79**(1), 85–92 (2003)
47. Yan, H., et al.: Flame-retardant performance of polystyrene enhanced by polyphenylene oxide and intumescent flame retardant. Polym.-Plast. Technol. Eng. **53**(4), 395–402 (2014)
48. Xiong, Z., et al.: Green and scalable fabrication of core-shell biobased flame retardants for reducing flammability of polylactic acid. ACS Sustain. Chem. Eng. **7**(9), 8954–8963 (2019)
49. Chow, W.S., Teoh, E.L., Karger-Kocsis, J.: Flame retarded poly(lactic acid): a review. Express Polym. Lett. **12**, 396–417 (2018)
50. Ma, X., et al.: Fabrication of highly efficient phenylphosphorylated chitosan bio-based flame retardants for flammable PLA biomaterial. Carbohyd. Polym. **287**, 119317 (2022)
51. Tawiah, B., et al.: Flame retardant poly(lactic acid) biocomposites reinforced by recycled wool fibers—thermal and mechanical properties. Express Polym. Lett. **13**, 697–712 (2019)
52. Réti, C., et al.: Flammability properties of intumescent PLA including starch and lignin. Polym. Adv. Technol. **19**(6), 628–635 (2008)
53. Yang, H., et al.: Lignin-derived bio-based flame retardants toward high-performance sustainable polymeric materials. Green Chem. **22**(7), 2129–2161 (2020)
54. Mokhena, T.C., et al.: Flame retardancy efficacy of phytic acid: an overview. J. Appl. Polym. Sci. **139**(27), e52495 (2022)
55. Mokhena, T.C., et al.: Mechanical properties of fire retardant wood-plastic composites: a review. Express Polym Lett **15**, 744–780 (2021)
56. Wang, M., et al.: Bio-based flame retardants to polymers: a review. Adv. Indus. Eng. Polym. Res. **6**(2), 132–155 (2023)
57. Mokhena, T.C., John, M.J.: Cellulose nanomaterials: new generation materials for solving global issues. Cellulose **27**(3), 1149–1194 (2020)
58. Mokhena, T.C., Jacobs, N.V., Luyt, A.S.: Nanofibrous alginate membrane coated with cellulose nanowhiskers for water purification. Cellulose **25**(1), 417–427 (2018)
59. Mokhena, T.C., et al.: Mechanical properties of cellulose nanofibril papers and their bionanocomposites: a review. Carbohyd. Polym. **273**, 118507 (2021)
60. Mokhena, T.C., John, M.J.: Esterified cellulose nanofibres from saw dust using vegetable oil. Int. J. Biol. Macromol. **148**, 1109–1117 (2020)
61. Shen, D., et al.: The overview of thermal decomposition of cellulose in lignocellulosic biomass. Cellulose-Biomass Convers. 193–226 (2013)
62. Costes, L., et al.: Bio-based flame retardants: when nature meets fire protection. Mater. Sci. Eng. R. Rep. **117**, 1–25 (2017)
63. Zhang, D., et al.: Flame-retardant properties and mechanism of polylactic acid-conjugated flame-retardant composites. Front. Chem. **10**, 894112 (2022)
64. Wu, K., et al.: Flame retardancy and thermal degradation of intumescent flame retardant starch-based biodegradable composites. Ind. Eng. Chem. Res. **48**(6), 3150–3157 (2009)
65. Yang, Y., et al.: Nanocarbon-based flame retardant polymer nanocomposites. Molecules **26**(15), 4670 (2021)

66. Wang, X., et al.: Flame retardancy and thermal degradation of intumescent flame retardant poly (lactic acid)/starch biocomposites. Ind. Eng. Chem. Res. **50**(2), 713–720 (2011)
67. Wang, X., et al.: Flame retardancy and thermal degradation of intumescent flame retardant poly(lactic acid)/starch biocomposites. Ind. Eng. Chem. Res. **50**(2), 713–720 (2011)
68. Wang, J., et al.: Improved flame-retardant properties of poly(lactic acid) foams using starch as a natural charring agent. Ind. Eng. Chem. Res. **53**(4), 1422–1430 (2014)
69. Maqsood, M., Seide, G.: Improved thermal processing of polylactic acid/oxidized starch composites and flame-retardant behavior of intumescent non-wovens. Coatings **10**(3), 291 (2020)
70. Fahami, A., et al.: Mica-based multilayer nanocoating as a highly effective flame retardant and smoke suppressant. ACS Appl. Mater. Interfaces. **12**(17), 19938–19943 (2020)
71. Zhang, S., et al.: The novel application of chitosan: effects of cross-linked chitosan on the fire performance of thermoplastic polyurethane. Carbohyd. Polym. **189**, 313–321 (2018)

Alginate as Flame Retardant: Synthesis, Structure, Properties, and Applications

L. K. Letsitsa, A. Kaleni, N. Jafta, M. J. Mochane, L. Theys, L. K. Lebelo, T. C. Mokhena, and M. T. Motloung

Abstract Alginate serves as a structural component of brown algae and assist with the shape of the plant. Alginate is a hydrophilic anionic polysaccharide derived from brown algae, bacteria and has been attracting attention as a promising flame-retardant material in recent years. Alginate is an unbranched polysaccharide composed of two linear copolymer blocks, L-guluronic acid (G) and D-mannuronic acid (M), which provides better thermal stability and flame retardancy to polymer matrices. These properties enhance alginate's high temperature resistance, making it a strong choice for use in flame retardant applications. Alginate is a biodegradable material that acts as a flame retardant by releasing water vapor when exposed to heat or flame, helping the material cool and preventing the spread of fire. Alginate flame retardants typically produce very little smoke when exposed to fire, thus contributing to improved fire safety. The versatility of alginate is also demonstrated by its compatibility with a wide variety of materials, allowing easy integration into fibres, coatings, composites, and foams. Incorporation of alginates into these matrices significantly improves the fire performance of the resulting materials, making them an attractive solution for improving fire safety across a wide range of applications. As research continues to advance, the incorporation of alginate into matrix materials is expected to increase and advance fire safety progress and promote sustainable practices in all industries. This chapter discusses the flame retardancy of alginate incorporated in

L. K. Letsitsa · A. Kaleni · N. Jafta · M. J. Mochane (✉) · L. Theys · M. T. Motloung
Department of Life Sciences, Central University of Technology, Bloemfontein, Free State, South Africa
e-mail: mochane.jonas@gmail.com

L. K. Lebelo
Occupational Health Division, School of Public Health, Faculty of Health Sciences, University of the Witwatersrand, Johannesburg, South Africa

T. C. Mokhena
DSI/Mintek-Nanotechnology Innovation Centre, Advanced Materials, Mintek, 2125 Randburg, South Africa

M. J. Mochane
Department of Chemistry, College of Science, University of South Africa, Johannesburg, South Africa

T. C. Mokhena et al. (eds.), *Biomaterials as Green Flame Retardants*, Engineering Materials, https://doi.org/10.1007/978-981-97-6871-4_3

various polymer matrices, and the structure of alginate that makes it an excellent flame-retardant filler.

Keywords Alginate flammability · Synthesis · Structure · Intumescent flame-retardant

1 Introduction

Algae are believed to be among the oldest life forms on Earth, dating back to the pre-Cambrian period. Marine algae have remained largely unchanged for millions of years after using light for photosynthesis. In contrast, other plants evolved stems, leaves, and roots after using light for photosynthesis [1]. Brown algae cell walls naturally contain an anionic polymer called alginate [2–4]. Alginate serves as a structural component in brown algae, helping to form the plant's shape. The gel matrix of alginate between the cells provides both sturdiness and pliability to the plant, allowing it to withstand water currents [4]. This is similar to cellulose in land plants. Figure 1 shows images of the algal species, including (A) the entire Padina boergesenii frond and (B) a diagram of the natural habitat of the NIOF site. The NIOF site is home to several marine plant species, including Turbinaria triquetra, Hormophysa cuneiformis, Dictyota ciliolata, and Sargassum. Each of these species can be observed in their natural habitat at the NIOF site, and whole fronds can also be observed [5].

The alginate composition varies among different seaweeds and it ranges from 20 to 60% in dry matter [2]. While alginate can also be produced by bacteria, all

Fig. 1 The seaweed species [6]

commercial alginate is derived from algae biomass. Alginates are hydrophilic anionic polysaccharides obtained from brown seaweed or bacteria [6–8]. These substances have varying chemical and physical properties due to their structural differences. Sodium alginate, the raw material, is produced by converting salts harvested from brown seaweed cell walls (class Phaeophyceae) into sodium, magnesium, calcium, strontium, and barium salts [5, 7]. One of the key characteristics of alginate is their ability to enhance flame retardancy. Herein, the chapter discusses history of algae, its properties, comparison to flame retardant fillers and its incorporation to various polymer matrices in relation to their flammability.

2 History of Alginate and a Brief Discussion on the Synthesis

Alginic acid, a natural polymer obtained from brown seaweed, was first discovered, and patented by British chemist E.C.C. Stanford on January 12, 1881 [1–4]. Stanford continued his research on the substance and helped to determine its chemical structure. His patent describes a process for extracting alginate from algae by soaking it in water or a diluted acid solution, then using sodium carbonate to extract it and finally precipitating the alginate by adding acid. Figure 2 shows a simplified diagram of the steps involved in extracting alginate from seaweed. The first step involves breaking down the plant cell wall with an acid or alkali. The seaweed biomass matrix is then extracted with sodium carbonate to obtain water-soluble alginate. After filtering the extract, a chemical called sodium or calcium chloride is added to make the alginate separate from the liquid. This alginate can then be changed into alginic acid by adding hydrochloric acid and further purification and conversion, a water-soluble sodium alginate powder is obtained [9, 10]. Alginate can be recovered from solution in three different ways: sodium alginate, calcium alginate, and alginic acid [5].

Krefting was the first to produce a purified form of alginic acid in 1896 and 1898 [1]. By 1927, alginates were available for commercial use. Today, around 40,000 tons of alginates are produced worldwide each year. They are used in various industries including food, pharmaceuticals, cosmetics, and dentistry due to their unique properties [11–13].

3 Selective Studies on Sources of Alginate

3.1 Algal Sources

The main source of alginates for commercial purposes, is the brown marine algae, sourced from the coastal region. There are various sources of whereby the brown algae is extracted from i.e., *Macrocystis pyrifera*, *Ascophyllum nodosum*, and *Laminaria*

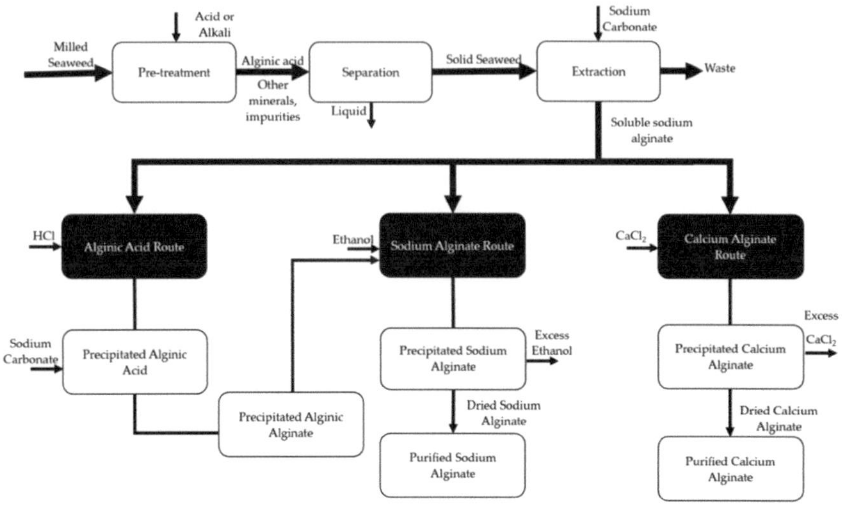

Fig. 2 The process of extracting alginate from seaweed [5]

hyperborean [3]. There are also other sources for extraction of alginates such as *Laminaria digitata*, *Ecklonia maxima* and *Lamina japonica*. Belattmania et al. [13] reported on the isolation and characterization of alginates found at the Atlantic Coast of Morocco. There were 8 species of brown seaweed that were harvested from the region i.e., *Bifurcaria bifurcata*, *Sargassum muticum*, *Lamnaria ochroleura*, *Fucus guiryl*, *Saccorhiza polyschides*, *Fucus vesiculosus f. volubilis*, and *Cystoseira humilis*. The alginate yields of the 8 species were in the range of 2.7% to 27.5% dw. The highest yields were recorded for *S. muticum*, *C. tamariscifolia*, *C. humilis* and *F. vesiculosus* with an excess of 18% dw. The identification of the species was analysed by Fourier transform InfraRed (FTIR) spectroscopy with reference to commercial sodium alginate (Fig. 3).

Generally, the extracted species were found to be similar to that of the commercial sodium alginate (see Fig. 4). The peak found at the range of 1600–1610 cm^{-1} was assigned to the O–C–O carboxylate asymmetric stretching. Furthermore, the reported bands in the region of 1400–1428 cm^{-1} were associated with the C–OH deformation vibration. Trica et al. [14] extracted alginates from the brown seawood (*Cystoseira barbata*) produced from the Romanian black sea. *Cystoseria* is known as a polyphyletic genus (brown seaweed) which belong to the family of Sargassaceae which is mostly found in the Mediterranean Sea and eastern Atlantic Ocea [14, 15]. The extraction of alginate yielded 19 ± 1.5% (w/w), which was fine powder. The produced alginate was characterized by NMR in comparison to sodium alginate. General the NMR signals for alginates is recognized in three signals i.e., signal I is associated with anomeric proton, which denote the guluronic acid (G-1), signal II is assigned to the anomeric proton (M-1) and signal III is attributed to the proton H-5 guluronic acid (G-5).

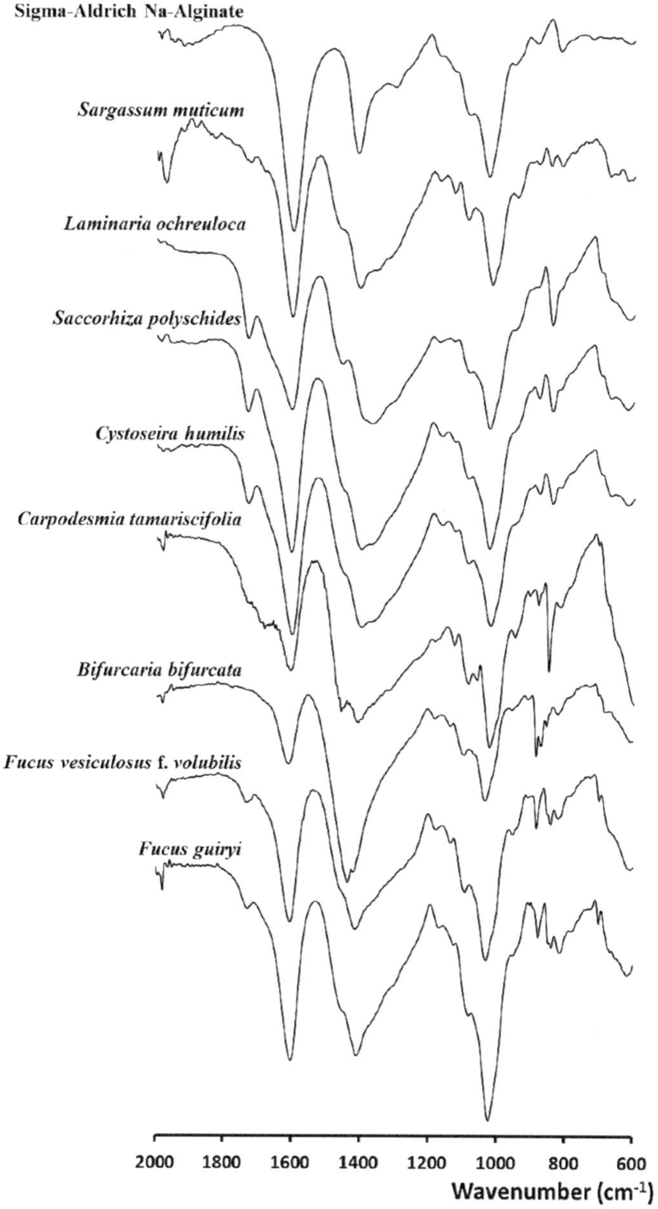

Fig. 3 FTIR spectra of all investigated samples [13]

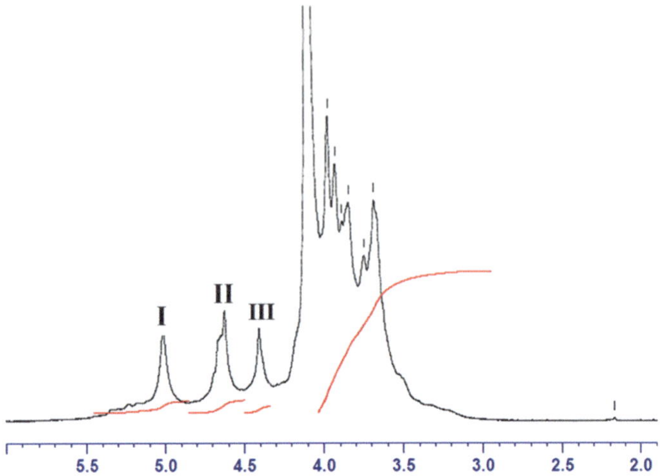

Fig. 4 1H NMR of the extracted alginate [14]

Flórez-Fernández and co-workers [16] reported on a green method for extraction of alginate from Sargassum muticum brown seaweed by utilizing ultrasound-assisted technique. Figure 5 depicts a typical example of extraction using ultrasound-assisted technique. The main aim of using this method is to minimize the utilization of chemicals, higher temperatures, and longer times of extractions. The precipitated alginate after ultrasound assisted extraction was analysed by FTIR-ATR and ^{1}H NMR. Sonification time was found to be key on the molar mass, rheology, and block structure. Sonification time used during extraction was found to decrease the lowest molecular weight i.e., <80 kDA. The strength of the alginate gels was found to vary, sonification conditions of <30 min resulted in alginate gels with intermediate strength.

4 Structure of Alginate and Its General Uses

Alginate is a biopolymer that is naturally found in brown seaweed and is made up of complex polysaccharides as defined in the abstract section. It is present as insoluble salts like calcium, magnesium, and sodium [5]. The chemical structure of alginic acid is complex and varies depending on the species from which it is extracted. These differences are thought to result from variations in the chain length of the polysaccharide [17, 18]. The chemical structure of polysaccharides is determined by factors such as their monosaccharide composition, sequence of monosaccharides, configuration and position of glycosidic linkages, solubility, and rheological properties [6]. Alginates are unbranched polysaccharides composed of two linear copolymer blocks, L-guluronic acid (G) and D-mannuronic acid (M) [8, 12, 19]. These blocks

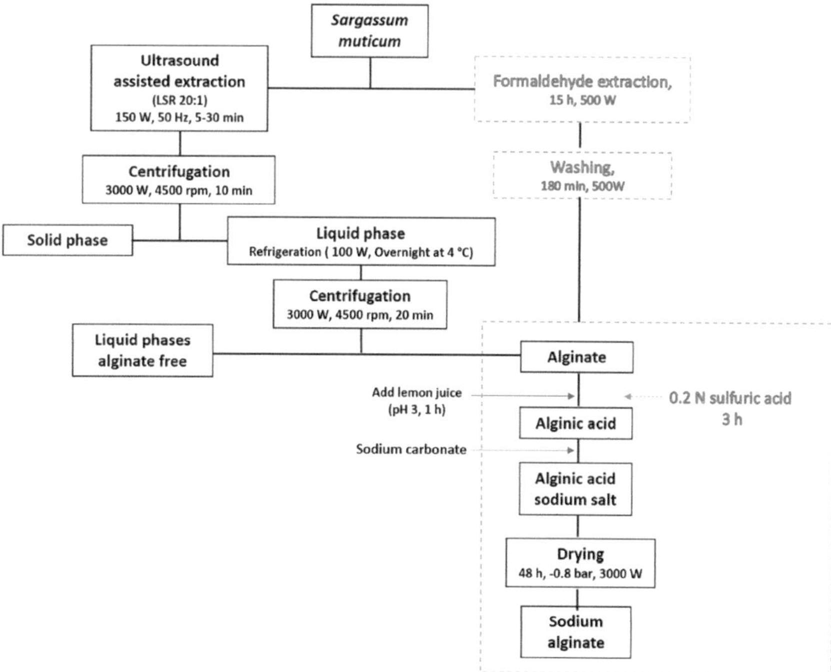

Fig. 5 Schematic drawing of the alginate extraction by using ultrasound-assisted technique [16]

are randomly arranged along the alginate chain of the GG, MM, and GM blocks. Alginates are composed of 1,4-D-mannuronic acid (M) and 1,4-L-guluronic acid (G) linked in 1–4 bonds [3, 4, 11, 18]. Figure 6 shows the structural alginates and shows how these monomers form homogeneous blocks (poly G or poly M) or heterogeneous blocks (MG). Alginate block copolymer structures containing D-mannuronic acid and L-guluronic acid can be determined by fractional [17]. The composition and sequence of mannuronic and guluronic acid blocks can vary among different alginate-producing species, resulting in differences in their alginate compositions.

Alginate gels are compatible with living tissues and are commonly used in the food industry to improve and maintain the structure of food products [7, 21]. They also been applied in the field of biomaterials for delivering drugs [7]. In addition to these uses, alginates have a wide range of other applications, such as textile coatings, paper sizing agents, mould release agents, and materials that resist flames. Alginate fibres, typically made from calcium alginate, are highly flame-retardant with a 34.0 oxygen index (LOI). This property, along with its non-toxicity, biocompatibility, hydrophilicity, and excellent formability, makes alginate a valuable material of choice [11, 18] in flame resistance applications. There is growing interest in alginates as a sustainable source of raw materials for a range of products, particularly hydrogel-polymer composites with applications in many fields [7].

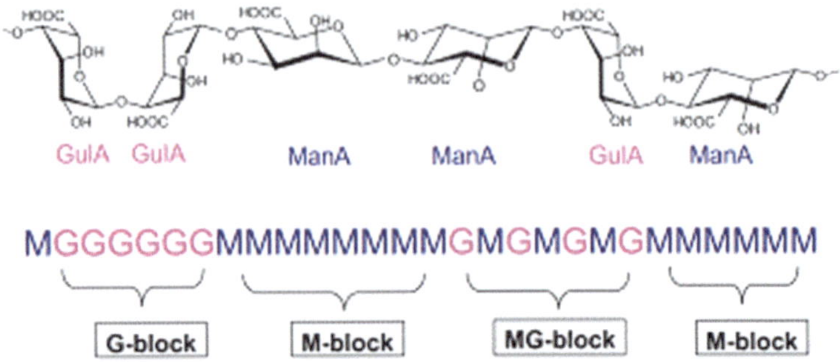

Fig. 6 Structural features of alginates [20]

5 Properties of Alginate

The properties of the alginate are discussed in relation to 4 parameters i.e., molecular weight, solubility, hydrogels, and stability. Alginates containing divalent and trivalent cations have a tendency of being insoluble in water due to the presence of the –COO- anion functional group[22]. However, it is worth mentioning that the alginates with monovalent such as Na^+, K^+, and NH_4^+ were found to be soluble in hot as well as cold water. Furthermore, the pH was found to be another component of importance in the solubility of the alginates. For an example, the pH of the alginate's solution has to be in a certain value in order for carboxylate groups to be deprotonated. The pH importance in alginate solution is the one ranging between 4–5, while the pH range of 5–11 does not affect the viscosity of the solution. The structure of alginate(s) was found to destabilize at pH of approximately 1 and a pH greater than 10 [23]. Alginates have a tendency of forming hydrogels, consisting of crosslinked polymer in a matrix in the form water. The crosslinking of alginate with G block contents results in better strength, ductility, hardness, and heat resistance when compared with the M block contents [24, 25]. The formation of the gel depends on crosslinking through three process i.e., diffusion, internal and cooling gelation. During the diffusion process, ions that diffused through the polymer chains are easily attracted by the G blocks. This is allowing a speedy gelation of higher extent. The *In-situ* gelation is the most common method of gelation which is associated with the atomic-scale mixing of ions in the alginate. The process take place by utilizing metal chlorides salts and nitrates. Another important factor is the thermoreversible hydrogels which is characterized by two types, i.e., gelation occurs during cooling and when heating them, hydrogels liquefy [26]. The reverse is also occurring during the thermogelling process, whereby the upon heating hydrogels are formed and upon cooling there is a liquid that is formed, and typical example of such systems is the solution of polymethacrylic acid [27]. The stability of the alginates is very important, because alginates are suitable candidate in flame retardancy applications. Alginic acid, which is the main component in the synthesis of alginates degrades faster than the calcium alginate and ammonium alginates. The stability of the alginates depends on various factors such as extent of crosslinking, graft polymerization, and the type of intercalating cation [28]. Various techniques have been utilized in order to examine the molecular weight of the alginates. Techniques such light scattering, viscosity measurements, sedimentation and diffusion were utilized to determine the molecular weight [29]. It was reported that the molecular weights of pure alginates range between 10 000–600 000 g/mol. Futhermore, it is reported that high molecular weight and high viscosity are designed by utilizing concentrated alginates solution with low molecular weight.

6 Alginate Applications as a Flame-Retardant Filler

Alginates are used in a wide range of industrial applications, including biotechnology, pharmaceuticals, chemicals, textiles, packaging, construction, food, aquaculture, dentistry, papermaking, crafts, and leisure [9, 30, 31]. Alginate is used in chemical engineering for emulsions, dispersions, wastewater treatment, and separation of oil from water [32]. In the textile industry, it enhances flame-retardant properties and is often used in protective clothing, upholstery, and curtains [31, 33]. Alginate can also be formulated into coatings for surfaces such as wood, paper, and metals to provide flame retardancy [34]. It has been applied as a flame-retardant filler in construction materials such as insulation, sealants, and adhesives. In addition to roof and wall insulation, alginate coatings are also suitable for timber frames, panels, and asphalt shingles. It can also be used indoors in the form of wall panels, ceiling panels, mouldings, countertops, tiles, fire doors, etc. Alginate coatings are suitable for furniture and vehicle interiors [7]. Table 1 illustrates various industrial uses of alginate.

There is a strong need for fire-resistant fibres and fabrics for various purposes like clothing, upholstery, bedding, and transportation[35]. However, concerns about the negative effects on health and the environment from using flame-retardant chemicals on textiles have grown [36]. Alginate materials have been effectively used to create fire resistant fibres through a wet spinning process that does not require harmful solvents. Calcium ions (Ca^{2+}) are often used for this purpose, along with various types of polyvalent metal ions, mainly because they are less toxic, inexpensive, and cause moderate gelation [36]. Calcium alginate fibers sometimes aren't very good at resisting salts and detergents because the calcium ions in them can be swapped out with other smaller ions [36]. To overcome their limitations, other types of metal ions like aluminium, copper, and barium are incorporated into the alginate since they stick well with the alginate. Figure 7 illustrates various application of the alginate, specifically for flame retardancy.

Alginates as film-forming materials can be combined with organic and inorganic additives to form flame-retardant coatings as matrix resins. Alginate can also be used as the negatively charged component of layered assemblies to construct flame-retardant coatings containing positively charged polyelectrolytes such as chitosan, branched polyethyleneimines, clays, carbon materials, and metal ions [7, 36]. To achieve satisfactory flame retardancy with a layer-by-layer deposition technique,

Table 1 Summary of the application and specific uses of alginates in industries [35]

Industry	Specific use
Biotechnology, bioengineering, biomedicine, and clinical	Dressings for wounds and burns
	Heavy metal chelator
	Scaffolding in tissue engineering
	3D bio-printing
Pharmaceutical industry	Controlled release
	Cancer therapy
Chemical, textile, packaging, and construction industry	Textile inks,
	Detergents,
	Adhesives
	Welding
	Building soundproofing
	Biodegradable containers
Food and drinks	Beverages and ice
	Paper industry binders
Paper industry	Binder
Aquaculture	Food supplements
Leisure industry	Cosmetics
Arts and crafts	Textile inks

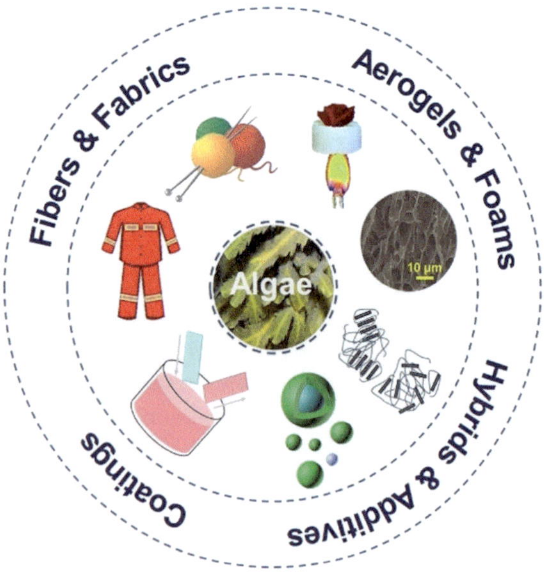

Fig. 7 An illustration of a flame-resistant alginate application [36]

many bilayers, such as 20/30, are usually deposited [36, 37]. Alginate, an anionic polymer, has an excellent ability to form carbides, so this process is slow and tedious and can cause significant weight gains of over 10%. This can ultimately cause deterioration of textiles' hand-feel, strength, and drape [37]. Since alginate fibres are polysaccharide materials capable of forming char, they are often combined with other polysaccharide fibres to form flame-retardant fabrics and fibres. Bio-based blended fibres and fabrics were combined with alginate fibres [33, 38] and the resultant product was found to be flame retardant. Alginates, which are plentiful and renewable polymers derived from biological sources and are inherently fire-resistant, have gained significant interest for creating flame-resistant aerogels [38, 39]. The combination of alginate and clay in these aerogels results in extremely lightweight structures with remarkable fire resistance properties.

7 Alginate as a Flame-Retardant Filler in Polymer-Based Materials

In recent years, the outstanding flame-retardant behavior of alginate and alginate derived nanoparticles has caught substantial awareness and infiltrated the polymer industry. Many researchers have focused their attention on studying alginates as eco-friendly reinforcing additives to polymer-based materials for advanced applications [7, 40, 41]. Cone calorimetry technique has been the preferred technique to analyze the the flame retardancy of alginates in polymer composites. This technique primarily measures the decline rate in oxygen concentration of combustible gases of a specimen at a standard heat flux rate of 10–100 kW/m^2 [7]. In order to successfully evaluate flame retardancy using the cone calorimetry technique, various parameters are studied and reported, such as:

- Time to ignition (TTI) which refers to: (1) time of combustion (TOF) and mass loss during combustion.
- Heat release rates (HRR's) which refers to: (1) peak heat release rate (PHRR), (2) total heat release rate (THRR), and (3) fire growth rate index (FGRI).
- Smoke density.
- Oxygen and carbon dioxide concentrations.
- Total smoke released (TSR); and
- Exhaust gas flow rate.

Pallmann et al. [42] utilized ammonium polyphosphate, phosphorylated and ion exchanged sodium alginate, and dipentaerythritol to formulate an intumescent flame retardant (IFR) additive for reinforcement of polypropylene (PP). The cone calorimetry results revealed that neat PP had a PHRR of 782 kW/m^2 after 190 s with the TTI starting at 15 s. The minimum HRR was observed at 709 kW/m^2 with a sharp decline after 240 s. This can be seen as an indication of the quick combustion rate of neat PP. The incorporation of IFR's into the PP matrix was observed to almost double the TTI to 27 s. The PHRR is seen at 208 kW/m^2 after 70 s which is attributed to the effective char formation by the IFR. Moreover, The SEM results support this statement as PP/IFR reinforced polymer composites were reported to have a smooth surface with no irregular curvatures compared to neat PP composites. Additionally, observation of the surface residue of the PP/IFR composites showed that there was no disintegration of the char layer formed by IFR's, indicating successful barrier against pyrolysis gases and oxygen [42]. Figures 8 and 9 depicts photographic images of neat PP and PP/IFR before and after cone calorimetric test.

Considering Pallmann et al. [42] results, it can be said that combining alginate with other flame-retardant fillers or additives can further enhance its flammability resistance in polymer composites. Kundu et al. [43] and Zhang et al. [44] utilized chitosan in addition to sodium alginate and alginate coated ammonium polyphosphate (APP) to reinforce polyamide 66 (PA 66) fabric and polylactic acid (PLA),

Fig. 8 Neat PP before (**a**) and after (**b**) cone calorimetric testing [47]. [Pallmann et al. 2018. Journal of Applied Polymer Science. Wiley]

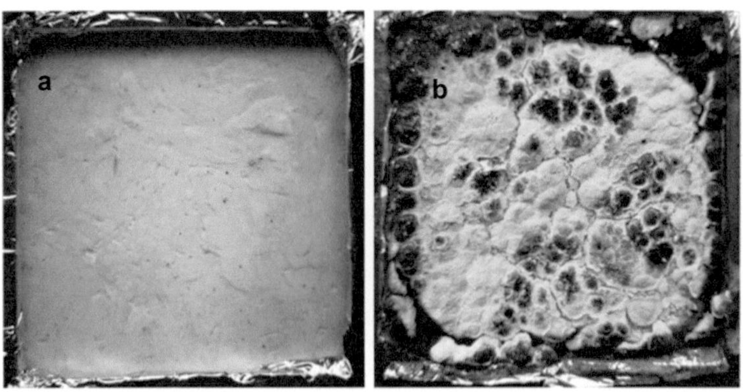

Fig. 9 PP/IFR before (**a**) and after (**b**) cone calorimetric testing [42]. [Pallamnn et al. 2018. Journal of Applied Polymer Science. Wiley]

respectively. The authors examined the effect of the fillers on the flammability resistance of PA 66 and PLA using cone calorimetry and UL-94 techniques. The UL-94 technique utilizes a Bunsen burner to ignite a specimen for 10 s, then the time taken for the specimen to self-extinguish is recorded. A flame is applied for a second time and time to extinguish was recorded. Three significant parameters are considered when analysing a sample using the UL-94 technique: (1) Time taken by the specimen to self-extinguish, (2) melt dripping of the specimen onto cotton pad placed underneath the specimen and (3) whether the dripping causes the cotton pad to ignite. Kundu et al. [43] used chitosan (CS), phytic acid (PA) and oxidized sodium alginate (OSA) in a quadralayer fashion denoted as $(CS\text{-}PA\text{-}CS\text{-}OSA)_n$, where “n” represents the number of quadralayers, to enhance the flammability resistance of PA 66. The vertical burning test (UL-94) results showed that PA 66 samples treated with $(CS\text{-}PA\text{-}CS\text{-}OSA)_{10}$ and $(CS\text{-}PA\text{-}CS\text{-}OSA)_{15}$ effectively reduced melting dripping of the specimen. The PHRR values were reduced for all samples treated with (CS-PA-CS-OSA) and the lowest PHRR was observed with $(CS\text{-}PA\text{-}CS\text{-}OSA)_5$ [43]. Zhang et al. [44] fabricated a core–shell fire resistant additive to PLA where the core is APP and the shell is alginate (AA) and chitosan (CS) bilayer (BL). The consequent fire retardant is APP@CS@AA-nBL (where “n” is also the number of bilayers) in a PLA matrix. All samples (except neat PLA) were reinforced with 10 wt% of the APP@CS@AA-nBL in a PLA matrix having 1-3BL. The authors reported that at 10 wt% APP@CS@AA-3BL, the composites exhibited the best flammability resistance having the upmost reduction in PHRR by 23.1% compared to neat PLA. Furthermore, 10 wt% APP@CS@AA-3BL sample achieved a V-0 UL-94 rating which qualifies it as a flame-retardant composite [44]. The use of eco-friendly fillers such as chitosan and alginate in polymer composites are an indication of the progress in research towards green technology. Further to this, other eco-friendly and biodegradable polymer reinforcing materials has been explored by researchers such as John [45] in recent years. The author investigates the treatment of flax fibre with alginate as compatibilizers in order to reinforce PLA and PHBV composites. Figure 10 shows the preparation process of the samples using marine biomass derived alginates and flax nonwovens.

Cone calorimetry results showed that the untreated flax fabric had a TTI of 17 s and its PHRR reached 122 kW/m^2. For the alginate-treated flax fabric, the TTI was 10 s and the PHRR reached 109 kW/m^2. However, a more than 50% reduction in the total amount of smoke emitted was observed in alginate-treated samples. This result suggests that alginate can be used as an excellent smoke suppressant and an effective flame retardant filler for polymer matrices [45]. A flame-retardant composite based on alginate and natural polymer in the form cotton was produced by Wang, et al. [46]. The cotton and alginate fibers were prepared by the carding machine at the ratio of cotton/alginate of 4/1, 5/5 and 1/4. Based on the ratios, the samples were named as cotton4/alginate 1, cotton5/alginate5 as well as cotton/alginate 4. Neat cotton was found to catch fire more easily, and after 4 s the whole cotton was on fire (see Fig. 11). The flammability resistance of the blend was reduced with the addition of the alginate fibres into the system. Furthermore, the cotton5/alginate5 extinguished once and produced char residues.

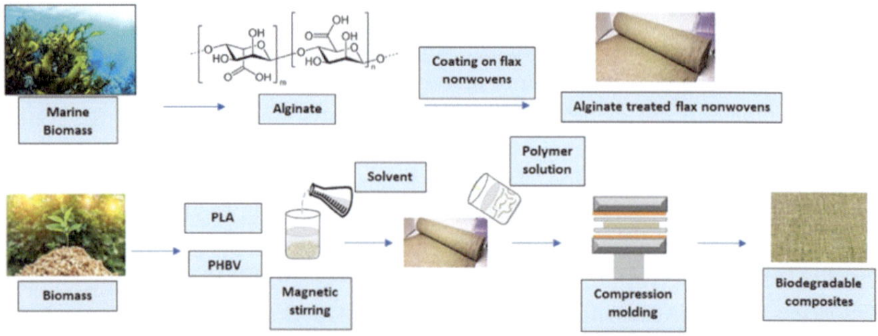

Fig. 10 Development process of composites [45] [Maya. 2022. Current Research in Green and Sustainable Chemistry. Elsevier. OPEN ACCESS]

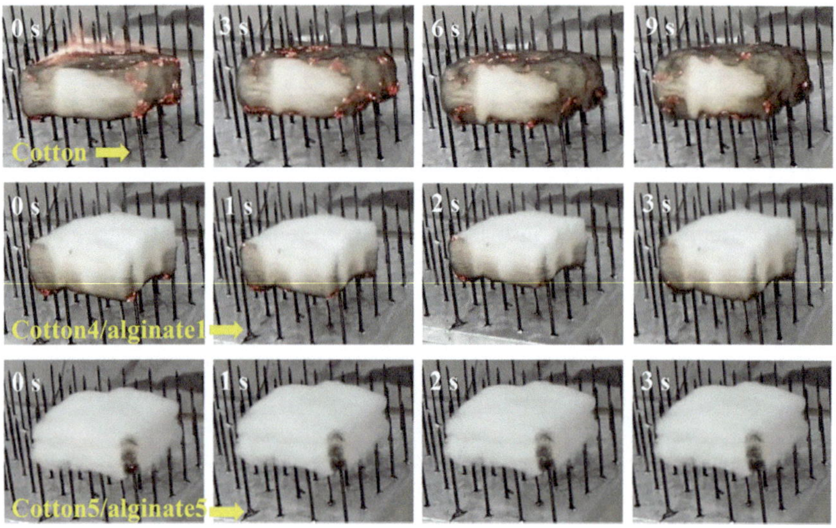

Fig. 11 Photos of the flammability test of the investigated samples at various times [46]

The flammability test was further investigated by cone calorimetry, whereby the neat cotton burned quickly with PHRR value of 95 kW/m^2, with a time to ignition value of 21 s [46]. The alginates fibres revealed a PHRR of 47 kW/m^2 with a TTI value of 100 s (see Fig. 12). The incorporation of the alginate into the cotton reduced the PHRR by 9.47%, 30.53% and 56.84%, for cotton4/alginate1, cotton5/alginate5 and cotton1/alginate4, respectively. The formation of char residues in the presence of alginates in blends prevents heat from entering the system and as a result lower HRR values. The carbon dioxide production (CO_2P) and total smoke production values of the cotton/alginates were lower that the neat cotton. This is evidence that the alginates fibres are capable of reducing smoke during fire.

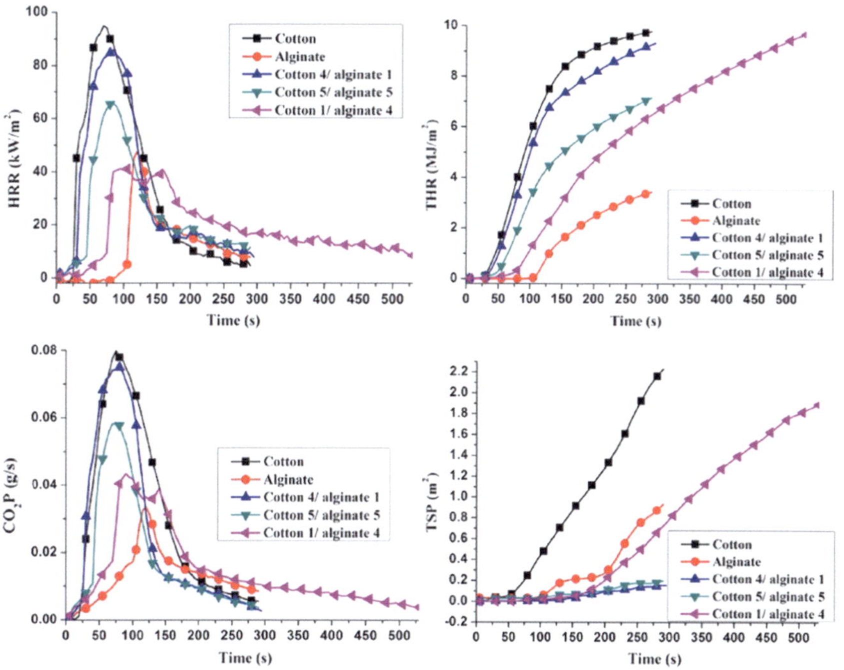

Fig. 12 **a** HRR, **b** THR, **c** CO_2P and **d** TSP of all the investigated samples [46]

8 Conclusion and Future Recommendations

Alginate-based flame retardants are a promising filler(s) to improve the fire resistance of polymeric materials. Their unique structure enables efficient dispersion within the polymer matrix, resulting in an improved flame-retardant property. Alginate's ability to form a charred protective layer reduces heat and mass transfer and slow the spread of fire. Alginate flame retardants have low toxicity, biodegradable, and environmentally friendly, making them a desirable alternative to traditional halogenated or toxic flame retardants. They can be incorporated into various polymers such as polyethylene, polypropylene, and polyurethane to improve fire resistance without significantly affecting other material properties. The addition of alginates within polymer matrices was shown to have improved the flame retardancy of various polymer matrices. The ability of alginates in improving the flame retardancy was due to the formation of effective and compact char residues, and such chars are known to prevent heat from entering the system and block volatile gases from leaving the system. As an environmentally friendly flame-retardant filler, there is a need in the market for fabrication of more flame-retardant green hybrid natural fiber biocomposites for sustainable advanced applications.

References

1. Mathis, H. A.: The History of Alginate Chemistry. Shock Vib. Dig. **25**, 10–14 (1993)
2. Pereira, L., Cotas, J.: Introductory chapter: alginates—a general overview. Alginates - Recent Uses This Nat. Polym. 1–16 (2020). https://doi.org/10.5772/intechopen.88381
3. Hasnain, M.S., Jameel, E., Mohanta, B., Dhara, A.K., Alkahtani, S., Nayak, A.K.: Alginates: Sources, Structure, and Properties. INC (2020)
4. Draget, K.I.: Alginates. Handb. Hydrocoll. Second Ed. 807–828 (2009). https://doi.org/10.1533/9781845695873.807
5. Saji, S., Hebden, A., Goswami, P., Du, C.: A brief review on the development of alginate extraction process and its sustainability. 1–20 (2022)
6. Rashedy, S.H., Abd El Hafez, M.S.M., Dar, M.A., Cotas, J., Pereira, L.: Evaluation and characterization of alginate extracted from brown seaweed collected in the red sea. Appl. Sci. **11** (2021). https://doi.org/10.3390/app11146290
7. Kabir, I.I., Sorrell, C.C., Mofarah, S.S., Yang, W., Yuen, A.C.Y., Nazir, M.T., Yeoh, G.H.: Alginate/polymer-based materials for fire retardancy: synthesis, structure, properties, and applications. Polym. Rev. **61**, 357–414 (2021). https://doi.org/10.1080/15583724.2020.1801726
8. Kijjoa, A., Sawangwong, P.: Marine drugs. Mar. Drugs. 73–82 (2004)
9. Lee, K.Y., Mooney, D.J.: Alginate: properties and biomedical applications. Prog. Polym. Sci. **37**, 106–126 (2012). https://doi.org/10.1016/j.progpolymsci.2011.06.003
10. Yong, K., Mooney, D.J.: Progress in polymer science alginate: properties and biomedical applications. Prog. Polym. Sci. **37**, 106–126 (2012). https://doi.org/10.1016/j.progpolymsci.2011.06.003
11. Kong, Q.S., Wang, B.B., Ji, Q., Xia, Y.Z., Guo, Z.X., Yu, J.: Thermal degradation and flame retardancy of calcium alginate fibers. Chinese J. Polym. Sci. (English Ed.) **27**, 807–812 (2009). https://doi.org/10.1142/S0256767909004527
12. Rosiak, P., Latanska, I., Paul, P., Sujka, W., Kolesinska, B.: Modification of alginates to modulate their physic-chemical properties and obtain biomaterials with different functional properties (2021)
13. Belattmania, Z., Kaidi, S., Atouani, S.E., Katif, C., Bentiss, F., Jama, C., Reani, A., Sabour, B., Vasconcelos, V.: Isolation and FTIR-ATR and 1H NMR characterization of alginates from the main alginophyte species of the atlantic coast of Morocco. Molecules **25**, 1–9 (2020). https://doi.org/10.3390/molecules25184335
14. Trica, B., Delattre, C., Gros, F., Ursu, A.V., Dobre, T., Djelveh, G., Michaud, P., Oancea, F.: Extraction and characterization of alginate from an edible brown seaweed (*Cystoseira barbata*) harvested in the Romanian Black Sea. Mar. Drugs. **17** (2019). https://doi.org/10.3390/md17070405
15. Marin, O.A., Timofte, F.: Atlasul macrofitelor de la litoralul românesc (2011)
16. Flórez-Fernández, N., Domínguez, H., Torres, M.D.: A green approach for alginate extraction from *Sargassum muticum* brown seaweed using ultrasound-assisted technique. Int. J. Biol. Macromol. **124**, 451–459 (2019). https://doi.org/10.1016/j.ijbiomac.2018.11.232
17. Guo, X., Wang, Y., Qin, Y., Shen, P., Peng, Q.: Structures, properties and application of alginic acid: a review. Int. J. Biol. Macromol. (2020). https://doi.org/10.1016/j.ijbiomac.2020.06.180
18. Zheng, C., Sun, Y., Cui, Y., Yang, W., Lu, Z., Shen, S., Xia, Y., Xiong, Z.: Superhydrophobic and flame-retardant alginate fabrics prepared through a one-step dip-coating surface-treatment. Cellulose **28**, 5973–5984 (2021). https://doi.org/10.1007/s10570-021-03890-y
19. Cattelan, G., Guerrero Gerbolés, A., Foresti, R., Pramstaller, P.P., Rossini, A., Miragoli, M., Caffarra Malvezzi, C.: Alginate formulations: current developments in the race for hydrogel-based cardiac regeneration. Front. Bioeng. Biotechnol. **8** (2020). https://doi.org/10.3389/fbioe.2020.00414
20. Zdiri, K., Elamri, A., Erard, A., Salaun, F.: Alginate-based bio-composites and their potential applications (2022)

21. Jeong, C., Kim, S., Lee, C., Cho, S., Kim, S.B.: Changes in the physical properties of calcium alginate gel beads under a wide range of gelation temperature conditions. Foods **9** (2020). https://doi.org/10.3390/foods9020180
22. Kabir, I.I., Sorrell, C.C., Mofarah, S.S., Yang, W., Yuen, A.C.Y., Nazir, M.T., Yeoh, G.H.: Alginate/polymer-based materials for fire retardancy: synthesis, structure, properties, and applications. Polym. Rev. **61**, 357–414 (2021). https://doi.org/10.1080/15583724.2020.1801726
23. Ribeiro, A.J., Silva, C., Ferreira, D., Veiga, F.: Chitosan-reinforced alginate microspheres obtained through the emulsification/internal gelation technique. Eur. J. Pharm. Sci. **25**, 31–40 (2005). https://doi.org/10.1016/j.ejps.2005.01.016
24. Paredes Juárez, G.A., Spasojevic, M., Faas, M.M., de Vos, P.: Immunological and technical considerations in application of alginate-based microencapsulation systems. Front. Bioeng. Biotechnol. **2**, 1–15 (2014). https://doi.org/10.3389/fbioe.2014.00026
25. Skjåk-Bræk, G., Grasdalen, H., Smidsrød, O.: Inhomogeneous polysaccharide ionic gels. Carbohydr. Polym. **10**, 31–54 (1989). https://doi.org/10.1016/0144-8617(89)90030-1
26. Yoshioka, H., Mori, Y., Tsukikawa, S., Kubota, S.: Polym. Adv. Technol. **9**, 155–158 (1998)
27. Eliassaf, J., Silberberg, A.: The gelation of aqueous solutions of polymethacrylic acid. Polymer (Guildf). **3**, 555–564 (1962). https://doi.org/10.1016/0032-3861(62)90103-9
28. da Silva, T.L., Vidart, J.M.M., da Silva, M.G.C., Gimenes, M.L., Vieira, M.G.A.: Alginate and sericin: environmental and pharmaceutical applications. Biol. Act. Appl. Mar. Polysaccharides. (2017). https://doi.org/10.5772/65257
29. Fujihara, M., Nagumo, T.: Determination of molecular weights of algianates by agarose gel filtration (1989)
30. Puscaselu, R.G., Lobiuc, A., Dimian, M., Covasa, M.: Alginate: from food industry to biomedical applications and management of metabolic disorders. Polymers (Basel). **12**, 1–30 (2020). https://doi.org/10.3390/polym12102417
31. Zhang, Y., Li, T.T., Shiu, B.C., Lin, J.H., Lou, C.W.: Multifunctional sodium Alginate@urushiol fiber with targeted Antibacterial, acid corrosion resistance and flame retardant properties for personal protection based on wet spinning. Appl. Surf. Sci. **584**, 152573 (2022). https://doi.org/10.1016/j.apsusc.2022.152573
32. Gady, O., Poirson, M., Vincent, T., Sonnier, R., Guibal, E., Gady, O., Poirson, M., Vincent, T., Sonnier, R., Guibal, E., Gady, O., Poirson, M., Vincent, T., Sonnier, R., Guibal, E.: Elaboration of light composite materials based on alginate and algal biomass for flame retardancy: preliminary tests To cite this version: HAL Id : hal-02906438 Elaboration of light composite materials based on alginate and algal biomass for flame retar (2020)
33. Li, S.Q., Tang, R.C., Yu, C.B.: Flame retardant treatment of jute fabric with chitosan and sodium alginate. Polym. Degrad. Stab. **196**, 109826 (2022). https://doi.org/10.1016/j.polymdegradstab.2022.109826
34. Shen, J., Liang, J., Lin, X., Lin, H., Yu, J., Wang, S.: The flame-retardant mechanisms and preparation of polymer composites and their potential application in construction engineering. Polymers (Basel). **14** (2022). https://doi.org/10.3390/polym14010082
35. Hurtado, A., Aljabali, A.A.A., Mishra, V., Tambuwala, M.M., Serrano-Aroca, Á.: Alginate: enhancement strategies for advanced applications. Int. J. Mol. Sci. **23** (2022). https://doi.org/10.3390/ijms23094486
36. Xu, Y.-J., Qu, L.-Y., Liu, Y., Zhu, P.: An overview of alginates as flame-retardant materials: pyrolysis behaviors, flame retardancy, and applications. Carbohyd. Polym. **260**, 117827 (2021). https://doi.org/10.1016/j.carbpol.2021.117827
37. Kundu, C.K., Hossen, M.T., Islam, T., Mollick, S., Song, L., Hu, Y.: Flame retardant coatings from bio-derived chitosan, sodium alginate, and metal salts for polyamide 66 textiles. ACS Omega **7**, 30841–30848 (2022). https://doi.org/10.1021/acsomega.2c02466
38. Li, P., Wang, Q.Z., Wang, B., Liu, Y.Y., Xu, Y.J., Liu, Y., Zhu, P.: Blending alginate fibers with polyester fibers for flame-retardant filling materials: thermal decomposition behaviors and fire performance. Polym. Degrad. Stab. **183**, 109470 (2021). https://doi.org/10.1016/j.polymdegradstab.2020.109470

39. Xu, Y.J., Qu, L.Y., Liu, Y., Zhu, P.: An overview of alginates as flame-retardant materials: pyrolysis behaviors, flame retardancy, and applications. Carbohydr. Polym. **260**, 117827 (2021). https://doi.org/10.1016/j.carbpol.2021.117827
40. Chen, W., Liu, Y., Liu, Y., Wang, Q.: Preparation of alginate flame retardant containing P and Si and its flame retardancy in epoxy resin. J. Appl. Polym. Sci. **134** (2017). https://doi.org/10.1002/app.45552
41. Yu, Z., Liu, J., He, H., Ma, S., Yao, J.: Flame-retardant PNIPAAm/sodium alginate/polyvinyl alcohol hydrogels used for fire-fighting application: preparation and characteristic evaluations. Carbohydr. Polym. **255**, 117485 (2021). https://doi.org/10.1016/j.carbpol.2020.117485
42. Pallmann, J., Ren, Y.L., Mahltig, B., Huo, T.G.: Phosphorylated sodium alginate/APP/DPER intumescent flame retardant used for polypropylene. J. Appl. Polym. Sci. **136**, 1–10 (2019). https://doi.org/10.1002/app.47794
43. Kumar Kundu, C., Wang, W., Zhou, S., Wang, X., Sheng, H., Pan, Y., Song, L., Hu, Y.: A green approach to constructing multilayered nanocoating for flame retardant treatment of polyamide 66 fabric from chitosan and sodium alginate. Carbohydr. Polym. **166**, 131–138 (2017). https://doi.org/10.1016/j.carbpol.2017.02.084
44. Zhang, Y., Xiong, Z., Ge, H., Ni, L., Zhang, T., Huo, S., Song, P., Fang, Z.: Core-shell bioderived flame retardants based on chitosan/alginate coated ammonia polyphosphate for enhancing flame retardancy of polylactic acid. ACS Sustain. Chem. Eng. **8**, 6402–6412 (2020). https://doi.org/10.1021/acssuschemeng.0c00634
45. John, M.J.: Biobased alginate treatments on flax fibre reinforced PLA and PHBV composites. Curr. Res. Green Sustain. Chem. **5**, 100319 (2022). https://doi.org/10.1016/j.crgsc.2022.100319
46. Wang, B., Ping, L.P., Xu, Y.-J., Jiang, Z.-M., Dong, C.-H., Liu, Y., Zhu, P.: Bio-based, nontoxic and flame-retardant cotton/alginate blended fibres as filling materials: thermal degradation properties, flammability and flame-retardant mechanism. Compos. B Eng. **194**, 108038 (2020). https://doi.org/10.1016/j.compositesb.2020.108038

Phytic Acid: A Novel Phosphate Bio-Based Flame Retardant

S. T. Sikhosana, Mokgaotsa Jonas Mochane, N. J. Malebo, T. C. Mokhena, T. G. Mofokeng, and E. R. Sadiku

Abstract The development and application of sustainable flame-retardant agents have become a research hotspot due to the urgent need for eco-friendly materials and the demand for a sustainable environmental protection. Phytic acid has gained popularity as a natural substance that is present in plant seeds with the potential to be a bio-based flame retardant. Its distinct structure, which consists of six phosphate groups, connected to a myoinositol core, confers inherent flame retardancy, making it a suitable candidate for fire safety in polymeric materials. Phytic acid can be added to polymer matrices through physical mixing, chemical grafting, or covalent bonding. This alters the thermal and mechanical properties of the resulting flame-retardant composite, as well as its fire performance. Ongoing research aims to maximize the potential of phytic acid in different polymeric systems, with the goal of creating safe, eco-friendly, and sustainable materials for fire safety applications in fields, such as: construction, automotive, electronics, and textiles.

Keywords Phytic acid · Biobased filler · Phosphate · Flame retardancy · Synergy

---

S. T. Sikhosana · M. J. Mochane (✉) · N. J. Malebo
Department of Life Sciences, Central University of Technology, Bloemfontein, Free State, South Africa
e-mail: mochane.jonas@gmail.com

T. C. Mokhena
DSI/Mintek-Nanotechnology Innovation Centre, Advanced Materials, Mintek, Randburg 2125, South Africa

T. G. Mofokeng
Centre for Nanostructures and Advanced Materials, DSI-CSIR Nanotechnology Innovation Centre, Council for Scientific and Industrial Research, Pretoria 0001, South Africa

E. R. Sadiku
Institute of NanoEngineering Research (INER), Department of Chemical, Metallurgical and Materials Engineering, Tshwane University of Technology, RSA, Pretoria 0001, South Africa

M. J. Mochane
Department of Chemistry, College of Science, University of South Africa, Johannesburg, South Africa

T. C. Mokhena et al. (eds.), *Biomaterials as Green Flame Retardants*, Engineering Materials, https://doi.org/10.1007/978-981-97-6871-4_4

1 Introduction

Flame-retardants refer to a class of materials that are specifically created to decrease the flammability of other materials [1]. Flame retardant materials are widely used to enhance fire safety and adhere to fire safety rules and regulations in various products and materials, such as: plastics, textiles, electronics, furniture, and building materials. These materials have, the abilities to halt the combustion process, create a protection heat shield, and emit gases that extinguish flames when exposed to high temperatures. The importance of these materials cannot be overstated, since they play a critical role in increasing fire safety across various settings, including homes, workplaces, public areas, and numerous industries. They provide occupants with additional time for effective evacuation and firefighting efforts and this extra time has been proven to be extremely crucial [1–4]. Some industries are governed by stringent fire safety rules that necessitate the use of flame-retardant materials for certain applications. Manufacturers can use flame-retardants to comply with certain laws and standards [5]. Since most polymer materials burn easily, they are usually mixed with flame retardants to ensure their safe usage [1, 6]. The addition of flame-retardants is crucial in improving the fire resistance of flammable polymers. These materials help lower the potential for polymer combustion and the production of toxic fumes, which are essential for the safe development and use of new materials [3, 7, 8]. In the past, halogenated fire-retardant fillers were commonly used because of their effectiveness in preventing fires. However, they have several significant drawbacks and environmental concerns. Halogenated compounds remain in the environment for a long time after their disposal and they can release harmful substances into the air and water, leading to pollution and posing long-term risks to ecosystems. Additionally, some fire-retardant fillers containing halogens, e.g., chlorinated or brominated compounds, have been found to contribute to ozone depletion, if released into the atmosphere. As a result, they have been subject to regulatory restrictions and have been phased out or banned [3, 5, 8–10]. Lately, there has been a shift towards creating and utilizing halogen-free substitutes for flame-retardant fillers. Both the industry and academia are making efforts to develop more ecofriendly and sustainable flame-retardant materials [11]. These alternatives are better for the environment, and they pose fewer health risks, while providing sufficient and effective fire protection. One effective method for achieving the necessary flame-retardant properties is to incorporate additives that contain nitrogen, phosphorous, or silica. Research on various bio-based materials, such as: chitosan, lignin, phytic acid, polydopamine, tannic acid, and *b*-cyclodextrin, has gained considerable attention from different bio-based research groups. Due to its intrinsic phosphate-based flame-retardant characteristics and its renewable origin, phytic acid has emerged as a particularly fascinating alternative among these choices [10–15]. The purpose of this chapter is to explore the potential of phytic acid in creating sustainable and high-performing flame-retardant materials. The chapter discusses the history of phytic acid and its reinforcement with polymers and/or polymer/fiber composites.

2 History: Phytic Acid: A Novel Phosphate Bio-Based Flame Retardant and Its Applications

Pfeffer was, in 1872, the first to recognise Phytic acid, which is also known as inositol hexakisphosphate (IP6), inositol polyphosphate, or phytate when in the salt form Pfeffer [15] (1872), and in 1903, Posternak [16] introduced the word "Phytine". Nassar et al. [17] (2021) reported that in 1914, Anderson [18], (1914) was the first scientist to describe the structure of IP6. Subsequently, in 1969, Johnson and Tate [19] confirmed this structural depiction by using the nuclear magnetic resonance spectroscopy Nassar et al. [17] (2021), reported that IP6 is a saturated cyclic acid and the phosphate ester of inositol, which has a high density of negative charges due to its six phosphate groups that become partially ionized at a physiological pH, where the negative charges are counter-balanced by cations. Nassar et al. [17] (2021) further explained that IP6 is in abundance in plants and that it has a significant nutritional role, acting as the principal storage form of phosphorus in many plant tissues, especially bran and seeds. Reddy et al. [20] (1982) and Schlemmer et al. [21] (2009) further explained that IP6 is a source of myoinositol as in animal cells since it has a high concentration, ranging from 10 to 100 μM in Mammalian cells and this depends on factors, such as the cell type and developmental stage. Schlemmer et al. [21] (2009) and Gupta et al. [22] (2015), have reported that due to the IP6 strong binding affinity to dietary minerals, it also serves as a potent nutrient. However, other studies have indicated that when iron or zinc binds to IP6, insoluble precipitates are formed, contributing to the deficiencies of these elements in people whose diets rely on foods for their mineral intake. In the year 1983, Graf [23], played a significant role in researching the characteristics and potential applications of IP6. Notably, Graf was among the pioneering researchers to provide an extensive account of the metal chelation and antioxidant properties [17, 23] that are associated with IP6. Graf's research highlighted [17, 23] the merits of IP6, including its cost-effectiveness, inert nature, non-toxicity, and its abundance. Moreover, it could be easily obtained from various plant sources, through the relatively straightforward procedures [9]. Despite the limited existing research, as per Nassar et al. [17] (2021) report, there are indications of IP6's potential in oral care applications. Graf [23] (1983) demonstrated that IP6 possesses the capability to enhance the physical and chemical attributes of dental cements when used as an additive. Prosser et al. [24] (1983) study, concluded that the resultant cement low vulnerability to early attack by water and acid, as well as the better adhesion to enamel when compared with dentine, are due to the lower mineral content of the latter. The mechanical properties of zinc phosphate cements [16], also improved with the addition of IP6. The study demonstrated that by increasing the IP6 concentration from 0 to 2%, resulted in a doubling of the compressive strength [3–17]. Substituting a portion of phosphoric acid with between 3 and 5% IP6, yielded the maximum compressive strength achievable [17]. Gbureck et al. [25] (2004), reported that the addition of IP6 alone, triggered a swift reaction and brought about a short and controllable setting time. By adjusting the ratio of IP6 to phosphoric acid or by manipulating the water content, allowed for achieving a more practical setting time.

Ongoing clinical trials are actively working on advancing these attributes [16, 25]. In 2017, a study by Meininger et al. [26] (2017) highlighted that IP6 exhibited a setting-retarding effect on di-calcium phosphate cements. This effect was crucial to meet the clinical requirement for cement handling time, which typically sets in less than a minute in an environment without retardants. Meininger et al. [26] (2017) suggested that IP6 achieves this outcome by absorbing di-calcium phosphate crystals within the cement, thereby slowing down the crystal growth and reducing the maximum temperature reached during the setting process. In the context of brushite cements, IP6 was utilized as a retarder to achieve the highest cement strength, hence resulting in values surpassing those attained with retardant-free cement [26]. Furthermore, Meininger et al. [26] (2017), emphasized the fact that cements incorporating with IP6, displayed a notably improved compatibility with osteoblast cells in comparison to citric-acid-containing cement. Furthermore, research that was conducted by Hurle et al. [27] in 2018 focused on the influence of IP6 on the hydration mechanism and setting kinetics of brushite cements [13]. Their findings were consistent with the observations of Meininger et al. [26, 27] (2017), demonstrating that controlled concentrations of IP6 acted as a retarder for the cement setting reaction. This effect translated to enhanced mechanical performance, as reported by Hurle et al. [27] (2018), resulting in a cement that was composed of a crystalline structure, known as monetite. In 2015, phytic acid (IP6) was introduced as a natural chelating agent [28] to assist in overcoming the deleterious effects of Ethylenediaminetetaacetic acid (EDTA). It has shown various benefits in medicine and the food industry as a preservative [28]. It is claimed to possess anti-cariogenic and anti-plaque effects because of its inherent ability to prevent dissolution of enamel [28]. Moreover, it was reported that IP6 is more biocompatible and less cytotoxic than EDTA with considerable smear layer removal ability [14]. In a recent report, Afzan et al. [28] (2020) indicated that due to these advantages, 1% IP6 was proposed as an alternative chelating agent. Properties, such as: its effect on microhardness, its removal ability of in-tracanal medication, effect on dentine bond, chelating efficacy, calcium loss from root dentine, effect on dental pulp stem cells, effect on calcium silicate-based cements, antimicrobial effect and its role in regenerative endodontics, have been reported previously by Afzan et al. [28] (2020). However, literature on its erosive potential and smear removal ability of a 1% IP6, was reported [29]. Kalçay [29], (2018) reported on a better erosion performance of dentine with a 1% IP6 when compared to a 17% EDTA. Nassar et al. [17] reported equal or better smear removal of IP6 when compared to 17% EDTA. In contrast, however, Jagzap et al. [30] (2017) reported a less effective smear removal by IP6. This contrary evidence warrants a study on IP6 with a known pH. Therefore, the objective of this study was to compare the erosive potential and smear layer removal ability of a pH = 3, 1% IP6 with 17% EDTA, by using both as final irrigants at different time intervals [28]. Boot et al. (2021) reported that the phytic acid extraction methods may be applied to various types of grain, as well as biological and vegetable sources [31]. The extraction method includes phytic acid solubilization and acidic dissociation from related proteins or mixed salts [31]. Phytic acid is largely ionized under physiological pH conditions, thereby, becoming negatively charged and thus, reacting easily with positively charged substances, e.g., proteins

[31]. Based on the forementioned, different phytic acid extraction conditions have been studied by using different acid solutions [31]. In 1970, Wheeler and Ferrel [32] conducted the first studies on the investigation of organic phosphorus extraction from wheat and wheat fractions and they found out that 0.3 mol L^{-1} trichloroacetic acid (50 mL), under mechanical shaking for 30 min or occasional manual swirling for 45 min, was more extraction-efficient than 6 mol L^{-1} hydrochloric acid, extracting ~88% of total phosphorus. There has been recent potential industry application, i.e., the use of phytic acid for imparting long-lasting flame-retardant properties to fabrics [32]. Bloot et al. [17–31] (2021) reported on the different approaches, based on the use of phytic acid as a bio-based phosphorus flame retardant that have been applied to develop fabrics, such as: cotton, wool, silk, cellulosic fibers, and nonwoven fabric with flame retardant properties or heat and smoke suppressing ability [33–39]. In conclusion, the historical path of flame-retardant development has seen remarkable shifts towards the environmentally friendly and sustainable solutions. The emergence of phytic acid as a novel phosphate bio-based flame retardant, represents a significant milestone in this journey. With its roots in traditional practices and its application in modern research, phytic acid offers a unique blend of historical wisdom and scientific innovations.

3 Structure and Properties

Phytic acid (PA) is a natural plant, known as Myo-inositol-1,2,3,4,5,6-hexakisphosphate, and Fig. 1 shows its structure. It can be found in different forms because it possesses 12 interchangeable protons that enable it to interact with multivalent cations and positively charged proteins. One of the several myo-inositol (Ins) phosphorylated derivatives, is classified as a cyclic alcohol that offers the carbon backbones that are required for phytic acid production. The chemical formula for phytic acid is: $C_6H_{18}O_{24}P_6$, and it has a molar mass of 660.04 g mol. The phosphate groups are axially connected to carbons 1, 3, 4, 5, and 6 and equatorially bonded to carbon 2, according to X-ray crystallographic study [40–44].

The calcium salt of PA is called phytate, while the calcium/magnesium salt is known as phytin. In accordance with the physiological pH and the metal ions available, phytic acid can exist as free acid, phytate, or phytin. Most of the work done on

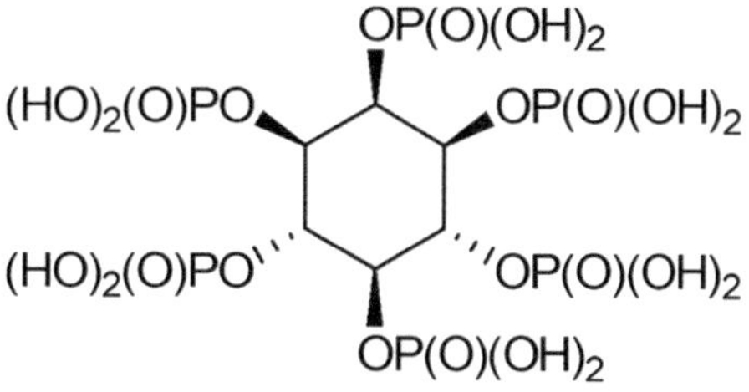

Fig. 1 Basic structure of phytic acid

phytic acid systems, has utilized interchangeably, these types of phytic acid. Inositol and inorganic phosphates are produced when phytic acid is completely hydrolyzed. Because the phytic acid molecule is inert and extremely stable, it may be kept as a solid for months or years in neutral or alkaline aqueous solutions, without producing any by-products of breakdown [40–42]. Phytic acid exhibits a significant affinity with polyvalent cations in the declining stability order, shown below: Cu^{2+} is followed by Zn^{2+}, Ni^{2+}, Co, Mn, Fe, and Ca. Such cations are chelated by phytic acid, resulting in the formation of water-insoluble phytate salts. The most prevalent mono- and divalent cations, are K^{+}, Mg^{2+}, and Ca^{2+}. Protein and starch can interact with phytic acid, either directly or indirectly. For many years, phytic acid was regarded as an anti-nutrient. However, several research, using both human and animal models, have shown how phytic acid can prevent a variety of diseases. In the eukaryotic species, phytic acid is everywhere, and lower in concentrations of myo-inositol with fewer phosphate groups (IP1-5) and are important for controlling crucial cellular processes, such as: cell division, cellular differentiation, exocytosis, and endocytosis [40, 43–46]. Table 1 illustrates the physiochemical properties of phytic acid.

4 Phytic Acid/Polymer Composites

To evaluate the effectiveness of flame-retardant properties on polymers, various tests, such as the: limiting oxygen index (LOI), vertical burning test (UL 94), thermogravimetric analysis (TGA) and the cone calorimeter tests, are conducted. The LOI measures the minimum oxygen needed to support the burning of a polymer; it happens in a controlled atmospheric environment [47]. High values of LOI, results in the best performance of flame retardancy. The UL 94 test measures the flammability of plastic materials. UL 94 has various ratings that are used to explain its results. These ratings include, viz: V-0 (flame extinguishes within 10 s with no flame drippings), V-1 (flame burns out within 30 s with no flame drippings) and V-2 (burns out within 30 s with drippings). For the best flame retardancy results, the material needs to attain a V-0 UL 94 rating [48]. The thermogravimetric analysis measures the changes in weight (thermal degradation) of a material when heated in controlled atmospheric conditions [49]. The monitored changes in weight of a sample are then, recorded down so that they can be plotted against time and/or temperature in order to derive/generate the TGA curve [3]. The cone calorimeter tests are used to measure the fire response behaviors of materials [50]. It relies on the fact that the heat release from the combustion of a material is dependent on the oxygen needed to sustain the combustion. It measures parameters, such as: the heat release rate (HRR), peak heat release rate (PHRR), total heat release (THR), temperature of maximum heat release rate (T_{max}), ignition time, flame spread, smoke production, and other parameters related to the combustion process of a material [50, 51]. A recent study on phytic acid (PA) conducted by Cheng et al. [39] (2016), showed that PA greatly improved the flame-retardant properties of polylactic acid (PLA) fabric. The untreated PLA fabric exhibited an LOI value of 26.3% and the PLA that was treated with 250 g L^{-1}

Table 1 Physiochemical properties of phytic acid (Chemicalbook.com)

Molecular formula	$C_6H_{18}O_{24}P_6$
Molar mass	660.04 g/mol
Density	1.432 g/mL at 25 °C
Melting point	<25°C
Boling point	105 °C
Flash point	673.9 °C
Water solubility	Miscible
Solubility	It is easily soluble in water, ethanol and acetone and it is almost insoluble in ether, benzene and chloroform
Vapor pressure	0.039 Pa at 60 °C
Appearance	It is composed of 98% white powder and 70% liquid that is light yellow to light brown slurry colour
Specific Gravity	1.282
Merck	14,7387
BRN	2,201,952
pKa	1.13 ± 0.10 (predicted)
Storage condition	2–8 °C
Stability	Stable. It is incompatible with oxidizing agents that are strong
Sensitive	Thermal sensitivity
Refractive index	n20/D 1.4
MDL	MFCD00082309
Physical and chemical properties	Trait light yellow to light brown serous liquid Soluble in water, ethanol, and acetone, almost insoluble in ether, benzene, and chloroform

(PLA-250) had a higher value of 36.1%, which clearly indicated an improvement of flame-retardant efficacy. Figure 2 shows the vertical burning tests of PLA fabrics that have been treated with 0,100 and 250 g L^{-1} of PA. The black color represents the average length of the char with respect to the PLA fabrics. It can be observed that the burnt area of the fabrics became smaller with an increase in the content of PA. The cone calorimeter tests, showed that the pHRR and the HRC values decreased with an increase in the PA content. The pHRR value of the untreated fabric was 463.6 W g^{-1}, while the treated fabrics of PLA-100 and PLA-250 were 329.7 and 279 W g^{-1}, respectively. From these outcomes, it can be deduced that the treated fabric can reduce the flame spread and exhibit a low fire risk when exposed to fire. The TGA showed that the onset temperature for pure PLA is 309 °C compared to the treated fabric of PLA-100 and PLA-250 331 °C and 323 °C respectively. The char residue increased with an increase in PA content in PLA. These results indicate that PA served as an effective char forming agent, which greatly improved the flame-retardant properties of PLA.

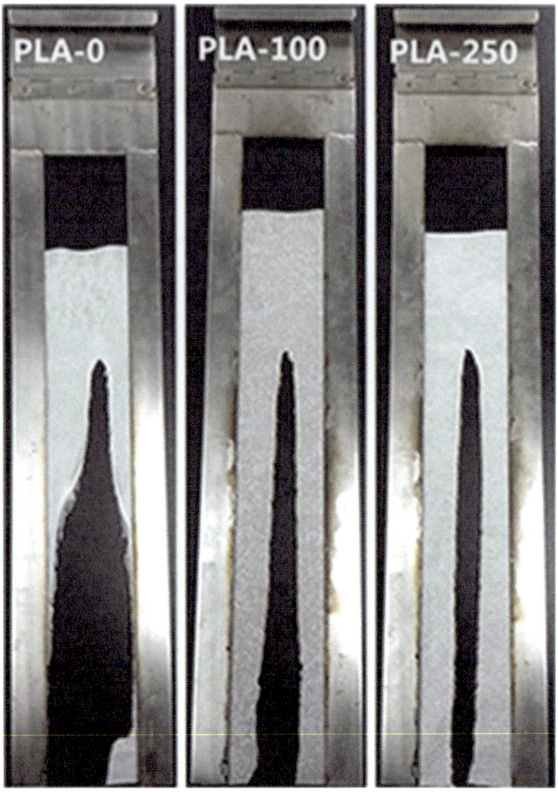

Fig. 2 Vertical burning tests of PLA treated with 0, 100 and 250 PA [39]

Gao et al. [52] (2019) conducted a study on the effectiveness of phytic acid salt that was prepared via the salt formation reaction between phytic acid and piperazine (PHYPI) as a flame retardant in a polypropylene (PP) matrix. The mechanism of PP/PHYPI is illustrated in Fig. 3 and the dehydration of phytic acid as well as the breakdown of ionic bonds in PHYPI, happened at the same time during the initial phase. This led to the production of non-flammable volatile substances, e.g., H_2O. Significantly, some thermally stable structures, such as P-N–C and P-O-C, were produced in this stage, which resulted in enabling the charring process of PP/PHYPI composite [52, 53], which clearly indicated the success and suitability of PA as a flame retardant in PP.

Phytic acid, as a flame retardant, can be applied in more than one industry, and it can be applied in plastics, wool, wood, cotton, and various types of materials [54]. Wool and cotton are two naturally occurring fabrics that have been extensively, used in upholstery, clothing, and industrial applications. These fabrics are mostly used because they provide softness and comfort, can absorb moisture, are biocompatible and they possess breathability properties. However, like most materials, they have low thermal stability, and they are fire hazard-prone when exposed to flames [54]. A recent study of PA and wool demonstrated that the wool's LOI values increased with

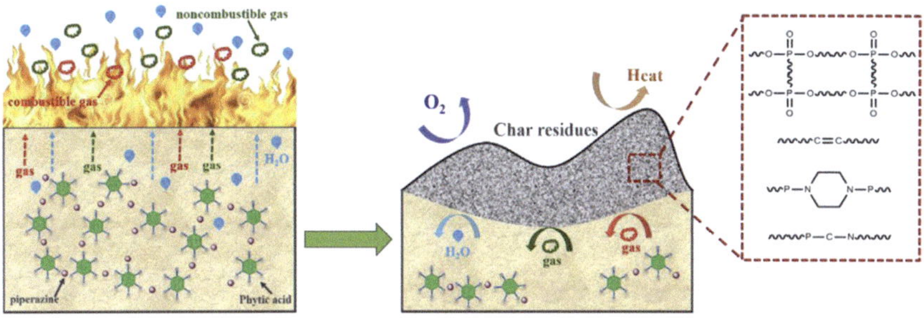

Fig. 3 The mechanism of PHYPI on the flame retardancy of PP [52]

an increase in the content of PA [55]. Wool with PA exhibited a higher LOI value of ~35.2% when compared to the untreated wool that recorded an LOI value of ~23.6%. This study employed the pyrolysis combustion flow calorimetry (PCFC), which is a testing technique that measures the heat release rate and combustion qualities of materials, during pyrolysis and combustion processes. The peak heat release rate of the untreated wool was ~131.7, while for the wool-40, wool-80 and wool-120, the values were: 89.8, 80.7 and 76.6 W/g, which exhibited reductions of 31.8%, 38.7% and 41.7%, respectively. According to the data, it can be concluded that PA significantly improved that flame retardancy of wool [55]. Polyamide 66 is a synthetic fibrous material that is broadly used by civilians and the military because of its ability to absorb and retain dye, good mechanical properties, and resistance to chemicals and shrinkage [47]. However, its application in various industries, is limited because its flammability and melt-dripping tendencies, when exposed to flames or fire, which pose danger to human health and properties [47]. Liu et al. [47] reported on the flammability properties of phytic acid (labelled IP_6)/polyamide 66 (PA) fabric via the pad dry method (see Fig. 4).

Table 2 shows the LOI results for polyamide 66 that has been coated with phytic acid. The untreated PA fabric had the lowest LOI value of 19.6% when compared to the fabrics that were coated with phytic acid (IP6), which recorded the highest LOI value of 32.3%. It can also be observed that the LOI value is directly proportional to the content of IP6 content that was added; hence, an increase in the content of IP6, resulted in an increase in the LOI value, indicating an improvement in the flame retardancy properties of the material under test.

The vertical burning test revealed that the char length of the untreated PA fabric was 18.1 with an associated dripping during combustion process, and it was also reported that the 15IP6-PA fabric had a char length of ~21.4 with no dripping. This can be attributed to the possibility that the melt dripping that occurred during the combustion of the polymer substrate, redirected the heat and fire away from the sample. However, the 25 and 35 IP_6-PA samples showed excellent vertical burning tests results, with a char length of ~12.5 and ~11.9, respectively and with no dripping, effectively proving the fact that phytic acid can improve the flammability characteristics of polyamide 66.

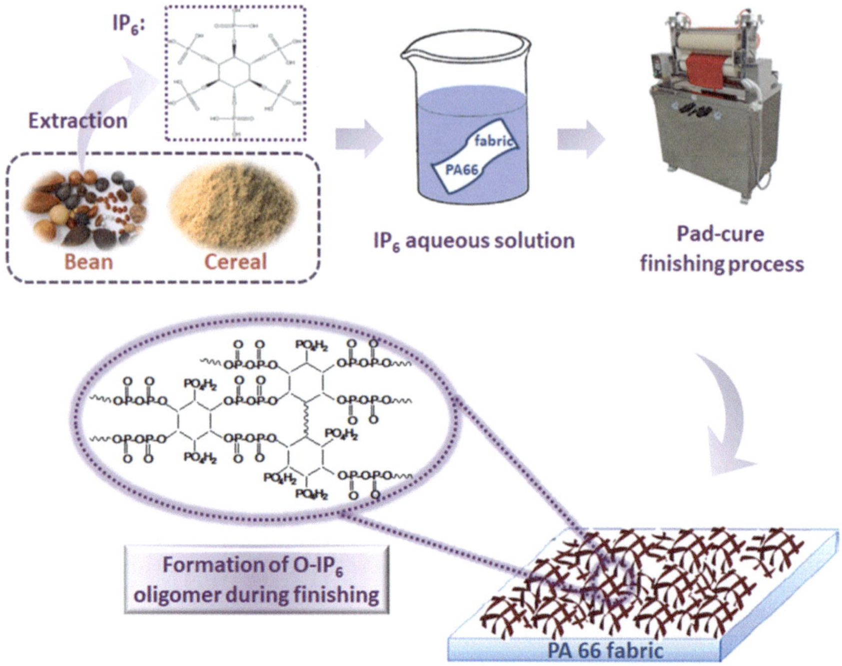

Fig. 4 Step by step scheme displaying method of preparation for polyamide 66 incorporated in phytic acid [47]

Table 2 LOI values for IP_6/PA fabric blends [47]

Sample	IP_6 concentration/%	LOI value/%
Untreated PA fabric	0	19.6
15IP_6-PA fabric	15	26.3
25IP_6-PA fabric	25	31.8
35IP_6-PA fabric	35	32.3

5 The Synergy of Phytic Acid with Other Flame-Retardant Fillers

An improvement in the flammability of polymers by incorporating green flame-retardant fillers is a step forward, but the use of a single filler material may not meet or guarantee the universally acceptable safety standards. The addition of supporting additives to the phytic acid systems, can enhance the protective ability of these materials, thereby, resulting in a synergistic effect. By combining two or more flame-retardant fillers, the flammability resistance of the resultant product can significantly, be enhanced. Synergism can result in a more robust flame-retardant system, which is critical in areas, such as: construction, electronics, transportation, and textile manufacturing, where fire safety is important. The combination of flame-retardant fillers can prevent or delay ignition and hence, reduce the generation of the associated harmful gases and smokes during combustion. The use of different (hybrid) flame-retardant fillers together, can result in superior formulations that require low contents of the individual filler loadings, resulting in cost savings, while preserving or even increasing fire performance [12, 56–58]. The incorporation of non-combustible fillers to a polymer, not only decreases the flammability of the polymer, but it can also increase the heat capacity, thermal conductivity, reflectivity, and emissivity of the material. Phosphorus-based flame retardants, e.g., phytic acid can act in both the gas and condensed phases during a fire out-brake [4, 12, 58]. The ability of phosphorous to adopt a range of oxidation states, allows the phosphorus-based flame-retardants to exhibit different flame-retardant mechanisms, i.e., both the gas and condensed phases. This means that they can work in the gaseous phase by creating phosphorous radicals, such as $PO^{\cdot}$, $PO_2^{\cdot}$, and $HPO_2^{\cdot}$, which trap the highly reactive radicals $H^{\cdot}$ and $OH^{\cdot}$, while in the condensed phase phosphorous promotes char formation [13]. Due to its characteristic decomposition nature at elevated temperatures (200 °C), phytic acid reduces the flammability of materials by absorbing heat and releasing water vapor, thereby diluting combustible gases. Additionally, phytic acid promotes the formation of a protective char layer on the surface of the polymer and hence, hindering the accessibility of oxygen and further inhibiting combustion [13, 56]. As an emerging environmentally friendly phosphorous-based biomass flame retardant, phytic acid has an obvious developmental potential in the flame-retardancy of polymers. Figure 5 illustrates the interaction of phytic acid with other flame retardant fillers [53]. However, recent studies have revealed the fact that the use of phytic acid alone as a flame-retardant additive, does not work well. When used alone, phytic acid induces the decomposition of the composites and, in some cases, increases their hygroscopic behaviours. In addition, its flame-retardant efficiency is low, and its application spectrum is limited [53, 59, 60]. To improve its flame-retardant abilities, phytic acid is commonly used in conjunction with other materials. For instance, improved properties have been reported in composites made of PLA, phytic acid, and lignin combinations [59]. The presence of both phytic acid and lignin effectively, limited the thermal degradation of PLA and composite hygroscopic tendency. Not only that, but they also managed to significantly, reduce ~44%, the pHRR in the cone calorimeter test, thereby achieving a V-2 classification in the UL-94 test. Furthermore, certain combinations have been proven to enhance mechanical properties, e.g., increasing elongation-at-break from a mere 3.1%, to an impressive 12.6%.

To create efficient flame-retardant formulations, it is critical to understand and maximize their combinations [57]. Synergistic combinations of flame-retardant fillers often, act through different mechanisms to address the various stages of the fire triangle, viz: heat, fuel, and oxygen (Fig. 6). This comprehensive approach provides a more effective barrier against fire. The ability to mix and match various flame-retardant fillers, allows for the tailoring of fire performance to suit specific application requirements and regulatory standards [61]. Phytic acid is a strong acid that exhibits six negatively charged phosphate groups in its structure and often, it chelates and/or

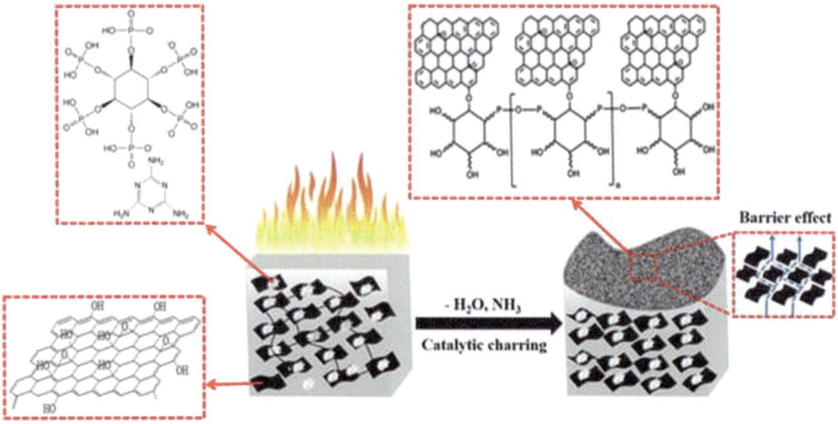

Fig. 5 Schematic illustration of flame-retardant mechanism. GO and PA derivatives as synergistic flame retardant. Reproduced from [60]. Open access

forms stable complexes with metal ions to improve the flammability properties the material [53]. Some examples of the metal-based fire retardants that have been used with phytic acid, include those containing zinc, iron, or aluminium compounds, which have been studied through various techniques, including spectrophotometry, potentiometry, and NMR in conjunction with computational data analysis. The stability and solubility of the ion–phytate complexes, are influenced by the specific cation, pH value, phytate-to-cation molar ratio, and the presence of other compounds in the solution. These compounds have been paired with various polymer matrices for their efficacies as flame-retardants [53, 57, 62].

Nonetheless, the most common and the frequently explored systems are those of phytic acid and nitrogen-containing compounds, fondly known as the P-N system [8, 13]. When nitrogen and phosphorous are combined, the two elements have a synergistic effect that greatly improves the flame retardance efficiency of the resulting product. The synergism between phytic acid and nitrogen-containing compounds, exhibits excellent flame-retardant properties, high thermal stability, high LOI values, and low smoke production. Mokhena et al. [56] (2022) reported on the different nitrogen-containing substances, including melamine, amino acids, DNA, and proteins, that have proven to be ideal P-N systems, adequate for the enhancement of the overall fire resistance performance of the resulting products. Depending on the polymer matrix, the efficiency of the synergy between phytic acid and other flame-retardant additives can vary. Different polymers have different combustion characteristics, thermal properties, and interactions with flame retardant compositions/chemicals [63]. Phytic acid and its derivatives as composite modifiers, are commonly used to improve the fire safety of EP, cotton, PLA, PP, EVA, and other materials [8, 57]. Zhou et al. [64] reported on the flame retardancy characteristics of phytic acid-functionalized graphene oxide (PA-GO), incorporated into the rigid

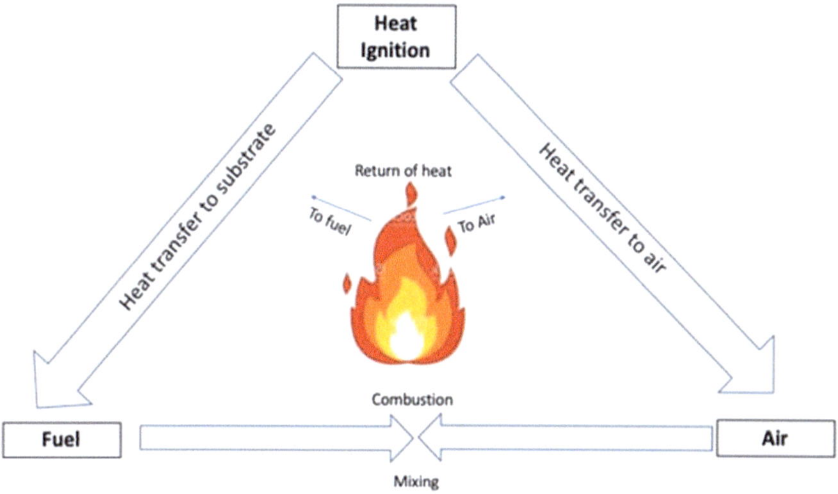

Fig. 6 A self-sustaining combustion cycle (i.e., Emman's fire triangle) for flammable materials [61]

polyurethane foam. Figure 7 illustrates a synthetic route for PA-GO flame retardant filler.

The flammability properties of PA-GO, in synergy with expandable graphite (EG) and ammonium phosphate incorporated in rigid polyurethane foam (RPUF), were analysed for their limiting oxygen index (LOI), by using the UL-94 vertical combustion tester and cone calorimeter. The LOI of the neat RPUF was ~18.4%, however with the incorporation of the 12 wt% IFR, i.e., 9 wt% EG and 3.0 wt% GO, increased the LOI value to ~25.7%. There was a reduction in the LOI value (viz 25.9%) of the hybrid composite consisting of 0.3 wt% GO together with 12 wt%, which is associated with poor dispersion of the fillers within the RPUF. From the cone calorimeter

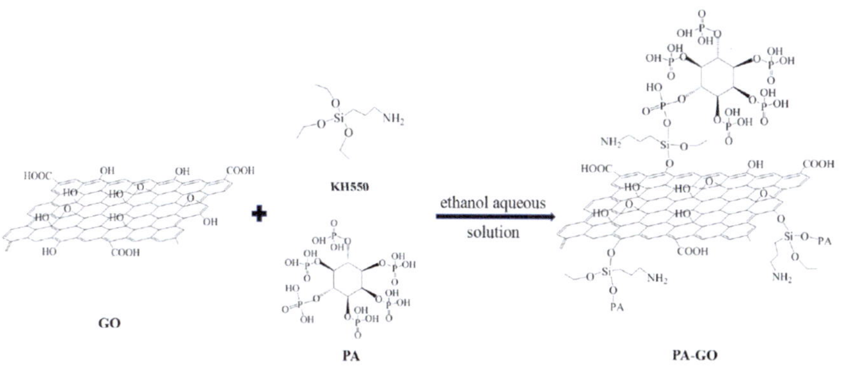

Fig. 7 Schematic drawings of the formation of PA-GO [64]. Open access

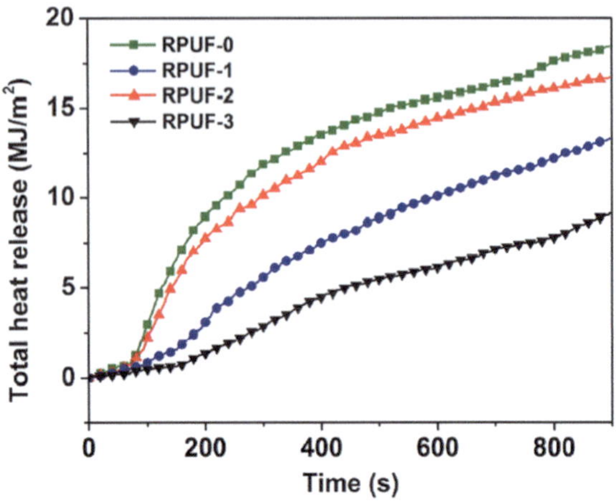

Fig. 8 Total heat release (THR) of all the investigated samples [64]. Open access

results, the PHRR of neat RPUF was recorded to be 84.32 kW/m^2 and the PHRR of the RPUF-1 (9.0 wt% EG + 3.0 APP) was recorded to be 35.94. A noticeable reduction was observed for the RPUF-3 (9.0 wt% APP + 0.3 PA-GO), with a reduction of ~18.71 kW m^{-2} in the PHRR. The spread of fire was relatively slow in the RPUF, which was associated with the effective synergy between PA-GO and an enhanced dispersion within the RPUF matrix. Similar results were obtained in the THR study, whereby the RPUF-3 samples showed a lower THR value (viz., 9.16 MJ m^{-2}) when compared with the RPUF-0 with a THR value of 18.46 MJ m^{-2} (Fig. 8).

Furthermore, Wang et al. [65] reported on the synergistic effect of phytic acid and melamine, incorporated into epoxy resin (EP). It is known that the melamine can produce non-flammable nitrogen-containing compounds, which can inhibit flammability. Neat epoxy (EP) recorded an LOI value of 24%, whereas the LOI value of the EP/PAMA (bio-based organic–inorganic hyperbranched hybrid) (EP-6), reached a value of 29.7%. Neat epoxy recorded a PHRR value of 1050.5 kW m^{-2}, while the EP-6 samples showed a reduction of 62.3% (PHRR = 395.6 kW m^{-2}) (Fig. 8). The THR of neat epoxy was observed to be 101 MJ/m^2, while the addition of the flame-retardant filler, reduced the value to 91 MJ m^{-2} (Fig. 9). This is associated with the incorporation of nano layered PAMA, which reduced the flammability of the system. The PAMA formed a continuous and compact device, which can prevent heat from entering the systems and also block the volatile products from leaving the system, and as a result, enhanced the flame resistance of the composite.

6 Phytic Acid Fibre Hybrid Systems

Hybridization opens-up new alternatives for improving the function of composite materials, particularly in sophisticated applications, by increasing the composites toughness or impact resistance. In addition, hybrid composites provide more design

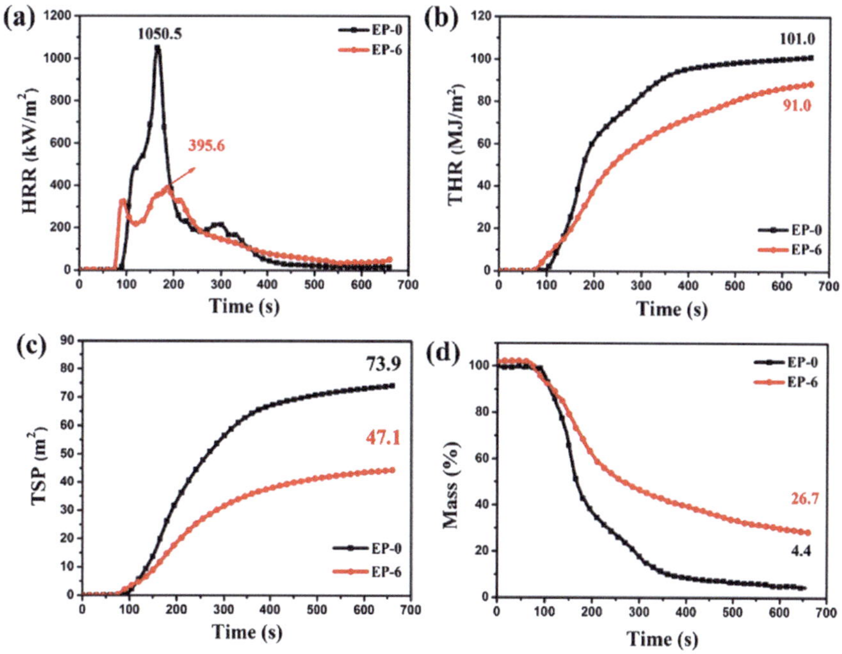

Fig. 9 The **a** HRR, **b** THR, **c** TSP and **d** of the investigated samples [65]

freedom than the non-hybrid composites, resulting in a synergetic effect that no single material can equal. The synergetic effect can be achieved in a variety of ways, including fiber selection, suitable fiber combination and orientation, and hybrid system interaction. Hybridization is a design technique for fibre-reinforced composite materials that involves combining two or more fibres of different groups within a single matrix to manipulate the desired properties [66]. They may be made from an admixture of natural and synthetic fibres, synthetic and synthetic fibres, or natural fibre and carbonaceous materials. Owing to their diverse properties of low production cost, high strength-to-weight ratio, and simple manufacturing process, hybrid natural fibre composite materials are useful in various structural and engineering applications. The utilization of hybrid natural fibre polymer composites in advanced applications, is usually hindered by their high flammability. High cellulose levels can enhance the risks of high flammability, while high lignin levels are responsible for char formation [67]. It is noteworthy that inorganic fillers, e.g., phytic acid can potentially affect, not only the combustion properties, but also the general properties of the hybrid system. Apart from their potential flame-retardant capabilities, phytic acid and its derivatives have also been utilized as modifiers to increase dispersion and consequently, the overall qualities of the final products [68]. The unique molecular structure of phytic acid, adds to its remarkable chelating ability to interact, in a variety of ways, with metallic ions or cationic polymers. The rich phosphate groups in phytic acid can induce mineralization of material surfaces to build a robust biomineralized skeleton and hence, improving a composite performance. Phytic acid can thus, bridge biopolymers and filler reinforcement via a significant interfacial chelation and an efficient electrostatic attraction, thereby, resulting in a strong biopolymer-based composite [69]. Zhang et al. [70] (2021) employed phytic acid to modify hemp fiber (HF-P) as a green flame retardant for vehicle parts containing unsaturated polyester resins (UPR). Phytic acid, in combination with melamine cyanurate, were used to modify hemp fiber to enhance the flame resistance of the resulting composites. The flammability properties of the composites were analysed with the LOI test and cone calorimeter test. The LOI value of the neat UPR was recorded to be 19.5% (Fig. 10), while the addition of 3% hemp fiber slightly, reduced the value to 19.2% (Fig. 10). The reduction in LOI value is an indication of the high flammability of the composite system. The synergy of phytic acid and melamine cyanurate, revealed an LOI value of 26.3% (Fig. 10).

The cone calorimeter test of the investigated samples recorded a PHRR value of 751.3 kW m^{-2} and with the incorporation of the hemp fiber, raised the PHRR value to 756.9 kW m^{-2} [70]. The incorporation of MCA and hemp fiber-phytic acid into the neat UPR, reduced the peak heat release rate (PHRR) to 237.2 kW m^{-2} (see Fig. 11a). Furthermore, the UPR/MCA/HF-P composite showed a lower THR value (Fig. 11b). The improvement in flammability resistance was associated with the HF-P, acting as a skeleton for the expansion of the carbon layer formed from the MCA and the matrix. The expanded and continuous char layer, can prevent the pyrolysis products from entering the UPR matrix and thereby, enhancing the flammability resistance.

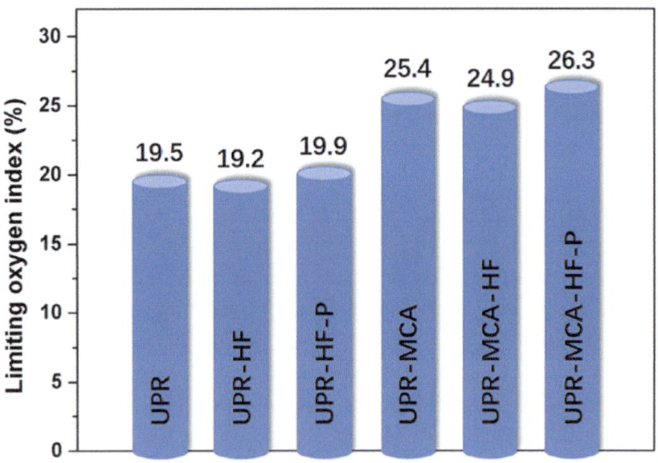

Fig. 10 Limiting oxygen index (LOI) of the investigated samples [70]. Open access

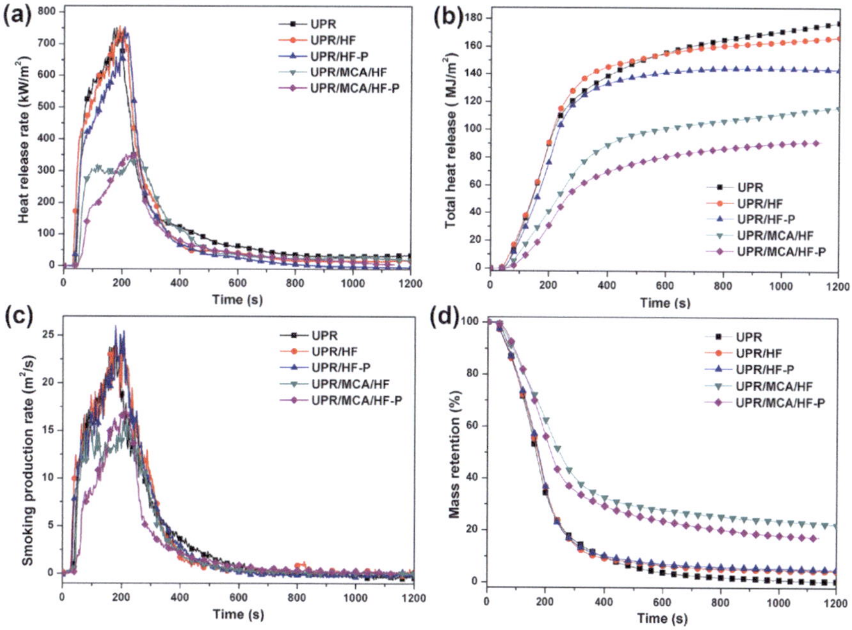

Fig. 11 **a** HRR, **b** THR, **c** SPR and **d** mass retention of all the investigated samples [70]. Open access

7 Conclusions

The use of phytic acid as a flame-retardant agent is an area of ongoing research. One of the primary benefits of using phytic acid as a flame retardant, lies in its eco-friendliness nature: it is derived from renewable plant sources and it is biodegradable, making it a more sustainable option in comparison to some traditional chemical flame retardants. Although research into phytic acid as a bioflame retardant is still in its rudimentary stages, phytic acid has been studied for use as a flame retardant in polymers and other materials. Its potential ascribes to its ability to form char and release phosphorous compounds when exposed to heat, which helps to reduce the flammability of materials. The potential synergies between phytic acid and other flame-retardant compounds, have also been studied. So far, the most important or prominent synergies are the P-N complexes. By combining phytic acid with other additives or materials could enhance its effectiveness in reducing flammability, and in some cases, a UL-94 V rating of V0, has been achieved. There are concerns, however, over its compatibility with some matrices and only limited studies on its use in fibre–fibre hybrid systems are available. If research continues to expand and it shows promising results, there could be potential applications in various industries, such as: construction, textiles, electronics, and automotive, where flame-resistant materials are of necessity and required.

References

1. Babrauskas, V., Fuoco, R., Blum, A.: Flame Retardant Additives in Polymers, pp. 87–118. Elsevier eBooks (2014). https://doi.org/10.1016/b978-0-444-53808-6.00003-2
2. Shen, J., et al.: The flame-retardant mechanisms and preparation of polymer composites and their potential application in construction engineering. Polymers **14**, 82–105 (2021)
3. Hull, T.R., Witkowski, A., Hollingbery, L.: Fire retardant action of mineral fillers. Polym. Degrad. Stab. **96**, 1462–1469 (2011)
4. Trovato, V., et al.: How to address flame-retardant technology on cotton fabrics by using functional inorganic sol–gel precursors and nanofillers: flammability insights, research advances, and sustainability challenges. Inorganics (Basel) **11**, 306–306 (2023)
5. Yang, Y., Díaz Palencia, J.L., Wang, N., Jiang, Y., Wang, D.-Y.: Nanocarbon-based flame retardant polymer nanocomposites. Molecules **26**, 4670–4701 (2021)
6. Kim, Y., Lee, S., Yoon, H.: Fire-safe polymer composites: flame-retardant effect of nanofillers. Polymers **13**(4), 540–585 (2021)
7. Krishnadevi, K., Selvaraj, V.: Development of halogen-free flame retardant phosphazene and rice husk ash incorporated benzoxazine blended epoxy composites for microelectronic applications. New J. Chem. **39**, 6555–6567 (2015)
8. Yang, G., et al.: Synthesis of a novel phosphazene-based flame retardant with active amine groups and its application in reducing the fire hazard of Epoxy Resin. J. Haz. Mater. **366**, 78–87 (2019)
9. Chen, Y., Xu, L., Wu, X., Xu, B.: The influence of nano ZnO coated by phosphazene/triazine bi-group molecular on the flame retardant property and mechanical property of intumescent flame retardant poly (lactic acid) composites. Thermochim. Acta **679**, 178336 (2019)
10. Mazela, B., Anyelkis, B., Grze´skowiak, W.: Expandable graphite as a fire retardant for cellulosic materials—a review. Forests **11**, 755–765 (2020)
11. Ahmad, M., et al.: Phosphazene cyclomatrix network-based polymer: chemistry, synthesis, and applications. ACS Omega **7**, 28694–28707 (2022)
12. Wang, L., Hu, X., Mao, Z., Wang, J., Wang, X.: Phosphorylated metal–organic framework for reducing fire hazards of poly(methyl methacrylate). Polymers **14**, 4871–4871 (2022)
13. Velencoso, M.M., Battig, A., Markwart, J.C., Schartel, B., Wurm, F.R.: Molecular firefighting—how modern phosphorus chemistry can help solve the challenge of flame retardancy. Angew. Chem. Int. Ed. **57**, 10450–10467 (2018)
14. He, S., et al.: Fully Bio-based phytic acid–basic amino acid salt for flame-retardant polypropylene. ACS Appl. Polym. Mater. **3**, 1488–1498 (2021)
15. Pfeffer, E.: The inositol phosphates: chemical synthesis and biological significance. Jahrb. Wiss. Bot. **8**, 429 (1872)
16. Posternak, S.: Sur un nouveau principe phospho-organique d'origine végétale, la phytine. Compt. Rend. Soc. de Biol. **55**, 1190–1192 (1903)
17. Nassar, M., et al.: Phytic acid: properties and potential applications in dentistry. Front. Mater. **8**, 638909 (2021)
18. Anderson, R.J.: A contribution to the chemistry of phytin. J. Biol. Chem. **17**, 171–190 (1914)
19. Johnson, L.F., Tate, M.E.: Structure of "phytic acids." Can. J. Chem. **47**, 63–73 (1969)
20. Reddy, N.R., Sathe, S.K., Salunkhe, D.K.: Phytates in legumes and cereals. Adv. Food Res. **28**, 1–92 (1982)
21. Schlemmer, U., Frølich, W., Prieto, R.M., Grases, F.: Phytate in foods and significance for humans: food sources, intake, processing, bioavailability, protective role and analysis. Mol. Nutr. Food Res. **53**, S330–S375 (2009)
22. Gupta, R.K., Gangoliya, S.S., Singh, N.K.: Reduction of phytic acid and enhancement of bioavailable micronutrients in food grains. J. Food Sci. Technol. **52**, 676–684 (2015)
23. Graf, E.: Applications of phytic acid. J. Am. Oil Chem. Soc. **60**, 1861–1867 (1983)
24. Prosser, H.J., Brant, P.J., Scott, R.P., Wilson, A.D.: The cement forming properties of phytic acid. J. Dent. Res. **62**, 598–600 (1983)

25. Gbureck, U., Barralet, J.E., Spatz, K., Grover, L.M., Thull, R.: Ionic modification of calcium phosphate cement viscosity. Part I: hypodermic injection and strength improvement of apatite cement. Biomaterials **25**, 2187–2195 (2004)
26. Meininger, S., et al.: Phytic acid as alternative setting retarder enhanced biological performance of dicalcium phosphate cement in vitro. Sci. Rep. **7**, 558–610 (2017)
27. Hurle, K., et al.: Hydration mechanism of a calcium phosphate cement modified with phytic acid. Acta Biomater. **80**, 378–389 (2018)
28. Afshan, Z., Jat, S.A., Khan, J.A., Hasan, A., Rehman Qazi, F.U.: Erosive Potential of 1% phytic acid on radicular dentine at different time intervals. Eur. Endod, J. **1**, 28–34 (2020)
29. Kalçay, M., Tinaz, A.C.: Effects of different concentrations of phytic acid on smear layer removal and erosion. Atatürk Üniversitesi Diş Hekimliği Fakültesi Dergisi. **28**, 341–347 (2018)
30. Jagzap, J.B., et al.: Effectiveness of three different irrigants-17% ethylenediaminetetraacetic acid, Q-MIX, and phytic acid in smear layer removal: a comparative scanning electron microscope study. Contemp. Clin. Dent. **8**, 459 (2017)
31. Bloot, A.P.M., Kalschne, D.L., Amaral, J.A.S., Baraldi, I.J., Cristiane Canan, C.: A review of phytic acid sources, obtention, and applications. Food Rev. Int. **39**, 73–92 (2021)
32. Wheeler, E.L., Ferrel, R.E.: A method for phytic acid determination in wheat and wheat fractions. Cereal Chem. **48**, 312–320 (1971)
33. Cheng, X.W., Tang, R.C., Guan, J.P., Zhou, S.Q.: An eco-friendly and effective flame retardant coating for cotton fabric based on phytic acid doped silica sol approach. Prog. Org. Coatings. **141**, 105539 (2020)
34. Thota, S., et al.: Covalent functionalization of cellulose in cotton and a nylon-cotton blend with phytic acid for flame retardant properties. Cellulose **27**, 11–24 (2020)
35. Zhou, Q., Chen, J., Zhou, T., Shao, J.: In situ polymerization of polyaniline on cotton fabrics with phytic acid as a novel efficient dopant for flame retardancy and conductivity switching. New J. Chem. **44**, 3504–3513 (2020)
36. Cheng, X.W., Guan, J.P., Kiekens, P., Yang, X.H., Tang, R.C.: Preparation and evaluation of an eco-friendly, reactive, and phytic acid-based flame retardant for wool. React. Funct. Polym. **134**, 58–66 (2019)
37. Cheng, X.W., Guan, J.P., Yang, X.H., Tang, R.C.: Improvement of flame retardancy of silk fabric by bio-based phytic acid, nano-TiO_2, and polycarboxylic acid. Prog. Org. Coatings. **112**, 18–26 (2017)
38. Liu, X., Zhang, Q., Cheng, B., Ren, Y., Zhang, Y., Ding, C.: Durable flame retardant cellulosic fibers modified with novel, facile and efficient phytic acid-based finishing agent. Cellulose **25**, 799–811 (2018)
39. Cheng, X.W., Guan, J.P., Tang, R.C., Liu, K.Q.: Phytic acid as a bio-based phosphorus flame retardant for poly(lactic acid) nonwoven fabric. J. Clean. Prod. **124**, 114–119 (2016)
40. Bloot, A.P.M., Kalschne, D.L., Amaral, J.A.S., Baraldi, I.J., Canan, C.: A review of phytic acid sources, obtention, and applications. Food Rev. Int. **39**, 73–92 (2023)
41. Oatway, L., Vasanthan, T., Helm, J.H.: Phytic acid. Food Rev. Int. **17**, 419–431 (2001)
42. Feil, B.: Phytic acid. J. New Seeds. **3**, 1–35 (2001)
43. Sarkhel, S., Roy, A.: Phytic acid and its reduction in pulse matrix: structure–function relationship owing to bioavailability enhancement of micronutrients. J. Food Process Eng. **45**, 14030 (2022)
44. Wang, R., Guo, S.: Phytic acid and its interactions: contributions to protein functionality, food processing, and safety. Compr. Rev. Food Sci. Food Saf. **20**, 2081–2105 (2021)
45. Nissar, J., Ahad, T., Naik, H.R., Hussain, S.Z.: A review phytic acid: as antinutrient or nutraceutical. J. Pharmacogn. Phytochem. **6**, 1554–1560 (2017)
46. Hídvégi, M., Lásztity, R.: Phytic acid content of cereals and legumes and interaction with proteins. Period. Polytech. Chem. Eng. **46**, 59–64 (2002)
47. Liu, W., et al.: A bio-based flame retardant coating used for polyamide 66 fabric. Prog. Org. Coatings. **156**, 106271 (2021)
48. Marti, J., Idelsohn, S.R., Oñate, E.: A finite element model for the simulation of the UL-94 burning test. Fire Technol. **54**, 1783–1805 (2018)

49. Ng, H.M., et al.: Thermogravimetric analysis of polymers. Encycl. Polym. Sci. Technol. 1–29 (2018). https://doi.org/10.1002/0471440264.pst667
50. Schartel, B., Hull, T.R.: Cone calorimeter—a tool for measuring heat release rate. Fire Mater. **31**, 327–354 (2007)
51. Schartel, B., Bartholmai, M., Knoll, U.: Some comments on the use of cone calorimeter data. Polym. Degrad. Stab. **88**, 540–547 (2005)
52. Gao, Y.Y., Deng, C., Du, Y.Y., Huang, S.C., Wang, Y.Z.: A novel bio-based flame retardant for polypropylene from phytic acid. Polym. Degrad. Stab. **161**, 298–308 (2019)
53. Liu, Y., et al.: Recent advances in biomass phytic acid flame retardants. Polym. Test. **124**, 108100 (2023)
54. Sykam, K., Försth, M., Sas, G., Restás, Á., Das, O.: Phytic acid: A bio-based flame retardant for cotton and wool fabrics. Ind. Crops Prod. **164**, (2021)
55. Cheng, X.W., Guan, J.P., Chen, G., Yang, X.H., Tang, R.C.: Adsorption and flame retardant properties of bio-based phytic acid on wool fabric. Polymers (Basel) **8**, (2016)
56. Mokhena, T.C., et al.: Flame retardancy efficacy of phytic acid: an overview. J. Appl. Polym. Sci. **139** (2022)
57. Wang, M., et al.: Bio-based flame retardants to polymers: a review. Adv. Indus. Eng. Polym. Res. **6**, 132–155 (2022)
58. Özer, M.S., Gaan, S.: Recent developments in phosphorus based flame retardant coatings for textiles: synthesis, applications and performance. Prog. Org. Coat. **171**, 107027 (2022)
59. Costes, L., Laoutid, F., Brohez, S., Delvosalle, C., Dubois, P.: Phytic acid–lignin combination: a simple and efficient route for enhancing thermal and flame retardant properties of polylactide. Eur. Polym. J. **94**, 270–285 (2017)
60. Yang, P., et al.: A novel self-assembled graphene-based flame retardant: synthesis and flame retardant performance in PLA. Polymers **13**, 4216 (2021)
61. Malucelli, G.: Sol-gel and layer-by-layer coatings for flame-retardant cotton fabrics: recent advances. Coatings **10**, 333 (2020)
62. Marolt, G., Gričar, E., Pihlar, B., Kolar, M.: Complex formation of phytic acid with selected monovalent and divalent metals. Front. Chem. **8**, 582746 (2020)
63. Wilén, C.-E., Pfaendner, R.: Improving weathering resistance of flame-retarded polymers. J. Appl. Polym. **129**, 925–944 (2013)
64. Zhou, X., et al.: Study on the flame retardancy of rigid polyurethane foam with phytic acid-functionalized graphene oxide. Molecules **28**, 6267–6267 (2023)
65. Wang, P., et al.: Facile fabrication of biobased P N C-containing nano-layered hybrid: preparation, growth mechanism and its efficient fire retardancy in epoxy. Polym. Degrad. Stab. **159**, 153–162 (2019)
66. Nurazzi, N.M., et al.: Review on mechanical performance of hybrid natural fiber polymer composites for structural applications. Polymers **13**, 2170 (2021)
67. Phonphuak, N., Chindaprasirt, P.: Types of waste, properties, and durability of pore-forming waste-based fired masonry bricks. Eco-Efficient Masonry Bricks Blocks (2015). https://doi.org/10.1016/b978-1-78242-305-8.00006-1
68. Mochane, M.J., et al.: Recent progress on natural fiber hybrid composites for advanced applications: a review. Express Polym. Lett. **13**, 159–198 (2019)
69. Jin, S., et al.: Phytic acid-assisted fabrication for soybean meal/nanofiber composite adhesive via bioinspired chelation reinforcement strategy. J. Hazard. Mater. **399**, 123064–123064 (2020)
70. Zhang, J., Zhang, H., Wang, X., Zhang, M.: Modified hemp fiber combined with melamine cyanurate for enhancing fire safety and mechanical properties of unsaturated polyester resins for automobile parts. AIP Adv. **11**, 125329 (2021). https://doi.org/10.1063/5.0076024

The Preparation and Characterization of Tannic Acid: A Novel Bio-Based Flame Retardant

K. P. Matabola, M. F. Bambo, K. I. Matshetshe, and M. Kgatle

Abstract Man-made polymers have contributed significantly in the economy and modern world. Despite having played a significant role, there are serious drawbacks linked with the use of these plastics in the industries. One of the setbacks is their increased amount of flammability which limit their full potential. As a result, the development of efficient flame retardants for these plastics became a priority. At present, a vast number of the flame retardants poses a serious threats to the environment which warranted the development of sustainable and environmentally friendly bio-based flame retardant. This chapter will focus on the progress made on using tannic acid and its composites as a bio-based flame retardant material applied to different polymers. Finally, conclusions and perspectives are briefly described.

Keywords Tannic acid · Flame retardant · Flammability · Polymers · Composites

1 Introduction

Polymeric materials are widely employed in mainstream compounds precisely on the grounds of being lighter, flexibility, and lower cost. That being said, the exhaustive use of these plastics will require the improvement of properties like thermal, antimicrobial, and mechanical properties. Most importantly, they are susceptible to heat and fire, viz, the mechanical properties weaken at increased temperatures and toxic components are discharged during the process of combustion, which restricted their full applications [1, 2]. As a result, a lot of researches have been carried out to

K. P. Matabola (✉)
Department of Water and Sanitation, University of Limpopo, Private Bag X 1106, Sovenga 0727, South Africa
e-mail: kgabo.matabola@ul.ac.za

M. F. Bambo · K. I. Matshetshe · M. Kgatle
DSI/Mintek—Nanotechnology Innovation Centre, Advanced Materials Division, Mintek, Private Bag X 3015, Randburg 2125, South Africa

M. F. Bambo
Department of Natural Sciences, Vaal University of Technology, Vanderbijlpark, South Africa

T. C. Mokhena et al. (eds.), *Biomaterials as Green Flame Retardants*, Engineering Materials, https://doi.org/10.1007/978-981-97-6871-4_5

subdue their flammability. These days, fire retardancy of polymer-based materials has turned out to be a significant aspect of their applications [3]. Initially, the halogenated compounds were mostly exploited as a flame retarding additives [4]. However, they tend to release harmful halogen-containing gases that endanger the human's health. That said, a variety of options like phosphorus-containing compounds [5] and nano-additives, such as clays [6–8], carbon nanotubes [6, 9] and graphenes [10–12] have been recommended as suitable FR alternatives. The heating of the phosphorus-containing compounds results in the production of a phosphoric acid, which consequentially causes a flame retarding impediment. On the other hand, nano-additives are capable of improving the mechanical properties of matrix as a fillers as well as a flame retardancy. In harmony with phosphorus containing compounds, they take up part as a layer blocking off flammable gases. However, the increased loadings of these additives tend to weaken the initial mechanical and thermal properties of polymer matrix [13–15]. These imitations can only be addressed by using different kind of additives.

Lately, the bioresource by-products in place of petroleum-based chemicals have been utilised for making the bio-based flame retardants precisely on account of them presenting much less carbon footprint as compared to fossil carbon. Moreover, the interest in the bioresource is largely motivated by their reasonable price, readily available and the increasing environmental interest. Amongst the different bio-based materials, tannins, which are polyphenolic compounds extracted from plants, have been studied for preparation of a slew of bio-based flame retardants. They have interesting features, such as high chemical and thermal stability as well as low thermal conductivity because of their inherent aromatic structure. These outstanding features make tannins a suitable candidate in the preparation of thermally insulating and flame retardant materials. Moreover, tannins can be divided into condensed tannins and hydrolyzable tannins [16, 17]. The condensed tannins are composed of repetitive flavonoid units, while the hydrolyzable tannins are saccharides partly or fully substituted by gallic or ellagic acid. By virtue of their molecular structure containing a plenty of aromatic rings and hydroxyl groups, both the covalent and noncovalent interactions are permitted, which affect the mechanical properties of the epoxy resins [18, 19]. During the breakdown of materials comprising of phenol groups, phenoxy radicals could quell the oxygen free radicals devoid of phosphorus elements in the polymer structure. Moreover, they have propensity to form high char when burned by fire, [20] resulting into a high yield of char layer, or protective layer, which stop heat, oxygen, and flammable gases [21]. The synchronised effect of features of tannins can improve not just flame retardancy but also mechanical properties of polymeric materials.

Among tannins, this chapter concentrates on hydrolysable tannin or tannic acid (TA) precisely on the basis that it has less conjugated and easy structure as compared to the condensed tannins. Moreover, it is soluble in organic solvents which enable easy preparation of TA-based polymer composites [22]. This chapter provides the current state of the tannic acid-based flame retardants. First up, the preparation methods are briefly highlighted followed by their resultant properties. Then, the thermal behaviour as well as the mechanisms are highlighted. It will culminate with the conclusions and future perspectives.

2 Preparation Methods for the Polymer-Tannic Acid Composites

Recognizing that there is a need for increased flame retardancy in materials, a variety of methods has been utilized to prepare these class of materials. This section will give a summary of the methods used to prepare, in particular, the polymer/TA-based flame retardants. Table 5.1 summarized recent studies based on the use of TA as green FR for various polymeric materials. The choice of the preparation method is often dependent on the polymeric materials. For instance, melt mixing is often utilized when thermoplastic polymers are used to prepare the flame-retarded materials.

The process most frequently used to create flame-retarded composites is melt mixing [3, 23, 24, 26, 28–30]. In this instance, the molten polymer was exposed to severe shear from rotors moving at a certain speed while at a higher temperature. The polymer can be mixed with FR prior introduction into the melt mixer or the FR can be introduced into the molten polymer in the melt mixer. The commonly used secondary method after melt mixing is melt compression. In this case, the pellets obtained from the melt mixer are pressed under desired pressure at the temperatures above the melting temperature of the polymeric material to afford sheets for flame resistance measurements. For instance, Laoutid et al. [3] prepared poly(lactic acid) (PLA)/ phytic acid (PA)/ranched polyethyleneimine (PEI)-TA composite using Brabender internal mixer with temperature kept at 160 °C and rotor speed of 70 rpm for 7 min. The as-prepared composites were then compression moulded at 160 °C through Agila PEZO hydraulic press as per a precise pressure program. Similarly, PLA/TA-based composites were prepared using melt mixing followed by compression moulding. PLA/boric acid (BA)-TA- melamine (MEL) (3, 4, 5%) composites were prepared by

Table 5.1 Preparation methods for polymer/TA-based flame retardants

Polymer	Additive	Methods	References
PLA	20 wt% Phyt/PEI-TA	Melt processing	[3]
PLA	BA-TA-MEL (3,4,5%)	Melt processing	[23]
PLA	20 wt% TA-oMMT	Melt processsing	[24]
Nylon 6	TA, TAT	Compression moulding	[25]
PVC	MH@PASn@TAZn	Hot pressing	[26]
PU	TA@RGO	Electrospinning	[27]
NR	APP/TA	Layer-by-layer assembly	[28]
PLA	5 wt% chitosan/TA@5wt% HNT	Hot compaction	[29]
PLA	LC-PA/TA (6,8,10 wt%)	Melt processing	[30]

hot pressing at 2, 6, 10, and 15 MPa for 2 min in a vulcanizer pre-heated at 17 °C for 3 min succeeded by cold pressing for 5 min. The hot pressing preceded melt blending of various contents of BA-TA-MEL with PLA in a ThermoHaake Rheomixer (Polylab, Germany) with a roller speed of 60 rpm, at 175 °C for 5 min. The sheet was then cut into desired sizes for further tests [23]. Laoutid et al. [24] prepared the PLA/TA-organomodified montmorillonite (OMMT) composites by the compression molding set at 160 °C using an Agila PE20 hydraulic press in accordance with a precise pressure program. Prior to that, the mixing of tannis and OMMT within the liquefied PLA was attained in a Brabender internal mixer at 160 °C for 7 min mixing at 70 rpm.

Elsewhere, the Nylon 6-TA and Nylon 6- chemically modified TA (TAT) composites were prepared utilizing a heated compression moulding press (Model 4389, Carver Inc.) [25]. The top and bottom platen temperature was set 222 °C while the compression force was kept at about 2 metric tons. The sample size was approximately $6.4 \times 12.7 \times 0.02$ cm^3 (width × length × thickness). The pure Nylon 6 sheet sample was also made to be tested as a control sample.

Meng et al. [26] prepared the Polyvinyl chloride/magnesium hydroxide@tin phytate@zinc tannate (PVC/MH@PASn@TAZn) composites using the hot compaction method. Firstly, the starting materials were mixed thoroughly on a two-roll mill (140 °C for 10 min) and then transferred into a mould. Afterwards, the composite film was obtained via hot pressing at 160 °C and 10 MPa for 8 min. Ultimately, the resultant sheet was cut into test pieces with required sizes for studying flame retardant and mechanical properties.

Besides melt blending processing techniques, electrospinning technique was also reported as suitable technique to fabricate TA-based composites [27]. The basic idea behind electrospinning is to first dissolve the polymer in an appropriate solvent before electrospinning. In this instance, a reservoir or needle is filled with the polymer solution before being exposed to high voltage. Surface tension in the solution is overcome by the applied voltage to create a jet that elongates to form fibers. The resultant fibres have diameters ranging between few nanometres to few microns. In their study, Kim et al. [27] prepared the polyurethane (PU)-TA@ reduced graphene oxide (rGO) composite nanofibers using the electrospinning process. A 10 wt% PU solution was prepared in dimethylformamide (DMF). Then, different loadings of TA@RGO were added into neat PU solutions at 1, 3, and 5 wt%, and stirred for 6 h at 25 °C. The resultant nanofibers were produced at an applied voltage of 18 kV, feed rate of 0.5 mL h^{-1}, tip–to-collector distance of 15 cm and a 21G needle gauge.

Li et al. [28] prepared natural rubber (NR)/ammonium phosphate (APP)/TA functionalized graphene (TGE) composite by the layer-by-layer assembly method. The N98 (5 phr), ZnO (5phr), APP (variable), N330 (30 phr) were successfully added to the TGE/NR compound and mixed for 8 min. The latter mixer was discharged onto a two roll mill set at 40 °C, where the sulfur (1 phr), CZ (1.2 phr) and DM (0.6 phr) were added.

3 Flame Retardant Properties of Polymer-Tannic Acid Composites

Polymer-tannic acid composites have gained significant interest in materials science due to their valuable applications in various fields [31]. These composites have the capability to retard or halt the expansion of fire, making them highly useful in various fields, such as building construction, electrical technology, and transportation. The increasing demand for safer and more environmentally friendly alternatives has driven research into natural, bio-based compounds like tannic acid (TA), which is derived sustainably from many plants [32]. TA enhances the flame retardancy of polymers by enhancing the creation of a protective char layer during combustion, while also providing excellent thermal stability and antioxidant properties [33]. By incorporating TA, polymer composites exhibit improved thermal stability and fire retardancy, making them appealing for diverse applications.

The behaviour of polymer-TA composites under heat exposure is interesting. When subjected to heating, TA undergoes dehydration reactions, leading to the development of a carbon-rich char layer on the composite's surface [34]. This charred layer serves as an insulating barrier, preventing further degradation and reducing the release of flammable gases, significantly slowing down the material's burning rate. Additionally, TA releases non-flammable gases like water vapour and carbon dioxide during pyrolysis, which dilute flammable gases and hinder flame propagation. Moreover, TA plays a dual role as both a charring agent and a radical scavenger due to its high phenolic content [35]. Its radical scavenging ability interrupts chain reactions during combustion by capturing free radicals, thus suppressing flame spread. Therefore, the incorporation of TA effectively enhances the flame retardancy of polymers through char formation and inhibition of radical-mediated combustion reactions.

The properties of flame retardancy in polymer-TA composites are examined using various techniques [36]. These techniques include the evaluation of factors, such as ignitability, flame spread rate, and heat release to estimate the flame retardant properties of a sample. These tests include limiting oxygen index (LOI), cone calorimetry, microscale combustion calorimetry (MCC), Underwriters Laboratories test standard (UL-94), and thermogravimetric analysis (TGA) [36]. In this part, the focus will be on discussing the properties measured with the cone calorimeter with that of LOI and UL-94 and their significance in understanding the combustion process and effectiveness of flame retardants. The cone calorimeter is an established fire testing instrument that employs oxygen consumption calorimetry to quantitatively measure the inherent flammability of materials. It follows the standard technique outlined by ASTM E-1354/ISO 5660 [37]. Over time, the cone calorimeter has become more popular as an efficient tool for analyzing various reaction-to-fire properties during combustion. These properties include time to ignition (TTI), heat release rate (HRR) and mass loss and other relevant parameters [38]. Only a few of the metrics will be covered in this chapter in order to better understand the flammability behaviour of composite armour. For example, De la Cruz observed enhanced flame-retardant properties in polyvinyl alcohol/TA composites with just 10 wt% TA resulting in a 35% reduction in peak heat release rate (pHRR) [39]. The heightened efficiency of flame retardancy observed with higher concentrations of TA in the TA-diglycidyl ether of bisphenol A (TA-DGEBA) thermosets was verified to stem from TA's charring effect and possibly its capacity to quench radicals derived from phenolic compounds as observed by Kim, et al. [40]. Consequently, this investigation into the direct interaction between TA and an epoxy monomer presents a promising avenue for crafting flame-retardant epoxy thermosets suitable for applications in different areas. Moreover, incorporating TA into polymers offers additional advantages, including improved mechanical strength and thermal stability, as demonstrated by Wang et al. in polylactic acid/TA composites [41].

To address compatibility and dispersion issues when mixing TA with polymers, researchers have developed innovative strategies such as creating nanostructures or employing crosslinking techniques [41]. In the research conducted by Korey et al. [42], the integration of TA into the thermoset matrix was explored. The behaviour of the molecular system hinged on the concentration of TA, wherein lower concentrations led to surface functionalization of the molecule, while higher concentrations resulted in its integration into the network via cross-linking. While TA exhibited effective performance as an intumescent agent, it did not lead to a reduction in the heat release rate. The outcomes of this study indicated that the effectiveness of TA's fire retardancy ability in epoxy is influenced independently by the external (hydrophilic surface-phenol groups) and internal (D-glucose and attached phenyl groups) structural regions of the TA molecule. Ensuring the structural integrity of both these regions proves pivotal for the synergistic fire retardancy behaviour of the molecule [42].

Price et al. [43] devised and assessed composites based on TA in contrast to control composites containing pentaerythritol (PER), a common component associated with intumescent char formation. Across all TA systems, the time to failure underwent an extension from seconds or minutes to an impressive duration exceeding 15 min (up to 27 min). Notably, the TA-based composites exhibited notably lower values in peak heat release (108 vs. 211 kW m^{-2}), total heat release (24.4 vs. 37.2 MJ m^{-2}), and fire growth rates (1.27 vs. 2.43 kW m^{-2} s^{-1}) when compared to PER composites. This research, in comparison to earlier studies utilizing TA composites, introduces TA-containing composites that offer significantly heightened fire protection. This marks a substantial advancement in the realms of fire protection and packaging industries. Wang et al. [44] explored the use of bio-based TA as a reactive flame retardant for the preparation of polyurethane elastomers, which exhibited improved flame retardancy and mechanical properties.

The incorporation of TA into polymer composites has been extensively studied, showcasing its effectiveness in enhancing flame retardancy and overall material performance through various mechanisms and strategies. These findings provide promising avenues for the development of safer and more efficient flame-retardant materials for various applications. The flame retardant mechanism of a new expansive flame retardant system (ACT) system in the NR matrix also involves a heat-stabilized clay sheet acting as a barrier against flammable gas in the condensed phase, catalytic charring, and ammonia release, contributing to remarkable flame-retardant performance in NR composites [45].

Most studies on polymer-TA composites focus on enhancing flame retardancy through physical means, like forming protective chars, but few explore the influence of chemical modifications on their behaviour under fire conditions. Such investigations could offer new possibilities for further improving their performance. For instance, chemically modifying TA might enhance their thermal stability or alter their decomposition products during pyrolysis, directly affecting flame retardancy. Xia et al. [25] conducted research using chemically modified TA (mTA) as a stable char-forming flame retardant additive for Nylon. The Nylon 6 blends with 15 wt% mTA exhibited significantly lower pHRR (52% reduction) and heat release capacity (HRC) (36% reduction), along with enhanced char yield (about 9%). The presence of mTA caused certain primary amide/amine and secondary amide species in Nylon 6 to degrade into less flammable moieties like ammonia, while mTA's aromatic moieties enhanced char formation in the condensed phase, reducing flammability. In flame tests, Nylon 6 blends with 15 wt% mTA showed reduced values after flame time and self-extinguishing behaviour, and the oxygen index increased from 22 to 25%. This research provides new insights into modifying natural polyphenols to create effective melt-processable flame retardant additives for Nylon.

Xia et al. [22] also reported a single-step chemical modification process that overcame TA's limitations while utilizing its beneficial properties, resulting in TA terephthalate (TAT) through crosslinking with terephthaloyl chloride. TAT exhibited enhanced thermal stability up to 230 °C with 30% higher char yield and very low heat release capacity (<80 J/g-K) compared to pristine TA. Crosslinked phenolic species in TAT enhanced char formation in the condensed phase, making it suitable as a flame retardant coating for polymers like Nylon 66. TAT-coated Nylon 66 fabric significantly impeded flame propagation, displaying quick self-extinguishing behaviour and reduced char length in vertical flame tests. Morphological characterization and thermal studies confirmed the beneficial effects of char formation on flame retardancy.

Kim et al. [27] synthesized graphene nanoparticles coated with TA as non-halogen flame retardants and electrospun them with polyurethane to make multifunctional composite nanofibers. The addition of bio-based, non-halogen flame retardant resulted in improved flame retardant, antimicrobial, and mechanical properties of the composite nanofibers. For instance, at 5 wt% of flame retardant, the pHRR was reduced by ~31.08%; and ~500% and ~135% improvements were observed in antimicrobial activity and Young's modulus, respectively, when compared with neat polyurethane fibers.

Different strategies have been explored to modify TA and its thermal degradation pathway to improve its charring effect [24]. For instance, mixing TA with organomodified montmorillonite (OMMT) resulted in a reduction of the pHRR in PLA-based compositions. Another strategy involved associating TA with a biosourced phosphorous-based compound, metallic phytate salt, to enhance TA's flame retardant effect. The third approach chemically modified TA through a chemical grafting of phosphoric acid groups, which showed the most effective flame-retardant effect, but with a reduction in PLA molecular weight. Table 5.2 listed some notable flame retardant properties of polymer-tannic acid composites.

Some other parameters from cone calorimeter measurements: heat release capacity (HRC), total heat release (THR), fire growth rate (FIGRA), smoke extinction area (SEA), total smoke production (TSP).

In a study by Lin et al. [28], a new intumescent flame retardant (IFR) system consisting of ammonium polyphosphate (APP)/TA functionalized graphene (TGE) presented synergistic flame retardant and smoke suppression effects for natural rubber (NR). The AGT system changed the decomposition behaviour of NR, forming a highly graphitized and phosphorous-containing carbonaceous structure char layer on the composites' surface, resulting in enhanced flame retardancy and smoke suppression properties. Furthermore, TGE improved the mechanical properties of the NR composite effectively when used as part of the AGT system.

Moreover, other studies have highlighted the significance of different types of interactions between TA and polymers in achieving other enhancements [46, 47]. Notably, hydrogen bonding interactions between the hydroxyl groups in TA and functional groups within polymers play a prominent role. These interactions promote better dispersion of TA within the polymer matrix and strengthen the interface between filler and matrix, resulting in improved mechanical properties and flame resistance. The growing body of evidence clearly indicates the promising potential of polymer-TA composites as superior flame retardant materials. However, further investigation is necessary to fully comprehend the impact of varying parameters, such as processing methods, polymer type, and TA proportions, on the final characteristics of these composites.

From a commercial standpoint, these composites offer economic benefits by utilizing natural substances like TA instead of expensive synthetic compounds traditionally used to enhance flame resistance. Additionally, being derived from renewable sources adds a sustainability factor, making these materials even more appealing for manufacturers striving for greener production practices.

It can be said that while existing literature provides a solid scientific foundation supporting the effectiveness of TA based polymer composites as flame retardants, there is scope for further exploration, particularly in understanding the microstructure-property relationships in these materials. Advanced analytical techniques such as scanning electron microscopy (SEM) and transmission electron microscopy (TEM) could shed light on the optimization strategies aimed at achieving maximal property enhancements with minimal additive usage. Polymer-TA composites provide a glimpse into a future where safety concerns related to flammable materials can be effectively addressed without compromising on environmental considerations or financial feasibility inherent in any industrial application scenario.

Table 5.2 Flame retardant properties of polymer-tannic acid composites

Sample	FR content (%)	TTI (s)	pHRR (kW/m^2)	THR (MJ/m^2)	HRC (J/gk)	FIGRA	Mass loss (g)	SEA (m^2/kg)	TSP (m^2)	References
0 wt% TA	0	96.0 ± 14.4	670.0 ± 93.8	63.0 ± 5.7	–	–	–	–	–	[42]
0.1 wt% TA	0.1	105.0 ± 13.7	96.0 ± 14.4	64.0 ± 9.6	–	–	–	–	–	
1.0 wt% TA	1.0	80.0 ± 8.8	96.0 ± 14.4	75.0 ± 6.7	–	–	–	–	–	
3.0 wt%-TA	3.0	114.0 ± 9.1	96.0 ± 14.4	80.0 ± 9.6	–	–	–	–	–	
5.0 wt% TA	5.0	84.0 ± 6.9	96.0 ± 14.4	74.0 ± 8.1	–	–	–	–	–	
8.0 wt% TA	8.0	116.0 ± 12.8	96.0 ± 14.4	83.0 ± 11.7	–	–	–	–	–	
TA composite	1	14.0 ± 0.9	108.0 ± 6.7	24.4 ± 0.57	–	1.27 ± 0.07	12.8 ± 1.2	–	–	[43]
Control	0	–	436 ± 10	37.7 ± 0.6	–	–	–	812 ± 8	10.5 ± 0.3	[46]
T-CFRP0.8	0.8	–	400 ± 28	27.4 ± 0.6	–	–	–	701 ± 48	8.3 ± 0.6	
T-CFRP1.0	1.0	–	382 ± 22	27.6 ± 0.6	–	–	–	662 ± 9	7.9 ± 0.2	
T-CFRP1.2	1.2	–	356 ± 13	26.8 ± 2.2	–	–	–	634 ± 32	7.6 ± 0.2	
Neat PU	0	–	340.75 ± 1.98	29.32 ± 0.79	529.69 ± 28.86	–	–	–	–	[62]
PU-TA	1	–	295.40 ± 6.33	24.45 ± 0.67	428.5 ± 17.62	–	–	–	–	
PU-TA@RGO1	1	–	272.26 ± 10.36	23.20 ± 0.67	377.63 ± 7.0	–	–	–	–	
PU-TA@RGO3	3	–	249.94 ± 9.39	22.28 ± 0.62	372.01 ± 8.71	–	–	–	–	
PU-TA@RGO5	5	–	235.23 ± 3.78	21.72 ± 0.42	356.81 ± 11.15	–	–	–	–	

4 Thermal Behaviour of Tannic Acid Based Polymer Composites

The thermal decomposition pathway and stability of polymers are one of the crucial indicators to determine their flame retardancy. Tannic acid (TA)-based polymer composites materials are one of the commonly studied and utilized flame retardant materials. The development of flame-retardant composites using tannic acid and its derivatives have shown to produce flame retardant materials with enhanced thermal stability, char formation and char thermal stability [24, 25, 48]. In this unit, we examine the impact of tannic acid and tannic acid derivatives flame-retardant materials on the thermal behaviour of various polymer composites using thermogravimetric analysis (TGA). Figure 5.1 shows the TGA thermogram of the pristine TA run under air and nitrogen atmosphere. The curves display similar trend between 100 and 400 °C, thus signifying that the nature of gas used has no impact on the thermal behaviour of the TA at temperatures below 400 °C [24, 48, 49]. However, at temperatures above 400 °C, the TGA curves of the TA were different; under air flow there is almost no residue observed, while under nitrogen flow there was a char formation amounting to approximately 28% [24]. The weight loss observed between 220 and 400 °C was attributed to the release of 1, 2, 3-benzenetriol and carbon dioxide emanating from the thermal decomposition of the outer layer gallic acid groups on the TA (Fig. 5.1). The formation of the char under nitrogen atmosphere at higher temperature above 400 °C could be assigned to the swelling effect of the cross-linked inner gallic acids groups in the TA [48].

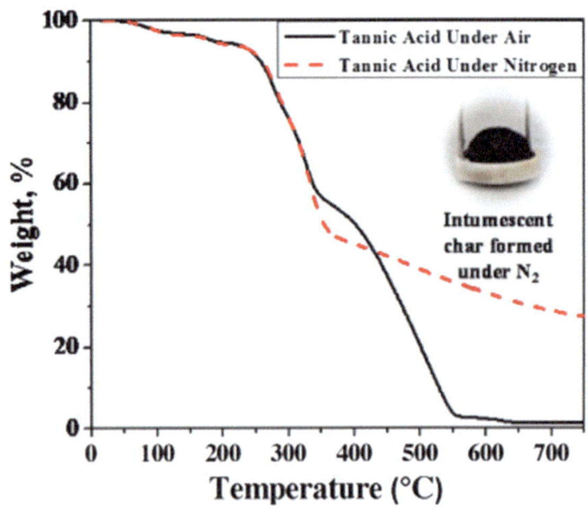

Fig. 5.1 TGA curves of tannic acid under air and nitrogen at 20 °C/min [49]

4.1 PLA Embedded Pristine TA

Tannic acid has been purported as an excellent charring material as confirmed by TGA analysis in Fig. 5.1, and it has been extensively studied as a flame retardant agent in many polymer composites formulations. In this subsection, we discuss the thermal properties of pristine TA incorporated PLA composite with respect to unmodified PLA and TA (Fig. 5.2). Undoubtedly, the incorporation of TA into PLA had a considerable influence on the thermal behaviour of the formed composite. Compared to the pristine PLA, the TA/PLA composite exhibited fast thermal decomposition at temperatures between 100 and 400 °C, attributable to the quick decomposition of the TA into highly flammable compounds, which in turn accelerate or fuel the degradation of the PLA [24, 48]. However, under pyrolytic conditions at temperatures above 400 °C, the thermal stability of the TA/PLA composite was relatively improved when compared to the pure PLA. This could be due to the charring effect of TA under pyrolytic conditions. Nonetheless, the formed char exhibited relatively poor thermal stability, evidenced by continually decomposition at temperatures above 500 °C [24]. Furthermore, the presence of TA into PLA induces premature decomposition of the formed composite. Therefore, there is still a need to improve thermal properties of the pristine TA based PLA in order to enhance its thermal stability, char forming, char thermal stability and ultimately its flame resistance properties. More recently, various studies have transitioned towards the use of functional TA obtained by blending or chemical functionalization with other functional materials with desired properties for flame retardancy [24, 25, 50].

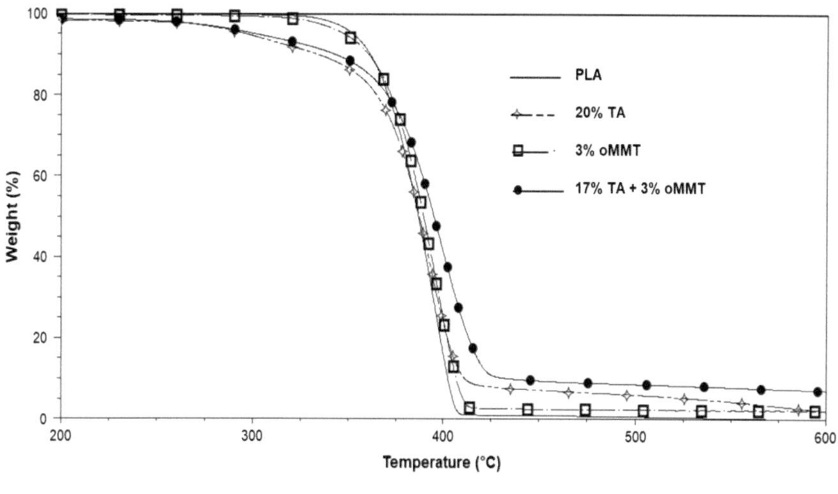

Fig. 5.2 TGA thermograms of pristine TA, OMMT nanoclay and TA combined with OMMT nanoclay incorporated PLA composites under nitrogen [24]

4.2 TA/Organo-Modified Montmorillonite (OMMT) Nanoclay Blend Based Poly (Lactide) (PLA) Composites

Chemical modification and/or blending of TA with various compounds present a great strategy to promote and enhance the char forming and char thermal stability of the TA based polymer composites. Herein, we delve into the thermal behaviour of TA blended with OMMT nanoclay incorporated PLA composite in respect to the composites of individual TA and OMMTs. Similar to TA/PLA composites, the TGA curves of TA/OMMTs PLA composites (Fig. 5.2) displayed significant weight losses between 300 and 450 °C, which correspond to premature thermal decomposition induced by volatile benzyl alcohol degradation product released from TA. This shows that the use of oMMTs in conjunction with TA did not increase the thermal degradation pathway of the PLA composites at temperature below 400 °C. Nonetheless, the starting degradation temperature of TA/OMMTs composite was slightly shift to approximately 300 °C from 296 °C of the pristine TA PLA composites. Meanwhile, between 450 and 600 °C, the TA/OMMTs composite exhibited improved thermal stability compared to individual TA and OMMTs based composites as confirmed by the lower mass loss (ML) and formation of the char layer (Fig. 5.2). At higher temperature above 400 °C, the combination of TA and OMMTs favoured the formation of high amount of a char residue amounting to 7.2% at 600 °C higher than 2% and 2.5% for individual TA and OMMTs based PLA composite, respectively. Compared to 2.4%/°C mass loss rate (MLR) for pristine PLA, the PLA blended with 17%TA/3% OMMT exhibits low average mass loss rate of 1.1%/°C (Table 5.3). These results demonstrate that OMMTs combined with TA hold great potential as a char forming agent for the development of flame-retardant polymer composite with improved char thermal stability [24].

4.3 Phosphorylated TA (TA-POH)/PLA Composite

Phosphorous compounds are vital constituents in flame retardant materials, which are well-known of their ability to enhance char formation and char thermal stability via acid-catalysed cross-linking condensation reactions [50, 51]. Therefore, chemical grafting of phosphorous molecules on the TA structure may change or influence the thermal properties of the resulting composites [24]. In this regard, the polymer composites embedded TA chemically grafted with phosphorous-based compounds "phytate salts" are anticipated to exhibit advanced char formation and char thermal stability. Figure 5.3 shows the TGA thermograms of pristine PLA and composites of PLA containing 20% phosphorylated TA (TA-POH) and 20% TA, respectively. Undesirable, the results show that between 280 and ~420 °C, the PLA incorporated 20% TA-POH exhibited faster early thermal decomposition compared to pure PLA and 20% TA loaded PLA as evidenced by lower onset temperature (Fig. 5.3 and Table 5.3) [24]. This could be attributed to synergist premature catalytic degradation

Table 5.3 The thermal behaviour of pristine polymer and their corresponding tannic acid based polymer composites under N_2

Polymer material	Composite additives	Polymer composite		Pure polymer		MMLR, %/ °C		Char yield at maximum high temperature °C, %		References
		T_{onset},[a], °C	T_{max}[b], °C	T_{onset}	T_{max}	Composite	Pure polymer	Composite	Pure Polymer	
PLA	20 wt% TA	296	391	312	372	2.1	2.1	2	0	[24]
PLA	17 wt% TA + 3 wt% oMMT	300	400	312	372	1.1	2.1	7.2	0	[24]
PLA	20 wt% TA-POH	286	350	312	372	2.7	2.1	6.0	0	[24]
Nylon 6	15 wt% TA	322	479	411	469	1.2	2.0	6.6	1.2	[25]
Nylon 6	15 wt% TAT	332	453	411	469	0.9	2.0	8.9	1.2	[25]
PLA	5 wt% BA-TA-MEL	~340	392	353	389	2.5	3.2	4	0.5	[54]
PLA	20 wt% Phyt/PEI-TA	366	417	347	390			12.5	0	[3]
EP	4 wt% PHDT@FeCo-LDH	329	392[c]	368	404[c]			22.7	14.1	[55]
EP	15 wt% APP@TA/ adenine/Ni	341	387[c]	368	402[c]			27.0	16.8	[35]
EPS	20 wt% PER/APP/ TA@GE	260	524 (T_{max2}), 792 (T_{max3})	283	412	0.42	1.95	14.6	~0	[56]
Cotton	20 wt% TA/0.25 wt% NaOH	220	~363	265	361	~1.15	~2.5	~20	4.7	[20]
Polyacrylonitrile	2 wt% TA-MoS_2	330	333 (T_{max1}), 445 (T_{max3})	316	326 (T_{max1}), 443 (T_{max2})	~0.7	~1.2	49.7	42.8	[57]

(continued)

Table 5.3 (continued)

Polymer material	Composite additives	Polymer composite		Pure polymer		MMLR, %/ °C		Char yield at maximum high temperature °C, %		References
		$T_{onset,}$[a], °C	T_{max}[b], °C	T_{onset}	T_{max}	Composite	Pure polymer	Composite	Pure Polymer	
Polyurethane	5 wt%/TA/reduced graphene oxide	≤250	~480	310	~480			~15	~8	[27]
Poly(vinyl chloride)	10 wt% MH@PASn@TAZn	258	468	256	461	2.45	2.78	20.13	8.7	[26]

Note [a]T_{onset} refers to the temperature at which 5% weight loss occurs; [b]T_{max} refers to the temperature at which maximum weight loss occurred; [c]$T_{50\% \pm 0.1}$ refers to temperature at which 50% mass loss occurred; MMLR means Maximum Mass Loss Rate; MH@PASn@TAZn refers to $Mg(OH)_2$@tin phytate@zinc tannate; blank—No data reported

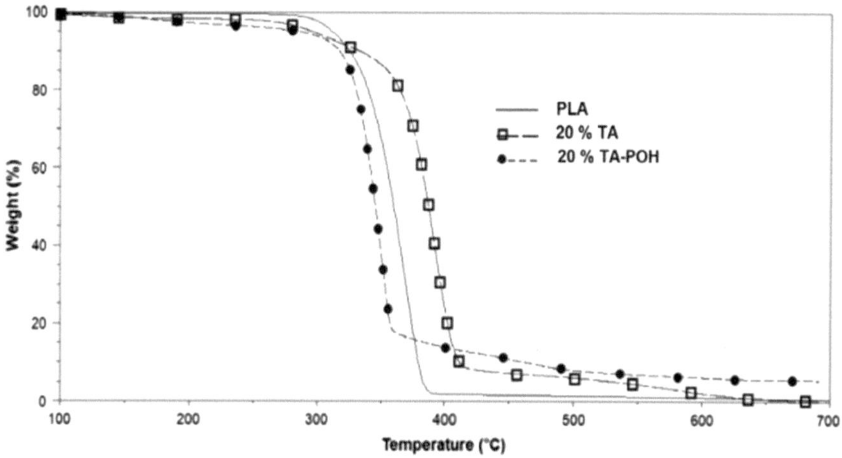

Fig. 5.3 TGA thermograms of PLA, and 20% TA and 20 TA-POH incorporated PLA composite under N_2 [24]

emanating from dehydration effect of the phosphoric acid groups and release of volatile degradants by the TA. Moreover, the average mass loss rate (2.7%) of the PLA/20% TA-POH composite was relatively high as compared to pristine PLA and PLA modified with pristine 20% TA (Table 5.3). Nonetheless, although TA-POH was observed to adversely influence the thermal stability of the PLA composite, it has prompted high char yield amounting to 6% at 600 °C (Table 5.3) as well as better char thermal stability.

4.4 Boric Acid (BA)-Tannic Acid (TA)-Melamine (MEL) PLA Composite

Inspired by the distinct physiochemical properties and thermal properties of boric acid (BA) tannic (TA) and melamine (MEL) as flame retardant agents, this unit reports on the bio-based tannic acid derivative (tribrid BA-TA-MEL) incorporated PLA composite to understand their synchronized or synergistic influence on the thermal characteristics of the resultant PLA composite [52–54]. The combination of the BA and MEL with TA is anticipated to result in a composite with distinctive and desirable degradation pathway desirable for flame retardancy [54]. TGA curves for PLA loaded BA-TA-MEL in comparison to pristine PLA are presented in Fig. 5.4 and Table 5.3. Similar to PLA composites discussed hitherto, the BA-TA-MEL/PLA composites displayed the high weight loss rate at temperatures less than 410 °C, which may be attributed to the premature thermal degradation of BA-TA-MEL at relatively lower temperatures. However, the presence of BA-TA-MEL into PLA composites decreases the average mass loss rate and promoted the creation high char yield calculated to

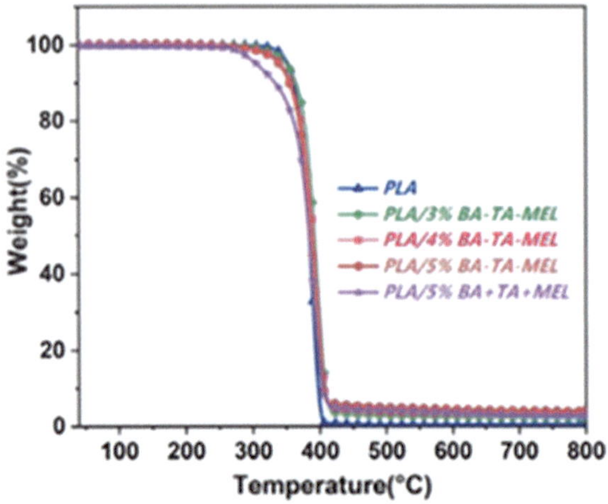

Fig. 5.4 TGA curves of pure PLA and PLA/BA-TA-MEL composite under N_2 [54]

be 4.0% (Figs. 5.4 and 5.3). The synergistic effect evolving from the combination of individual components of the BA-TA-MEL composites created a larger protective layer that covers the polymer during thermal decomposition.

4.5 PLA Incorporated Phytic Acid/Polyethyleneimine/Tannic Acid Complex (Phyt/PEI-TA) Composite

Figure 5.5 illustrates the impact of the tannic acid derivative (20 wt% Phyt/PEI-TA) on the thermal stability of the PLA utilizing TGA analysis under N_2. Positively, modification of PLA with Phyt/PEI-TA decreases the rate of weight loss and mass loss of PLA. The onset temperature (initial degradation temperature) of the PLA was increased from 347 to 366 °C for Phyt/PEI-TA blended PLA composites, thus signifying improved thermal stability (Figs. 5.3 and 5.5). Furthermore, the Phyt/PEI-TA displayed excellent char forming properties evidenced by the high char yield (12.5%) and char stability for the PLA composites compared to 0% char yield for pristine PLA at 700 °C (Table 5.3) [3]. It is clear that the use of Phyt/PEI-TA changed the degradation pathway of the PLA composite from noncharring to charring material, thus demonstrates a great potential as an additive for development of superior flame retardant material [3].

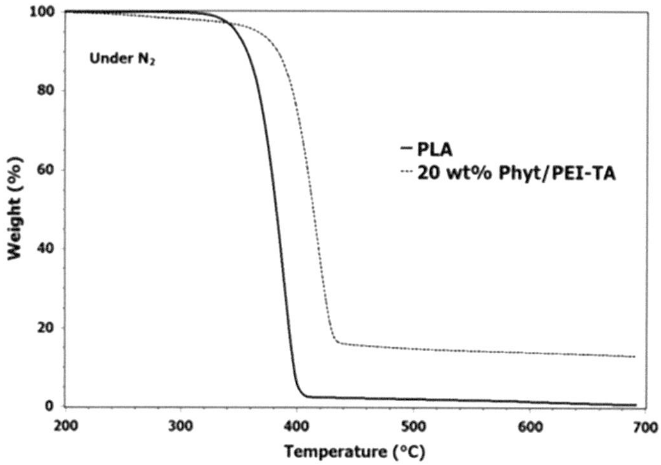

Fig. 5.5 TGA curves of PLA and PLA containing 20 wt% Phyt/PEI-TA under nitrogen (20 °C/min) [3]

4.6 TA and Terephthaloyl Cross-Linked Tannic Acid (TAT) Based Nylon 6 Composites

Nylon is among the main crucial polymers utilized in various industries including but not limited to automotive, textile and electronics. In complementary to its excellent mechanical properties, superior thermal stability and charming forming ability of this polymer are of paramount importance for the above mentioned applications. The processing of nylon requires non-toxic flame retardant additives that are thermally stable at high temperatures. Tannic acid based flame retardant material are among the materials that have been explored to improve the thermal behaviour of the nylon [25]. Herein, we evaluate the thermal behaviour of chemically modified TA "terephthaloy cross-linked TA (TAT)" incorporated nylon 6 composites versus polymer composite containing unmodified TA. Figure 5.6 shows the TGA thermograms of the nylon 6 blended with pristine TA and chemically modified TA (TAT) composites acquired by isothermal TGA experiment at 235 °C under air. Notably, 15 wt% TAT incorporated nylon 6 exhibited superior thermal stability with only ~2% weight loss when compared to the 15 wt% pristine TA incorporated nylon 6 with 5% weight loss [25].

To further understand thermal degradation behaviour of nylon 6 incorporated TA or TAT, the overall thermal stability analysis was carried out from 200 to 750 °C under nitrogen (Fig. 5.7). Both maximum 15 wt% TA and 15 wt% TAT loaded Nylon 6 composites exhibited faster degradation between 200 and 500 °C, compared to composites with lower loadings [25]. However, above 500 °C, 15 wt% TA and 15 wt% TAT based nylon composite presented high char forming and excellent char thermal stability compared to their respective composites with lower loadings. Compared to 15 wt% pristine TA loaded nylon 6 composite which exhibited maximum weight

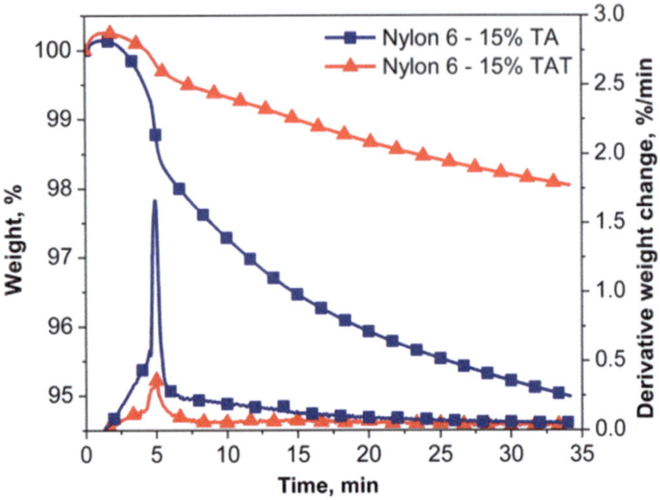

Fig. 5.6 Thermal stability of Nylon 6—TA/TAT (15 wt% loading) blends at Nylon 6 processing temperature (235 °C) [25]

loss rate of 1.2%/°C and approximately 1.2% char yield, 15% TAT loaded nylon 6 composite showed better thermal stability with maximum weight loss rate of 0.9%/°C and about 8.9% char yield (Fig. 5.7 and Table 5.3). In comparison to nylon 6, 15 wt% TA and 15 wt% TAT loaded nylon reduced maximum weight loss rate by 40% and 55%, respectively. This results demonstrated that chemical modification of TAT open alternative ways to further improve the FR properties TA based nylon composites.

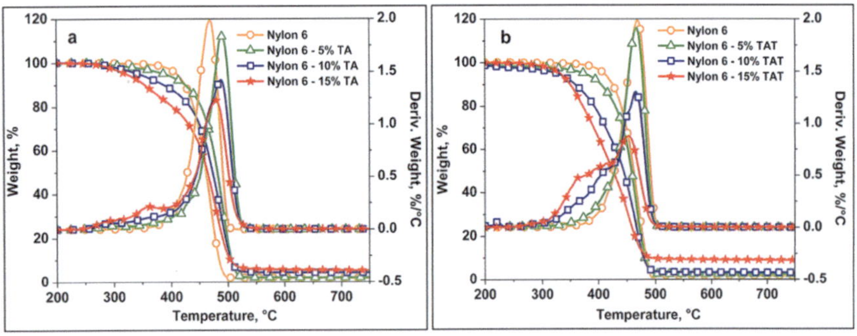

Fig. 5.7 TGA thermograms of **a** Nylon 6-TA blends and **b** Nylon 6-TAT blends [25]

4.7 Other Tannic Acid Based Polymer Composites

In addition to the afore-mentioned tannic acid based bio-polymer composites, thermal properties of other bio-polymer (e.g., epoxy resin, cotton, nylon) modified with tannic acid derivatives are also summarized in Table 5.3. For example, we explored the thermal properties of epoxy resin (EP) polymer composites prepared using polyphosphazene nanospheres (PHDT@FeCo-LDH) which was are prepared via polycondensation of tannic acid, hexachlorocyclotriphosphazene (HCCP) and 4,40-dihydroxybiphenyl followed by surface modification with FeCo-Layered double hydroxides (FeCo-LDH) [55]. The result showed that the incorporations of the functional nanospheres into EP reduced the initial decomposition temperature ($T_{5\%}$) and 50% mass loss temperature ($T_{50\%}$) (Table 5.3). This could be attributed to the catalytic effect of the embedded nanospheres which increased the rate of premature degradation of the EP. Remarkably, the EP composites presented high char yield, char thermal stability and low maximum mass loss rate (39.8% less) compared to pure EP, which could be ascribed to the carbonization by the PHDT as a results of the char-forming ability of the TA and formation of dehydrating compounds (e.g., Phosphoric) through degradation of the HCCP molecules in the PHDT [55]. Meanwhile, the LDH on the surface of the microsphere also provided further carbonization effect on EP during thermal degradation. Another similar study, reported EP matrix embedded with FR ammonium polysphophate particles grafted with tannic acid, adenine and Ni^{2+} (APP@TA/adenine/Ni) composite, and thermal properties are summarized in Table 5.3 [35]. The reported composite also demonstrated low maximum mass loss rate and high char yield when compared to pure polymer.

Another different study, reported on expanded polystyrene (EPS) modified with a FR composite of phenolic epoxy resin (PER)/ammonium polyphosphate (APP)/ tannic acid grafted graphene (TA@GE). The result showed that the modified EPS composite exhibited completely different degradation pathway comprised of three decomposition steps when compared to the unmodified EPS with only one degradation step [56]. Like previous polymer composites, the EPS composites showed better thermal stability, char forming and char thermal stability with respect to unmodified EPS. The thermal behavior of this systems are summarized in Table 5.3. Tannic acid can be modified with many other materials to develop advanced flame retardant polymer composite with enhanced thermal behavior [34]. Other than the already mentioned polymeric composite, Table 5.3 further gives other examples of reported tannic acid based polymer composites and summary of their thermal properties.

5 Flame Retardant Mechanism of Tannic Acid Modified Composites

Tannic acid, a polyphenol, exhibits a fire-resistant behavior that is related to its char-forming abilities. A study by Price et al. [43] burned puck containing tannic acid composites to test their fire resistant behaviour. Upon burning of the composites, a significant amount of char is formed (Fig. 5.8). The increased carbon content coupled with polyphenol hydroxyl structure of tannic acid make it suitable for use as a biobased flame retardant. The flame retardant mechanism of tannic acid can occur either in the gas or condensed phases or a combination of the two processes [29, 58, 59]. Furthermore, the degradation of tannic acid occurs under nitrogen or air in the condensed or gas phases [21]. The polymer pyrolysis process is the main method that produces the flame retardant mechanism of the condensed phase, which involves a protective char formation [58, 60]. The flame retardancy of the gas phase is accomplished by the termination of radical chemistry involved in combustion [58].

The evaluation of the thermal degradation and char formation of tannic acid showed that, significant degradation of tannic acid occurred at a temperature of ~350 °C. The production of char under nitrogen gas displayed typical intumescence [21]. Tannic acid possesses flame retardant qualities due to its ability to burn and produce graphite while deactivating radicals created by the combustible material [27]. In addition, the study by Kim et al. [27] also evaluated the generation of heat during combustion and flame retardancy using the peak heat release rate (PHRR) and total heat release (THR) values of the materials. The observed values of PHRR and THR for polyurethane (PU) and composites with reduced graphene oxide (rGO) and tannic acid are presented in Table 5.4. These values demonstrated that the release of heat was greater for neat PU, and the values dropped when RGO and tannic acid were added. The incorporation of tannic acid flame retardant coating in the composite allowed

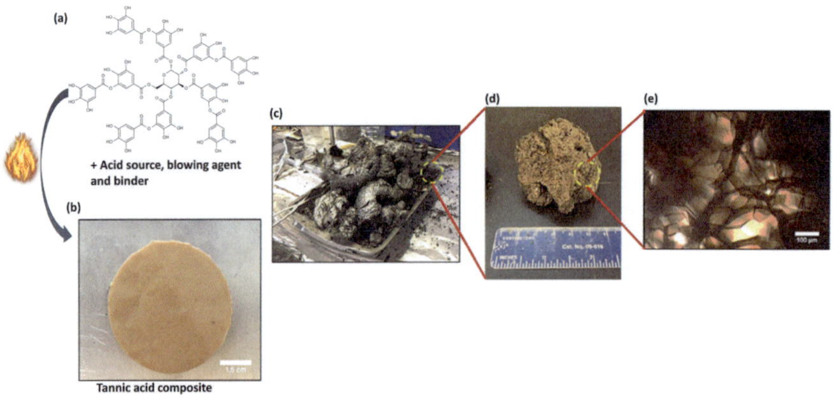

Fig. 5.8 **a** Chemical structure of tannic acid, **b** unburnt tannic acid composite puck, **c** burnt tannic acid tannic acid composite puck, **d** cross-sectional image of tannic acid char and **e** optical microscopy image of a 2 mm thick tannic acid char under transmitted light, scale bar = 100 μm [43]

Table 5.4 Data summary from cone calorimeter tests for neat PU, and PU-TA@rGO with different concentrations of rGO [27]

Sample	PHR (W/g)	THRR (kJ/g)
Neat PU	340.75 ± 1.98	29.32 ± 0.79
PU-TA	295.40 ± 6.33	24.45 ± 0.67
PU-TA@RGO1	272.26 ± 10.36	23.20 ± 0.67
PU-TA@RGO3	249.94 ± 9.39	22.28 ± 0.62
PU-TA@RGO5	235.23 ± 3.78	21.72 ± 0.42

TA = tannic acid

for the development of continuous and compact char layer during condensation step, which inhibits the transmission of second- to third-order heat.

Xia et al. [21] performed the analysis of flame retardancy of tannic acid and the IR peaks for tannic acid char are also summarized in the Fig. 5.9. The figure shows that tannic acid decomposition takes place between 300 and 400 °C. The tannic acid char FTIR analysis at the temperature of 400 °C shows a considerable shift because of the loss of hydroxyl (benzene triol) and ester (carbon dioxide) groups [21]. The bulk of the gallic acid units in the outer layer are thought to have broken down and been liberated in the gaseous phase at this temperature because of the ester bond cleavage. Higher temperatures do not affect tannic acid's inner layer's stability, and at 700 °C, it cross-linked to form intumescent carbonaceous char [61]. Thus, because only carbonaceous char is left at 800 °C, the majority of IR absorption from tannic acid vanishes. The formation of char coating is a result of pyrolysis of tannic acid, which is considered a char-forming agent and blowing agent during burning process [29]. The char layer that forms during combustion can offer defense against mass and heat transfer [29].

5.1 *Summary of the Proposed Flame Retardant Mechanism of Tannic Acid*

Tannic acid flame retardant action produces intumescent char in the condensed and gas phases, which is crucial for applications that involve protection against fire [22]. Nevertheless, these processes both take place throughout a complicated process that involves numerous simultaneous reactions. In the flame retardant mechanism of tannic acid composites, a protective carbon layer will form on the polymer's surface due to the dehydration and carbonization of the flame retardant [60]. The carbon layer is crucial to the ability of polymers to resist flame. The composite's carbon layer becomes denser and has a higher degree of graphitization as a result of the flame retardant addition [60].

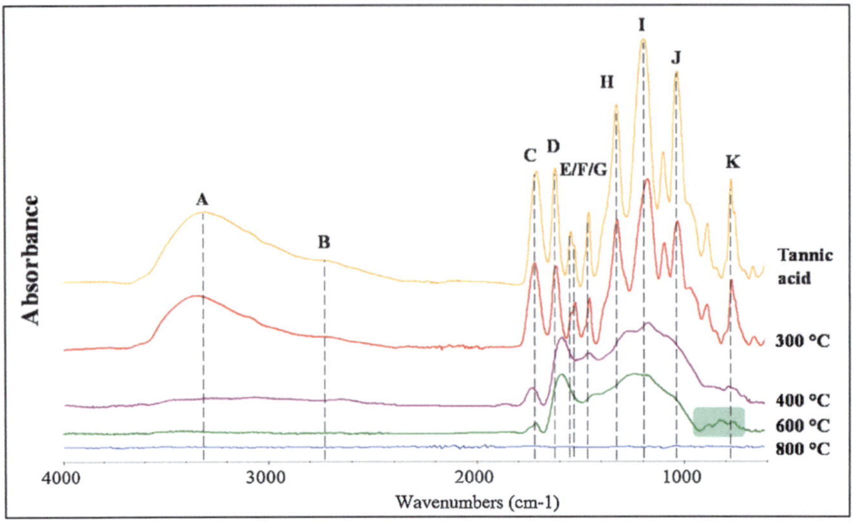

Fig. 5.9 ATR-FTIR of the condensed phase study of tannic acid A. (3325 cm^{-1}, –OH, H-bonding) B. (2710 cm^{-1}, –CH asymmetric stretch) C. (1698 cm^{-1}, C=O stretch) D. 1606 cm^{-1}, aromatic C–O symmetrical stretch) E/F/G. (1532/1518/1443 cm^{-1}, aromatic C=C), H/I. (1307/1177 cm^{-1}, aromatic C–O asymmetrical stretch) and J. (1016 cm^{-1}, C–O–C stretch) K. 754 cm^{-1}, C–H out-of-plane deformation). [21]

5.2 *Condensed Phase Flame Retardant Mechanism*

According to reports, tannic acid acts in the condensed phase by creating a useful char layer [62]. It is anticipated that char-forming technologies will lessen the quantity of dangerous volatile chemicals released during combustion [62]. The initial step in the condensed phase flame retardant mechanism includes the breakdown of the polymer by the tannic acid leading to the prominent flow of the polymer reducing the impact of the escaping flame. The degradation of tannic acid in the condensed phase shows the cleavage of the ester linkage, which releases benzene diol/triol [21]. Through the dehydration process of the created terminal alcohols, cross-linked or carbonized structures are produced at higher temperatures [62]. This happens, for instance, as a result of the flame retardant's dehydrating effect creating carbon–carbon double bonds in the polymer and crosslinking of the aromatic rings [21, 63]. This is then followed by charring, where the carbonization of tannic acid during combustion creates a comparatively stable carbon layer structure [60]. It appears that this thermal insulating layer can both prevent interaction between combustible volatiles and external oxygen; and delay the external heat radiation from the burning area to the polymer [63]. These can lessen the release of flammable gases and pyrolysis as they provide heat cooling effects [29].

5.3 Gas Phase Flame Retardant Mechanism

In the gas phase mechanism, there is a discharge of non-combustible gases, which can stop the burning cycle. In the pyrolysis of tannic acid, the primary products are phenols (1,2,3-benzene triol and 1,2- benzene diol), gases (carbon dioxide and carbon monoxide) and gaseous water [21, 22, 57, 63]. The combustion leads to the breakdown of the cross-linked phenoxy species, which was indicated by the production of carbon monoxide [22]. The phenoxy compounds in the combustion process have the ability to scavenge oxygen-containing radicals in the gaseous phase, which inhibits the free radical chain reaction of combustion and aids in gaseous phase flame retardation put an end to the chain reaction [20, 60]. In this case, the high-energy fire radicals are trapped in the flame zone. In addition, the abundant carbon dioxide produced is capable of preventing heat transfer and diluting the combustible gas [64]. These processes use cross-linking and cyclizing to create a carbonaceous protective layer. Thus, in the gas phase, the flame is extinguished by the formation of radical scavengers.

6 Concluding Remarks and Perspectives

This chapter reports on the recent progress of the novel bio-based (TA) flame retardant additives for polymers. The materials were brought in as a result of the negative impacts brought on by the traditional flame retardants on both the human health and the ecosystem. Interestingly, the flame retardants behaviours and thermal stability of the polymer composites became positively impacted by the addition of the bio-based flame retardants. Despite the considerable achievements, some limitations were observed that need to be addressed. The first being the poor interface between the additive and the polymer matrix of these bio-based compounds for it resulted in the low flame retardancy behaviour. Further, most of the bio-based flame retardants displayed lower efficiency particularly at lower loadings while the higher loadings weakened the mechanical properties of the resultant composites. Also, the thermal stability and the char residual in the combustion were also negatively affected. The industrial development of these bio-based flame retardants are constrained by their high costs and the intricate manufacturing processes.

In addition, some research areas in respect to bio-based flame retardants need to be further developed: The quest for new and different kinds of bio-based materials that can be employed as flame retardants ought to be pursued. Notwithstanding these challenges, the bio-based route is still an encouraging and tenable approach in real industrial applications.

Acknowledgements The authors would like to thank the Department of Science and Innovation (DSI)/Mintek Nanotechnology Innovation Centre, South Africa, for financial support.

References

1. Yang, H., Yu, B., Song, P., Maluk, C., Wang, H.: Surface-coating engineering for flame retardant flexible polyurethane foams: a critical review. Compos. Part B Eng. **176**, 107185 (2019). https://doi.org/10.1016/j.compositesb.2019.107185
2. He, W., Song, P., Yu, B., Fang, Z., Wang, H.: Flame retardant polymeric nanocomposites through the combination of nanomaterials and conventional flame retardants. Prog. Mater. Sci. **114**, 100687 (2020). https://doi.org/10.1016/j.pmatsci.2020.100687
3. Laoutid, F., Vahabi, H., Shabanian, M., Aryanasab, F., Zarrintaj, P., Saeb, M.R.: A new direction in design of bio-based flame retardants for poly(lactic acid). Fire Mater. **42**, 914–924 (2018). https://doi.org/10.1002/FAM.2646
4. Lenoir, D., Kampke-Thiel, K.: Formation of polybrominated dibenzodioxins and dibenzofurans in laboratory combustion processes of brominated flame retardants. Am. Chem. Soc. Nelson 377–392 (1995). https://doi.org/10.1021/bk-1995-0599.ch025
5. Zhang, W., Li, X., Yang, R.: Pyrolysis and fire behaviour of epoxy resin composites based on a phosphorus-containing polyhedral oligomeric silsesquioxane (DOPO-POSS). Polym. Degrad. Stab. **96**, 1821–1832 (2011). https://doi.org/10.1016/j.polymdegradstab.2011.07.014
6. Lee, S.K., Bai, B.C., Im, J.S., In, S.J., Lee, Y.S.: Flame retardant epoxy complex produced by addition of montmorillonite and carbon nanotube. J. Ind. Eng. Chem. **16**, 891–895 (2010). https://doi.org/10.1016/j.jiec.2010.09.014
7. Becker, C.M., Gabbardo, A.D., Wypych, F., Amico, S.C.: Mechanical and flame-retardant properties of epoxy/Mg–Al LDH composites. Compos. Part A Appl. Sci. Manuf. **42**, 196–202 (2011). https://doi.org/10.1016/J.COMPOSITESA.2010.11.005
8. Suihkonen, R., Nevalainen, K., Orell, O., Honkanen, M., Tang, L., Zhang, H., Zhang, Z., Vuorinen, J.: Performance of epoxy filled with nano- and micro-sized magnesium hydroxide. J. Mater. Sci. **47**, 1480–1488 (2012). https://doi.org/10.1007/s10853-011-5933-6
9. Im, J.S., Lee, S.K., In, S.J., Lee, Y.S.: Improved flame retardant properties of epoxy resin by fluorinated MMT/MWCNT additives. J. Anal. Appl. Pyrolysis **89**, 225–232 (2010). https://doi.org/10.1016/J.JAAP.2010.08.003
10. Kim, M.J., Jeon, I.Y., Seo, J.M., Dai, L., Baek, J.B.: Graphene phosphonic acid as an efficient flame retardant. ACS Nano **8**, 2820–2825 (2014). https://doi.org/10.1021/nn4066395
11. Yu, B., Shi, Y., Yuan, B., Qiu, S., Xing, W., Hu, W., Song, L., Lo, S., Hu, Y.: Enhanced thermal and flame retardant properties of flame-retardant-wrapped graphene/epoxy resin nanocomposites. J. Mater. Chem. A. **3**, 8034–8044 (2015). https://doi.org/10.1039/c4ta06613h
12. Gu, J., Liang, C., Zhao, X., Gan, B., Qiu, H., Guo, Y., Yang, X., Zhang, Q., Wang, D.Y.: Highly thermally conductive flame-retardant epoxy nanocomposites with reduced ignitability and excellent electrical conductivities. Compos. Sci. Technol. **139**, 83–89 (2017). https://doi.org/10.1016/j.compscitech.2016.12.015
13. Wang, C.S., Shieh, J.Y.: Synthesis and properties of epoxy resins containing bis(3-hydroxyphenyl) phenyl phosphate. Eur. Polym. J. **36**, 443–452 (2000). https://doi.org/10.1016/S0014-3057(99)00091-9
14. Hergenrother, P.M., Thompson, C.M., Smith, J.G., Connell, J.W., Hinkley, J.A., Lyon, R.E., Moulton, R.: Flame retardant aircraft epoxy resins containing phosphorus. Polymer (Guildf). **46**, 5012–5024 (2005). https://doi.org/10.1016/J.POLYMER.2005.04.025
15. Ribera, G., Mercado, L.A., Galià, M., Cádiz, V.: Flame retardant epoxy resins based on diglycidyl ether of isobutyl bis(hydroxypropyl)phosphine oxide. J. Appl. Polym. Sci. **99**, 1367–1373 (2006). https://doi.org/10.1002/app.22325
16. Arbenz, A., Avérous, L.: Chemical modification of tannins to elaborate aromatic biobased macromolecular architectures. Green Chem. **17**, 2626–2646 (2015). https://doi.org/10.1039/c5gc00282f
17. Shirmohammadli, Y., Efhamisisi, D., Pizzi, A.: Tannins as a sustainable raw material for green chemistry: a review. Ind. Crops Prod. **126**, 316–332 (2018). https://doi.org/10.1016/j.indcrop.2018.10.034

18. Qian, L., Feng, F., Tang, S.: Bi-phase flame-retardant effect of hexa-phenoxy-cyclotriphosphazene on rigid polyurethane foams containing expandable graphite. Polymer (Guildf). **55**, 95–101 (2014). https://doi.org/10.1016/j.polymer.2013.12.015
19. Mu, X., Yuan, B., Hu, W., Qiu, S., Song, L., Hu, Y.: Flame retardant and anti-dripping properties of polylactic acid/poly(bis(phenoxy)phosphazene)/expandable graphite composite and its flame retardant mechanism. RSC Adv. **5**, 76068–76077 (2015). https://doi.org/10.1039/c5ra12701g
20. Nam, S., Condon, B.D., Xia, Z., Nagarajan, R., Hinchliffe, D.J., Madison, C.A.: Intumescent flame-retardant cotton produced by tannic acid and sodium hydroxide. J. Anal. Appl. Pyrolysis **126**, 239–246 (2017). https://doi.org/10.1016/j.jaap.2017.06.003
21. Xia, Z., Singh, A., Kiratitanavit, W., Mosurkal, R., Kumar, J., Nagarajan, R.: Unraveling the mechanism of thermal and thermo-oxidative degradation of tannic acid. Thermochim. Acta **605**, 77–85 (2015). https://doi.org/10.1016/j.tca.2015.02.016
22. Xia, Z., Kiratitanavit, W., Facendola, P., Thota, S., Yu, S., Kumar, J., Mosurkal, R., Nagarajan, R.: Fire resistant polyphenols based on chemical modification of bio-derived tannic acid. Polym. Degrad. Stab. **153**, 227–243 (2018). https://doi.org/10.1016/j.polymdegradstab.2018.04.020
23. Qiu, S., Sun, J., Li, H., Gu, X., Fei, B., Zhang, S.: A green way to simultaneously enhance the mechanical, flame retardant and anti-ultraviolet aging properties of polylactide composites by the incorporation of tannic acid derivatives. Polym. Degrad. Stab. **196**, 109831 (2022). https://doi.org/10.1016/j.polymdegradstab.2022.109831
24. Laoutid, F., Karaseva, V., Costes, L., Brohez, S., Mincheva, R., Dubois, P.: Novel bio-based flame retardant systems derived from tannic acid. J. Renew. Mater. **6**, 559–572 (2018). https://doi.org/10.32604/JRM.2018.00004
25. Xia, Z., Kiratitanavit, W., Facendola, P., Yu, S., Kumar, J., Mosurkal, R., Nagarajan, R.: A bio-derived char forming flame retardant additive for nylon 6 based on crosslinked tannic acid. Thermochim. Acta. **693** (2020). https://doi.org/10.1016/j.tca.2020.178750
26. Meng, W., Dong, Y., Li, J., Cheng, L., Zhang, H., Wang, C., Jiao, Y., Xu, J., Hao, J., Qu, H.: Bio-based phytic acid and tannic acid chelate-mediated interfacial assembly of $Mg(OH)_2$ for simultaneously improved flame retardancy, smoke suppression and mechanical properties of PVC. Compos. Part B Eng. **188** (2020). https://doi.org/10.1016/j.compositesb.2020.107854
27. Kim, Y.N., Ha, Y.M., Park, J.E., Kim, Y.O., Jo, J.Y., Han, H., Lee, D.C., Kim, J., Jung, Y.C.: Flame retardant, antimicrobial, and mechanical properties of multifunctional polyurethane nanofibers containing tannic acid-coated reduced graphene oxide. Polym. Test. **93** (2021). https://doi.org/10.1016/j.polymertesting.2020.107006
28. Li, L., Liu, X., Shao, X., Jiang, L., Huang, K., Zhao, S.: Synergistic effects of a highly effective intumescent flame retardant based on tannic acid functionalized graphene on the flame retardancy and smoke suppression properties of natural rubber. Compos. Part A Appl. Sci. Manuf. **129** (2020). https://doi.org/10.1016/j.compositesa.2019.105715
29. Jin, X., Xiang, E., Zhang, R., Qin, D., Jiang, M., Jiang, Z.: Halloysite nanotubes immobilized by chitosan/tannic acid complex as a green flame retardant for bamboo fiber/poly(lactic acid) composites. J. Appl. Polym. Sci. **138** (2020). https://doi.org/10.1002/app.49621
30. Qiu, S., Li, Y., Qi, P., Meng, D., Sun, J., Li, H., Cui, Z., Gu, X., Zhang, S.: Improving the flame retardancy and accelerating the degradation of poly (lactic acid) in soil by introducing fully bio-based additives. Int. J. Biol. Macromol. **193**, 44–52 (2021). https://doi.org/10.1016/j.ijbiomac.2021.10.119
31. Alongi, J., Han, Z., Bourbigot, S.: Intumescence: tradition versus novelty. A comprehensive review. Prog. Polym. Sci. **51**, 28–73 (2014). https://doi.org/10.1016/j.progpolymsci.2015.04.010
32. Zhang, W., Roy, S., Ezati, P., Yang, D.P., Rhim, J.W.: Tannic acid: a green crosslinker for biopolymer-based food packaging films. Trends Food Sci. Technol. **136**, 11–23 (2023). https://doi.org/10.1016/j.tifs.2023.04.004
33. Kim, H., Panda, P.K., Sadeghi, K., Seo, J.: Poly (vinyl alcohol)/hydrothermally treated tannic acid composite films as sustainable antioxidant and barrier packaging materials. Prog. Org. Coatings. **174**, 107305 (2023). https://doi.org/10.1016/j.porgcoat.2022.107305

34. Wang, M., Yin, G.Z., Yang, Y., Fu, W., Díaz Palencia, J.L., Zhao, J., Wang, N., Jiang, Y., Wang, D.Y.: Bio-based flame retardants to polymers: a review. Adv. Ind. Eng. Polym. Res. **6**, 132–155 (2023). https://doi.org/10.1016/j.aiepr.2022.07.003
35. Sui, Y., Dai, X., Li, P., Zhang, C.: Superior radical scavenging and catalytic carbonization capacities of bioderived assembly modified ammonium polyphosphate as a mono-component intumescent flame retardant for epoxy resin. Eur. Polym. J. **156**, 110601 (2021). https://doi.org/10.1016/j.eurpolymj.2021.110601
36. Norouzi, M., Zare, Y., Kiany, P.: Nanoparticles as effective flame retardants for natural and synthetic textile polymers: application, mechanism, and optimization. Polym. Rev. **55**, 531–560 (2015). https://doi.org/10.1080/15583724.2014.980427
37. Babrauskas, V.: Development of the cone calorimeter—a bench-scale heat release rate apparatus based on oxygen consumption (1982)
38. Morgan, A.B., Bundy, M.: Cone calorimeter analysis of UL-94 V-rated plastics. Fire Mater. 257–283 (2007). https://doi.org/10.1002/fam
39. De la Cruz, L.G., Abt, T., León, N., Wang, L., Sánchez-Soto, M.: Ice-template crosslinked PVA aerogels modified with tannic acid and sodium alginate. Gels **8** (2022). https://doi.org/10.3390/gels8070419
40. Kim, Y.O., Cho, J., Yeo, H., Lee, B.W., Moon, B.J., Ha, Y.M., Jo, Y.R., Jung, Y.C.: Flame retardant epoxy derived from tannic acid as biobased hardener. ACS Sustain. Chem. Eng. **7**, 3858–3865 (2019). https://doi.org/10.1021/acssuschemeng.8b04851
41. Wang, X., Cai, W., Ye, D., Zhu, Y., Cui, M., Xi, J., Liu, J., Xing, W.: Bio-based polyphenol tannic acid as universal linker between metal oxide nanoparticles and thermoplastic polyurethane to enhance flame retardancy and mechanical properties. Compos. Part B Eng. **224**, 109206 (2021). https://doi.org/10.1016/j.compositesb.2021.109206
42. Korey, M., Johnson, A., Webb, W., Dietenberger, M., Youngblood, J., Howarter, J.: Tannic acid-based prepolymer systems for enhanced intumescence in epoxy thermosets. Green Mater. **8**, 150–161 (2019). https://doi.org/10.1680/jgrma.19.00061
43. Price, E.J., Covello, J., Paul, R., Wnek, G.E.: Tannic acid based super-intumescent coatings for prolonged fire protection of cardboard and wood. SPE Polym. **2**, 153–168 (2021). https://doi.org/10.1002/pls2.10043
44. Wang, J., Zhou, S., Zhang, Q., Lu, G.P., Lin, Y.: Tannic acid as cross-linker and flame retardant for preparation of flame-retardant polyurethane elastomers. React. Funct. Polym. **181**, 105454 (2022). https://doi.org/10.1016/j.reactfunctpolym.2022.105454
45. Wang, J., Wang, X., Zhou, Z., Liu, X., Xu, M., Zhao, F., Zhao, F., Li, S., Liu, Z., Li, L., Zhao, S.: Flame-retardant effect of tannic acid-based intumescent fire-retardant applied on flammable natural rubber. RSC Adv. **12**, 29928–29938 (2022). https://doi.org/10.1039/d2ra04682b
46. Kim, Y.O., Cho, J., Kim, Y.N., Kim, K.W., Lee, B.W., Kim, J.W., Kim, M., Jung, Y.C.: Recyclable, flame-retardant and smoke-suppressing tannic acid-based carbon-fiber-reinforced plastic. Compos. Part B Eng. **197**, 108173 (2020). https://doi.org/10.1016/j.compositesb.2020.108173
47. Bhakare, M.A., Lokhande, K.D., Bondarde, M.P., Dhumal, P.S., Some, S.: Dual functions of bioinspired, water-based, reusable composite as a highly efficient flame retardant and strong adhesive. Chem. Eng. J. **454**, 140421 (2023). https://doi.org/10.1016/j.cej.2022.140421
48. Yang, M., Guo, W., Liu, S., Zhang, B., Chen, Y., Wang, Y.: Highly stretchable gamma-irradiated poly (vinyl alcohol)/Tannic acid composite hydrogels with superior transparency and antibacterial activity. J. Polym. Res. **28**, 1–13 (2021). https://doi.org/10.1007/s10965-021-02777-7
49. Xia, Z., Singh, A., Kiratitanavit, W., Mosurkal, R., Kumar, J., Nagarajan, R.: Unraveling the mechanism of thermal and thermo-oxidative degradation of tannic acid. Sens. Actuators B. Chem. **605**, 77–85 (2015). https://doi.org/10.1016/j.tca.2015.02.016
50. Brosse, N., Pizzi, A.: Tannins for wood adhesives, foams and composites. In: He, Z. (ed.) Bio-Based Wood Adhesives, p. 366. CRC Press, Boca Raton (2017)
51. Costes, L., Laoutid, F., Brohez, S., Delvosalle, C., Dubois, P.: Phytic acid–lignin combination: a simple and efficient route for enhancing thermal and flame retardant properties of polylactide. Eur. Polym. J. **94**, 270–285 (2017). https://doi.org/10.1016/J.EURPOLYMJ.2017.07.018

52. Wu, Q., Cui, X., Mu, C., Sun, J., Gu, X., Li, H., Zhang, S.: Toward a new approach to synchronously improve the fire performance and toughness of polylactic acid by the incorporation of facilely synthesized ammonium polyphosphate derivatives. Compos. Part A Appl. Sci. Manuf. **150**, 106595 (2021). https://doi.org/10.1016/J.COMPOSITESA.2021.106595
53. Lesaffre, N., Bellayer, S., Fontaine, G., Jimenez, M., Bourbigot, S.: Revealing the impact of ageing on a flame retarded PLA. Polym. Degrad. Stab. **127**, 88–97 (2016). https://doi.org/10.1016/J.POLYMDEGRADSTAB.2015.10.019
54. Qiu, S., Sun, J., Li, H., Gu, X., Fei, B., Zhang, S.: A green way to simultaneously enhance the mechanical, flame retardant and anti-ultraviolet aging properties of polylactide composites by the incorporation of tannic acid derivatives. Polym. Degrad. Stab. **196** (2022). https://doi.org/10.1016/j.polymdegradstab.2022.109831
55. Sui, Y., Sima, H., Shao, W., Zhang, C.: Novel bioderived cross-linked polyphosphazene microspheres decorated with FeCo-layered double hydroxide as an all-in-one intumescent flame retardant for epoxy resin. Compos. Part B Eng. **229**, 109463 (2022). https://doi.org/10.1016/J.COMPOSITESB.2021.109463
56. Shao, X., Du, Y., Zheng, X., Wang, J., Wang, Y., Zhao, S., Xin, Z., Li, L.: Reduced fire hazards of expandable polystyrene building materials via intumescent flame-retardant coatings. J. Mater. Sci. **55**, 7555–7572 (2020). https://doi.org/10.1007/s10853-020-04548-z
57. Peng, H., Wang, D., Fu, S.: Tannic acid-assisted green exfoliation and functionalization of MoS2 nanosheets: significantly improve the mechanical and flame-retardant properties of polyacrylonitrile composite fibers. Chem. Eng. J. **384** (2020). https://doi.org/10.1016/j.cej.2019.123288
58. Hobbs, C.E.: Recent advances in bio-based flame retardant additives for synthetic polymeric materials (2019)
59. Wang, Z.H., Zhang, A.N., Liu, B.W., Wang, X.L., Zhao, H.B., Wang, Y.Z.: Durable flame-retardant cotton fabrics with tannic acid complexed by various metal ions. Polym. Degrad. Stab. **201**, 109997 (2022). https://doi.org/10.1016/j.polymdegradstab.2022.109997
60. Wang, M., Yin, G., Yang, Y., Fu, W.: Bio-based flame retardants to polymers: a review. Adv. Ind. Eng. Polym. Res. **6**, 132–155 (2023). https://doi.org/10.1016/j.aiepr.2022.07.003
61. Makhlouf, G., Abdelkhalik, A., Ameen, H.: Synthesis of a novel highly efficient flame-retardant coating for cotton fabrics with low combustion toxicity and antibacterial properties. Cellulose **28**, 8785–8806 (2021). https://doi.org/10.1007/s10570-021-04076-2
62. Laoutid, F., Vahabi, H., Shabanian, M., Aryanasab, F., Zarrintaj, P., Saeb, M., Laoutid, F., Vahabi, H., Shabanian, M., Aryanasab, F., Zarrintaj, P.: A new direction in design of bio-based flame retardants for poly (lactic acid). Fire Mater. **42**, 914–924 (2020)
63. Chen, Y., Zeng, Y., Wu, Y., Chen, T., Qiu, R., Liu, W.: Flame-retardant and recyclable soybean oil-based thermosets enabled by the dynamic phosphate ester and tannic acid. ACS Appl. Mater. Interfaces **15**, 5963–5973 (2023). https://doi.org/10.1021/acsami.2c21279
64. Zhang, A.N., Liu, B.W., Zhao, H.B., Wang, Y.Z.: Eco-friendly and durable flame-retardant coating for cotton fabrics based on dynamic coordination of Ca^{2+}-tannin acid. Prog. Org. Coatings. **170**, 106964 (2022). https://doi.org/10.1016/j.porgcoat.2022.106964

Recent Advances on the Use of Chitin and Its Derivatives as Flame Retardants for Different Polymeric Materials

A. Mtibe, L. Hlekelele, M. J. John, V. M. Khumalo, and T. C. Mokhena

Abstract There is an increase in the usage of plastic materials in various applications and due to their potential risk to burn easy and widespread the flames, there is a necessity the use of flame retardants to reduce the risk of fire hazard. Conventional flame retardancy such as halogenated flame retardants have been used but showed drawbacks. Due to environmental awareness and health concerns the use of halogenated flame retardants have been outlawed in most developed countries which opened an opportunity for development of ecofriendly flame retardancy. This present study highlights the use of chitin and chitin derivatives as flame retardants. The modification of chitin and chitin derivatives is discussed. The mechanism of flame retardants is also thoroughly discussed in this study. In addition, current status and new developments of chitin-based flame retardants are also discussed.

Keywords Chitin · Chitosan · Flame retardant · Modification · Mechanism

1 Introduction

Fire hazard has become a serious issue in recent years due to the use of combustible materials such as plastic. Plastic materials are widely used in the modern life in variety of applications such as aerospace, automotive, agricultural, and food packaging. This is due to their outstanding properties such as light weight, good mechanical properties, low price, and processability [1]. These materials when exposed to external

A. Mtibe (✉) · L. Hlekelele · M. J. John
Centre for Nanostructures and Advanced Materials, DSI-CSIR Nanotechnology Innovation Centre, Council for Scientific and Industrial Research, Pretoria, South Africa
e-mail: AMtibe@csir.co.za

M. J. John · V. M. Khumalo
Department of Chemistry, Nelson Mandela University, Port Elizabeth, South Africa

T. C. Mokhena
DSI/Mintek—Nanotechnology Innovation Centre, Advanced Materials, Mintek, Randburg 2125, South Africa

T. C. Mokhena et al. (eds.), *Biomaterials as Green Flame Retardants*, Engineering Materials, https://doi.org/10.1007/978-981-97-6871-4_6

source of heat they burn quickly and release large amounts of smoke and toxins which contains obnoxious gases that causes global warming in the long run.

Researchers and engineers have been working on finding solutions to prevent fire from happening or to reduce the impact caused by fires. Various solutions to prevent the impact of fires such as smoke alarms and emergency exits were proposed since the late 1900s [2]. It is worth noting that these systems do not prevent fires and they are expensive. However, fire still cause death although those system are in place. The suitable approach to reduce the risk of fire, to prevent the combustion of material as well as preventing the release of smoke and toxic gases was proposed. This approach involves the incorporation of flame-retardant additives into polymeric material. Different flame retardants such as minerals, halogenated, phosphorus containing compounds, nitrogen containing compounds, silicon containing compounds as well as nanomaterials have been previously investigated [3–5]. Flame retardants such as halogenated compounds have shown drawbacks i.e. during combustion they produce smoke and emit toxic gases which cause environmental pollution. Due to environmental awareness and health concerns the use of these materials was discontinued in most if not all the developed countries [4]. Therefore, there is a necessity to develop ecofriendly flame retardants.

Biobased flame retardants have garnered a considerable amount of interest as a promising replacement for conventional flame retardants. This is due to their abundantly available, sustainability, biodegradability, and environmental friendliness [6–8]. Chitin and its derivatives are among the widely used biobased flame retardants due to their due to its cationic structure, outstanding biological properties and inexpensive [9–12]. Also, chitin is a char forming agent during combustion which acts as a protective layer.

This book chapter is aimed at presenting recent developments in the valorization of chitin and its derivatives as flame retardant additives for polymeric materials. The recent developments of chitin-based flame-retardant additives will be discussed. Chitin cannot provide adequate flame retardancy alone, however, supplementing with other flame retardants will be also discussed.

2 Chitin and Chitosan

Chitin is regarded as the second most available abundantly natural occurring polymer on earth after cellulose. It is a linear polymer with higher molecular weight. Unlike cellulose, which is obtained from plant based, chitin is the major constituent of some living organisms [6]. This natural occurring polysaccharide is generally found in the cell wall of insects, fungi, and yeast, also in the crabs and shrimp shell of crustaceans as well as in the exoskeletons of arthropods [6, 13, 14]. Shells consists of 20–30% chitin, 30–40% proteins, 30–50% calcium carbonate and <1% lipids and astaxanthin [10]. The process of removing chitin from the shells involves four strategies. (1) Deproteinization by treating shells with dilute sodium hydroxide (NaOH). (2) Demineralization of deproteinated shell by treating with dilute hydrochloric acid

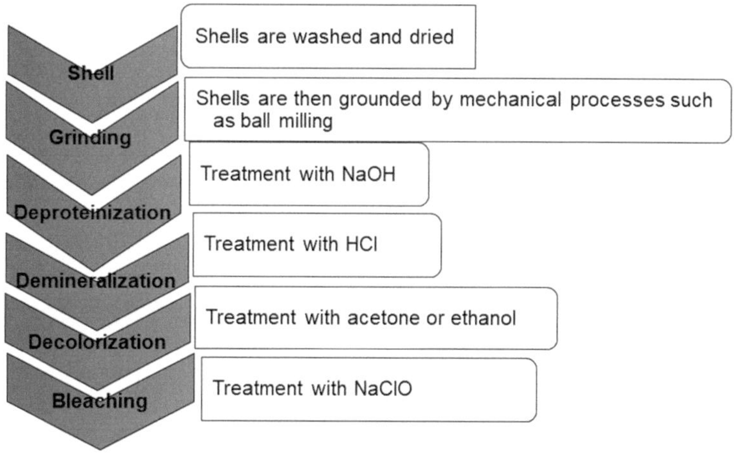

Fig. 6.1 Extraction of chitin from shell

(HCl). (3) Decolorization by extracting lipids and astaxanthin pigments with organic solvents such as acetone and ethanol. Lastly, bleaching with sodium chlorite (NaClO) thus resulting in white product [10]. The extraction of chitin from the shells is depicted in Fig. 6.1.

It can also be generated from waste sources such as mushroom waste (Agaricus bisporus which is more consumed in United States of America (USA)). In the study reported by Desai et al. [15] it was reported that mushroom waste can generates more than 1000 metric tons of fungal chitin per annum. The major advantages of chitin are their high modulus and large number of nitrogen and hydroxyl groups which can be modified by P_2O_5 to enhance their flame retardancy properties [11, 16]. Chitin can be converted to chitosan by deacetylation process. For instance, Mathaba and Daramola [17] produced chitosan from chitin obtained from seashells. In their study, the production process was performed in three stages namely: (1) deproteinization process. This process was achieved by treating crushed seashell with 6 wt.% NaOH for 2 h. (2) Demineralization method. Demineralization was performed by treating deproteinized seashell with 6 wt.% HCl for 2 h. (3) Deacetylation. In this process, deproteinized and demineralized chitin from seashell was treated again with NaOH at various loadings (20–60 wt.%) and various temperatures (80–120 °C). Fourier transform infrared (FTIR) confirmed the production of chitosan with the present of alcohol, alkaline as well as amine groups at absorption bands of 3500 cm^{-1}, 2400 cm^{-1} and 2800 cm^{-1} and 1450 to 1480 cm^{-1}, respectively. The highest degree of deacetylation and yield of chitosan produced in this study was achieved at 40 wt.% NaOH and 100 °C temperature. Degree of deacetylation depends on the deacetylation processing parameters such as reaction time, temperature and concentration of alkaline solution which can determine whether deacetylation of chitin is complete or partial [17]. Partially deacetylated chitin suggests that the degree of acetylation is approximately 50% or lower whereas, fully deacetylated chitin indicates that the degree of acetylation is 50% or more [17]. Degree of acetylation of chitin can positively or negatively influences the properties (mechanical, physical, biological as well as chemical) of the resultant chitosan as it suggests that the number of free amino groups exposed after deacetylation process. It is worth noting that high degree of acetylation of chitin results in the expansion of the applications of chitosan. For instance, a high degree of acetylation could result in many positively charged groups in the backbone of chitosan, thus expand chitosan applications such as antimicrobial activities for the growth of a variety of fungi, yeasts, and microorganisms [6]. The proposed structure of the production of chitosan from chitin is depicted in Fig. 6.2. Chitosan in its chemical structure, it comprises of large number of amine and hydroxyl groups making it easier to be modified by chemical reactions.

There has been growing interest in the use of chitosan in various research fields such as flame retardants applications. For instance, the flame retardancy and thermal decomposition behaviour of chitosan reinforcing thermoplastic polyurethane (TPU) was investigated [19]. The results suggested that the incorporation of chitosan at low loading of about 2 wt.% into TPU reduces heat release rate, smoke production and total smoke release by 65.9%, 79.4% and 54.2%, respectively in comparison to neat TPU. Chitosan also attracted considerable attention due to their thermal stability.

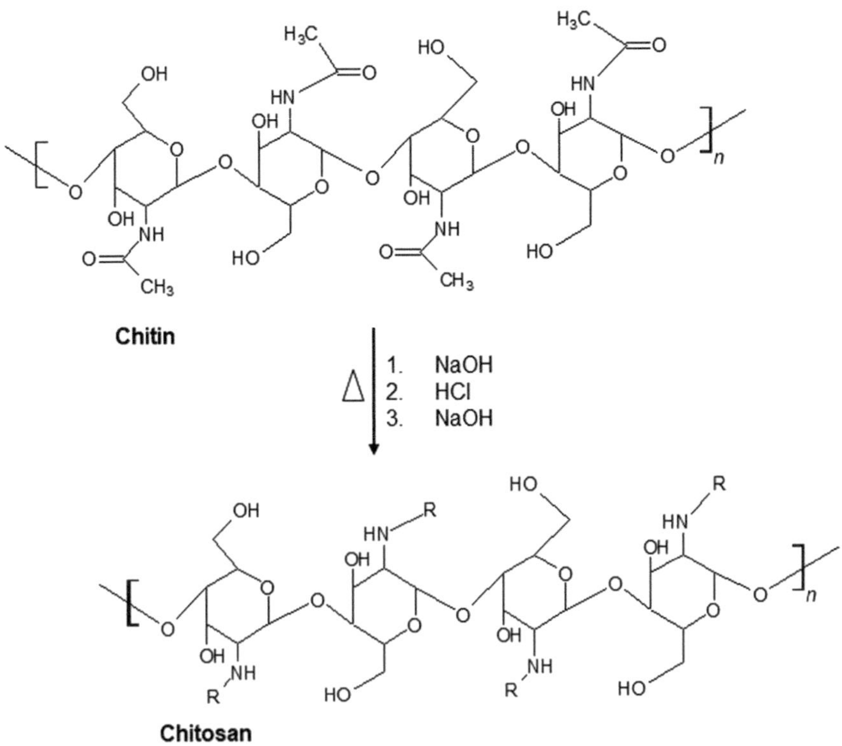

Fig. 6.2 Structure of chitin and chitosan (R = H or Acetyl group (Ac) (Ac < 50%) [18]

Maqsood and Seide [4] and Chukwunwike and Okafor [20] reported that chitosan has three decomposition steps which correspond to the evaporation of moisture at around 130 °C, the second step is depolymerization and deacetylation of chitosan at the temperatures between 260 and 360 °C. Lastly, the char residue of about 30% is formed at temperatures around 500 °C. Chitosan is regarded as a non-toxic polycation biodegradable and biocompatible polymer consisting of *N*-acetyl-D-glucosamine and D-glucosamine monomeric units. Other advantages include low cost, excellent biological activity, and nonallergenic [21]. D-glucosamine content in the chitosan depends on the degree of deacetylation process of chitin. As a result, chitosan is among the very few polymers consisting of a primary amino group in the backbone. As above mentioned, chitosan is a char forming polymer which make it a suitable material to be used as a flame retardant. Although chitosan demonstrated flame retardancy properties, it has limited applications. For example, it does not provide adequate flame retardancy when used alone [7]. Therefore, it must be supplemented with various additives by modification process [9, 22]. Zhang et al. [23] studied the effect of chitosan supplemented with phytic acid on the flame retardancy of ethylene vinyl acetate copolymer (EVA) based composites. The combination of chitosan and phytic acid led to the improved char content and decreased in the peak of heat release and total heat release. The increase in char content indicates that the resultant composites had excellent intumescent effect. Furthermore, it was also reported that the incorporation of these materials in EVA improves its tensile modulus.

3 Modification of Chitin

Modification of chitosan is carried out in order to meet or exceed their performance as flame retardants. Previous studies have revealed that the most effective modification method is chemical modification to form chitosan derivatives [24–36]. Previous study has demonstrated that functionalization of chitosan as a component can be achieved by modifying with the phosphorus-based material to improve their flame retardancy of material [28]. In that study, Carboxymethyl chitosan was used to encapsulate melamine polyphosphate and investigated their flame retardancy. Then, the functionalized chitosan together with melamine polyphosphate were incorporated in TPU to fabricate a composite material. It was noticed that when 15 wt.% of carboxymethyl chitosan encapsulated melamine polyphosphate was incorporated into TPU the LOI was 29.4% and UL-94 rating reaches V-0. In addition, the introduction of functionalized chitosan encapsulated with melamine polyphosphate in TPU significantly decreased peak heat release by 65.9%, total heat release by 24.1%, the total smoke release by 87.3 and carbon dioxide production by 60.6%. In the recent studies reported by Zhou et al. [25] and Chen et al. [37] chitosan was chemically modified to form chitosan derivatives and investigated their flame retardancy. 9,10-dihydro-9-oxa-10-phospha-phenanthrene-10-oxide (DOPO) was used to supplement chitosan derivative and they were compounded into epoxy resin to produce composites. In the case of Zhou et al. [25] chitosan was chemically modified with p-nitro benzaldehyde.

As anticipated, the incorporation of both chitosan derivatives and DOPO into epoxy resins improves char forming and the limiting oxygen index (LOI) reaches 34% as well as UL-94 reaches V-0. The smoke production and total heat release were significantly decreased. This trend was noticed as well as in epoxy resin composite reinforced with unmodified chitosan and DOPO [38]. There were no significant changes observed in terms of tensile strength, flexural strength, and transparency in comparison to neat epoxy resin. Similar studies were conducted by Wang et al. [29] and Ma et al. [30]. In case of Wang et al. [29], they prepared chitosan derivative containing sulfur element (CSI-TE), mixed with aluminum hypophosphite (AHPP) and reinforce polylactic acid (PLA) with the aim of improving flame retardancy. The LOI of PLA composites reinforced with 3.75 wt.% CS-TE and 11.25 wt.% APP was 28.5% and UL-94 rating reaches V-0. Likewise, the previous study reported by Zhou et al. [25], the smoke production as well as the peak of heat release were reduced, and the resultant composites demonstrated good anti-dripping behaviour. It was reported that the addition of chitosan derivative and AP into PLA increased char content and decreased tensile strength.

Other studies have demonstrated that the incorporation of inorganic materials such as clay, titanium dioxide (TiO_2), ferric sulfate hydrate and other nanoparticles into chitosan to improve their flame retardancy [8, 12, 22, 39–42]. The addition of montmorillonite (MMT) clay in chitosan aerogels have significantly improved the flame retardancy with the LOI reaches 85% [43]. Chitosan derivatives combined with modified MMT clay has also been used as a flame retardant to delay the combustion of TPU and linear low-density polyethylene (LLDPE) [40, 42, 44]. In the case of Liu et al. [44], the authors added a loading of 1 wt.% organically modified exfoliated montmorillonite (PMT) and 9 wt.% chitosan modified with aluminum hypophosphite (AHP) into TPU. During the combustion of the TPU composites the char was formed and well dispersed which act as a protecting agent of underlying TPU matrix from heat as shown in Fig. 6.3. In addition, the composites exhibited an improved flame retardancy.

Similar performance was observed in the case TPU composite reinforced with modified MMT and chitosan derivatives. Chitosan derivatives were fabricated by crosslinking chitosan with bis-(4-formylphenyl)-phenyl-phosphonate (ABPO). The crosslinked chitosan was combined with APP and modified MMT and used as flame retardant materials. The results displayed that TPU composites increased LOI. The peak heat release rate, total smoke release, and total heat release decreased as shown depicted in Fig. 6.4. Also, the char content was higher than virgin TPU. Similar behaviour was observed when other inorganic materials such as TiO_2 and ferric sulfate hydrate [8, 22].

Another method which has recently received research hotspot is direct surface coating or layer by layer technique. Chitosan based surface coating has been studied as a potential material that can reduce the flammability [45–47]. Chitosan based gel has been fabricated using facile approach whereby chitosan was mixed with polyvinyl alcohol-based macromonomer (PVAM), acrylic acid (AA) and copper ion (Cu(II)) [46]. As shown in Fig. 6.5, surface coating with chitosan-based gel displayed an improved flame retardancy of the material.

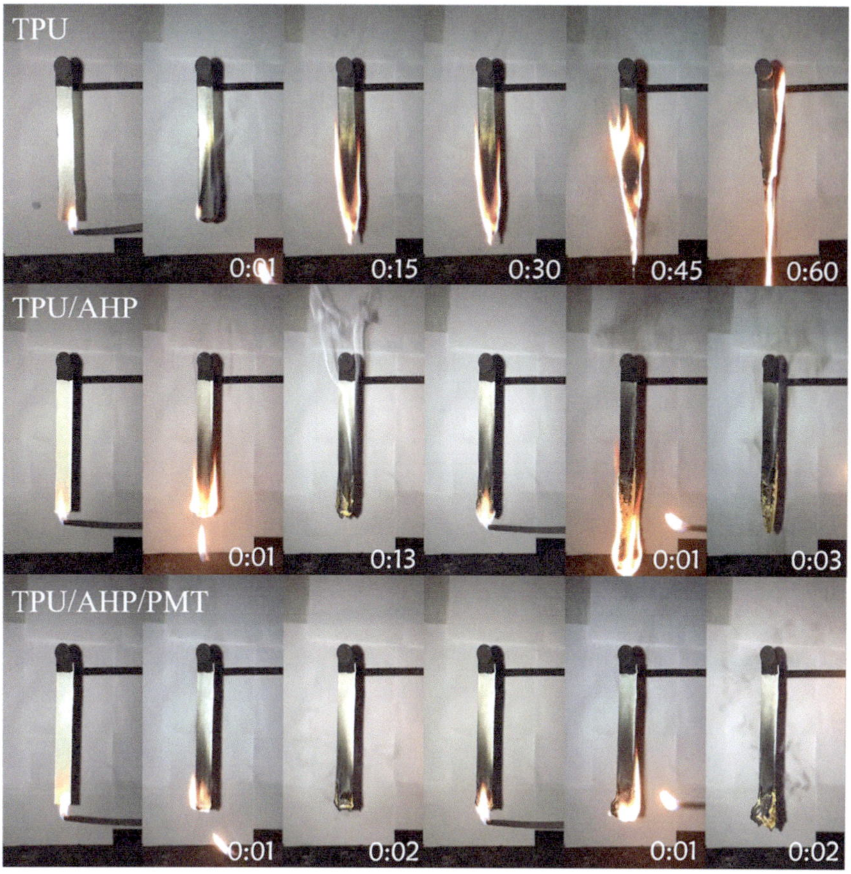

Fig. 6.3 Picture of TPU and TPU composites during UL-94 rest (reproduced from [44] with permission

Layer by layer technique is achieved by modifying the surface of the material such as polymeric material, foam or fabric by selecting chitosan which has a positive surface charge and other materials which has a negative surface charge [3, 13]. Chitosan is usually selected because of its flame retardancy properties. Research has demonstrated that anionic nanoparticles such as clay can be paired with cationic chitosan to produce material with improved flame retardancy properties. In the study presented by Laufer et al. [39], thin film made from chitosan and MMT was fabricated via layer-by-layer technique which allows the formation of multilayer films. 10 bilayers of chitosan-clay nanocoating were deposited in the polyurethane (PU) foam. The surface of the coated foam was characterized by uniform nanostructure as shown in Fig. 6.6 while the uncoated foam had a smooth surface.

Both coated and uncoated foam were exposed to heat. It was noticed that uncoated foam was completely destroyed after heat exposure for 30 s whereas coated foams

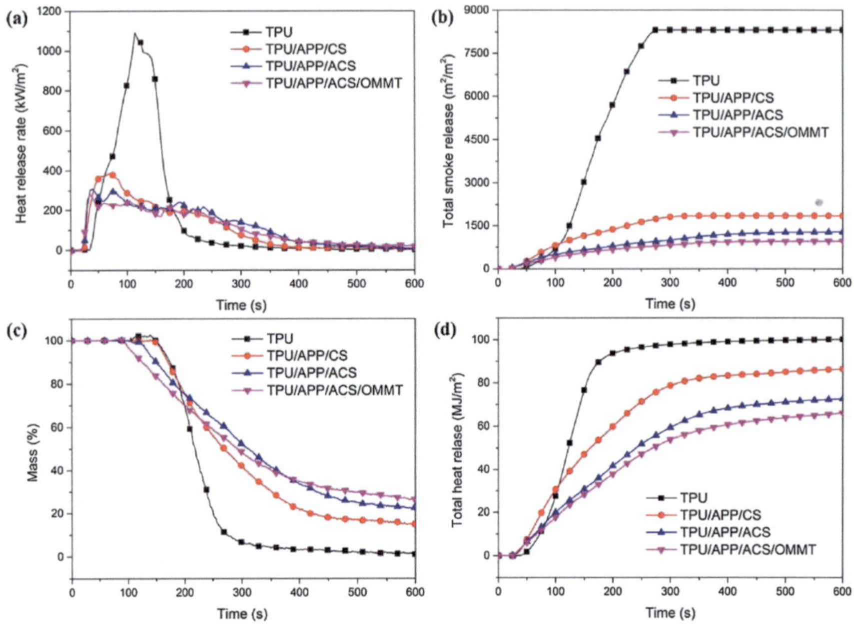

Fig. 6.4 **a** Heat release rate, **b** total smoke release, **c** mass loss, and **d** total heat release of TPU and TPU composites (reproduced from [42] with permission)

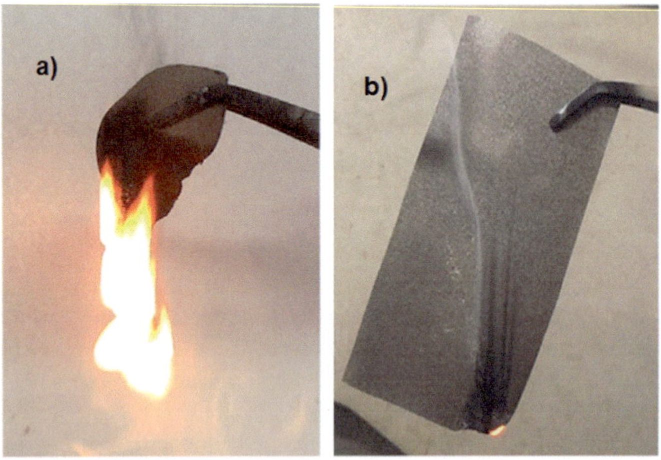

Fig. 6.5 **a** Uncoated material and **b** chitosan based gel coated material (reproduced from [46] with permission)

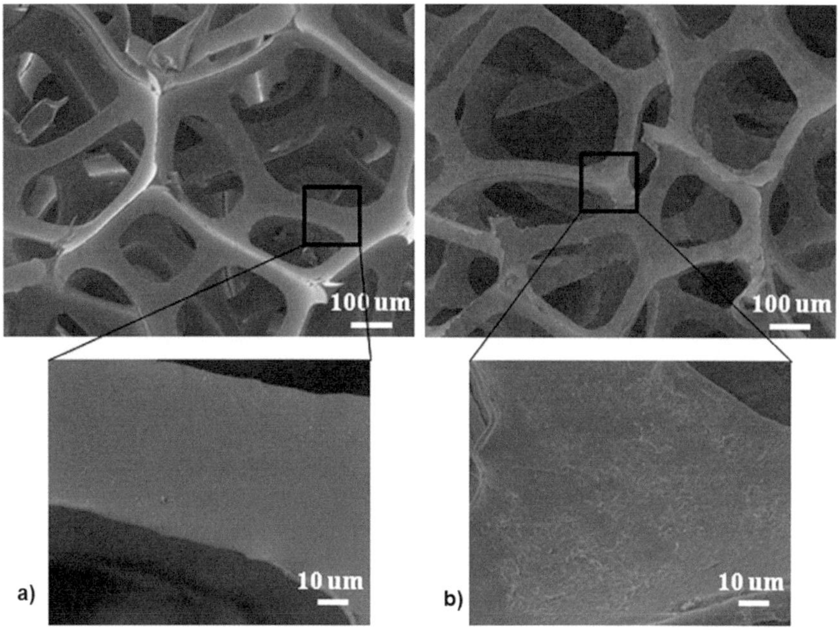

Fig. 6.6 Scanning electron microscopy (SEM) of **a** uncoated PU foam and **b** coated foam with 10 bilayers chitosan and MMT (reproduced from [39] with permission)

with 10 bilayers chitosan and MMT retained the shape with a slight collapse after the heat exposure. During the combustion process, the layer of char was formed which acts as a protective barrier. Moreover, the peak of heat release was reduced by 52% in comparison to uncoated foams. Deng et al. [41] compared the effectiveness of nanocoatings on flexible silicone foams using bilayers of chitosan/APP and chitosan/MMT. They reported that chitosan/MMT nanocoatings was more effective when compared to chitosan APP because they provide more barrier on smoke release and produce more denser layer of char during combustion process. Even though this nanocoating is more effective, both nanocoatings are low cost and environmentally benign. Similar trend was observed when different inorganic materials such as TiO_2 was used investigated [48]. Chitin multilayer as nanocoatings with flame retardancy properties was also investigated [16, 49]. In the study reported by Batool and coworkers [16], chitin was phosphorylated using phosphorus pentoxide (P_2O_5) followed by fabrication of multilayer consisting of poly (vinyl alcohol) (PVA), phosphorylated chitin, graphene oxide nanosheets and laponite. It is well known that PVA burns easily when exposed to heat. However, upon the incorporation of phosphorylated chitin, graphene oxide nanosheets and laponite the flammability of PVA decreased. The multilayer fabricated in this study displayed strong flame retardancy properties and it was noticed that these films did not show dripping behaviour when they were exposed to heat. In addition, the resultant films exhibited good mechanical properties. These outstanding properties of the films can expand their applications transportation, food packaging, construction, and textiles.

Chitosan can be also directly coated the surface of the polymeric material with the aid of flame retardants such as phosphorus pentoxide, APP and phytic acid to improve their flame retardancy [50–53]. Kundu et al. [50] used layer by layer technique to coat chitosan and phosphorylated chitosan on polyamide 66 (PA66) fabric surfaces along with poly-acrylate sodium (PAS). Only 1 wt.% of cationic chitosan and anionic (1 wt.% phosphorylated chitosan + 0.25 wt.% PAS) were added in PA66 to reduce peak release rate and improve LOI which showed good flame retardancy properties. Layer by layer technique was used to coat chitosan and phosphorylated cellulose (PCL) on the cotton fabric. About 0.5 wt.% and 2 wt.% of PCL was used to fabricate 20 multi-layers. The coating demonstrated a significant reduction in peak heat release rate and total heat release and increased char content. Chitosan derivatives have been also used as surface coating applying layer by layer technique [52, 53]. Chitosan derivatives were prepared by modifying chitosan with hypo phosphorous acid and subsequently deposited onto polyester cotton (PTCO) blends using the Layer-by-Layer assembly technique to form 20 bilayers [52]. Fourier transform infrared spectrometry (FTIR) and X-ray photoelectron spectroscopy (XPS) were used to confirm the modification of chitosan and the results confirmed the modification of chitosan. The introduction of chitosan derivatives improves thermal stability of the fabric. In addition, the fabrication of chitosan derivatives and subsequently coating fabrics produced an intumescent flame-retardant system. The modification of chitosan with hypo phosphorous acid increases the phosphorus content thereby improving flame retardancy. In the recent study [53], adenosine triphosphate/chitosan 5 bilayers bio-coatings were assembled by layer-by-layer method. It was observed that this biocoatings can make polyurethane (PU) foam to obtain excellent flame retardancy. It was also reported that the application of biocoatings on the foam does not damage the structure of the foam. The deposition of 5 bilayers biocoatings enhanced LOI to 35% from 17% and drastically reduced peak heat release and total smoke release by 42% and 30.6%, respectively. Furthermore, the addition of biocoatings did not significantly affect the mechanical properties of PU.

4 Mechanism of Flame Retardants

It is well known that polymeric materials are susceptible to external sources of heat. Upon exposure to sufficient flame, polymers undergo thermal decomposition. During combustion process, polymers absorb heat after a while the bonds between polymer chains break and release volatiles as fuel to the heat. It has been reported that most of the polymers decompose between 200 and 450 °C and leave little char residues to protect them from fully combustion [4]. For this reason, scientists and researchers have been investigating the most suitable flame-retardant material that will prolong the ignition of the material. Flame retardants played a very crucial role in the field of polymer composites. The effectiveness of flame retardants is assessed by measuring heat release rate, flame speed rate, ignition of the material as well as formation of smoke and toxic gases [4]. Various flame retardants such as minerals, halogenated, phosphorus containing compounds, nitrogen containing compounds, silicon containing compounds as well as nanomaterials have been studied. These materials prevent the fire to occur or slowing down the spreading of fire [5]. The choice of flame retardancy depends on thermal properties of polymers, chemical and physical properties of flame retardants, fire hazard scenario, processing methods, environmental stability, sustainability and in some case recyclability [54]. In addition, as mentioned above, flame retardants are categorized based on their protection mechanisms. There are three approaches of flame retardants that are used protect polymeric materials; (1) gaseous phase inhibition. In this approach flame retardants react with polymer at a gas phase during combustion process. The volatile emitted from polymer decomposition reacts with oxygen in the air [5, 55]. The widely used flame retardants that are operating in a gaseous phase are halogenated based flame retardants. Their mechanism involves releasing halogens radicals i.e. $Cl^{\cdot}$ and $Br^{\cdot}$ which react with hydrogen and hydroxyl group to inhibit free radicals involved in the combustion process. Even though these materials were effective at low loading in polymeric materials, during combustion they produce smoke and emit toxic gases which cause environmental pollution. However, Halogenated free flame retardants were then developed to address the shortcomings of halogenated based flame retardants. (2) Condensed phase. Phosphorus-based compounds are widely used flame retardants. During combustion, polymeric based materials are dehydrated and decarbonized to form a protective cover/layer of char of the polymeric. The formation of char protects the material from complete combustion. Furthermore, the formation of char slows down the decomposition rate of material by reducing the emission of volatiles by-products [5]. Chitin and chitosan-based flame-retardant materials have recently garnered considerable attention. They follow condensed phase approach. The mechanism involves ring opening during combustion process and produces a dense char layer. Chitosan contains amino group however, during combustion at high temperatures, amino group is converted into ammonium thereby diluting the available oxygen in the system [3]. (3) Quench and cool system. Mineral hydrated fillers such as aluminum tri-hydroxide (ATH) and magnesium hydroxide ($Mg(OH)_2$) are widely in this approach [4]. Mineral fillers break down to water molecules when subjected to higher temperatures which dilute volatiles released during combustion. The major drawback of mineral hydrated fillers is their effectiveness depends on high loadings which results in a decrease in mechanical properties thus limiting their wide range of applications.

5 Current Status and New Developments of Chitin-Based Flame Retardants

Chitin is one of abundant available natural polysaccharide that can be obtained from the shells of crustaceans, insects and fungi [11, 26, 56]. It has been used in different fields such as agriculture, food packaging, cosmetics, and biomedical. This results from its attractive features such as biocompatibility, biodegradability, and antimicrobial activity. Recently, chitin and its derivatives received much attention as potential flame retardant for polymers, fabrics and many more. Chitin and its derivatives were found to be suitable carbon source that facilitate the formation of stable compact char on the surface of the flammable materials including polymers, fabrics, and wood. The resulting char act as shield that protect the underlying material from heat and oxygen and inhibit the release of flammable product during combustion. The hybridization of chitin and its derivatives with other materials such as phosphorus, nitrogen and/or metal compounds has been the main subject of research to further improve their flame retardant property and forming multifunctional materials [56, 57]. For instance, Zhang et al. [11] phosphorylated chitin nanofibrils using phosphorus pentoxide (P_2O_5) to enhance flame retardant of paper. The authors pointed out that LOI values increased with increase in the phosphorus content, i.e. increase in phosphorylated chitin nanofibrils. In addition, total heat release (THR), peak heat release rate (PHRR) and temperature of PHHR decreased with increase in phosphorylated chitin nanofibrils (Table 6.1). The neat paper burned completely while the presence of phosphorylated chitin led to no ignition from 1 to 7 s with self-extinguishing behavior (Fig. 6.7a, b). The tensile strength and stiffness under dry conditions increased by 50%, whereas under wet conditions the mechanical strength doubled that of neat paper. This was attributed to the strong interaction between chitin nanofibrils with their high specific surface and cellulose fibers.

Chitin and its derivatives such as chitosan, carboxymethyl chitosan, chitosan phosphate, and chitosan nanofibers have been used as flame retardant of various

Table 6.1 Flammability results for phosphorylated chitin nanofibrils [11]

Sample	Phosphorus content (%)	PHRR (W g^{-1})	Temperature of PHRR (°C)	THR (kJ g^{-1})	LOI value (%)
Control paper	–	343	379	12	17
5% P3C1N-16	0.20	313	319	8.9	20
10%5% P3C1N-16	0.40	278	308	7.2	23
20% P3C1N-16	0.80	193	296	4.8	27
30% P3C1N-16	1.2	129	286	3.4	30

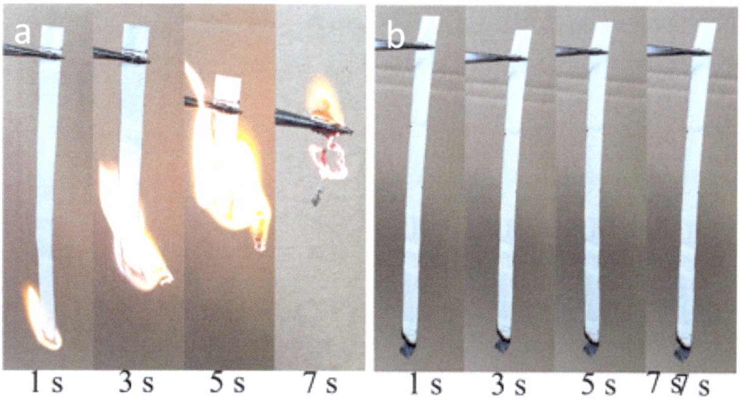

Fig. 6.7 Burning behavior of **a** neat paper and **b** 30% P3C1N-16-loaded paper [11]

polymeric materials [26, 58–60]. The inclusion of these materials into polymers such as polyethylene, polypropylene, polyethyelene terephthalate and thermoplastic polyurethane was not only to improve their fire resistance but to improve the overall properties including thermal stability, and mechanical properties. The chitin-based materials has been employed in wood fiber-based and polymer-based packaging materials. These materials have advantages over the synthetic counterparts including environmental benign, biodegradable, and renewable. They also have beneficial features such as antimicrobial, antioxidant, and barrier properties to the packaging materials. The modification of chitosan through phosphorylation and phosphoramidation using phenylphosphoryl dichloride and tetraethylenepentamine, respectively [30]. The resultant flame retarded PLA composites exhibited excellent fire resistance performance when compared to neat PLA. Elsewhere, chitosan phosphate acrylate (CPA) was added into epoxy acrylate (EA) to enhance flame retardant property. The LOI value increased from 21 to 26% with addition of 20% CPA [61]. The PHHR decreased from 332 to 145 Wg^{-1} and THR from 21 to 11 kJ g^{-1} with the addition of of 20% CPA. From Real-Time Infra-Red (RT-IR) and thermogravimetric-Infra-Red (TGA-IR) it was concluded that the presence of CPA facilitated the generation of stable char that inhibited the release of flammable volatized products. In addition, the thermal stability increased with increase in CPA content.

On the other hand, the synthesis of metal-based flame retardants such as aluminum, magnesium, and zinc from chitin and its derivatives. The resulting flame-retardants form metal oxides or hydroxides during combustion, serving as heat sinks, oxygen scavengers, and smoking suppression agents. They also found to facilitate charring while improving mechanical performance of the resultant materials.

Yet blending chitin-based FRs with other biomaterials especially to form polyelectrolyte complex as green FR has been the major topic to preserve the unique features of these materials. Alginate, tannic acid and others have been blended with chitin-based materials to develop well-balanced materials that have elements that are highly inflammable. The main method for preparation is often layer-by-layer assembly.

6 Summary and Perspective

Based on the summary and discussion, it can be concluded that chitin and its derivatives as flame-retardants are constantly improving and emerging endlesslessly. The conception of using nitrogen-containing materials especially from natural-based materials never fade in the flame retardant field. The modification of these natural materials overcome their intrinsic disadvantages. The grafting, hybridization and encapsulation are the commonly employed methods to enhance compatibility between chitin-based FRs and flame retarded materials. The less complicated modification techniques to afford thermally stable products are essential to develop high efficiency flame-retardants. In recent published papers, chitin and derivatives treated with phosphorus-containing compounds are mainly reported to enhance fire resistance performance of polymers, fabrics and wood-based products. The ease functionalization of chitin and its derivatives because of the presence of large number of functionalities makes it possible to develop and design FRs that are environmentally benign and sustainable. In general, there are two distinctive styles that explains the flame retardant mechanisms of chitin-based FRs. Under the guideline of these mechanisms, there is a huge possibility of developing chitin-based FRs that can be sustained and constantly updated to meet increasingly stringent environmental and commercial needs.

Chitin-based FRs as green FRs have a huge potential to be applied in various fields because of their fire resistance capability, high smoke suppression and compatibility with most treated materials. Although, the use of biomaterials as alternative FRs is a new concept that still need more research to qualify some of the obtained fire resistance performance, the modifications of these FRs need to be careful considered in order to preserve their valuable features when compared to conventional synthetic FRs. The modifications have to be cheaper and eco-friendlier such that these materials still stay competitive from ecological and economical viewpoints.

References

1. Ignatyev, I.A., Thielemans, W., Vander, B.: Recycling of polymers: a review. Chem Sus Chem **7**(6), 1579–1593 (2014). https://doi.org/10.1002/cssc.201300898
2. Watson, D.A.V., Schiraldi, D.A.: Biomolecules as flame retardant additives for polymers: a review. Polymers (Basel) **12**(4), 849 (2020). https://doi.org/10.3390/polym12040849
3. Wang, M., Yin, G., Yang, Y., Fu, W.: Bio-based flame retardants to polymers: a review. Adv. Ind. Eng. Polym. Res. **6**, 132–155 (2023). https://doi.org/10.1016/j.aiepr.2022.07.003
4. Maqsood, M., Seide, G.: Biodegradable flame retardants for biodegradable polymer. Biomolecules **10**, 1038 (2020). https://doi.org/10.3390/biom10071038

5. Taib M.N.A.M. et al.: Current progress of biopolymer-based flame retardant. Polym. Degrad. Stab. **205** (2022). https://doi.org/10.1016/j.polymdegradstab.2022.110153
6. Mokhena, T.C., Luyt, A.S.: Development of multifunctional nano/ultrafiltration membrane based on a chitosan thin film on alginate electrospun nanofibres. J. Clean. Prod. **156**, 470–479 (2017). https://doi.org/10.1016/j.jclepro.2017.04.073
7. Wang, G., Li, H., Shao, Y., Wu, X., He, Q., Niu, K.: Preparation method of biogenic phytic acid and chitosan—related epoxy resin flame retardant: a review. J. Polym. Res. **30**(7) (2023). https://doi.org/10.1007/s10965-023-03580-2
8. Zhou, C., et al.: Chitosan–Fe(III) complexes via green preparation toward flame retardant and high-mechanical-strength epoxy composites. ACS Sustain. Chem. Eng. **10**(40), 13453–13464 (2022). https://doi.org/10.1021/acssuschemeng.2c04181
9. Jin, X., Cui, S., Sun, S., Sun, J., Zhang, S.: The preparation and characterization of polylactic acid composites with chitin-based intumescent flame retardants. Polymers (Basel) **13**(20), 3513 (2021). https://doi.org/10.3390/polym13203513
10. Zeng, J., He, Y., Li, S., Wang, Y.: Chitin whiskers: an overview. Biomacromol **13**(1), 1–11 (2012). https://doi.org/10.1021/bm201564a
11. Zhang, T., Kuga, S., Wu, M., Huang, Y.: Chitin nanofibril-based flame retardant for paper application. ACS Sustain. Chem. Eng. **8**(33), 12360–12365 (2020). https://doi.org/10.1021/acssuschemeng.0c02016
12. Riehle, F., Hoenders, D., Guo, J., Eckert, A., Ifuku, S., Walther, A.: Sustainable chitin nanofibrils provide outstanding flame-retardant nanopapers. Biomacromol **20**(2), 1098–1108 (2019). https://doi.org/10.1021/acs.biomac.8b01766
13. Costes, L., Laoutid, F., Brohez, S., Dubois, P.: Bio-based flame retardants: when nature meets fire protection. Mater. Sci. Eng. R **117**, 1–25 (2017). https://doi.org/10.1016/j.mser.2017.04.001
14. Cabrera-Barjas, G. et al.: Value-added long-chain aliphatic compounds obtained through pyrolysis of phosphorylated chitin. Int. J. Biol. Macromol. **238** (2023). https://doi.org/10.1016/j.ijbiomac.2023.124130
15. Desai, K., Kit, K., Li, J., Davidson, P.M., Zivanovic, S., Meyer, H.: Nanofibrous chitosan non-wovens for filtration applications. Polymer (Guildf) **50**(15), 3661–3669 (2009). https://doi.org/10.1016/j.polymer.2009.05.058
16. Batool, S., Guo, W., Gill, R., Xin, W., Hu, Y.: Chitin based multi-layered coatings with flame retardancy an approach to mimic nacre: synthesis, characterization and mechanical properties. Carbohydr. Polym. **291** (2022). https://doi.org/10.1016/j.carbpol.2022.119488
17. Mathaba, M., Daramola, M.O.: Effect of Chitosan's degree of deacetylation on the performance of PES membrane infused with Chitosan during AMD treatment. Membranes (Basel). **10**(3) (2020)
18. Song, E.-H., Shang, J., Ratner, D.M.: Polysaccharides. Polym. Sci. A Compr. Ref. **9**, 137–155 (2012). https://doi.org/10.1016/B978-0-444-53349-4.00246-6
19. Jiao, C., Li, M., Chen, X., Li, S.: Flame retardancy and thermal decomposition behavior of TPU/chitosan composites. Polym. Adv. Technol. 1–11 (2019). https://doi.org/10.1002/pat.4752
20. Chukwunwike, S.A., Okafor, K.J.: A review on some selected bio-based (green) flame retardants. J. Eng. Technol. **8**(1) (2019)
21. Li, J., et al.: Chitosan natural polymer material for improving antibacterial properties of textiles. ACS Appl. Bio Mater. **4**(5), 4014–4038 (2021). https://doi.org/10.1021/acsabm.1c00078
22. Becenen, N., Erdogan, S.: Chitosan and nano-TiO_2 coating improves the flame retardancy of dyed and undyed denim fabrics by increasing the charring. J. Ind. Text. **51**, 1252S-1278S (2022). https://doi.org/10.1177/15280837221099632
23. Zhang, T., Yan, H., Shen, L., Fang, Z., Zhang, X., Wang, J.: Chitosan/phytic acid polyelectrolyte complex : a green and renewable intumescent flame retardant system for ethylene—vinyl acetate copolymer. Ind. Eng. Chem. Res. **53**(49), 19199–19207 (2014). https://doi.org/10.1021/ie503421f
24. Patankar, K.C. et al.: Fabrication of chitosan-based finishing agent for flame-retardant, UV-protective, and antibacterial cotton fabrics. Mater. Today Commun. **33** (2022). https://doi.org/10.1016/j.mtcomm.2022.104637

25. Zhou, C., Wang, X., Wang, J., Pan, Z., Zhou, H.: Epoxy resin modified with chitosan derivatives and DOPO: Improved flame retardancy, mechanical properties and transparency. Polym. Degrad. Stab. **199** (2022). https://doi.org/10.1016/j.polymdegradstab.2022.109931
26. Cui, H., Wu, N., Ma, X., Niu, F.: Superior intrinsic flame-retardant phosphorylated chitosan aerogel as fully sustainable thermal insulation bio-based material. Polym. Degrad. Stab. **207** (2023). https://doi.org/10.1016/j.polymdegradstab.2022.110213
27. Yu, H., Xu, X., Xia, Y., Pan, M., Zarshad, N., Pang, B.: Synthesis of a novel modified chitosan as an intumescent flame retardant for epoxy resin. e-Polymers **20**, 303–316 (2020). https://doi.org/10.1515/epoly-2020-0036
28. Liu, X., et al.: Effects of carboxymethyl chitosan microencapsulated melamine polyphosphate on the flame retardancy and water resistance of thermoplastic polyurethane. Polym. Degrad. Stab. **160**, 168–176 (2019). https://doi.org/10.1016/j.polymdegradstab.2018.12.019
29. Wang, Y., Yuan, J., Ma, L., Yin, X., Zhu, Z., Song, P.: Fabrication of anti-dripping and flame-retardant polylactide modified with chitosan derivative/aluminum hypophosphite. Carbohydr. Polym. **298** (2022). https://doi.org/10.1016/j.carbpol.2022.120141
30. Ma, X., Wu, N., Liu, P., Cui, H.: Fabrication of highly efficient phenylphosphorylated chitosan bio-based flame retardants for flammable PLA biomaterial. Carbohydr. Polym. **287** (2022). https://doi.org/10.1016/j.carbpol.2022.119317
31. Yang, Z., Li, H., Niu, G., Wang, J., Zhu, D.: Poly(vinylalcohol)/chitosan-based high-strength, fire-retardant and smoke-suppressant composite aerogels incorporating aluminum species via freeze drying. Compos. Part B **219** (2021)
32. Liu, X., Gu, X., Sun, J., Zhang, S.: Preparation and characterization of chitosan derivatives and their application as flame retardants in thermoplastic polyurethane. Carbohydr. Polym. **167**, 356–363 (2017). https://doi.org/10.1016/j.carbpol.2017.03.011
33. Hu, S., Song, L., Hu, Y.: Preparation and characterization of chitosan-based flame retardant and its thermal and combustible behavior on polyvinyl alcohol. Polym. Plast. Technol. Eng. **52**(4), 393–399 (2013). https://doi.org/10.1080/03602559.2012.754465
34. Cui, X., Wu, Q., Sun, J., Gu, X., Li, H., Zhang, S.: Preparation of 4-formylphenylboronic modified chitosan and its effects on the flame retardancy of poly(lactic acid). Polym. Degrad. Stab. **202** (2022). https://doi.org/10.1016/j.polymdegradstab.2022.110037
35. Xiao, Y., Zheng, Y., Wang, X., Chen, Z., Xu, Z.: Preparation of a chitosan-based flame-retardant synergist and its application in flame-retardant polypropylene. J. Appl. Polym. Sci. **131**(19) (2014). https://doi.org/10.1002/app.40845
36. Wang, J., et al.: Preparation of boron-containing chitosan derivative and its application as intumescent flame retardant for epoxy resin. Cellulose **30**(7), 4663–4681 (2023). https://doi.org/10.1007/s10570-023-05174-z
37. Chen, R., Luo, Z., Yu, X., Tang, H., Zhou, Y., Zhou, H.: Synthesis of chitosan-based flame retardant and its fire resistance in epoxy resin Rui. Carbohydr. Polym. **245**, 116530 (2020). https://doi.org/10.1016/j.carbpol.2020.116530
38. Zhou, P., Wang, J., Bai, T., Chevali, V., You, D., Huo, S.: Facile approach for preparing chitosan/phosphaphenanthrene-containing epoxy resins with superior flame retardancy and mechanical properties. ACS Appl. Polym. Mater. **5**(6), 4256–4265 (2023). https://doi.org/10.1021/acsapm.3c00452
39. Nanocoatings, F., Laufer, G., Kirkland, C., Cain, A.A., Grunlan, J.C.: Clay—chitosan nanobrick walls: completely renewable gas barrier and flame-retardant nanocoatings. ACS Appl. Mater. Interfaces **4**(3), 1643–1649 (2012). https://doi.org/10.1021/am2017915
40. Hassan, M., Nour, M., Abdelmonem, Y., Makhlouf, G.: Synergistic effect of chitosan-based flame retardant and modified clay on the flammability properties of LLDPE. Polym. Degrad. Stab. **133**, 8–15 (2016). https://doi.org/10.1016/j.polymdegradstab.2016.07.011
41. Deng, S., Liao, W., Yang, J., Cao, Z., Wang, Y.: Flame-retardant and smoke-suppressed silicone foams with chitosan-based nanocoatings. Ind. Eng. Chem. Res. **55**(27), 7239–7248 (2016). https://doi.org/10.1021/acs.iecr.6b00532
42. Zhang, S., Liu, X., Jin, X., Li, H., Sun, J., Gu, X.: The novel application of chitosan: effects of cross-linked chitosan on the fire performance of thermoplastic polyurethane. Carbohydr. Polym. **189**, 313–321 (2018). https://doi.org/10.1016/j.carbpol.2018.02.034

43. Jiang, X. et al.: Chitosan/clay aerogel: Microstructural evolution, flame resistance and sound absorption. Appl. Clay Sci. **228** (2022) https://doi.org/10.1016/j.clay.2022.106624
44. Liu, X., et al.: Enhancing the flame retardancy of thermoplastic polyurethane by introducing montmorillonite nanosheets modified with phosphorylated chitosan. Compos. Part A **119**, 291–298 (2019). https://doi.org/10.1016/j.compositesa.2019.02.009
45. Cho, W. et al.: Ion—complexed chitosan formulations as effective fire-retardant coatings for wood substrates. Polym. Degrad. Stab. **197** (2022). https://doi.org/10.1016/j.polymdegradstab.2022.109870
46. Liu, X., Xiao, C.: Formation of self-healable fire-retardant water-soluble chitosan/chemically cross-linked polyvinyl alcohol/Cu(II) gel. Environ. Technol. Innov. **20** (2020). https://doi.org/10.1016/j.eti.2020.101087
47. Makhlouf, G., Abdelkhalik, A., Ameen, H.: Preparation of highly efficient chitosan-based flame retardant coatings with good antibacterial properties for cotton fabrics. Prog. Org. Coatings **163** (2022). https://doi.org/10.1016/j.porgcoat.2021.106627
48. Pan, H., Ma, W., Zhang, Z.: Construction of layer-by-layer assembled green coating on titanate nanotubes to improve the flame retardancy of epoxy resin. J. Appl. Polym. Sci. **137**(44) (2020). https://doi.org/10.1002/app.49369
49. Pan, H., Wang, W., Pan, Y., Song, L., Hu, Y., Meow, K.: Formation of self-extinguishing flame retardant biobased coating on cotton fabrics via Layer-by-Layer assembly of chitin derivatives. Carbohydr. Polym. **115**, 516–524 (2015). https://doi.org/10.1016/j.carbpol.2014.08.084
50. Kundu, C.K., Wang, X., Song, L., Hu, Y.: Chitosan-based flame retardant coatings for polyamide 66 textiles: one-pot deposition versus layer-by-layer assembly. Int. J. Biol. Macromol. **143**, 1–10 (2019). https://doi.org/10.1016/j.ijbiomac.2019.11.220
51. Pan, H., Song, L., Ma, L., Pan, Y.: Layer-by-layer assembled thin films based on fully biobased polysaccharides: chitosan and phosphorylated cellulose for flame-retardant cotton fabric. Cellulose **21**(4), 2995–3006 (2014). https://doi.org/10.1007/s10570-014-0276-5
52. Liu, L., Pan, Y., Wang, Z., Hou, Y., Gui, Z., Hu, Y.: Layer-by-Layer assembly of hypophosphorous acid-modified chitosan based coating for flame retardant polyester-cotton blends. Ind. Eng. Chem. Res. **56**(34), 9429–9436 (2017). https://doi.org/10.1021/acs.iecr.7b02303
53. Song, H., Park, C.H., Jeong, S.H., Heo, J.H., Lee, J.H.: Synergistic adenosine triphosphate/chitosan bio-coatings on polyurethane foam for simultaneously improved flame retardancy and smoke suppression. ACS Appl. Polym. Mater. **5**(6), 4388–4399 (2023). https://doi.org/10.1021/acsapm.3c00514
54. Kovacevic, Z., Grgac, S.F., Bischof, S.: Progress in biodegradable flame retardant nano-biocomposites. Polymers (Basel) **13**(5), 741 (2021). https://doi.org/10.3390/polym13050741Academic
55. Reynolds, K.J., Zagho, M.M., Robertson, M., Qiang, Z., Nazarenko, S.: Environmental, health, and legislation considerations for rational design of nonreactive flame-retardant additives for polymeric materials: future perspectives. Macromol. Rapid Commun. **43**(24) (2022). https://doi.org/10.1002/marc.202200472
56. Chen, F., Wang, C., Liu, X., Wang, Z., Fang, L., Zhang, Y.: Flame - retardant properties of chitin liquefaction - based polyurethane foam. J. Polym. Res. **30**, 143 (2023). https://doi.org/10.1007/s10965-023-03521-z
57. Zhou, K., Tang, G., Gao, R., Guo, H.: Constructing hierarchical polymer@MoS2 core-shell structures for regulating thermal and fire safety properties of polystyrene nanocomposites. Compos. Part A **107**, 144–154 (2018). https://doi.org/10.1016/j.compositesa.2017.12.026
58. Kundu, C.K., Wang, X., Hou, Y., Hu, Y.: Construction of flame retardant coating on polyamide 6.6 via UV grafting of phosphorylated chitosan and sol–gel process of organo-silane. Carbohydr. Polym. **181**, 833–840 (2018). https://doi.org/10.1016/j.carbpol.2017.11.069
59. Liu, S., Kuan, C., Ke, C., Shen, M., Chiang, C.: A novel effective halogen-free biomass-based chitosan derived flame retardant through microencapsulation technology for improved flame retardancy, thermal stability of polyurethane. Polym. Degrad. Stab. **219** (2024). https://doi.org/10.1016/j.polymdegradstab.2023.110604

60. Wang, B., Luo, C., Zhu, P., Liu, Y., Xu, Y.: Facile construction of H_3PO_3-modified chitosan/montmorillonite coatings for highly efficient flame retardation of polyester–cotton fabrics. Prog. Org. Coatings **184** (2023). https://doi.org/10.1016/j.porgcoat.2023.107864
61. Hu, S., Song, L., Pan, H., Hu, Y., Gong, X.: Thermal properties and combustion behaviors of flame retarded epoxy acrylate with a chitosan based flame retardant containing phosphorus and acrylate structure. J. Anal. Appl. Pyrolysis **97**, 109–115 (2012). https://doi.org/10.1016/j.jaap.2012.06.003

Recent Developments on Flame Retardant Fibre-Reinforced Composite Materials

M. Dada and P. Popoola

Abstract The pursuit of fire safety and improved material properties has led to significant advancements in flame-retardant fiber-reinforced composite materials. This chapter highlights the most effective flame retardants for fiber-reinforced composites, how different flame retardants affect the mechanical properties of the composites, the latest developments in flame retardant fiber-reinforced composites, the potential applications of flame retardant fiber-reinforced composites, the environmental and health implications of using flame retardant fiber-reinforced composites, how the cost of flame retardant fiber-reinforced composites can be reduced, and the challenges in commercializing flame retardant fiber-reinforced composites. There is a need to significantly reduce flammability while maintaining structural integrity. These fire-resistant composites have gained research interest in industries like aerospace, automotive, and construction, where safety and durability are crucial. Hence, ongoing efforts and significant findings in flame-retardant fiber-reinforced composite materials need to be reported, highlighting their potential to contribute to a safer and more sustainable future.

Keywords Flame retardant · Fibre-reinforced composites · Fire safety · Polymers · Nanotechnology

1 Introduction

The growth in manufacturing industries has necessitated the development of stronger, stiffer, denser, and more cost-effective materials with improved sustainability [1–3]. Composite materials, which are made up of natural or synthetic fibers, have significant applications in various fields such as construction, mechanical, automobile, aerospace, biomedical, and marine. Research studies have shown that composites have a significant enhancement in structural, mechanical, and tribological properties

M. Dada (✉) · P. Popoola
Chemical, Metallurgical and Materials Engineering, Tshwane University of Technology, Pretoria 0183, South Africa
e-mail: dadamo@tut.ac.za

T. C. Mokhena et al. (eds.), *Biomaterials as Green Flame Retardants*, Engineering Materials, https://doi.org/10.1007/978-981-97-6871-4_7

[4–6]. However, concerns about plastic waste accumulation have led researchers to develop environmentally friendly materials with cleaner manufacturing processes. Various composite recycling processes have been developed to cope with the thousands of tons of composite waste generated annually. According to Datta et al. [7], mechanical recycling methods, including pulverization, pyrolysis, glycolysis, and HVF, are used to recycle materials. Additionally, natural fillers such as cellulose nanocrystals and natural fibers enhance material properties and reduce waste. Hybrid composites, which combine multiple fibers, are also created. Composites are categorized based on fiber length, with continuous fiber reinforcement composites having long fibers and discontinuous fiber reinforcement composites having short fibers [8–10].

Researchers have optimized composite materials for specific applications by altering their fiber content, orientation, and manufacturing processes [11]. The properties of fiber-reinforced composites (FRCs), including mechanical, physical, thermal, and electrical properties, are crucial for their intended use. Cellulosic fiber, known for its unique properties such as biocompatibility, biodegradability, and non-toxicity, is widely used in various industries [12–14]. It can be extracted from natural resources like rice husk, wheat husk, maize husk, pine needles, cotton, and jute, as well as waste biomass. Cellulosic fiber is cheap, plentiful, non-toxic, and can bond to the cement-based matrix, which improves properties like flexibility, toughness, flexural capacity, and impact resistance [15]. Fiber reinforced composites (FRCs) have become increasingly popular in various sectors due to their exceptional mechanical properties and lightweight structure. However, their fire safety remains a serious concern. Natural fibers as polymer reinforcement face two significant issues: incompatibility with hydrophobic polymers and high flammability. Chemical modification of fibers can enhance mechanical properties but increase flammability. Higher lignin concentrations retard flame propagation, while higher cellulose concentrations heighten flammability [16].

The growing need for flame-resistant materials has motivated researchers and industry experts to investigate and create effective flame retardants capable of protecting FRCs from the destructive effects of fire events. Several flame retardants have been studied throughout the years, each with its own set of benefits and drawbacks [17–20]. However, determining the most effective flame retardant for FRCs remains a difficulty. A comprehensive investigation to assess various flame retardant materials, both existing and novel, to determine their performance in preventing or retarding flame spread and combustion and to contribute to the advancement of fire safety in fiber reinforced composites, is necessary. Fiber-reinforced polymer composites consist of continuous fibers embedded in a polymeric resin matrix, with common fibers including carbon, glass, aramid, and basalt. FRP composites are applied to concrete as laminates, rods, or sheets [21]. Previous studies have established suitable methods and techniques for strengthening existing RC structures using FRPs and established guidelines for their application in critical environments and loading conditions. FRP is a noncorrosive, high strength-to-weight ratio, and durable material with excellent fatigue properties, lower relaxation losses, minimal disruption, and easy handling. Furthermore, it features irregular surfaces and electrical insulators without a magnetic signature [22]. FRP is 5 times lighter than conventional steel and has 8–10 times higher tensile strength than steel. However, the widespread use of fiber reinforced composites (FRCs) in a variety of sectors has emphasized the need for improved fire safety measures. Despite their superior mechanical qualities and lightweight attributes, FRCs are nonetheless vulnerable to fire events, which may have disastrous repercussions [23]. Fiber reinforced polymeric composite materials offer high strength, fatigue endurance, and thermal insulation. However, they have poor impact damage tolerance and are not suitable for high-demanding fire-safety and mechanical properties [24]. The use of nanofillers in composites has improved various properties such as thermal, mechanical, energy absorption, biodegradability, moisture absorption, tribology, and flame retardancy. Inorganic nano-filled NFRPCs have mechanical strength comparable to synthetic fibers and lower environmental impact, making them suitable for engineering applications. However, long-term durability in outdoor applications requires addressing challenges like moisture absorption and weather conditions and more research is needed. [25].

Composite materials made from natural fibers and recyclable materials like polylactic acid (PLA) and bamboo fiber (BF) are important for achieving carbon–neutral solutions. To improve the properties of these composites, the interface between BF and PLA must be carefully considered. To limit flammability, halogen-free flame retardants can be added to PLA, but this may negatively impact its mechanical characteristics. Carbon fiber reinforced polymer composites are also widely used due to their rigidity. Graphene fibers, with their strength and conductivity, have potential applications in lightweight conductive cables, supercapacitors, solar cell fabrics, and more [23]. Basalt fibre (BF) offers superior physical and mechanical qualities to fibreglass while being much less expensive than carbon fibre [26]. The addition of glass or carbon fibres to Kevlar fiber-reinforced composites (KFRCs) may improve their thermal characteristics [27]. Sisal fibre and hemp composites are commonly used in vehicle interiors and furniture upholstery due to their good tribological qualities. Hemp composites exhibit a 52% improvement in specific flexural strength compare to glass fibre (GF)-reinforced composites with a propylene matrix. Glass fibre reinforced polymer composites (GFRP) are also widely used in engineering applications due to their strength and durability [28]. To increase rigidity and density, glass fibres may be used. Thermoplastics and thermosets, for example, bond the fibres and transmit loads and stresses. The high cost of polymers has restricted their commercial application, although fillers may improve composite qualities while decreasing their value [29]. Because of their environmental resilience, specific strength, and stiffness, GFRP composites are extensively employed in the maritime and pipe sectors. Because of their small weight and great fatigue resistance, they are also employed in the aviation sector. The damping capabilities of randomly oriented coir fiber-reinforced polypropylene composites are superior to those of synthetic fiber-reinforced composites [30]. Chemical fibres have a wide range of technical uses, including aviation, vehicles, ships, sports equipment, and building. However, because of their brittleness, CF-based composites are prone to stress concentrations. Hybridization methods may address these concerns by substituting carbon fibre layers with ductile fibre layers, resulting in cost reductions as well as enhanced physical and mechanical qualities [31]. Basalt fibres are an effective option for reinforcing CFRP composites due to their eco-friendliness, affordability, and ease of processing. These fibres are produced from molten basalt-based igneous volcanic rock, making them superior to E-glass in terms of tensile and compressive qualities. Additionally, they are less expensive than carbon-based fibres [32]. Palm fibres (PFs) exhibit superior fiber-matrix interaction and a greater Young's modulus than homopolymers. Friction composites are made using abaca fibre (AF) reinforcement, which provides good wear resistance while decreasing density as fibre content increases [33].

According to the study, luffa fibres (LFs) can enhance the mechanical properties of composites by improving tensile, compressive, flexural, impact strength, and water absorption. However, thermoplastic composites reinforced with natural fibres have lower strength performance than thermoset composites [34]. Natural fibre composites are combined with a small amount of synthetic fibre to enhance recycling potential and design flexibility. The resulting hemp/glass fibre hybrid polypropylene composites exhibit better flexural strength, impact strength, and water absorption qualities [35]. Oil palm/kenaf fiber-reinforced epoxy hybrid composites display improved tensile and flexural capabilities, which makes them a viable option in the automobile industry. Fibre hybridization, which combines multiple types of fibre in a matrix, is a potential method for offsetting the drawbacks of one type while enhancing their properties [36]. Hybrid composite materials for infrastructure and civil applications are created using basalt fibres. Banana and jute fibres are also utilised in composites, resulting in higher tensile and compressive strength. Basalt fibres have been garnering attention due to their improved mechanical qualities, making them a promising alternative for the polymer sector [37]. The embedding or reinforcing of foreign elements in a host matrix can result in the creation of new and superior properties like increased elastic modulus, ductility, light weight, and flame retardancy [38]. While glass fibres (GFs) are frequently utilised because of their high strength, durability, thermal stability, resistance to impact, chemical, friction, and wear, machining GFRPs is time-consuming, difficult, and has a short tool life. Glass fibre (GF) is a low-cost, high-tensile-strength, corrosion-resistant, and insulating substance that is often employed as a reinforcing material to increase the mechanical strength of polymer matrix [39]. Polyamide 66 (PA66) and polyamide 6 (PA6) are two of the most common polyamides. Both are synthetic polymers derived from hexamethylene diamine and adipic acid monomers. Polyamide 66 (PA66) is a prominent engineering thermoplastic in today's industrial industries because of its design flexibility, corrosion resistance, high strength, dimensional stability, lightweight, reduced tooling and finishing loss, and other advantages. While GF is often used to increase tensile strength and thermal distortion temperature in PA6. The poor interface contacts between glass fibre (GF) and organic polymer matrix have limited the performance of glass fibre reinforced PA6 composites. Surface modification of GF is desired to improve compatibility, but integrating GF may result in toughness and ductility loss. Flame retardancy and melt-dripping are significant flaws in GF/PA6 composites, making it challenging to create them using PA6's low melt viscosity. Phosphorous flame retardants have been widely used to enhance their flame retardancy, but interfacial flame retardation is still significant due to the larger combustion rate of interface regions [40].

2 The Most Effective Flame Retardants for Fiber-Reinforced Composites

To improve flame retardancy, flame-retardant treatments have been developed to enhance thermal stability and fire resistance [41]. These treatments include Inorganic flame retardants, based on metal hydroxides like magnesium hydroxide (MDH) and alumina trihydrate (ATH), which suppresses combustion by releasing water vapor. **Organophosphorus flame retardants** treatment like **tributyl phosphate (TBP)**, Triphenyl phosphate (TPP), Octadecyl diphenyl phosphate (OPPD), Tris (2-chloroethyl) phosphate (TCP) and Tetrakis(hydroxymethyl) phosphonium chloride (TPHP) interrupt the combustion process by releasing non-flammable gases, forming a char layer, and affecting free radical production [42–44]. Halogenated flame retardants like **Chlorinated paraffins (CPs)**, **Firemaster BP-6**, **Polybrominated diphenyl ethers (PBDEs)**, Aroclors and **Brominated flame retardants (BFRs)** release halogen atoms but can be harmful to the environment. Nanoparticles improve material flame retardancy without significantly affecting properties. Nonetheless, bio-based flame retardants, made from renewable resources, are considered more environmentally friendly than traditional flame retardants [45–47]. The choice of fiber-reinforced polymer (FRP) materials for retrofitting structures depends on various factors. While carbon fiber-reinforced polymer (CFRP) is resistant to alkali and enhances concrete strength, it is more expensive and requires less resin [48]. Glass fiber-reinforced polymer (GFRP) is recommended for earthquake-strengthening projects due to its good energy absorption capacity. Hybrid FRPs, which combine CFRP and GFRP, offer high ductility, deformability, break resistance, and are inexpensive. BFR reinforcements are preferred over CFRP components due to their better strain control capability at failure. However, the selection of FRP for strengthening requires further research [49, 50]. To achieve a flame retardancy rating of UL94 V-0, researchers have used non-halogen flame retardants like melamine cyanurate (MCA), red phosphorus, and melamine polyphosphate (MPP). However, these non-halogen flame retardants have limitations due to their corrosiveness and toxicity. Free-halogen FRs like aluminum diethylphosphinate (AlPi), melamine polyphosphate (MPP), and red phosphorus have been used to weaken or solve the melt dripping problem. Melt dripping is the occurrence of molten plastic dripping from a burning material, causing fire to spread. Factors contributing to melt dripping include material type, fire temperature, material size and shape, and the presence of flame retardants. Larger and irregular materials are more likely to melt and drip, while larger, more uniform materials are less likely. Inorganic filler retardants like montmorillonite and silica nanocomposites have also been explored for their FR effects. Xu et al. [51] prepared montmorillonite/silica-based nanocomposites to investigate thermosetting phenolic resin properties using FT-IR, NMR, and other testing methods. Results showed a higher molecular weight, a higher glass transition temperature, and improved thermal decomposition performance compared to BPR-400. The synthesized montmorillonite/silica nanocomposites showed a reduced ablative rate and back temperature after 600 s. The authors stated that the material is expected to be used in solid rocket engine thermal protection materials due to the montmorillonite/silica nanocomposites.

Aluminum diethylphosphinate (AlPi) is a favorable free-halogen organophosphorous FR [52]. While 9,10-dihydro-9-oxa-10-phosphaphenanthrene-10-oxide (DOPO) derivatives are also promising nonhalogenated compounds for FR materials, DOPO-derivatives have been designed and applied to polyamides for outstanding flame retardancy. AlPi and DOPO-derivatives interrupt combustion by creating a protective layer and releasing phosphoric acid vapors, preventing oxygen, and acting as a fire suppressor by releasing phosphoric acid vapors. Inorganic nanometer FRs such as sepiolite and nanosilica can enhance the formation of a uniform char layer and improve the properties of AlPi and polyamide (PA). Similar effects can be observed with metal salts of ZnS and ZnB when used as synergists with AlPi for PA materials. The mechanical properties of composites are determined by characteristics like matrix and fibre properties, as well as adhesion strength. However, stress transfer is limited due to the low interaction between PA66 and GF. Scientists aim to develop biocomposites that are completely biodegradable by combining natural fibres with biodegradable polymers like polylactic acid, starch polymers, soybean polymers, and cellulosic polymers. These biocomposites provide benefits such as higher tensile strength, lower density, lower cost, and greater biodegradability [53–55]. Fiber biocomposites are eco-friendly materials made of natural fibers and a polymer matrix that provide strength, stiffness, toughness, and bonding. They have applications in industries like automotive, building, and aerospace due to their ecological and economic benefits. However, their use is limited by hydrophobic polymer matrices, sensitivity to heat during the compounding process, and flammability. Flame retardancy is becoming increasingly important, and two general approaches are additive and reactive. The flame-retardant property of biocomposites can be obtained through various methods, like blending polylactic acid with Ammonium polyphosphate, ramie fibers, undergoing flame-retardant treatment with APP, and blending PLA and ramie fibers. Grafting DOPO on the surface of glass fiber enhances tensile strength and flame retardancy. This addition, however, reduces the mechanical characteristics of PBT/GF [56–59].

To enhance the bonding between fibers and resins, compatibility must be improved, which can be achieved through PPC, a compatibilizer composed of maleic anhydride and diphenyl phosphate. This leads to improved mechanical properties. The combustion of fiber-reinforced composites differs from pure resin combustion due to the wick effect of the fiber, which transports fuel fragments to the gas phase. Coating DOPO and vinyl trialkoxy silane on the surface of BF has not been successful in making PLA/BF composites less flammable. A modified BF covered by a SiP hybrid aerogel was identified as BF–SiP. Flame retardants in FRCs have shown promise in terms of lowering flammability and limiting fire spread. However, the use of these chemicals may present complexity and trade-offs that may influence the mechanical characteristics of the composite. The effect of flame retardants on critical mechanical features such as strength, stiffness, flexural and tensile capabilities, impact resistance, and interfacial bonding was studied using rigorous experimental testing and characterization. Furthermore, the effect of flame retardants on the weight, density, and thermal stability of FRCs, since these parameters might have a big impact on the material's applicability for real-world applications [60, 43, 61–63].

3 The Effect of Flame Retardants on the Mechanical Properties of Fiber-Reinforced Composites

Flame retardant fiber-reinforced composites (FRFRCs) are widely used in various industries due to their high stiffness, strength, and fire-resistant properties. Different substances, including ZB, MnO_2, MMT, and SnO_2, can be used to create composites with high flame retardancy. Boron compounds like borax/boric acid and ZB, as well as phosphate compounds, such as mono and diammonium phosphates, and phosphate treatments, have been shown to offer minor improvements in flame performance. Wood fibers/Polypropylene (PP), APP, and silica are effective in creating FRs for wood fibers/PP composites, with silica synergizing with APP in wood fibers/PP composites. Wood/PP composites with APP and silica have smaller color changes compared to unfilled ones. APP and $Mg(OH)_2$ have been found to improve the flame performance of WPCs, with wool having a positive effect on improving flame retardancy. It has been discovered that sisal fibers/PP with APP and $Mg(OH)_2$ decrease burning rate and boost thermal stability without affecting mechanical qualities. APP, as well as the combination of APP and ZB, have been shown to improve flame retardancy in PP composites without compromising mechanical qualities. The incorporation of FRs improves the flame and heat retardancy of flax/epoxy laminates and balsa-flax/epoxy sandwich composites significantly. Flax fiber/PLA/TPS composites with APP addition have the highest vertical burning fire performance, with a flammability rating of V-0 indicating exceptional self-extinguishing qualities. Flax fibers/PBS with APP, DAP, and PA have been found to have a decisive improvement in flame retardancy properties without affecting the composite's mechanical properties or technological feasibility. Cotton fabric reinforced epoxy composites show improvements in thermal and flammability properties [64, 65]

Bazan et al. [66] discovered that combining natural fibers and basalt in flame retardant polypropylene (PP) can increase strength and stiffness while maintaining low flammability. The research found that a coconut fiber composite offers the best flame retardant qualities. Aging decreases impact strength and deformability but increases strength qualities. Water absorption-induced swelling of fibers can lead to increased strength qualities and fill gaps between fibers and matrix during manufacturing. The study aimed to determine if water-swelling fiber could fill thermal shrinkage gaps between jute fibers and the polypropylene matrix. Fiber reinforcement in Onyx matrix material can increase tensile and impact strength. High-strength high-temperature (HSHT) fiberglass can increase tensile strength by 305.95% and impact strength by ~549%. [67]. Katsoulis et al. [68] conducted research on the impact of flame retardants and nanoparticles on the flexural performance of glass fiber-reinforced epoxy composites. They found that increasing heat flux reduced the flexural stiffness and modulus values of radiant heat-damaged composites. Flame retardants were successful in increasing char formation but did not adequately inhibit fiber reinforcements. Meanwhile, Chen et al. developed CGF/PE composites with a novel intumescent flame retardant (IFR) system. The IFR system significantly improved flame retardancy and mechanical properties of the composites, and SEM scans showed a toughening impact on the fiber-matrix interface [69]. Petersen et al. [70] investigate the influence of alumina trihydrate (ATH) as a fire-resistant material on the mechanical characteristics of Glass Fiber-Reinforced Polymer (GFRP) materials. The control material was found to be the strongest, with a 3% weaker flexural strength. However, the 25% ATH sample showed a 19% decrease in compressive strength and a 25% decrease in shear and tensile strength compared to the control. The addition of ATH reduces overall strength and makes FRP more brittle. The study suggests that a 50% ATH loading has a greater influence on GFRP's mechanical characteristics, potentially enabling the creation of fire-resistant FRP systems. Increasing the ATH quantity did not affect stiffness, but it reduced strength and made the FRP fragile. Combining 25% ATH with isophthalic had no significant effect on stiffness and strength, but 50% ATH had a more significant effect. Toldy et al. [71] compared the flame retardant and mechanical performance of flame-retardant epoxy resin systems and composites to their reference systems. Carbon fibers significantly improved flame retardant performance, with outstanding LOI and vertical burning test (UL-94) values for both systems. However, the simultaneous application of flame retardant and carbon fibers did not lead to the predicted improvement in heat release outcomes. The addition of carbon fibers improved mechanical qualities, but polar flame retardants in composites reduced their mechanical qualities. The addition of 40% carbon fibers increased residual mass by 12.2% in the aliphatic system and 9.3% in the aromatic system, despite disregarding carbon fibre mass in the aliphatic and aromatic systems. A multilayer composite with a reference composite core and an intumescent epoxy resin coating layer was developed to increase flame retardant and mechanical performance simultaneously [72–74]. This synergistic combination prevented the composite from igniting under 50 kW/m^2 heat flux, thanks to the gas and solid phase effects of the flame-retardant coating and the heat distribution effect of the carbon fiber reinforcement. Zhao et al. [75] developed flame-retardant glass fibre reinforced epoxy composites (FGRE) by adding P-MA, a phenylphosphonate-based flame retardant, to GRE. The composites had a 33% LOI value and a V-0 rating in UL 94 testing. P-MA did not affect the hardness or interfacial strength of the epoxy matrix and had a weaker influence on flame retardancy in GRE than in EP. However, P-MA increased the interfacial properties between the epoxy matrix and the glass fibre. The investigation found that the interfacial property of FGRE was maintained at the same level as that of GRE. Yargici et al. [76] developed flame-retardant glass fiber reinforced epoxy composites (FGRE) by integrating a phenylphosphonate-based flame retardant. The composites had a high flame retardancy rating and demonstrated enhanced mechanical properties. The study analyzed the flame retardancy and mechanical properties of phosphorus-based flame retardants on virgin PLA and found that disphosphoric acid-based FR had the least detrimental effect on mechanical properties. With optimal FR and chain extender additives, the PLA-GF composite exhibited superior mechanical and flammability performance, providing an excellent bio-based plastic alternative for the durable appliance industry. These study findings will be crucial for developing next-generation multifunctional flame retardant fibre-reinforced composite materials. Pomazi et al. [76] developed flame-retardant (FR) epoxy gelcoats using ammonium polyphosphate (APP) and resorcinol bis(diphenyl phosphate) (RDP) for improved fire behavior. The study compared the fire performance of different composite and gelcoat manufacturing techniques. Carbon fiber-reinforced epoxy composites were produced through vacuum infusion and gelcoats were sprayed on their surfaces with the help of a diluent. The addition of RDP in the gelcoat affected the fire resistance of the gelcoat matrices. The vacuum infusion method produced lower heat release than damp compression molding, and spraying the gelcoat produced a more uniform surface. The best fire resistance was achieved when the composite base was made via vacuum infusion and the gelcoat was applied via spraying. RDP was found to be unnecessary in the vacuum-infused composite base. The compatibility between the matrices affected the adhesion between the composite matrix and gelcoat. Multifunctional flame-retardant epoxy gelcoats containing phosphorus-containing flame retardants were explored for industrial application. Gelcoats with RDP were sprayable without diluent, resulting in superior fire performance and decreased hardness. Spraying and vacuum infusion significantly reduced total heat release and peak heat release rate. The authors concluded that the composites with vacuum infusion and coating showed the highest overall fire resistance. While the composite base's inherent flame retardancy can be omitted. Fiber-reinforced polymer (FRP) composites offer weight reduction and exceptional properties like strength, stiffness, and corrosion resistance. They are used in various industries, but fire resistance is crucial. A thermosetting resin with UV-curable acrylic monomers and transparent aluminum trihydrate (ATH) flame-retardant additives was used to create flame-retardant UV-curable composites. A study by Perez-de-Eulate et al. [77] was conducted to determine the optimal additive percentage and matrix viscosity for the railway industry, as well as the compatibility of the additives with the matrix's and composite's combustion properties. The composition of ATH has a moderate impact on UV light transmission during curing. As the surface area increases, specimens require longer curing periods. Several UV flame-retardant resin formulations were created, with flame-retardant glass fiber reinforced composites using the highest concentration of ATH. The composite exhibited moderate flame spread, the absence of igniting droplets/particles, low heat emission, moderate haze, and hazardous gas emission. It meets the R1HL3 requirements of the EN 45,545-2 standard, making it suitable for domestic and exterior applications on massive vehicle components and rail vehicles. Javaid et al. [78] investigated the fire resistance and mechanical features of ultra-lightweight truss core-based structures treated with varied concentrations of ammonium polyphosphate (APP) as a flame-retardant additive. According to their results, a formulation containing 10% APP got UL-94 V-1 certification and a LOI value of 29%. The thermal stability of the modified epoxy compositions was high, with maximum shear, flexural, and compressive moduli of 872.5, 3814, and 152.3 MPa, respectively. Sandwich composites like this might be employed in the civil infrastructure and marine sectors. Higher APP concentrations enhanced the truss core-based sandwich composites' mechanical characteristics and fire retardancy, resulting in better fire resistance owing to the creation of a protective char layer on the specimen's surface. The study concluded that the concentration of APP was a key factor in enhancing the composites' mechanical properties. In their study, Vieille et al. [79] analyzed the effects of a phosphorus-based flame-retardant additive, Exolit, on the fire-behavior and mechanical properties of composite laminates made from unidirectional carbon fibers and polyether-ketone-ketone (PEKK) thermoplastic. The study found that Exolit delayed thermal degradation and maintained composite structural integrity, resulting in excellent mechanical properties. However, the positive effects were most prominent in orthotropic laminates with carbon fibers at a 0° orientation. Semi-isotropic laminates experienced moderate decreases in axial rigidity and strength compared to virgin specimens, while orthotropic laminates had significant decreases. In another study by Alao et al. [80], alkaline treatment and fire-retardant coating were investigated for producing biocomposites from frost-retted hemp fiber and PLA. The researchers found that treatment with a 5 wt.% NaOH solution for 1 h was the most effective in enhancing the material's strength and modulus values. The study emphasizes the possibility of integrating fiber surface treatment with fire-retardant coating to enhance the mechanical and fire protection capabilities of biocomposites, contributing to sustainable building design principles. Loganathan et al. [81] investigated the viscoelastic, fire-retardant, and thermal characteristics of MWCNT-modified phenolic composites reinforced with hybrid Cytostachys Renda (CR)/kenaf fibres (KFs). The hybrid composites exhibited the maximum thermal stability, according to thermogravimetric and differential scanning calorimetric studies. Multiwalled carbon nanotubes (MWCNT)-modified composites based on kenaf fibre demonstrated higher storage and loss moduli. The hybrid composites had the highest average rate of heat emission, the lowest total heat release rate, the average mass loss rate, the average fire growth rate index, and the highest average rate of fire performance index. The smoke density was less than 200, meeting FAR 25.853d requirements. For high-temperature aerospace and automotive applications, the hybrid composite sample (15/35; 15% CR and 35% K) was selected as the best composite. All of these fire resistant fibre reinforced composites have applications in aerospace, construction, automotive, transportation, electronic, and electrical fields.

4 Recent Developments in Flame Retardant Fiber-Reinforced Composites

A low-cost and eco-friendly method for producing a recyclable fire safety design includes using recycled polymer waste and PP fiber-reinforced materials. Flame retardant treatments and metal components can enhance their thermal stability and fire resistance. Incorporating FR compounds or micro/nano FR additives can increase the flame slope of the polymer matrix, but halogen-based flame retardants are harmful to the environment and have been prohibited. Composites made from palm oil waste and recycled materials are a promising alternative. Foiler metal laminates are commonly used in the manufacturing of hybrid natural fiber/metal polymer composites. The addition of aluminium layers to the composite structure enhances their tensile, flexural, and impact characteristics significantly [82]. Because of their excellent efficacy, halogen flame retardants are a preferred option for flame-resistant materials due to the fast development of technology. However, as environmental concerns grow, halogen flame retardants are being abandoned, leading to the need for halogen-free alternatives like nitrogen, phosphorus, bio-based, and nanometer infill flame retardants. Phosphorus flame retardants have become essential alternatives, but they have disadvantages such as poor compatibility, surface hygroscopicity, and potential toxic PH3 gas production. To improve phosphorus-based flame retardants, researchers must address their compatibility and dispersion. Encapsulation techniques can enhance compatibility and dispersion, while performance-enhancing materials can improve flame retardancy and mechanical properties of composites. The development of future flame retardants should prioritize environmental protection by utilizing bio-based materials that offer biodegradability. Synergistic flame-retardant systems are gaining popularity, with potential in binary, ternary, and multi-flame-retardant systems. There is a focus on developing "multi-in-one" flame-retardants to guarantee optimal performance and mechanical properties in flame-retardant composite materials [83].

Recyclable flame retardant thermosets (RFRs) have made significant advancements in recent years, but there are still challenges to overcome. Sustainability concerns, high processing costs, and environmental issues remain. The synthesis of RFRs often requires organic solvents, which can limit their industrial applications. Additionally, cytotoxicity of RFRs containing phosphorus, halogen, and silicon must be thoroughly analyzed. The majority of RFRs have complex synthesis routes, making it difficult to simplify the synthesis approach, enhance recycling and flame retardant efficacy, and suggest suitable routes for various requirements. The thermal stability and chemical resistance of RFRs are compromised due to lower bond energy of reversible dynamic covalent bonds. Recycling these materials can be challenging, and chemical recycling using vitrimer's network decomposition mechanism requires further research. RFRs can be used with natural construction elements and optimized for fire resistance and flame-retardant nanomaterials. Further research is needed to explore the safety concerns of dynamic bonds in fire-retardant thermosets of living organisms and ecosystems. Future research in the development of hyperbranched dynamic bonds and fire retardants is expected to produce eco-friendly, sustainable, and mechanically robust thermosetting materials that prioritize human life and living environments. Despite the challenges, this field of study offers promising possibilities [84]. The utilization of plant fibers as reinforcements in composite materials and structures has increased due to their unique chemical composition and hierarchical microstructures. Nanotechnology and hybridization techniques have been employed to improve the mechanical properties of composites, including acoustic absorption, attenuation, thermal isolation, and electromagnetic wave transmission. Plant fiber-reinforced composites (PFRCs) can be manufactured using the same techniques as man-made fiber-reinforced composites (MFRCs), but their high defect rates and low fiber volume fraction hinder their mechanical properties. Superior mechanical properties have been achieved using nanocellulose or cellulose nanofibrils, leveraging the hierarchical microstructure of plant fibers [85]. To enhance FR PLA, straightforward production and additive synthesis methods are recommended by Yang et al. [86]. Matching processing properties of polymer matrix with FRs and additives is crucial for flame-retardancy improvement. Miniaturization of FRs is essential for mechanical properties. Reinforcing FRs with fiber, reactive extrusion, and toughening with rubber has been attempted, but durability and mechanical strength negatively impact PLA's performance. A comprehensive toughening strategy is needed to achieve highly reinforced PLA without sacrificing stiffness.

In recent times, fiber reinforced composites require additives for flame retardancy, thermal and electrical conductivity, and other properties. Solid additives can compromise the structure and cause irregular particulate distribution. Multifunctional gelcoats, developed using unsaturated polyester and epoxy, offer water resistance, electric conductivity, and flame retardancy. This article discusses preparation methods, common defects, and testing procedures for multifunctional gelcoats [87]. Gelcoating methods include brush and roller application and spray application. Brush application is simple and has excellent air release and minimal styrene emission. It can be difficult to maintain even layer thickness and can be problematic for large surfaces. Spray application is superior due to higher productivity and lower Volatile organic compound (VOC) emission. It allows for a greater diversity of molds and requires fewer man-hours. However, spraying-compatible gelcoat systems can increase styrene emission. To minimize VOC emissions and maximize air release, spray particles should be as large as possible and sprinkling pressure should be minimal. The latest gelcoating techniques that could overcome the limitations of traditional methods include UV curable gelcoat systems, which are crucial in the overall gelcoating process. Thermoset resins turn from liquid to solid during the curing process, and photoinitiated curing of coatings is a primary area of research, due to its short curing periods and low VOC emissions. Photopolymerization curing requires strong irradiation, and various electromagnetic irradiation sources are used. Carbon nanomaterials (CNMs) have been used to enhance the efficacy of UV-curable coatings In-mold coating (IMC) processes provide a consistent gelcoat thickness while lowering VOC emissions.

The traditional open-mould gelcoating process struggles to achieve the 8-h average occupational exposure limit of 20 ppm for volatile organic compounds.

Harper's in-mould surfacing (IMS) and in-mould gelcoating process using a trilayer separator fabric (IMGC) both significantly reduce styrene emission levels. However, IMS still has issues with styrene leakage during shim removal. A study was conducted on a hand-painted gelcoat and the unique IMGC and IMS techniques. Electric vehicle (EV) technology is set to replace half of newly purchased ICE vehicles by 2030, with the goal of reducing greenhouse gas emissions and improving air quality. However, the success of EVs relies on their power sources being carbon–neutral, such as solar or wind power. Despite challenges like limited range, slow charging, battery damage, thermal events, increased weight, and end-of-life issues, a carbon–neutral vehicle fleet requires advancements in battery technology and lightweighting, as well as reliance on the electricity mix. Materials research can help mitigate thermal runaway and enhance sustainability, with the integration of renewable, recycled, and waste-recovered materials enabling the circular economy and reducing environmental impact. Polymers and composites provide an opportunity for the shift towards EV technology, with the development of lightweight, high-performance, and functional materials [88]. Researchers have worked hard to create green and ecologically friendly bio-based flame retardants, which have shown both academic and industrial promise. The primary technique of modification is incorporating flame retardant components such as nitrogen, phosphorus, and silicon into bio-based compounds. However, obstacles such as limited flame-retardant efficiency, low thermal stability, and impediments to industrialisation owing to immature methodologies, high prices, and complicated manufacturing processes confront the advancement. To overcome these challenges, further research should focus on open-source materials, deep development, and low-cost sustainable flame retardant exploration. Open-source materials can explore novel and diverse materials, while deep development involves functionalizing bio-based flame retardants with multifunctional chemical groups. Exploration of low-cost sustainable flame retardants may be accomplished by pursuing green techniques and integrating them with new energy sources to develop a complete, sustainable, safe, and high-tech sector [89].

Nanotechnology develops fire retardant fiber reinforced compositeswith enhanced flame retardancy and mechanical properties by incorporating nanoparticles. Natural fiber reinforced polymer composites offer superior mechanical properties, low density, low cost, and low solidity, making them advantageous in various sectors like automotive, buildings, and constructions. The use of nanofillers has improved the properties of composites, including thermal, mechanical, energy absorption, biodegradability, moisture absorption, tribology and flame retardancy. Inorganic nano-filled NFRPCs have mechanical strength comparable to synthetic fibers and lower environmental impact, making them suitable for engineering applications. However, challenges like moisture absorption and weather conditions need to be addressed for long-term durability in outdoor applications. Further research is necessary [90]. Nanomaterials research has shown significant potential in fire retardancy applications, with potential for commercial use in fire-retardant additives. Nano-dimensional clay materials have been used to improve mechanical properties of polymers, with thermal stability and combustion retardancy being recognized. Halloysite nanotubes (HNTs), a potential alternative for fire retardancy, has wide availability and commercial value due to its superior dispersion potential, high loading, and cation exchange capacity. Further research is needed to determine its distribution in polymers and its impact on fire retardancy behavior. The adoption of low-cost fire-retardant nanofillers like HNT is a game changer in the area, since HNT-reinforced polymers display improved heat stability and mechanical qualities even at low concentrations [91, 92]. In summary, recent advancements in fire retardant fibre reinforced composites focus on improving flame retardancy and their corresponding mechanical properties. Key research interests include developing new flame retardants, using nanotechnology to create barrier against heat and oxygen, and developing innovative processing methods to regulate flame retardant dispersion and improve interfacial adhesion between fibers and matrix.

5 Conclusion and Outlook

The environmental and physiological effects of using flame retardant fiber-reinforced composites (FRFRCs) are complex, influenced by factors such as the type of flame retardant used, the quantity used, and the method of disposal. Halogenated flame retardants, which are effective in preventing fires, can be hazardous to human and environmental health. Phenorus-based flame retardants, on the other hand, are generally considered less hazardous but have unknown long-term health effects. The quantity of flame retardant used in a FRFRC can also have environmental and health consequences. The disposal method of FRFRCs can also have environmental and health consequences. When disposed of in landfills, hazardous chemicals can be released, but recycled FRFRCs can help reduce their environmental impact. Halogenated flame retardants, composed of chlorine, bromine, or fluorine, are effective in preventing fires but can be hazardous to human and environmental health. Phosphorus-based flame retardants, containing phosphorus, are generally less hazardous but have unknown long-term health effects. FRFRCs discarded in landfills can release toxic compounds into the environment due to their degradation over time. Recycling FRFRCs reduces the environmental impact of these materials by extracting and using flame retardants in other products during the recycling process. The commercialization of flame-retardant fiber-reinforced composites encounters several challenges like higher costs, limited accessibility, potential performance issues, and regulatory approval requirements. These factors make it challenging to sell FRFRCs in certain markets and less suitable for non-flame-retardant applications. To reduce the cost of FRFRCs, manufacturers can use more efficient flame retardants, different processing techniques, recycled materials, and increased production volume. These methods can significantly impact the material's cost, reducing the amount of flame retardant used, and the material's flame retardancy. Using recycled materials can also help reduce production costs. In conclusion, recent advancements in flame-retardant fiber-reinforced composite materials have significantly improved fire safety and durability in various industries. These innovations have led to safer, more reliable products in aerospace, automotive, and construction. As research continues, more innovative solutions will be developed to combat fire hazards and enhance material performance. These advancements are crucial for a more resilient and fireproof future, emphasizing sustainability and safety.

References

1. Adeniyi, A.G., Onifade, D.V., Ighalo, J.O., Adeoye, A.S.: A review of coir fiber reinforced polymer composites. Compos. B Eng. **176**, 107305 (2019)
2. Morampudi, P., Namala, K.K., Gajjela, Y.K., Barath, M., Prudhvi, G.: Review on glass fiber reinforced polymer composites. Mater. Today: Proc. **43**, 314–319 (2021)
3. Elseify, L.A., Midani, M., Shihata, L.A., El-Mously, H.: Review on cellulosic fibers extracted from date palms (*Phoenix dactylifera* L.) and their applications. Cellulose **26**, 2209–2232 (2019)
4. Erden, S., Ho, K.: Fiber reinforced composites. In: Fiber Technology for Fiber-Reinforced Composites, pp. 51–79. Woodhead Publishing (2017)
5. Gururaja, M.N., Rao, A.H.: A review on recent applications and future prospectus of hybrid composites. Int. J. Soft Comput. Eng. **1**(6), 352–355 (2012)
6. Hajj, R., Brunel, R., Sonnier, R., Longuet, C., Ganachaud, F.: Silicone-recycled pyrolyzed fillers for enhanced thermal-and flame-resistant silicone elastomers. Polym. Degrad. Stab. **200**, 109947 (2022)
7. Datta, J., Głowińska, E., Włoch, M.: Mechanical recycling via regrinding, rebonding, adhesive pressing, and molding. In: Recycling of Polyurethane Foams, pp. 57–65. William Andrew Publishing (2018)
8. Jawaid, M., Siengchin, S.: Hybrid composites: a versatile materials for future. Appl. Sci. Eng. Progress **12**(4), 223–223 (2019)
9. Krauklis, A.E., Karl, C.W., Gagani, A.I., Jørgensen, J.K.: Composite material recycling technology—state-of-the-art and sustainable development for the 2020s. J. Compos. Sci. **5**(1), 28 (2021)
10. Mallick, P.K.: Fiber-Reinforced Composites: Materials, Manufacturing, and Design. CRC Press (2007)
11. Naser, M.Z., Hawileh, R.A., Abdalla, J.A.: Fiber-reinforced polymer composites in strengthening reinforced concrete structures: a critical review. Eng. Struct. **198**, 109542 (2019)
12. Liu, G., Xiong, Y., Zhou, L.: Additive manufacturing of continuous fiber reinforced polymer composites: design opportunities and novel applications. Compos. Commun. **27**, 100907 (2021)
13. Nunna, S., Chandra, P.R., Shrivastava, S., Jalan, A.K.: A review on mechanical behavior of natural fiber based hybrid composites. J. Reinf. Plast. Compos. **31**(11), 759–769 (2012)
14. Omran, A.A.B., Mohammed, A.A., Sapuan, S.M., Ilyas, R.A., Asyraf, M.R.M., Rahimian Koloor, S.S., Petrů, M.: Micro-and nanocellulose in polymer composite materials: a review. Polymers **13**(2), 231 (2021)
15. Prasad, D.S., Shoba, C.: Hybrid composites–a better choice for high wear resistant materials. J. Market. Res. **3**(2), 172–178 (2014)
16. Prasad, D.S., Shoba, C., Ramanaiah, N.: Investigations on mechanical properties of aluminum hybrid composites. J. Market. Res. **3**(1), 79–85 (2014)
17. Prashanth, S., Subbaya, K.M., Nithin, K., Sachhidananda, S.: Fiber reinforced composites-a review. J. Mater. Sci. Eng. **6**(03), 2–6 (2017)
18. Rajak, D.K., Pagar, D.D., Kumar, R., Pruncu, C.I.: Recent progress of reinforcement materials: a comprehensive overview of composite materials. J. Market. Res. **8**(6), 6354–6374 (2019)
19. Safri, S.N.A., Sultan, M.T.H., Jawaid, M., Jayakrishna, K.: Impact behaviour of hybrid composites for structural applications: a review. Compos. B Eng. **133**, 112–121 (2018)
20. Sienkiewicz, A., Czub, P.: Flame retardancy of biobased composites—research development. Materials **13**(22), 5253 (2020)

21. Wang, B., Zhong, S., Lee, T.L., Fancey, K.S., Mi, J.: Non-destructive testing and evaluation of composite materials/structures: a state-of-the-art review. Adv. Mech. Eng. **12**(4), 1687814020913761 (2020)
22. Yang, G., Park, M., Park, S.J.: Recent progresses of fabrication and characterization of fibers-reinforced composites: a review. Compos. Commun. **14**, 34–42 (2019)
23. Zheng, H., Zhang, W., Li, B., Zhu, J., Wang, C., Song, G., Wu, G., Yang, X., Huang, Y., Ma, L.: Recent advances of interphases in carbon fiber-reinforced polymer composites: a review. Compos. B Eng. **233**, 109639 (2022)
24. Zielecka, M., Rabajczyk, A., Pastuszka, Ł, Jurecki, L.: Flame resistant silicone-containing coating materials. Coatings **10**(5), 479 (2020)
25. Zweben, C.: Tensile strength of hybrid composites. J. Mater. Sci. **12**, 1325–1337 (1977)
26. Li, Y., Zhang, J., He, Y., Huang, G., Li, J., Niu, Z., Gao, B.: A review on durability of basalt fiber reinforced concrete. Compos. Sci. Technol. **225**, 109519 (2022)
27. Nasser, J., Lin, J., Steinke, K., Sodano, H.A.: Enhanced interfacial strength of aramid fiber reinforced composites through adsorbed aramid nanofiber coatings. Compos. Sci. Technol. **174**, 125–133 (2019)
28. Ren, G., Yao, B., Huang, H., Gao, X.: Influence of sisal fibers on the mechanical performance of ultra-high performance concretes. Constr. Build. Mater. **286**, 122958 (2021)
29. Domun, N., Kaboglu, C., Paton, K.R., Dear, J.P., Liu, J., Blackman, B.R., Liaghat, G., Hadavinia, H.: Ballistic impact behaviour of glass fibre reinforced polymer composite with 1D/2D nanomodified epoxy matrices. Compos. B Eng. **167**, 497–506 (2019)
30. Abu-Okail, M., Alsaleh, N.A., Farouk, W.M., Elsheikh, A., Abu-Oqail, A., Abdelraouf, Y.A., Ghafaar, M.A.: Effect of dispersion of alumina nanoparticles and graphene nanoplatelets on microstructural and mechanical characteristics of hybrid carbon/glass fibers reinforced polymer composite. J. Mater. Res. Technol. **14**, 2624–2637 (2021)
31. Nurazzi, N.M., Asyraf, M.R.M., Rayung, M., Norrrahim, M.N.F., Shazleen, S.S., Rani, M.S.A., Shafi, A.R., Aisyah, H.A., Radzi, M.H.M., Sabaruddin, F.A., Ilyas, R.A.: Thermogravimetric analysis properties of cellulosic natural fiber polymer composites: a review on influence of chemical treatments. Polymers **13**(16), 2710 (2021)
32. Xie, H., Yang, L., Zhang, Q., Huang, C., Chen, M., Zhao, K.: Research on energy dissipation and damage evolution of dynamic splitting failure of basalt fiber reinforced concrete. Constr. Build. Mater. **330**, 127292 (2022)
33. Alhijazi, M., Zeeshan, Q., Safaei, B., Asmael, M., Qin, Z.: Recent developments in palm fibers composites: a review. J. Polym. Environ. **28**, 3029–3054 (2020)
34. Alhijazi, M., Safaei, B., Zeeshan, Q., Asmael, M., Eyvazian, A., Qin, Z.: Recent developments in luffa natural fiber composites. Sustainability **12**(18), 7683 (2020)
35. Bhoopathi, R., Ramesh, M.: Optimization of drilling output responses of eggshell fillers reinforced hemp/glass fibres hybrid composites. Mater. Today: Proc. **46**, 3245–3250 (2021)
36. Ghori, S.W., Rao, G.S.: Mechanical and thermal properties of date palm/kenaf fiber-reinforced epoxy hybrid composites. Polym. Compos. **42**(5), 2217–2224 (2021)
37. Khan, M., Cao, M., Xie, C., Ali, M.: Effectiveness of hybrid steel-basalt fiber reinforced concrete under compression. Case Stud. Constr. Mater. **16**, e00941 (2022)
38. Kumar, A., Sharma, K., Dixit, A.R.: A review of the mechanical and thermal properties of graphene and its hybrid polymer nanocomposites for structural applications. J. Mater. Sci. **54**(8), 5992–6026 (2019)
39. Rajak, D.K., Wagh, P.H., Linul, E.: Manufacturing technologies of carbon/glass fiber-reinforced polymer composites and their properties: a review. Polymers **13**(21), 3721 (2021)
40. Thomason, J.L.: Glass fibre sizing: a review. Compos. A Appl. Sci. Manuf. **127**, 105619 (2019)
41. Zhao, W., Kundu, C.K., Li, Z., Li, X., Zhang, Z.: Flame retardant treatments for polypropylene: strategies and recent advances. Compos. A Appl. Sci. Manuf. **145**, 106382 (2021)
42. Lazar, S.T., Kolibaba, T.J., Grunlan, J.C.: Flame-retardant surface treatments. Nat. Rev. Mater. **5**(4), 259–275 (2020)
43. Liu, B.W., Zhao, H.B., Wang, Y.Z.: Advanced flame-retardant methods for polymeric materials. Adv. Mater. **34**(46), 2107905 (2022)

44. Liao, Y., Chen, Y., Wan, C., Zhang, G., Zhang, F.: An eco-friendly NP flame retardant for durable flame-retardant treatment of cotton fabric. Int. J. Biol. Macromol. **187**, 251–261 (2021)
45. Nazir, R., Parida, D., Borgstädt, J., Lehner, S., Jovic, M., Rentsch, D., Bülbül, E., Huch, A., Altenried, S., Ren, Q., Rupper, P.: In-situ phosphine oxide physical networks: a facile strategy to achieve durable flame retardant and antimicrobial treatments of cellulose. Chem. Eng. J. **417**, 128028 (2021)
46. Kundu, C.K., Gangireddy, C.S.R., Song, L., Hu, Y.: Flame retardant treatments for polyamide 66 textiles: analysis the role of phosphorus compounds. Polym. Degrad. Stab. **182**, 109376 (2020)
47. Li, S.Q., Tang, R.C., Yu, C.B.: Flame retardant treatment of jute fabric with chitosan and sodium alginate. Polym. Degrad. Stab. **196**, 109826 (2022)
48. Sharkey, M., Harrad, S., Abdallah, M.A.E., Drage, D.S., Berresheim, H.: Phasing-out of legacy brominated flame retardants: The UNEP Stockholm Convention and other legislative action worldwide. Environ. Int. **144**, 106041 (2020)
49. Uthaman, A., Xian, G., Thomas, S., Wang, Y., Zheng, Q., Liu, X.: Durability of an epoxy resin and its carbon fiber-reinforced polymer composite upon immersion in water, acidic, and alkaline solutions. Polymers **12**(3), 614 (2020)
50. Zhao, Q., Zhang, K., Zhu, S., Xu, H., Cao, D., Zhao, L., Zhang, R., Yin, W.: Review on the electrical resistance/conductivity of carbon fiber reinforced polymer. Appl. Sci. **9**(11), 2390 (2019)
51. Xu, Q., Zhang, Z., Li, Y., Zeng, F., Deng, P., Zhu, J., Su, Z.: Synthesis of thermosetting phenolic resin and its montmorillonite/silica nanocomposites under alkali-free conditions. J. Adhesion Sci. Technol. 1–11 (2022)
52. Liu, C., Li, P., Xu, Y.J., Liu, Y., Zhu, P.: Nickel alginate-enhanced fire safety of aluminum diethylphosphinate on epoxy resin. J. Appl. Polym. Sci. **140**(9), e53552 (2023)
53. Bifulco, A., Varganici, C.D., Rosu, L., Mustata, F., Rosu, D., Gaan, S.: Recent advances in flame retardant epoxy systems containing non-reactive DOPO based phosphorus additives. Polym. Degrad. Stab. **200**, 109962 (2022)
54. Varganici, C.D., Rosu, L., Bifulco, A., Rosu, D., Mustata, F., Gaan, S.: Recent advances in flame retardant epoxy systems from reactive DOPO–based phosphorus additives. Polym. Degrad. Stab. **202**, 110020 (2022)
55. Yang, Y., Wang, D.Y., Jian, R.K., Liu, Z., Huang, G.: Chemical structure construction of DOPO-containing compounds for flame retardancy of epoxy resin: a review. Prog. Org. Coat. **175**, 107316 (2023)
56. Asyraf, M.R.M., Ishak, M.R., Norrrahim, M.N.F., Nurazzi, N.M., Shazleen, S.S., Ilyas, R.A., Rafidah, M., Razman, M.R.: Recent advances of thermal properties of sugar palm lignocellulosic fibre reinforced polymer composites. Int. J. Biol. Macromol. **193**, 1587–1599 (2021)
57. Ramesh, M., Deepa, C., Kumar, L.R., Sanjay, M.R., Siengchin, S.: Life-cycle and environmental impact assessments on processing of plant fibres and its bio-composites: a critical review. J. Indus. Textiles **51**(4_suppl), 5518S–5542S (2022)
58. Ilyas, R.A., Sapuan, S.M., Harussani, M.M., Hakimi, M.Y.A.Y., Haziq, M.Z.M., Atikah, M.S.N., Asyraf, M.R.M., Ishak, M.R., Razman, M.R., Nurazzi, N.M., Norrrahim, M.N.F.: Polylactic acid (PLA) biocomposite: processing, additive manufacturing and advanced applications. Polymers **13**(8), 1326 (2021)
59. Asyraf, M.R.M., Ishak, M.R., Syamsir, A., Nurazzi, N.M., Sabaruddin, F.A., Shazleen, S.S., Norrrahim, M.N.F., Rafidah, M., Ilyas, R.A., Abd Rashid, M.Z., Razman, M.R.: Mechanical properties of oil palm fibre-reinforced polymer composites: a review. J. Market. Res. **17**, 33–65 (2022)
60. Niu, Q., Yue, X., Guo, Z., Yan, H., Fang, Z., Li, J.: Flame retardant bamboo fiber reinforced polylactic acid composites regulated by interfacial phosphorus-silicon aerogel. Polymer **252**, 124961 (2022)
61. Zhou, X., Qiu, S., Mu, X., Zhou, M., Cai, W., Song, L., Xing, W., Hu, Y.: Polyphosphazenes-based flame retardants: a review. Compos. B Eng. **202**, 108397 (2020)

62. Huo, S., Song, P., Yu, B., Ran, S., Chevali, V.S., Liu, L., Fang, Z., Wang, H.: Phosphorus-containing flame retardant epoxy thermosets: recent advances and future perspectives. Prog. Polym. Sci. **114**, 101366 (2021)
63. Dogan, M., Dogan, S.D., Savas, L.A., Ozcelik, G., Tayfun, U.: Flame retardant effect of boron compounds in polymeric materials. Compos. B Eng. **222**, 109088 (2021)
64. Saba, N., Jawaid, M., Alothman, O.Y., Inuwa, I.M., Hassan, A.: A review on potential development of flame retardant kenaf fibers reinforced polymer composites. Polym. Adv. Technol. **28**(4), 424–434 (2017)
65. Wang, S., Zhong, J., Gu, Y., Li, G., Cui, J.: Mechanical properties, flame retardancy, and thermal stability of basalt fiber reinforced polypropylene composites. Polym. Compos. **41**(10), 4181–4191 (2020)
66. Bazan, P., Salasińska, K., Kuciel, S.: Flame retardant polypropylene reinforced with natural additives. Ind. Crops Prod. **164**, 113356 (2021)
67. Prajapati, A.R., Dave, H.K., Raval, H.K.: An experimental study on mechanical, thermal and flame-retardant properties of 3D-printed glass-fiber-reinforced polymer composites. J. Mater. Eng. Perform. **30**, 5266–5277 (2021)
68. Katsoulis, C., Kandola, B.K., Myler, P., Kandare, E.: Post-fire flexural performance of epoxy-nanocomposite matrix glass fibre composites containing conventional flame retardants. Compos. A Appl. Sci. Manuf. **43**(8), 1389–1399 (2012)
69. Chen, J., Wang, J., Ding, A., Ni, A., Chen, H.: Flame retardancy and mechanical properties of glass fibre reinforced polyethylene composites filled with novel intumescent flame retardant. Compos. B Eng. **179**, 107555 (2019)
70. Petersen, M.R., Chen, A., Roll, M., Jung, S.J., Yossef, M.: Mechanical properties of fire-retardant glass fiber-reinforced polymer materials with alumina tri-hydrate filler. Compos. B Eng. **78**, 109–121 (2015)
71. Toldy, A., Szolnoki, B., Marosi, G.: Flame retardancy of fibre-reinforced epoxy resin composites for aerospace applications. Polym. Degrad. Stab. **96**(3), 371–376 (2011)
72. Chen, H., Wang, J., Ni, A., Ding, A., Sun, Z., Han, X.: Effect of novel intumescent flame retardant on mechanical and flame retardant properties of continuous glass fibre reinforced polypropylene composites. Compos. Struct. **203**, 894–902 (2018)
73. Khalili, P., Tshai, K.Y., Hui, D., Kong, I.J.C.P.B.E.: Synergistic of ammonium polyphosphate and alumina trihydrate as fire retardants for natural fiber reinforced epoxy composite. Compos. B Eng. **114**, 101–110 (2017)
74. Chen, J., Wang, J., Chen, H., Ni, A., Ding, A.: Synergistic effect of intumescent flame retardant and attapulgite on mechanical properties and flame retardancy of glass fibre reinforced polyethylene composites. Compos. Struct. **246**, 112404 (2020)
75. Zhao, X., Yang, L., Martin, F.H., Zhang, X.Q., Wang, R., Wang, D.Y.: Influence of phenylphosphonate based flame retardant on epoxy/glass fiber reinforced composites (GRE): flammability, mechanical and thermal stability properties. Compos. B Eng. **110**, 511–519 (2017)
76. Yargici Kovanci, C., Nofar, M., Ghanbari, A.: Synergistic enhancement of flame retardancy behavior of glass-fiber reinforced polylactide composites through using phosphorus-based flame retardants and chain modifiers. Polymers **14**(23), 5324 (2022)
77. Pérez-de-Eulate, N.G., Ares Elejoste, P., Goenaga, G., Urrutxua, M., Vallejo, F.J., Ballestero, J., Allue, A., Gómez-Alonso, J.L.: On the influence of flame-retardant additives on UV-curable thermosetting glass fiber-reinforced composites. Polymers **15**(1), 240 (2023)
78. Javaid, A., Hashmi, S.: Development of the fire-retardant truss core sandwich structures using carbon fiber reinforced epoxy composites. Fire Technol. **58**(3), 1493–1507 (2022)
79. Vieille, B., Coppalle, A., Le Pluart, L., Schuhler, E., Chaudhary, A., Rijal, B., Alia, A., Delpouve, N., Bourdet, A.: Influence of a flame-retardant on the fire-behaviour and the residual mechanical properties of C/PEKK composite laminates exposed to a kerosene flame. Compos. A Appl. Sci. Manuf. **152**, 106720 (2022)
80. Alao, P.F., Press, R., Kallakas, H., Ruponen, J., Poltimäe, T., Kers, J.: Investigation of efficient alkali treatment and the effect of flame retardant on the mechanical and fire performance of frost-retted hemp fiber reinforced PLA. Polymers **14**(11), 2280 (2022)

81. Loganathan, T.M., Sultan, M.T.H., Ahsan, Q., Jawaid, M., Naveen, J., Shah, A.U.M., Talib, A.R.A., Basri, A.A.: Thermal degradation, visco-elastic and fire-retardant behavior of hybrid Cyrtostachys Renda/kenaf fiber-reinforced MWCNT-modified phenolic composites. J. Therm. Anal. Calorim. **147**(24), 14079–14096 (2022)
82. Ilyas, R.A., Sapuan, S.M., Asyraf, M.R.M., Dayana, D.A.Z.N., Amelia, J.J.N., Rani, M.S.A., Norrrahim, M.N.F., Nurazzi, N.M., Aisyah, H.A., Sharma, S., Ishak, M.R.: Polymer composites filled with metal derivatives: a review of flame retardants. Polymers **13**(11), 1701 (2021)
83. Gao, J., Wu, Y., Li, J., Peng, X., Yin, D., Jin, H., Wang, S., Wang, J., Wang, X., Jin, M., Yao, Z.: A review of the recent developments in flame-retardant nylon composites. Compos. Part C: Open Access 100297 (2022)
84. Rashid, M.A., Liu, W., Wei, Y., Jiang, Q.: Review on intrinsically recyclable flame retardant thermosets enabled through covalent bonds. J. Appl. Polym. Sci. **139**(27), e52493 (2022)
85. Zhang, Z., Cai, S., Li, Y., Wang, Z., Long, Y., Yu, T., Shen, Y.: High performances of plant fiber reinforced composites—a new insight from hierarchical microstructures. Compos. Sci. Technol. **194**, 108151 (2020)
86. Yang, X.M., Qiu, S., Yusuf, A., Sun, J., Zhai, Z., Zhao, J., Yin, G.Z.: Recent advances in flame retardant and mechanical properties of polylactic acid: a review. Int. J. Biol. Macromol. 125050 (2023)
87. Pomázi, Á., Toldy, A.: Multifunctional gelcoats for fiber reinforced composites. Coatings **9**(3), 173 (2019)
88. Mohanty, A.K., Vivekanandhan, S., Tripathi, N., Roy, P., Snowdon, M., Drzal, L.T., Misra, M.: Sustainable composites for lightweight and flame retardant parts for electric vehicles to boost climate benefits: a perspective. Compos. Part C: Open Access 100380 (2023)
89. Wang, M., Yin, G.Z., Yang, Y., Fu, W., Palencia, J.L.D., Zhao, J., Wang, N., Jiang, Y., Wang, D.Y.: Bio-based flame retardants to polymers: a review. Adv. Indus. Eng. Polym. Res. (2022)
90. Mishra, T., Mandal, P., Rout, A.K., Sahoo, D.: A state-of-the-art review on potential applications of natural fiber-reinforced polymer composite filled with inorganic nanoparticle. Compos. Part C: Open Access 100298 (2022)
91. Ilyas, R.A., Nurazzi, N.M., Norrrahim, M.N.F.: Fiber-reinforced polymer nanocomposites. Nanomaterials **12**(17), 3045 (2022)
92. Rawtani, D., Satish, S., Rao, P.: Flame retardancy of nanocomposites with emphasis on Halloysite nanotubes. J. Ind. Eng. Chem. **125**, 1–13 (2023)

Wood Flour and Other Fiber Composites: Properties and Flame Retardancy

M. J. Phiri, M. M. Phiri, N. M. Mzinyane, and S. P. Hlangothi

Abstract Flame retardant materials prevent and slow down the spread of fireby hindering the combustion of the given material, and in some cases, the flame retardants create a protective coating on the surface of the burning material. The flame retardant can be used with the base material as an additive flame retardant or chemically grafted to the material as a reactive flame retardant. Most flame retardants release highly-toxic components which are harmful to the environment. Currently, an intriguing area of research work focusses on the synthesis and characterization of environmentally friendly flame retardants that are novel, efficient, have lower toxicity, and are bio-based materials. Natural plant-based fibers have been used in several composite applications including automotive, construction, and agriculture. The current work presents wood fiber-reinforced composites for use as flame-retardant products. The reaction mechanism of how the flame retardants work will be illustrated and the chemical, thermal, mechanical, and morphological properties of the flame-retardant composites will also be discussed.

Keywords Lignocellulosic materials · Wood fibres · Flame retardant property

1 Introduction

Wood fiber-reinforced polymeric composites have found application in mostly sound damping. Current research trends indicate an inclination to continuous growth in this industry and promise to keep increasing over the years to come. These materials are compounded by mixing the said natural fiber with the polymer matrix. The function of the fiber filler is mostly to improve the tensile strength, impart lightweight properties,

M. J. Phiri (✉) · N. M. Mzinyane
Department of Natural Sciences, Vaal University of Technology, Private Bag X021, Vanderbijlpark 1911, South Africa
e-mail: mohaup@vut.ac.za; mohaujphiri@gmail.com

M. M. Phiri · S. P. Hlangothi
Department of Chemistry, Nelson Mandela University, P.O. Box 7700, Gqeberha 6031, South Africa

T. C. Mokhena et al. (eds.), *Biomaterials as Green Flame Retardants*, Engineering Materials, https://doi.org/10.1007/978-981-97-6871-4_8

and noise reduction among others. The polymer matrix on the other hand facilitates uniform distribution of the fibers and uniform distribution of the load within the matrix. Additionally, the polymer matrix also aids in binding the composite components together and provides a protective coating for the fibers against environmental conditions. These materials are gradually replacing ceramic and metallic composites in some applications due to their low density, high stiffness, high strength, excellent resistance to weatherability, good thermal insulation, low thermal expansion, and high fatigue resistance. They are, however, associated with limitations that hinder their complete adoption. Among others, the disadvantage of the wood fiber reinforced composites is their poor fire resistance due to the high flammability of the polymer matrix with typical decomposition temperatures ranging from 300 to 400 °C. Additionally, the high amounts of heat released during the thermal decomposition of polymer, the release of toxic gases, smoke, and soot could potentially cause severe, delayed health issues or even death. Similar to the polymer matrix, wood fiber is also highly flammable [1]. This makes it impossible for these composites to be fully adopted across various industries, especially in the building sector.

Flame retardant additives used to date motivated research into improving the flame retardancy of wood fiber-based polymer composites. Several strategies have been considered to enhance the flame retardancy of these composites. The research trends point to the use of treatment methods for the fiber surface, the polymer, or the composite itself. The advantages and limitations associated with each procedure are also discussed. The enhancement is typically studied via different analytical methods, such as thermogravimetric analysis (TGA). For instance, the increase in the thermal decomposition of the resulting composite is often associated with the enhancement in flame retardant. Other techniques often used to measure flammability of the wood-reinforced composites have also been used including cone calorimeter, MCC, oxygen index meter, vertical or horizontal burning tests. To understand the mechanism that are involved during decomposition, other techniques such X-ray photoelectron spectroscopy (XPS), energy-dispersive X-ray (EDX), TG-FTIR and scanning electron microscope (SEM), are usually employed.

2 Types of Flame Retardants

Flame retardant chemicals have been employed since ancient times where the chemical used would control the diffusion rate of the flame or prevent continued ignition. Alum and vinegar were some of the first flame retardants and would be added to the wood to protect it against fire. There have been developments in flame retardant chemicals, however, the mechanism of action is still the same. Halogenated flame retardants, chlorine, and bromine-based, were developed in the 1930s [2] and organophosphorus flame retardants in the 1950s [3, 4]. While these chemicals have been phased out as flame retardants, these materials helped shape the landscape for the existing flame retardants since underlying chemicals concepts are still similar.

Flame retardants have two main classification types, namely, non-reactive and reactive types. The additive flame-retardant type can further be categorized into organic and inorganic [5, 6]. They are typically compounded with the polymer matrix. On the other hand, their inorganic counterparts are more popular due to their affordability and ease of functionalization into halogenated moieties and are used as fillers in the composites. The reactive flame retardants work by chemically modifying the structure of the polymer matrix with the flame-retardant components [7–12]. These flame retardants are categorized according to their compositions into 4 classes, namely, halogens, melamine, phosphate, and inorganic additives.

3 Mechanism of Flame Retardance

To understand the mechanism of flame retardance, it is important to look at the typical steps involved in the event of fire (Fig. 8.1). The initial step is ignition where the combustion starts to occur; this is subsequently accompanied by the fire propagation step. During this development stage, both the heat released, and the temperature of the fire increases significantly. At this point, the fire starts to spread rapidly, within a short time, producing excessive smoke, gases and heat, this is referred to as “flashover”. Thus, the fire becomes uncontainable and hence very dangerous [13–17].

Flame retardants work through various mechanisms to slow down or prevent the spread of fire. It is important to note that different flame retardants may employ a combination of these mechanisms to achieve fire resistance. The specific mechanism depends on the type and properties of the flame retardant, as well as the material. Some common mechanisms include the following:

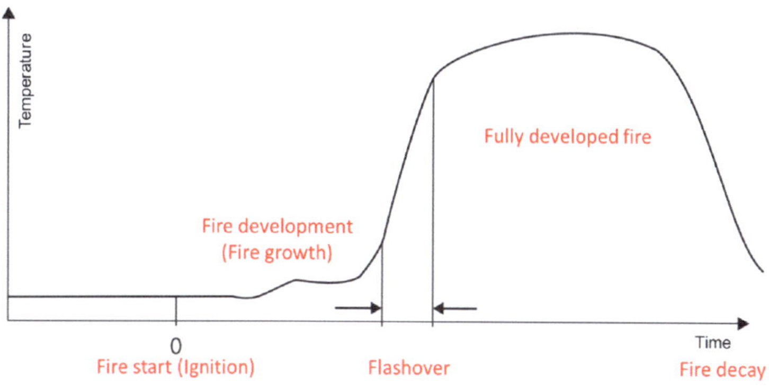

Fig. 8.1 The schematic of a fire situation. A typical example of fire will begin with fire ignition and its growth and continue to a well-defined and established fire until the decay stage is reached. The smoke, gases and heat generated from the fire will result in a flashover. Reprinted with permission from Dewaghe et al. [13]

1. Thermal insulation: flame retardants can create a barrier between the flame and a material, reducing the transfer of heat. This slows down the temperature rise and prevents the material from igniting or spreading the fire.
2. Chemical reaction: certain flame retardants release chemical compounds when exposed to high temperatures. These compounds can interrupt the combustion process by absorbing heat, and/or diluting the flammable gases. Others generate non-combustible gases (e.g., water or carbon dioxide) which essentially act as fire suppressants.
3. Formation of protective char layer: flame retardants can promote the formation of a thick, protective layer of char on the surface of the material. This char layer acts as a shield, insulating the underlying material from flames. It slows down the combustion process and reduces the fuel available for the fire.
4. Catalytic action: some flame retardants act as catalysts, promoting chemical reactions that hinder the combustion process. They may break down flammable gases or accelerate the formation of non-flammable compounds, thereby suppressing the fire.

Flame retardant modes of action differ. Additive flame retardants are suitable for high-temperature action via the deformation process. Thus, they are more effective at the temperature where combustion resumes. Contrarily, reactive flame retardants are integrated into the polymer structure via the polymerization process or via the post-polymerization treatment process [5]. Typical examples of halogenated flame retardants containing bromine are shown in Fig. 8.2 [18]. It is important to note that not all of the available organobromine flame retardants are affordable. Sometimes the flame-retardant compound may need to be functionalized in order to improve its compatibility with the polymer matrix. However, these processes must be cost-effective and should not hinder the release of the bromine under suitable fire conditions. Thus, bromine should not be released after complete decomposition of the polymer matrix and not too early prior to polymer decomposition.

The halogen-containing flame retardants (e.g., R-Br) work by blocking the combustion cycle of the fire. The halogen flame retardant undergoes homolytic cleavage releasing the bromine radical and the R-radical which consequently react with the combustion radicals, hydroxy radicals (HO·), and hydrogen radicals (H·), and hence obstructing their oxidation. These halogenated retardants are usually used in conjunction with antimony to increase the efficiency of fire retardancy. Antimony oxide facilitates the migration of halogens in the gas phase and together enables gas

$$R{-}Br \rightarrow R\cdot + Br\cdot \quad Br\cdot + R{-}H \rightarrow HBr + R\cdot \quad H\cdot + Br\cdot \rightarrow HBr R{-}Br \rightarrow R\cdot + Br\cdot \quad Br\cdot + R{-}H \rightarrow HBr + R\cdot \quad H\cdot + Br\cdot \rightarrow HBr$$

$$HO\cdot + HBr \rightarrow H_2O + Br\cdot \quad H\cdot + HBr \rightarrow H_2 + Br\cdot HO\cdot + HBr \rightarrow H2O + Br\cdot \quad H\cdot + HBr \rightarrow H2 + Br\cdot$$

$$SbBr_3 + 3H\cdot \rightarrow Sb + 3HBr SbBr3 + 3H\cdot \rightarrow Sb + 3HBr$$

$$Sb + HO\cdot \rightarrow SbOH Sb + HO\cdot \rightarrow SbOH$$

$$SbOH + HO\cdot \rightarrow SbO + H_2O SbOH + HO\cdot \rightarrow SbO + H2O$$

Fig. 8.2 Reaction steps of gaseous antimony oxide and the combustion radicals phase flame retardation/inhibition. The reaction steps involved between the volatile antimony species are shown in Fig. 8.2. These species act as effective radicals that disturb the combustion reaction cycle. The series of reactions is proposed in the multiple steps given below (Fig. 8.3).

Examples of the representative phosphorus compounds are shown in Fig. 8.4 [19]. To improve the char formation or char oxidation durability obtained in the presence of phosphorus flame retardant, other additives are usually added. Unlike brominated flame-retardant compounds, the existing versions of phosphorous flame retardants are much less; with only 9,10-dihydro-9-oxa-10-phosphaphenanthrene-10-oxide (DOPO) as the only commercially adopted [20–25].

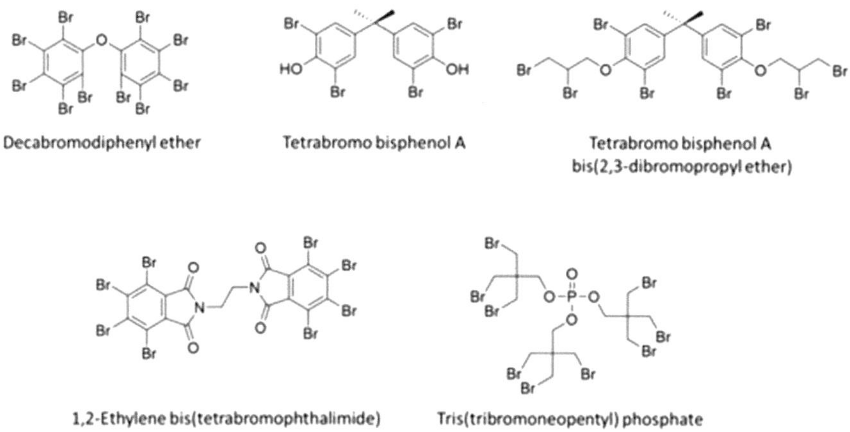

Fig. 8.3 Illustrative chemical structures of bromine-containing flame retardants with inhibitory properties on combustion reactions. Reprinted with permission from Morgan et al. [18]

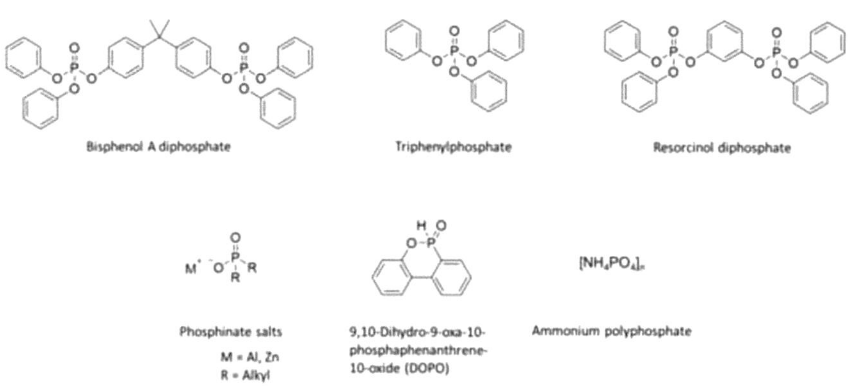

Fig. 8.4 Chemical structures of characteristic phosphorus-containing flame retardants, which are mostly used in the vapor phase, and not in the condensed phase. Reprinted with permission from Mihajlovic et al. [19]

4 Preparation and Fabrication of the Wood-Fiber Reinforced Composites

Wood-fiber reinforced composites have gained significant attention in various industries due to the unique characteristics of cellulose fibers obtained from wood [26, 27]. These lignocellulosic fibers offer distinct advantages, such as biodegradability and renewability, making them an attractive choice for composite materials. However, their hydrophilic nature poses challenges in achieving strong interfaces with hydrophobic polymeric matrices, leading to suboptimal mechanical properties. To overcome this limitation, various modifications can be employed to enhance the mechanical properties of these fibers and establish better compatibility with the matrix, ultimately improving the moisture resistance [28]. This chapter explores the methods to enhance the mechanical properties of wood fiber-reinforced composites and the different preparation techniques used in their manufacturing.

4.1 Surface Modification of Natural Fibers

To improve the mechanical properties of wood-fiber reinforced composites, surface modifications of cellulose fibers have proven to be effective approach. These modifications include chemical treatments to reduce hydroxyl groups, thereby enhancing compatibility with hydrophobic polymer matrices [28]. Furthermore, surface grafting of coupling agents can promote better adhesion between the fibers and the matrix material, resulting in improved mechanical performance.

4.2 Hybridization with Other Reinforcements

Hybridization with other reinforcing materials, such as carbon or glass fibers, can significantly enhance the mechanical properties of wood-fiber composites [29]. The combination of cellulose fibers with pre-impregnated carbon fabrics has shown great potential for achieving optimized lightweight construction, increased load-bearing capabilities, and greater design flexibility.

5 Preparation Techniques for Wood-Fiber Reinforced Composites

The choice of preparation method plays a crucial role in determining the final properties of the composite and its suitability for specific applications. Several methods are used for manufacturing wood fiber-reinforced composites, each offering unique advantages based on the desired product requirements.

5.1 Compression Moulding

Compression moulding stands out as a favoured technique for producing composites with improved mechanical properties, especially for applications in the automotive and aerospace industries [29]. This method involves precise temperature and pressure control to manufacture polymer matrix composites efficiently. Advantages include minimal fiber attrition, increased processing speed, and the ability to create complex shapes and designs [30].

Various factors influence the mechanical properties of natural fiber composites. Researchers, such as [31] have conducted experiments to explore the impact of these factors. For instance, Srinivasa et al. found that the flexural and impact strength of these composites increases with higher fiber loading percentages. Moreover, they observed a significant improvement in both flexural and impact strength when using treated areca fibers compared to untreated ones. Meanwhile, Saravanan and Dhurai investigated the impact strength of jute fiber-reinforced polypropylene composite under different conditions, such as temperature, pressure, and fiber treatment time, using the compression moulding process. In another study, Zampaloni et al. [32] produced kenaf-polypropylene natural fiber composites, and their findings revealed that using 30% and 40% fiber content by weight in the compression moulding process adequately reinforced the polypropylene powder, thereby enhancing its strength.

Therefore, compression moulding proves to be a promising technique to produce composites with superior mechanical properties. Researchers have shown that factors such as fiber loading percentage, fiber treatment, and moulding conditions play vital roles in determining the final properties of natural fiber composites. Hence, further exploration and optimization of these parameters, compression moulding can offer significant advancements in composite materials for various industrial applications.

5.2 Injection Moulding

This is an overview of the increasing importance of injection moulding in wood-fiber reinforced composites, emphasizing the significance of understanding materials and machine configuration to achieve successful outcomes. The method's rapid development is attributed to its widespread adoption across various industries, including automotive, hardware/appliances, medical, and packaging. The cyclic process of mould filling, cooling, and ejection enables efficient and high-speed production, making it a preferred choice for manufacturers.

A critical aspect in the injection moulding process is the intricate behaviour of materials during the fundamental stages. Understanding the physical phenomena involved and their correlation with the properties and performance of the final moulded part is essential for optimal results. Proper selection of machine configuration that suits the material type becomes imperative for successful outcomes. Furthermore, injection moulding exhibits remarkable versatility by accommodating a wide range of materials, including metals, glasses, elastomers, confections, and most commonly, thermoplastic and thermosetting polymers. This characteristic opens possibilities for designing and manufacturing diverse products, solidifying its position as a preferred manufacturing process across multiple industries [33].

To choose a stable fiber for composite processing, it is essential to assess the thermal stability of various fibers. Sun et al. (2010) [34] conducted a thermogravimetric analysis on Sisal and jute fibers. The findings revealed that Sisal fiber exhibited lower weight loss compared to jute fiber at temperatures below 250 °C, demonstrating that Sisal fiber had superior thermal stability within the composite processing temperature range (170–220 °C). While, Venkatesh et al. (2023) [35] achieved significant improvements in the thermal and mechanical properties of polypropylene hybrid composites. They utilized varying weight percentage ratios of glass/hemp fibers (0:35, 5:30, 10:25, and 15:20) combined with a 5 wt% compatibilizer using the injection moulding technique. The research highlighted that the combination of thermoplastic and natural fibers typically results in low mechanical strength and incompatibility, necessitating special treatment to effectively synthesize the natural fiber.

5.3 *Extrusion*

Extrusion is another commonly used technique for wood-fiber composites, offering continuous processing of profiles and products. By controlling the extrusion parameters, it is possible to achieve desired mechanical properties and fiber dispersion.

5.4 *Other Techniques*

Various other methods, such as Resin Transfer Moulding (RTM), Hot Pressing, Pultrusion, Hand Lay-up, and Spray-up, are available, each offering specific advantages for different applications.

6 Properties of Wood Fiber-Reinforced Composites

6.1 *Flammability and Thermal Properties of Wood Fiber-Reinforced Composites*

The most important characteristics of wood fiber-reinforced composites for flame retardance applications include flammability and thermal properties [36–38]. In the study by Gallo et al. [37], they mixed 30 wt.% of kenaf fiber with blends of poly(3-hydroxybutyrate-co-3-hydroxy-valerate) (PHBV) and poly(butylene adipate-co-terephthalate) (PBAT) in the presence of aluminum diethyl phosphinate as the reactive flame retardant agent. The composites were melted and mixed using a twin-screw extruder in a two-step process and in the first step, a layer of 10 wt.% of flame retardant was formed which was followed by the second layer of 30 wt.% of kenaf fiber. Cone calorimeter results showed that the fire growth index values of neat PHBV/PBAT blend, 10 wt.% flame retardant composites, and 30 wt.% kenaf fiber composites were 16.3, 8.5, and 7.7 kw/m^2, respectively. The fire growth index is the measure of the flammability property of the material and the results showed that kenaf fiber has more ability to hinder fire progression. The authors further suggested that the thermal decomposition of flame retardant aided in improving intermolecular interactions of kenaf fiber with the polymer matrix and the first layer of flame retardant trapped some carbon monoxide release which inhibited the fire growth while the second layer of kenaf fiber provided ventilation situation within the composites. Scanning electron microscopy (SEM) images confirmed that the kenaf fiber layer had porous structures with voids and cavities. Furthermore, thermogravimetric analysis (TGA) curves showed that the kenaf fiber stabilized the onset thermal decomposition temperature of the composites.

In a similar study, Subasinghe et al. [39] prepared polypropylene (PP) composites with 30 wt.% kenaf fiber, 20 wt.% ammonium phosphate (APP) as a flame retardant and 3 wt.% maleic anhydride grafted polypropylene (MAPP) as a compatibilizer. The composites were melted and mixed with wool fiber and UV colorant (UVC) as additive flame retardants in the twin-screw extruder at the temperature of 165–185 °C. Small-scale experiments using a cone calorimeter were used and results showed that the total heat release of the PP/KF composites decreased with the addition of the kenaf fiber into the polymer matrix due to the presence of lignin in the fiber and results are as shown in Fig. 8.5. The addition of the flame retardants further decreased the total heat release with wool fiber and UVC aiding in the hybridization process which improved the thermal stability of the PP/KF composites. In addition, TGA results supported the finding of cone calorimeter experiments as the onset thermal degradation temperature was delayed with the addition of flame retardants. The study suggested that the flame retardants created a porous layer upon thermal degradation that decreased the overall thermal degradation rate of PP/KF composites and released carbon dioxide which prevented oxygen access for the combustion reaction.

Zhang et al. [40] used bamboo fiber fabric material to examine the flame retardancy using ammonium salts of tetrakis(phosphonomethyl) arginine (ATTPMA) as a

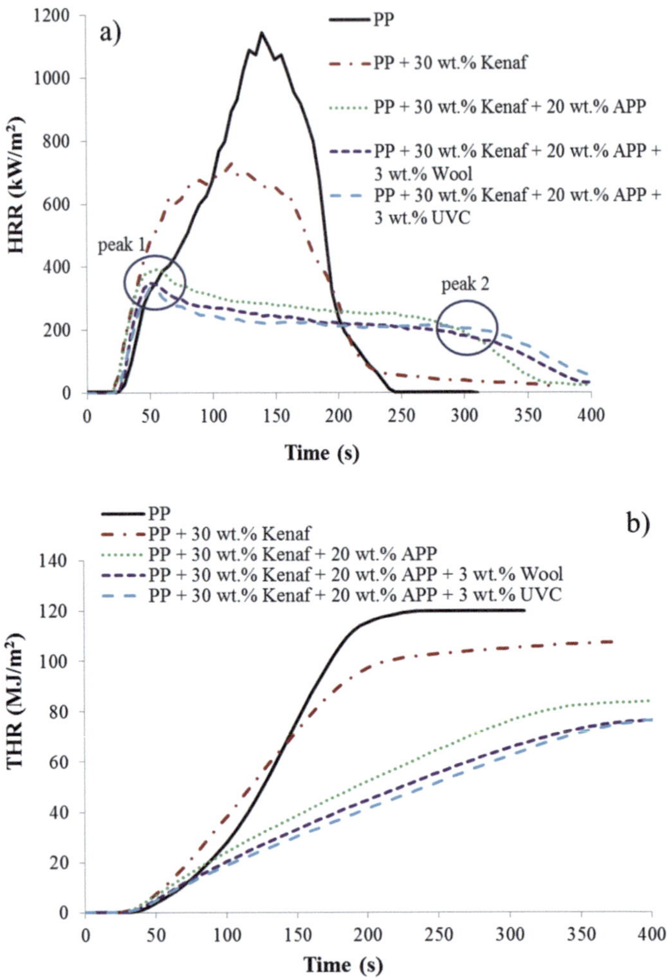

Fig. 8.5 Total heat release results for PP/KF composites with APP, wool fiber, and UVC flame retardants. Reprinted with permission from Subasinghe et al. [39]

reactive flame retardant. The cellulose of the bamboo fiber was grafted with different concentrations of 10, 20, and 30 wt.% ATTPMA as illustrated in Fig. 8.6. Based on the reaction mechanism, the bamboo fabric was immersed in an ATTPMA solution through physical adsorption, and cellulose functional groups in bamboo fiber were grafted through the phosphonate groups from the ATTPMA structure.

The classical vertical burning tests from the same study by Zhang and co-workers [40] showed improved flame retardancy properties of bamboo fiber fabric grafted with ATTPMA as illustrated in Fig. 8.7. As seen from this figure, the neat bamboo fiber fabric is completely burnt where bamboo fiber fabric grafted with ATTPMA flame retardant their degree of burning decreased from ~63 to ~48 mm with the

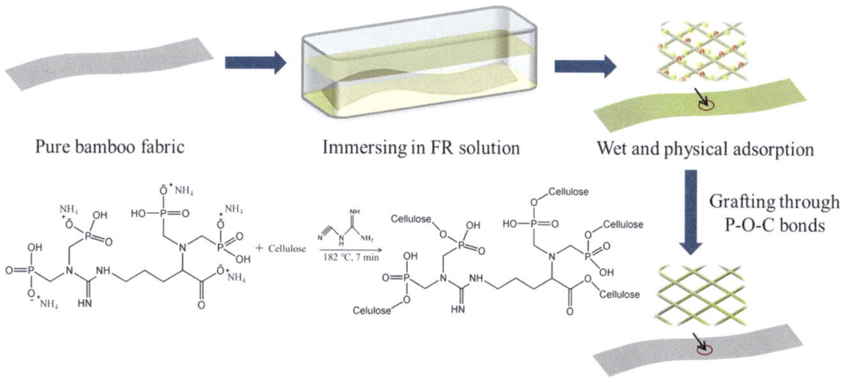

Fig. 8.6 Reaction mechanism for grafting of ATTPMA flame retardant to cellulose of the bamboo fiber fabric material. Reprinted with permission from Zhang et al. [40]

concentration of ATTPMA. These results were supported by limiting oxygen index values that increased from 20.3 to 50.2% with a higher concentration of ATTPMA solution. Limiting oxygen index indicates the ability of the material to prevent oxygen access into the material for the combustion reaction. In addition, cone calorimeter findings confirmed that 30 wt.% ATTPMA grafted bamboo fiber fabric has the lowest total heat release value as compared to other materials as there was a protective char layer that formed from the flame retardant. Moreover, TGA curves indicated that all three ATTPMA grafted bamboo fiber fabrics have two thermal degradation steps and this suggested that the flame retardant affected the decomposition process by decreasing the onset degradation temperature of the bamboo fiber fabric. The authors further assumed that the presence of char residue that increased with an increase in the amount of flame retardant has aided in the increase of the thermal stability of the overall bamboo fiber material. The char is assumed to absorb the heat generated during the combustion reaction.

With the aim to further understand the effect of flame retardant on the wood fiber-reinforced composites, Wang et al. [41] used 20 wt.% sisal fiber (SF) as a filler in PP composites while utilizing 22.0 wt.% ammonium phosphate with silane structure (Si-APP) as a flame retardant. Prior to the melt-mixing process, the sisal fiber surface was treated with 10 wt.% of sodium hydroxide solution to create functional groups for good adhesion interactions with the polymer matrix. PP/SF/Si-APP composites were prepared using a twin-screw extruder at a temperature of 175–180 °C and injection molded at the same temperature range. Cone calorimeter results indicated that the addition of sisal fiber increased the limiting oxygen index from ~18 to ~25% since the lignin and cellulose in sisal fiber produced char upon thermal degradation and it acted as the fire-resistance material. Moreover, the addition of the flame retardant has resulted in shorter time-to-ignition values for the PP/SF composites, and the total production of smoke gases has also increased. Their observations mean that the presence of flame retardant in the PP/SF created incomplete combustion which has hindered fire growth.

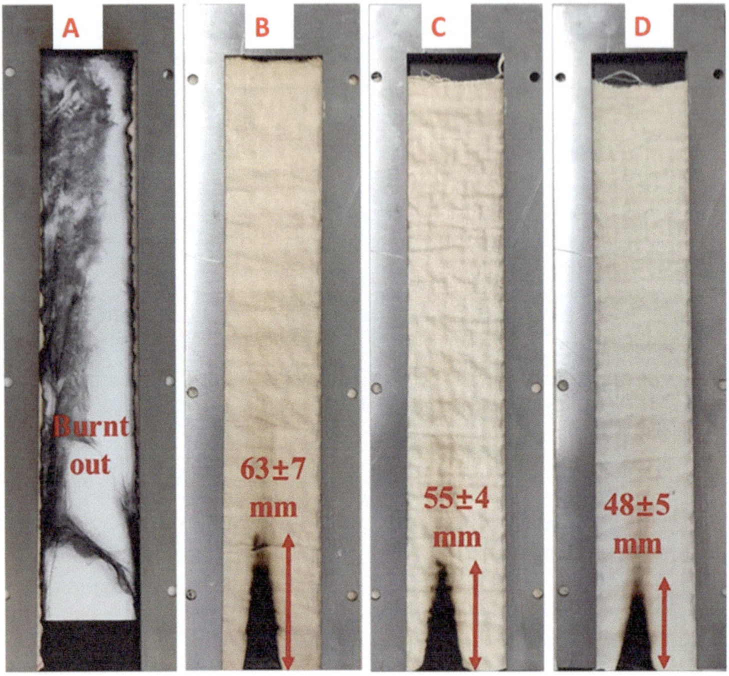

Fig. 8.7 Vertical burning tests for **a** neat bamboo fiber material, and bamboo fiber fabric grafted with ATTPMA flame retardant **b** 10 wt.%, **c** 20 wt.% and **d** 30 wt.%. Reprinted with permission from Zhang et al. [40]

Furthermore, the TGA results showed that the addition of sisal fiber and flame retardant decreased the onset thermal degradation temperature of the PP/SF composites to ~250 °C as compared to ~410 °C for the neat PP polymer. The char residue formed at this early degradation temperature is crucial in the condensed phase process of the flame retardant since it is used as a carbon-forming agent that further improves the thermal stability of the overall PP/SF composites. Based on the flammability and thermal stability studies on PP/SF composites, the authors have proposed flame retardance mechanisms of the Si-APP material as illustrated in Fig. 8.8. The reaction mechanism of Si-APP flame retardant was assumed to take place in both condensed and gaseous phases in the presence of Pentaerythritol (PER) as an additive. In the condensed phase, the flame retardant was thermally degraded to form a carbonaceous layer with sisal fiber and PP polymer matrix while in the gaseous phase, the flame retardant is thermally decomposed to release volatile organic compounds which inhibit the fire growth and enhance the limiting oxygen index property of the PP/SF composites.

The previous study by Wang and co-workers [41] concluded that different processing steps during melt-mixing could affect how the flame retardant interacts with the composite material. Another factor that affects the flame retardance

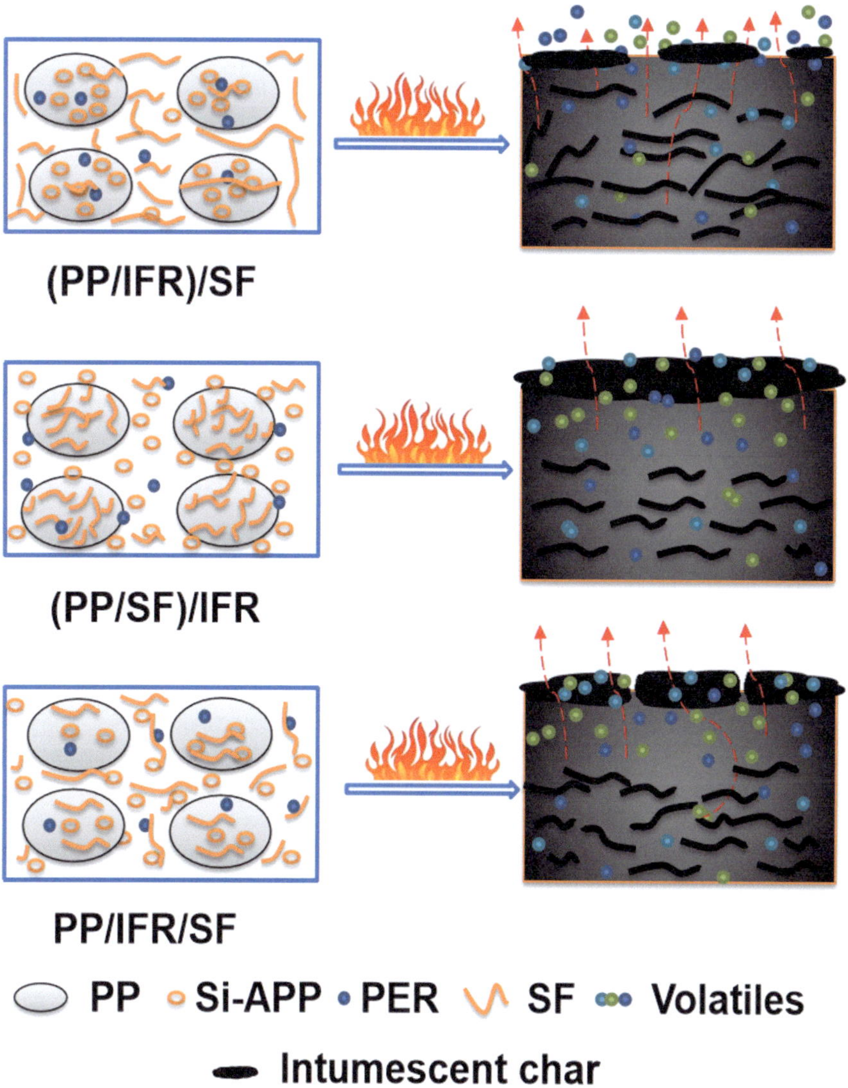

Fig. 8.8 Reaction mechanism of Si-APP flame retardant and PP/SF composites in the presence of PER additive. Reprinted with permission from Wang et al. [41]

behavior of the material is the type of flame retardant. Jeencham et al. [42] carried out a research study to evaluate the efficiency of three types of flame retardants and they prepared PP/SF composites with 30–40 wt.% of either ammonium phosphate (APP), magnesium hydroxide (MgOH) or zinc borate (ZB). The PP/SF composites with 30 wt.% fiber content were mixed in the presence of maleic anhydride PP as a compatibilizer. The internal mixer was operated at a temperature of 170 °C and a mixing time of 13 min, and samples were injection molded at the same temperature. Based on the horizontal burning rate experiments, sisal fiber was found to promote the flammability of the PP polymer matrix since the limiting oxygen index was decreased below 19% which is the detection limit of the instrument. However, the addition of all three flame retardants reduced the burning rate of PP/SF composites with APP showing the best results among the flame retardants.

The authors extended their study to determine the thermal stability of the PP/SF composites and they found that hemicellulose and cellulose in the sisal fiber thermally decomposed at much lower temperatures as compared to PP material to create char needed to increase the thermal stability of the PP/SF composites. The presence of lignin in the sisal fiber further increased the thermal decomposition temperature of PP material as presented in Table 8.1. Moreover, the TGA results showed that the addition of all three flame retardants increased the thermal stability of the PP/SF composites since when they are burnt, their char acts as a protective layer over the surface of the PP/SF composites. The char residue also acts as a great medium for the heat transfer process within the PP/SF composites.

In another study, Kandare et al. [43] prepared epoxy laminate composites with flax fiber content of 30% volume fraction by using the wet lay-up method in the presence of ammonium polyphosphate flame retardant. The results of the cone calorimeter experiments for the neat flax fiber/epoxy (FE) and flax fiber/epoxy with APP flame retardant (FE-APP) composites are presented in Fig. 8.9. The FE composite showed two maximum heat release rates of 590 and 616 kW/m^2. It was assumed that there were two layers formed during the heating process in the cone calorimeter. The first layer was assumed to be the temporary char layer from the thermal decomposition of bio-based polymer composites and the second layer was assigned to epoxy laminate material. On the other hand, FE-APP composites showed reduced heat release rates of 260 and 480 kW/m^2 for the first and second layers, respectively. The reduction in the rates was observed to be caused by a char-forming active ingredient in the flame retardant, and a carbonaceous layer was seen on the surface of the FE-APP

Table 8.1 Thermal properties of neat PP polymer, PP/SF composite with and without flame retardants

Sample name	T_5 (°C)	T_{50} (°C)	Char residue (%)
PP	408	458	0.00
PP/30SF	308	461	4.31
PP/30SF/40APP	322	474	12.25
PP/30SF/40MgOH	329	477	21.27
PP/30SF/40ZB	320	466	23.60
PP/30SF/30APP/10MgOH	322	472	18.97
PP/30SF/30APP/10ZB	316	472	19.54

Adopted and reprinted with permission from Jeencham et al. [42]
T_5 (°C) and T_{50} (°C) are the onset and maximum thermal degradation temperatures derived from TGA curves, respectively

composites as shown in Fig. 8.10. Furthermore, the authors found that the flame-out time for the FE-APP composites was longer than the FE composite, and those results confirmed the total heat release of the composites.

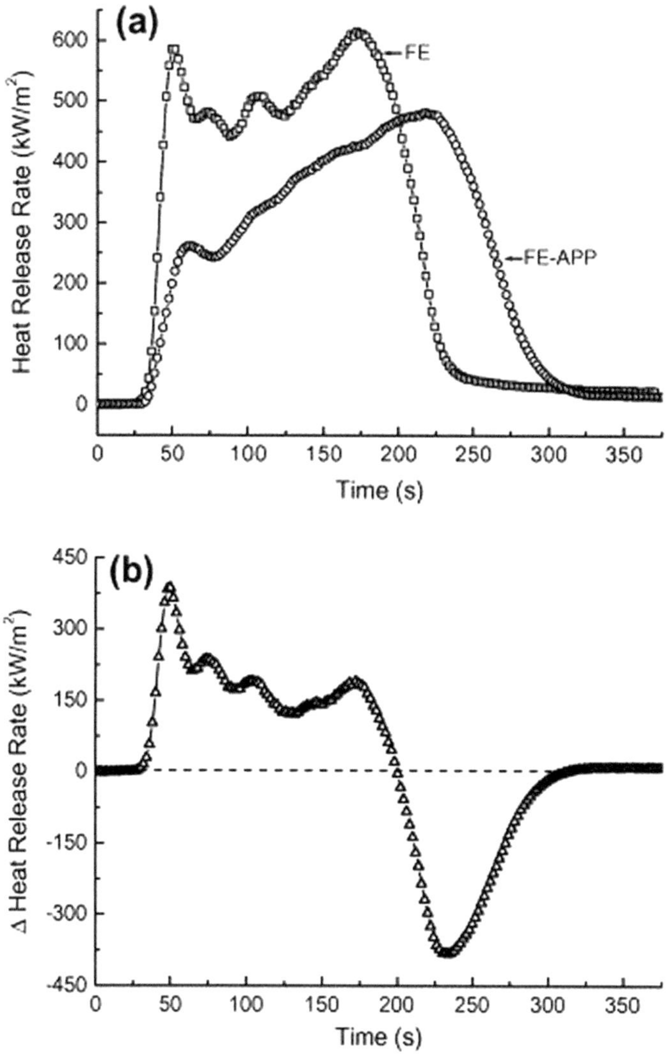

Fig. 8.9 Heat release rates for the neat flax fiber/epoxy (FE) and flax fiber/epoxy with APP flame retardant (FE-APP) composites. Reprinted with permission from Kandare et al. [43]

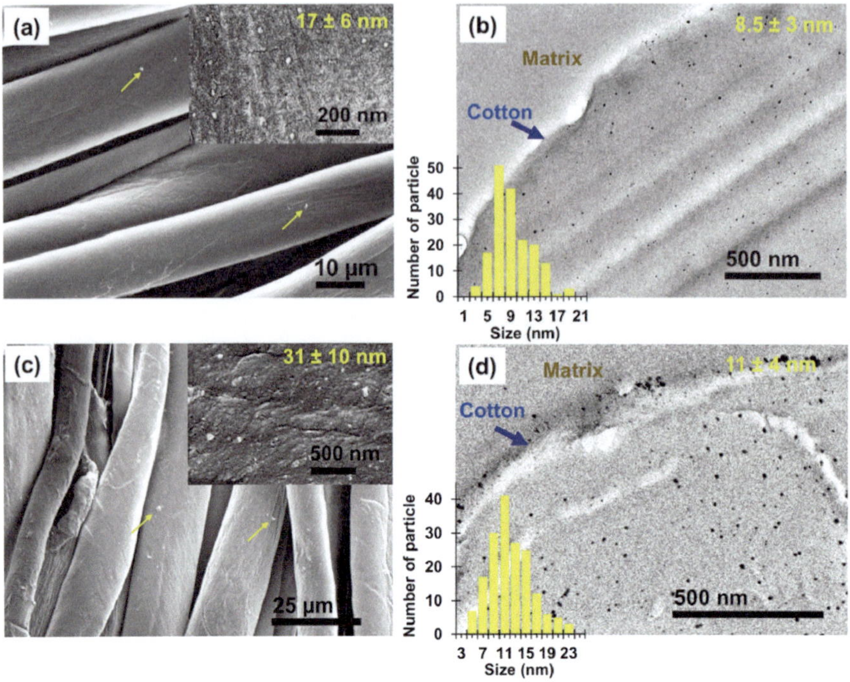

Fig. 8.10 Morphological properties of steam-cured cotton fiber fabric using TVPO as a flame retardant; SEM image (**a**) sample with 0.08% silver nanoparticles and its corresponding TEM image (**b**) and SEM image (**c**) sample with 0.75% silver nanoparticles and its corresponding TEM image (**d**). Reprinted with permission from Nazir et al. [44]

6.2 Morphological Properties of Wood-Fiber Reinforced Composites

The morphological properties of the wood fiber-reinforced composites are important for flame retardance application since they can determine the heat transfer process within the composite. In the work of Nazir et al. [44], the authors used cotton fiber fabric and grafted trivinyl phosphine oxide (TVPO) to the hydroxyl functional groups of the cellulose structure. They explored an innovative approach utilizing phosphine oxide physical networks created via a grafting reaction to impart flame retardancy properties of cotton fiber materials. The researchers demonstrated the effectiveness of the facile strategy, with a specific focus on SEM and transmission electron microscopy (TEM) analyses. SEM images showed that a phosphine oxide physical network has successfully formed on the surface of the cellulose as shown in Fig. 8.10. The network was distributed uniformly and densely throughout the cotton fabric structure, demonstrating effective covering of the cellulose fibers. The network structure functioned as a barrier by blocking oxygen and heat from reaching the cellulose substrate, improving the material's flame-retardant qualities. The SEM examination further illustrated the physical network's endurance by showing that it maintains its integrity even after being exposed to varied environmental conditions.

Moreover, TEM analysis provided valuable insights into the microstructure of the phosphine oxide physical networks. The images revealed a three-dimensional network morphology, with interconnected phosphine oxide clusters as illustrated in Fig. 8.10. The unique structure contributed to the enhanced flame retardant of the treated cellulose. The TEM analysis further confirmed the homogeneity and uniform distribution of the phosphine oxide networks throughout the cellulose matrix, highlighting their effective adhesion to the cellulose fibers. An important benefit of the phosphine oxide physical networks was their longevity which guarantees the treated cellulose's continued antibacterial and flame-retardant capabilities. For plant-based fiber materials, the study underlined the significance of endurance in achieving durable flame retardant and antibacterial treatments. Their approach was a promising one for dealing with the issues related to flame retardancy.

In an attempt to understand the effect of flame retardant on morphological properties of natural fiber/polymer composites, Subasinghe et al. [39] fabricated PP/kenaf fiber (KF) composites with 30 wt.% fiber content. In their study, they used 20 wt.% ammonium phosphate (APP) as a flame retardant and 3 wt.% maleic anhydride grafted polypropylene (MAPP) as a compatibilizer. The composites were further melted and mixed in the presence of wool fiber (WF) and UV colorant (UVC) as additive flame retardants in the twin-screw extruder.

The interfacial interactions among the components of PP/KF composites were studied using an SEM instrument and the results of the composites with different flame retardants are presented in Fig. 8.11. There were a lot of voids and cavities within the PP/KF composites, as shown in Fig. 8.11a, that resulted from the interaction of kenaf fiber and polymer matrix, flame retardant and the author also assumed that APP initiated the formation of new hydrogen bonds which contributed to more voids within the composite matrix. From Fig. 8.11b, there are few to no voids in the composite matrix due to wool fiber having long fibers which aided in cross-linking reactions among polymer components. On the other hand, the addition of UVC resulted in larger voids across the sample matrix and it was found that there was an improvement in the heat transfer process.

The type of wood fiber and polymer matrix plays an important role in flame retardancy of polymer composites. Kandola et al. [38] performed a study to evaluate the interaction of fiber type with different polymer materials. They prepared polypropylene (PP) and polylactic acid (PLA) composites with 40 wt.% fiber content of jute and sisal using a hot-press technique at a temperature of 190 °C for 2 min. The morphological analyses of the polymer/fiber composites were done with fractured samples after tensile testing using SEM and the images are presented in Fig. 8.12. There were fiber pull-out characteristics for the PP composites as shown in Fig. 8.12a, b which created holes and cavities. However, PLA composites had fewer fiber pull-out features, and the surface of the fracture was observed to be smooth with no cavities or voids. PLA composites were found to have better adhesion properties as compared to PP composites since PLA is an ester polymer with polar groups on its structure and the groups are compatible with the hydroxyl groups in the fiber cellulose structures.

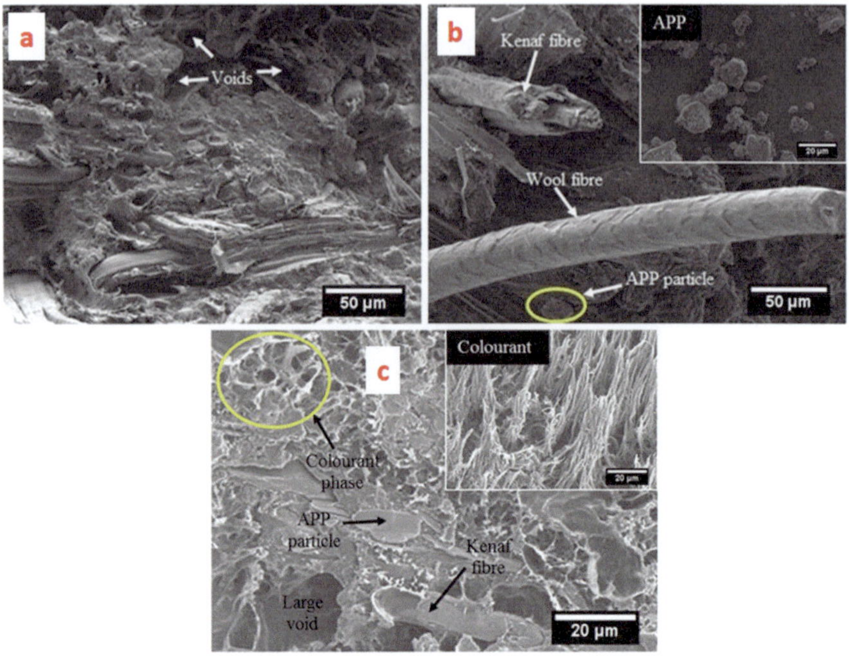

Fig. 8.11 SEM images of the PP/KF composites with different types of flame retardants, **a** PP/KK/APP, **b** PP/KK/APP/WF, and **c** PP/KF/APP/UVC. Reprinted with permission from Subasinghe et al. [39]

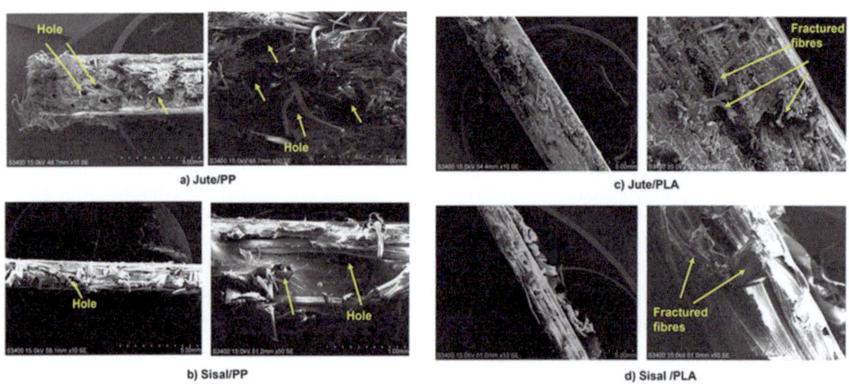

Fig. 8.12 SEM images of the fractured samples after tensile testing for the polymer/fiber composites, **a** Jute/PP, **b** Sisal/PP, **c** Jute/PLA, and **d** Sisal/PLA

The morphological characterization of polymer/fiber composites using SEM and TEM analyses was used to evaluate the effect of fiber content, flame retardant type, and polymer matrix on the fire retardancy application of the composites. SEM images provided information about the arrangement of polymer chains and the presence of nanoparticles within the polymer composite network. The SEM and TEM analysis confirmed the uniform distribution of nanoparticles, enhancing the material's fire-retardant properties. Furthermore, the study investigated the composites formed by incorporating fire retardant additives into the polymer composites. SEM and TEM analysis of the composites demonstrate the dispersion and integration of the additives within the material structure, leading to improved fire retardancy without compromising the mechanical properties.

7 Conclusions

The current study primarily investigated the flammability properties of biofiber, which are derived from plant-based natural sources. Several types of fibers, including jute, flax, hemp, kenaf, and sisal that have been used to reinforce the polymer composites. These fibers possess inherent advantages such as renewability, low density, and good mechanical properties, making them suitable for different applications.

Several treatment techniques were looked at to improve the fire resistance of these natural fibers. Alkalization, acetylation, grafting, and silane treatment are a few examples of the chemical processes used in these procedures. Coatings and surface modification of the fibers are examples of physical processes. It was emphasized that these treatments were beneficial in lowering flammability and increasing the thermal stability of wood fiber-reinforced polymer composites.

The study also provided SEM and TEM analyses of the wood fiber-reinforced polymer composites to allow for a detailed examination of the microstructure, fiber-matrix interface, and distribution of filler materials. These analyses helped in understanding the structural characteristics and provided insights into the fire behavior and flammability mechanisms of the materials.

In the development of wood-fiber reinforced polymer composites for flame retardancy application, fibers are typically mixed with a polymer matrix to achieve desired properties in the presence of flame retardant additives. The specific type of polymer used depends on the application, desired characteristics and commonly employed polymers include PP, PE, and PLA. The selection of the polymer matrix plays a crucial role in determining the flammability and mechanical properties of the resulting polymer composites.

Acknowledgements The authors would like to thank financial support from the Vaal University of Technology, Nelson Mandela University and the National Research Foundation (NRF), South Africa, NRF grant number: TTK23042597055.

References

1. Papadopoulos, A.M.: State of the art in thermal insulation materials and aims for future developments. Energy Build. **37**, 77–86 (2005). https://doi.org/10.1016/j.enbuild.2004.05.006
2. Alaee, M.: An overview of commercially used brominated flame retardants, their applications, their use patterns in different countries/regions and possible modes of release. Environ. Int. **29**, 683–689 (2003). https://doi.org/10.1016/S0160-4120(03)00121-1
3. Yang, J., Zhao, Y., Li, M., Du, M., Li, X., Li, Y.: A review of a class of emerging contaminants: the classification, distribution, intensity of consumption, synthesis routes, environmental effects and expectation of pollution abatement to organophosphate flame retardants (OPFRs). Int. J. Mol. Sci. **20**, 2874 (2019). https://doi.org/10.3390/ijms20122874
4. van der Veen, I., de Boer, J.: Phosphorus flame retardants: properties, production, environmental occurrence, toxicity and analysis. Chemosphere **88**, 1119–1153 (2012). https://doi.org/10.1016/j.chemosphere.2012.03.067
5. Laoutid, F., Bonnaud, L., Alexandre, M., Lopez-Cuesta, J.-M., Dubois, P.: New prospects in flame retardant polymer materials: from fundamentals to nanocomposites. Mater. Sci. Eng. R. Rep. **63**, 100–125 (2009). https://doi.org/10.1016/j.mser.2008.09.002
6. Bourbigot, S., Duquesne, S.: Fire retardant polymers: recent developments and opportunities. J. Mater. Chem. **17**, 2283 (2007). https://doi.org/10.1039/b702511d
7. Sun, Z., Hou, Y., Hu, Y., Hu, W.: Effect of additive phosphorus-nitrogen containing flame retardant on char formation and flame retardancy of epoxy resin. Mater. Chem. Phys. **214**, 154–164 (2018). https://doi.org/10.1016/j.matchemphys.2018.04.065
8. Zabihi, O., Ahmadi, M., Khayyam, H., Naebe, M.: Fish DNA-modified clays: towards highly flame retardant polymer nanocomposite with improved interfacial and mechanical performance. Sci. Rep. **6**, 38194 (2016). https://doi.org/10.1038/srep38194
9. Lazar, S.T., Kolibaba, T.J., Grunlan, J.C.: Flame-retardant surface treatments. Nat. Rev. Mater. **5**, 259–275 (2020). https://doi.org/10.1038/s41578-019-0164-6
10. Costes, L., Laoutid, F., Aguedo, M., Richel, A., Brohez, S., Delvosalle, C., et al.: Phosphorus and nitrogen derivatization as efficient route for improvement of lignin flame retardant action in PLA. Eur. Polym. J. **84**, 652–667 (2016). https://doi.org/10.1016/j.eurpolymj.2016.10.003
11. Gholamian, F., Nabiyouni, G., Ghanbari, D., Jalajerdi, R., Aminifazl, A.: Synergistic effect between Sb_2O_3 nanostructure and brominated compound on the flame retardant properties of the polymeric matrixes. High Temp. Mater. Processes (London) **32**, 125–132 (2013). https://doi.org/10.1515/htmp-2012-0039
12. Niu, L., Xu, J.-L., Yang, W.-L., Kang, C.-H., Ma, J.-Q., Su, J.-Q.: Synergistic effect between nano-Sb_2O_3 and brominated epoxy resin on the flame retardancy of poly(butylene terephthalate). Sci. Adv. Mater. **11**, 466–475 (2019). https://doi.org/10.1166/sam.2019.3434
13. Dewaghe, C., Lew, C.Y., Claes, M., Belgium, S.A., Dubois, P.: Fire-retardant applications of polymer–carbon nanotubes composites: improved barrier effect and synergism. In: Polymer–Carbon Nanotube Composites, pp. 718–745. Elsevier (2011). https://doi.org/10.1533/9780857091390.3.718
14. Cortés, D., Gil, D., Azorín, J., Vandecasteele, F., Verstockt, S.: A review of modelling and simulation methods for flashover prediction in confined space fires. Appl. Sci. **10**, 5609 (2020). https://doi.org/10.3390/app10165609
15. Cicione, A., Walls, R.S., Kahanji, C.: Experimental study of fire spread between multiple full scale informal settlement dwellings. Fire Saf. J. **105**, 19–27 (2019). https://doi.org/10.1016/j.firesaf.2019.02.001
16. Till, R.C., Coon, J.W.: Unwanted Fire and Fire Growth. Fire Protection, pp. 1–10. Springer International Publishing, Cham (2019). https://doi.org/10.1007/978-3-319-90844-1_1
17. Zong, R., Kang, R., Zhao, W., Tao, C.: Experimental Study and Model Analysis of Flashover in Confined Compartments. Fire Science and Technology 2015, pp. 649–657. Springer Singapore, Singapore (2017). https://doi.org/10.1007/978-981-10-0376-9_66

18. Morgan, A.B., Gilman, J.W.: An overview of flame retardancy of polymeric materials: application, technology, and future directions. Fire Mater. **37**, 259–279 (2013). https://doi.org/10.1002/fam.2128
19. Mihajlović, I.: Recent development of phosphorus flame retardants in thermoplastic blends and nanocomposites. 79–114 (2015). https://doi.org/10.1007/978-3-319-03467-6_4
20. Zilberman, J., Yoffe, D., Piotrowski, A., Singh, M.P., Suryadevara, K., Levchik, S.: Comparative study of reactive flame retardants based on 9,10-dihydro-9-oxa-10-phosphaphenanthrene 10-oxide. J. Fire Sci. **35**, 235–256 (2017). https://doi.org/10.1177/0734904117702157
21. Guo, Z., Wang, Z., Fang, Z.: Fabrication of 9,10-dihydro-9-oxa-10-phosphaphenanthrene-10-oxide-decorated fullerene to improve the anti-oxidative and flame-retardant properties of polypropylene. Compos. B Eng. **183**, 107672 (2020). https://doi.org/10.1016/j.compositesb.2019.107672
22. Wang, J.: Kinetic evaluation of 9,10-dihydro-9-oxa-10-phosphaphenanthrene-10-oxide as a flame retardant for epoxy resins. J. Macromol. Sci. Part B **59**, 542–550 (2020). https://doi.org/10.1080/00222348.2020.1753339
23. Yan, W., Yu, J., Zhang, M., Long, L., Wang, T., Qin, S., et al.: Novel flame retardancy effect of phenethyl-bridged DOPO derivative on epoxy resin. High Perform. Polym. **30**, 667–676 (2018). https://doi.org/10.1177/0954008317716525
24. Liu, M., Yin, H., Chen, X., Yang, J., Liang, Y., Zhang, J., et al.: Preliminary ecotoxicity hazard evaluation of DOPO-HQ as a potential alternative to halogenated flame retardants. Chemosphere **193**, 126–133 (2018). https://doi.org/10.1016/j.chemosphere.2017.10.142
25. Krivoshiev, B.V., Beemster, G.T.S., Sprangers, K., Cuypers, B., Laukens, K., Blust, R., et al.: Transcriptome profiling of HepG2 cells exposed to the flame retardant 9,10-dihydro-9-oxa-10-phosphaphenanthrene 10-oxide (DOPO). Toxicol Res (Camb) **7**, 492–502 (2018). https://doi.org/10.1039/C8TX00006A
26. Dai, D., Fan, M.: Wood fibres as reinforcements in natural fibre composites: structure, properties, processing and applications. Nat. Fibre Compos. 3–65 (2014). https://doi.org/10.1533/9780857099228.1.3
27. Zhang, L., Lv, S., Sun, C., Wan, L., Tan, H., Zhang, Y.: Effect of MAH-g-PLA on the properties of wood fiber/polylactic acid composites. Polymers (Basel) **9**, 591 (2017). https://doi.org/10.3390/polym9110591
28. Beg, M.D.H., Pickering, K.L.: Reprocessing of wood fibre reinforced polypropylene composites. Part I: effects on physical and mechanical properties. Compos. Part A Appl. Sci. Manuf. **39**, 1091–1100 (2008). https://doi.org/10.1016/j.compositesa.2008.04.013
29. Wulfsberg, J., Herrmann, A., Ziegmann, G., Lonsdorfer, G., Stöß, N., Fette, M.: Combination of carbon fibre sheet moulding compound and prepreg compression moulding in aerospace industry. Procedia Eng. **81**, 1601–1607 (2014). https://doi.org/10.1016/j.proeng.2014.10.197
30. Faruk, O., Bledzki, A.K., Fink, H.-P., Sain, M.: Biocomposites reinforced with natural fibers: 2000–2010. Prog. Polym. Sci. **37**, 1552–1596 (2012). https://doi.org/10.1016/j.progpolymsci.2012.04.003
31. Ku, H., Wang, H., Pattarachaiyakoop, N., Trada, M.: A review on the tensile properties of natural fiber reinforced polymer composites. Compos. B Eng. **42**, 856–873 (2011). https://doi.org/10.1016/j.compositesb.2011.01.010
32. Zampaloni, M., Pourboghrat, F., Yankovich, S.A., Rodgers, B.N., Moore, J., Drzal, L.T., et al.: Kenaf natural fiber reinforced polypropylene composites: a discussion on manufacturing problems and solutions. Compos. Part A Appl. Sci. Manuf. **38**, 1569–1580 (2007). https://doi.org/10.1016/j.compositesa.2007.01.001
33. Rabbi, M.S., Islam, T., Islam, G.M.S.: Injection-molded natural fiber-reinforced polymer composites–a review. Int. J. Mech. Mater. Eng. **16**, 15 (2021). https://doi.org/10.1186/s40712-021-00139-1
34. Sun, Z.-Y., Han, H.-S., Dai, G.-C.: Mechanical properties of injection-molded natural fiber-reinforced polypropylene composites: formulation and compounding processes. J. Reinf. Plast. Compos. **29**, 637–650 (2010). https://doi.org/10.1177/0731684408100264

35. Venkatesh, R., Kantharaj, I., Sasikumar, R., Kannan, C.R., Yadav, A., Karthigairajan, M., et al.: Thermal adsorption and mechanical behaviour of polypropylene hybrid composite synthesized by glass/hemp fibre via an injection moulding process. Adsorpt. Sci. Technol. **2023**, 1–7 (2023). https://doi.org/10.1155/2023/7450085
36. Yao, K., Song, C., Fang, H., Wang, F., Chen, L., Jiang, S., et al.: Freezing-extraction/vacuum-drying method for robust and fatigue-resistant polyimide fibrous aerogels and their composites with enhanced fire retardancy. Engineering **21**, 152–161 (2023). https://doi.org/10.1016/j.eng.2021.08.024
37. Gallo, E., Schartel, B., Acierno, D., Cimino, F., Russo, P.: Tailoring the flame retardant and mechanical performances of natural fiber-reinforced biopolymer by multi-component laminate. Compos. B Eng. **44**, 112–119 (2013). https://doi.org/10.1016/j.compositesb.2012.07.005
38. Kandola, B.K., Mistik, S.I., Pornwannachai, W., Anand, S.C.: Natural fibre-reinforced thermoplastic composites from woven-nonwoven textile preforms: mechanical and fire performance study. Compos. B Eng. **153**, 456–464 (2018). https://doi.org/10.1016/j.compositesb.2018.09.013
39. Subasinghe, A., Somashekar, A.A., Bhattacharyya, D.: Effects of wool fibre and other additives on the flammability and mechanical performance of polypropylene/kenaf composites. Compos. B Eng. **136**, 168–176 (2018). https://doi.org/10.1016/j.compositesb.2017.10.034
40. Zhang, Y., Liu, Y., Wen, H., Wang, Q.: Durable flame retardant and mechanism of bamboo fabric through grafting based on arginine. Cellulose **28**, 8187–8204 (2021). https://doi.org/10.1007/s10570-021-04056-6
41. Wang, Z., Feng, W., Ban, J., Yang, Z., Fang, X., Ding, T., et al.: Sisal-fiber-reinforced polypropylene flame-retardant composites: preparation and properties. Polymers (Basel) **15**, 893 (2023). https://doi.org/10.3390/polym15040893
42. Jeencham, R., Suppakarn, N., Jarukumjorn, K.: Effect of flame retardants on flame retardant, mechanical, and thermal properties of sisal fiber/polypropylene composites. Compos. B Eng. **56**, 249–253 (2014). https://doi.org/10.1016/j.compositesb.2013.08.012
43. Kandare, E., Luangtriratana, P., Kandola, B.K.: Fire reaction properties of flax/epoxy laminates and their balsa-core sandwich composites with or without fire protection. Compos. B Eng. **56**, 602–610 (2014). https://doi.org/10.1016/j.compositesb.2013.08.090
44. Nazir, R., Parida, D., Borgstädt, J., Lehner, S., Jovic, M., Rentsch, D., et al.: In-situ phosphine oxide physical networks: a facile strategy to achieve durable flame retardant and antimicrobial treatments of cellulose. Chem. Eng. J. **417**, 128028 (2021). https://doi.org/10.1016/j.cej.2020.128028

Fabrication and Characterization of Flame Retardant Nanocellulose-Based Materials

M. F. Bambo, K. P. Matabola, B. May, N. Chauke, and T. Munonde

Abstract High thermal conductivity and flame resistance in nanocellulose-based composites provide a double benefit of efficient heat dissipation and decreased fire risks in a variety of materials. Future generations could benefit greatly from the development of adaptable, ecologically friendly fire retardant technologies because to this dual capability. Recently, nanocellulose has attracted a lot of interest and use, largely because of its high aspect ratio, large specific surface area, adaptable modifiability, strong mechanical strength, renewable nature, and biodegradability. These qualities make nanocellulose an excellent starting point for creating fire-resistant materials. This chapter explores the characteristics of materials made of nanocellulose-based materials as well as their prospective uses in a variety of fields. It includes a discussion of the difficulties involved in developing flame retardant properties in nanocellulose-based and provides an overview of research projects aimed at advancing flame retardant nanocellulose-based materials solutions. The techniques

M. F. Bambo (✉)
DSI/Mintek Nanotechnology Innovation Center, Advanced Materials Division, Mintek, Randburg 2194, Johannesburg, South Africa
e-mail: mokaeb@mintek.co.za

K. P. Matabola
Department of Water and Sanitation, University of Limpopo, Private Bag X1106, Sovenga 0727, Polokwane, South Africa

B. May
Digby Wells Environmental, Turnberry Office Park, Grosvenor Rd, Bryanston 2191, Johannesburg, South Africa

N. Chauke
Mintek, Advanced Materials Division, Catalysis Group, Randburg 2194, Johannesburg, South Africa

T. Munonde
Institute for Nanotechnology and Water Sustainability (iNanoWS), College of Science, Engineering, and Technology, University of South Africa, Florida Campus, 28 Pioneer Avenue, Roodepoort 1709, Johannesburg, South Africa

M. F. Bambo
Department of Natural Sciences, Vaal University of Technology, Vanderbijlpark, South Africa

T. C. Mokhena et al. (eds.), *Biomaterials as Green Flame Retardants*, Engineering Materials, https://doi.org/10.1007/978-981-97-6871-4_9

used to create nanocellulose-based materials are highlighted, covering their preparation processes, distinctive properties, and the variables that affect them. Additionally, this work illuminates the variety of uses for flame retardant nanocellulose-based in a wide range of industrial applications.

Keywords Flame retardancy · Biomaterials · Nanocellulose · Filler · Nanomaterials · Composites

1 Introduction

Flame retardant materials have captured the attention of the scientific community for their potential utility across diverse industries like construction, automotive, aerospace, and textiles [1–3]. To qualify as effective flame retardants, certain characteristics must be exhibited, these include the capacity for endothermic decomposition to absorb heat and release inert substances, the ability to scavenge free radicals and suppress chemical combustion, the promotion of heightened heat or mass flux to shield materials, reduction of mass transfer or heat flux by forming a protective layer, and alteration of the chemical degradation pathway for polymers [4–6]. The enhancement of flame retardant composites can be achieved through protective coatings or the inclusion of reinforcing agents. Progress has been notable in the development of novel, environmentally friendly, cost-effective, and sustainable materials with remarkable attributes and functionalities. This development is largely spurred by concerns over greenhouse gas emissions, dwindling crude oil reserves, and the escalating demand for advanced materials in diverse high-end applications [7, 8].

Recent times have witnessed the emergence of nanocellulose-based materials as promising alternatives to conventional flame retardants, attributed to their distinctive qualities including substantial mechanical strength, low thermal conductivity, effective oxygen barrier properties, and exceptional biodegradability [9, 10]. Furthermore, these materials offer potential solutions to petroleum-derived polymers due to their abundant availability, renewability, biodegradability, biocompatibility, versatile modification capacities, adaptability, and diverse morphologies. This contributes to mitigating the depletion of non-renewable natural resources [11, 12]. However, streamlining the production of nanocellulose-based flame retardants for industrial-scale manufacturing remains a challenge. Critical steps involve fabricating and characterizing these innovative flame-retardant materials, thus gaining insights into their underlying mechanisms and optimizing their attributes for practical applications.

This chapter delves into the attributes of nanocellulose and its possible applications across various fields. It outlines the hurdles associated with the flame retardant capabilities of such materials, the research endeavours focused on the advancement of flame retardant nanocellulose-based substances, the assortment of flame retardant additives suitable for nanocellulose, and the diverse fabrication techniques employed for developing flame retardant nanocellulose-based materials. Moreover, the characterization techniques applicable to flame retardant nanocellulose-based materials are expounded upon, alongside the manifold applications these materials find use in.

2 Background

The prevention of materials from catching fire and reducing the rate of flame spreading is an essential requirement for many commercial products. Thus, flame retardants are incorporated into materials to delay or stop combustion processes [13, 14]. Developing flame retardant materials presents several challenges, which researchers strive to overcome to create effective and sustainable solutions. Some of the common challenges include: (i) balancing/maintaining the flame retardancy and material properties as a result of incorporating flame retardant additives into materials, which could negatively impact some material properties such as mechanical strength, flexibility, and transparency. (ii) Environmental and health concerns have been raised with many traditional flame retardant additives, such as halogenated compounds. (iii) Ensuring proper compatibility and uniform dispersion of flame-retardant additives within the material matrix can be challenging. Poor dispersion can lead to the concentration of additives in localized regions, resulting in uneven flame retardancy performance and potential material degradation. (iv) Flame retardant materials should maintain their fire resistance properties over time, even under various environmental conditions. Ensuring the durability and long-term stability of flame retardant additives within the material matrix is crucial to provide reliable and consistent protection against fire hazards. (v) Developing flame retardant materials that are cost-effective and economically viable for large-scale production is a challenge. Some flame retardant additives can be expensive, which may limit their widespread adoption, particularly in price-sensitive applications. (vi) Meeting the regulatory requirements and standards for flame retardancy in different industries and regions can be complex. It requires continuous testing, evaluation, and adherence to evolving fire safety regulations to ensure that the developed materials meet the necessary standards. (vii) Translating laboratory-scale flame retardant materials to large-scale manufacturing can present technical and logistical challenges. Scaling up production processes while maintaining consistent performance, cost-effectiveness, and quality control is a significant consideration [15, 16].

Addressing these challenges requires a multidisciplinary approach involving materials science, chemistry, engineering, and environmental considerations. Ongoing research focuses on developing innovative flame retardant materials, exploring alternative flame retardant technologies that display cost-effectiveness and employ green approaches.

Cellulose is an abundant renewable polymer that can be extracted from various natural sources such as plants, wood, algae, and bacteria. Cellulose is a high molecular weight polymer comprising numerous ß-D-glucopyranose units linked by ß-1.4

linkages [17]. These features have been extended by fabricating nano-sized cellulose (nanocellulose), which exhibits unique physiochemical, optical, and magnetic features compared to their bulk counterparts [18, 19]. Evidently, in the last two decades, nanocellulose materials have been prominent in several applications by substituting synthetic polymers such as poly(vinyl) alcohol (PVA), and polyurethane (PU) in applied materials owing to their excellent properties such as high porosity and surface area, low immunogenicity, stiffness, low toxicity, and biodegradability [20–23].

Nanocellulose are nanometre cellulose-based objects that are categorized into two major classes, namely: (i) nanostructure materials, (i.e. cellulose microcrystals and microfibrils) and (ii) nanofiber (i.e. cellulose nanofibers, nanocrystals, and bacterial cellulose) [24–26]. These different nanocellulose classes present different features that resulted in various applications including polymer composite reinforcing, energy storage, biomedicine, Pickering emulsifiers, wastewater treatment, wood adhesives, and flame retardants [18, 20, 25, 27, 28]. Cellulose microcrystals are crystalline powders obtained via chemical, enzymatic, and acid hydrolysis treatment of non-woody lignocellulosic biomass. The most commonly applied treatment is acid hydrolysis due to its low cost and rapid reaction times. During the treatment, the extent of polymerization is reduced and more crystalline particles are obtained [29–31]. Their particle size varies in micrometers and displays properties such as water insolubility, high crystallinity, mechanical strength, and good thermal stability as a result they have been widely applied in building materials and as reinforcing agents in nanocomposites [32, 33]. Cellulose microfibrils (CMFs) are packs of a large number of elementary fibril units isolated from lignocellulosic biomass via mechanical defibrilization and an alkaline steam explosion followed by oxidation, ultra-sonication, and hydrolysis [30, 34]. CMFs are flexible and transparent materials that exhibit a large surface area with high numbers of hydroxyl groups and are thermally stable. They have been applied in the medical, automotive, puckering emulsions, and electronics fields [34, 35]. Cellulose nanofibrils (CNFs) are also referred to as nanofibrous cellulose or nanofibrillated cellulose. They are flexible entangled network structures that are 20–100 nm in width and >10 000 nm in length [36, 37]. CNFs can be produced via chemical, mechanical, enzymatic treatments and a combination of the treatment of lignocellulosic biomass [38, 39]. Their unique thermal, mechanical, and optical properties have landed them in energy storage materials, smart electronics, biomedical materials, and food ingredients [39, 40]. Cellulose nanocrystals are also known as nanocrystalline cellulose, rodlike cellulose crystals, and cellulose whiskers/nanowhiskers. Cellulose nanocrystals are highly crystalline (54–88% crystallinity index) cylindrical rod-like nanoparticles about 4–70 nm in width, 100–6000 nm in length, and less flexible than CNFs. They are commonly produced via acid hydrolysis using different acids (e.g. sulphuric acid, hydrochloric acid etc.) and sometimes via a combination of enzymatic and physiochemical treatments [41–45]. Cellulose nanocrystals display high surface area, and acceptable thermal stability and exhibit easily modifiable functionality, making them suitable candidates for applications, such as biosensing, papermaking, biocomposites, Pickering emulsifiers, and fire retardancy [18, 24]. Bacterial cellulose is also known as bacterial nanocellulose, bio-cellulose, and microbial cellulose. They are commonly applied in biomedical uses [46, 47]. They differ from CNFs and nanocrystals because they are produced from bacterial species [48, 49]. Bacterial cellulose usually consists of ultrafine, pristine, ribbon-like nanofibers with widths of 20–100 nm and micrometers in length. They display better crystallinity and hydrophilicity than CNFs and CNCs [50]. Their fabrication preludes the harsh conditions employed during CNFs and CNCs but it is considered expensive due to the synthesis media used [48].

3 Flame Retardant Additives for Nanocellulose

The distinctiveness of nanocellulose-based materials arises from their amalgamation of nanocellulose with other beneficial compounds (such as macromolecular polymers and nanomaterials with dimensions of 0D, 1D, and 2D). This fusion leads to the enhancement of inherent physical characteristics (such as optical, mechanical, and thermal properties) and the introduction of diverse functionalities, encompassing electrical, plasmonic, luminescent, and magnetic attributes [51, 52].

Generally, nanocellulose displays low fire resistance owing to its chemical composition (carbon, hydrogen, and oxygen), which makes it easily combustible, producing char [19]. Consequently, the modification of nanocellulose through polymerization [51–53], oxidation [54–57], esterification [54, 58], silylation [56, 59, 60], and the incorporation of nanocellulose with flame-retardant additives such as phosphorus compounds [61–63], borates/boric acid [62–66] and inorganic nanofillers such as aluminum hydroxide, magnesium hydroxide [67–69], ZnO/TiO_2 nanoparticles [27, 70], carbon-based materials (e.g. graphene, carbon nanotubes) and clay nanostructured materials has improved the flame retardancy of nanocellulose and its derived products, making them more resistant to ignition, flame spread, and heat release [71]. The char produced during the combustion of nanocellulose within the resulting fire-retardant nanocomposite acts as a physical barrier against heat and oxygen to the underlying material, inhibiting further combustion [19].

3.1 Organic Compounds

Phosphorus-based fire retardancy additives are phosphorus compounds, such as [*N*-methylol-3-(dimethylphosphono) propionamide] (PCP, also known as Pyrovatex CP), phosphoric acid (PA), triallyl phosphate (TAP), diammonium phosphate (DAP), tributyl phosphate (TBP) and triallyl phosphoric triamide (TPT) [62]. They act as a coating agent for nanocellulose and work by releasing phosphorus radicals (phosphoric and polyphosphoric acids) during combustion which dehydrate cellulose and increase the rate of char formation. This disrupts the radical chain reactions and inhibits the spread of flames [62]. They are compatible with nanocellulose materials as they effectively bind via the exposed hydroxyl groups of nanocellulose. Phosphorus-based additives were intended to replace halogen-based additives since they exhibit lower toxicity [61]. Zhang et al. displayed a two-step fabrication process for a series of phosphorylated CNFs (P-CNFs), which involved phosphorylation treatment and mechanical grinding [63]. The CNFs were extracted from bamboo and wood pulps and then incorporated with diammonium hydrogen phosphate $(NH_4)_2$ (HPO_4) and $(NH_4)_2$ (HPO_4)/urea mixtures. The fabricated P-CNFs exhibited improved mechanical properties compared to the pristine CNFs such as high tensile strength (115.9 MPa), work of fracture of 41.8 MJ/m^{-3}, elongation at break of 53.1%, and Young's modulus of 2.5 GPa, demonstrating the formation of dense, homogenized polymer networks. The surface characterization and flame tests showed enhanced flame resistance with the P-CNFs compared to the pristine CNFs, which was attributed to the heat and oxygen barrier provided by the P_xO_y and carbon protective layer of the P-CNFs [63]. In another development, Salmeia et al. investigated the flammability of cellulose fibres produced from cotton and peat and the effect of a series of nitrogen-containing phosphorus compounds on their flame retardancy [61]. The study reported that the flammability properties of the cellulose fibers were dependent on the chemical composition. Peat-based cellulose fibers exhibited better fire retardancy compared to cotton-based cellulose, which was evident in the slow-burning and production of more char, attributed to the presence of lignin in its structure. The authors found that smaller phosphorus compounds and bis-phosphoramidates exhibited greater condensed phase flame retardancy on both cellulose fiber materials compared to larger phosphorus compounds and mono-phosphoramidates [61].

3.2 Boron Compounds

Boric acid and borate salts are commonly used as flame-retardant additives for nanocellulose-based materials due to their effectiveness in enhancing fire resistance. These additives work through a combination of physical and chemical mechanisms to suppress combustion and reduce flammability. Upon exposure to heat during combustion, boric acid and borate salts release water vapour and form boron oxide (B_2O_3) as a by-product. The boron oxide acts as a protective barrier, inhibiting the transfer of heat and oxygen to the underlying material. This barrier effect retards the propagation of flames and reduces the rate of burning [64]. Boric acid and borate salts can catalyze the charring process during combustion. They facilitate the conversion of flammable volatiles and decomposition products into charred residues. This char formation creates a stable and insulating layer on the material's surface, limiting the availability of fuel and oxygen, thus reducing the fire's intensity and spread. They generally have low toxicity, making them safer for certain applications where human exposure is a concern. These additives are often cost-effective and commercially available, making them accessible for various industrial applications. For example, zinc borate is considered a clean flame retardant additive that promotes carbon formation and the suppression of smoke. However, sodium borate/ boric acid and zinc borate/boric acid mixture is considered a safer fire retardant additive in terms of smoke production during combustion [65]. Lee et al. studied the effects of boric acid concentration on the flammability of cellulose-lignin composite fibers [66]. The study showed that the cellulose-lignin composite fibres produced no char while an increase in boric acid concentration from 1 to 6% resulted in enhanced char formation from 19.4% to 29.32% in air and 32.285 to 42.14% in nitrogen atmosphere. This demonstrated the catalyzing effect of boric acid on char formation during combustion.

3.3 Inorganic Compounds

3.3.1 Inorganic Hydroxides Minerals

Aluminium hydroxide and magnesium hydroxide are commonly used as mineral-based flame-retardant additives in nanocellulose composites. These additives release water vapor when exposed to heat, which cools the material and dilutes flammable gases. They are cost-effective, readily available, and have low toxicity. However, these additives may require higher loading levels to achieve desired flame retardancy [67–73]. Yuan et al. demonstrated good transparency, improved mechanical properties, and reduced flammability of cellulose aerogels after incorporating aluminium hydroxide nanoparticles [67]. In another report, Han et al. displayed in situ synthesis of magnesium hydroxide-based nanoparticle cellulose gel nanocomposite from waste cotton fabrics for flame retardancy applications [68]. The nanocomposite presented highly porous networks and good flame retardancy properties. Similar results were reported when cellulose-based aerogels were incorporated with aluminium hydroxide nanoparticles [69].

3.3.2 Nanomaterials

Metal nanoparticles such as titanium dioxide (TiO_2) or zinc oxide (ZnO) can be used as flame retardant additives. These nanoparticles can act as physical barriers or catalytic agents, inhibiting the combustion process and reducing flame propagation. Nanoparticle additives offer unique properties and can enhance flame retardancy in nanocellulose composites. They can act as physical barriers, catalytic agents, or heat absorbers. However, the dispersion and compatibility of nanoparticles in nanocellulose matrices can be challenging. There may be concerns related to nanoparticle toxicity and environmental impact [70–76]. The in situ synthesis of TiO_2-modified CNCs aerogel was demonstrated by Luo et al. [71]. Although the TiO_2@CNCs aerogel experienced a decline in crystallinity, the flame retardancy and thermal insulation properties were enhanced compared to the CNCs aerogel [71]. Bajwa et al. showed that the incorporation of ZnO nanoparticles coated the CNCs in high-density polyethylene (HDPE) polymer decreased the total smoke release the in the HDPE [27]. TEM images showed well-dispersed ZnO nanoparticles within the HDPE and are expected to perform well in flammability and mechanical strength tests [27].

3.3.3 Carbon Materials

Carbon material such as graphene and graphene derivatives, a two-dimensional carbon material has also been investigated as a fire retardant additive for nanocellulose-based materials. Graphene-based flame retardant systems have shown promise due to the unique combination of graphene's high thermal conductivity, barrier properties, specific area, and catalytic effects [72]. The flame retardancy of most carbon materials is centered around the char formation, which depends on structure and size [72, 77, 78].

3.3.4 Clay Materials

Clay materials such as montmorillonite clay (MMT) have shown the potential to improve the mechanical properties of polymeric foams by Medina et al. [79]. They fabricated MMT-based fire retardant nanocomposite foam using CNFs as a polymer phase via freeze-casting technique. Then studied the mechanical and fire retardant properties after the incorporation of PVA at varied amounts. The study showed the incorporation of PVA enhances the mechanical properties and optimum mechanical and flame retardancy properties were achieved at 15–30% PVA content with a 1:1 ratio of MMT: CNFs [79]. On the other hand, clay materials have been reported to improve the graphitization of char when combined with carbon material [80]. Wicklein et al. demonstrated a fire-retardant nanocomposite consisting of nanocellulose, sepiolite (SEP, magnesium silicate clay material), and graphene oxide [81]. Burning tests showed that the nanocomposite exhibited good fire retardancy properties but performed poorly in the absence of graphene or with little SEP, demonstrating some synergism between SEP and graphene oxide (GO) [81].

4 Fabrication Methods of Flame Retardant Nanocellulose-Based Materials

Cellulose is known as a fibrous polysaccharide which constitutes most parts of the plant cell wall (cellulose Iβ), which helps plants remain stiff and upright, with cellulose Iα mainly found in green algae and bacteria [81]. The linear chains of the cellulose have been reported to assemble into nanoscale fibrils held together through hydrogen bonding showing both crystalline and amorphous regions [82]. The formed nanocellulose fibrils can easily interact with other materials through Van de Waals, hydrophobic and electrostatic interactions to form nanocellulose-based materials or nanocellulose-nanocomposites [83–85]. The abundancy of cellulose from various wood pulp production have made nanocellulose-based materials attractive for a wide variety of applications as a practical and viable sustainable solution [82]. Though CNF are the most popular cellulose nanomaterials prepared from plants, CNCs and bacterial cellulose (BCs), as well as particles of amorphous nanocellulose and cellulose nanoyarns have also been reported [86, 87]. These cellulose types differ on the morphology and preparation method. Various methods for the fabrication of nanocellulose have been reported which includes chemical, mechanical, chemical–mechanical, and bio-enzymatic [88]. Due to their unique properties such as good hydrophilicity, chirality, easy surface modification, rich functional groups and biocompatibility, nanocellulose have been employed for a variety of applications such as water treatment, pickering emulsifiers, wood adhesives, biomedical applications and as flame retardant materials [88–91]. However, for the flame retardant applications, various inorganic and organic flame retardants additives have been fabricated into the nanocellulose structure, resulting in flame retardant nanocellulose-based materials [92, 93]. The synthetic pathway for nanocellulose synthesis and nanocellulose-composites is shown in Scheme 9.1. The electrospinning, hydrothermal, cross linking, solution casting, and sol–gel have been used to fabricate the nanocellulose and the flame retardant, which are then combined through either evaporation, supercritical or freeze drying (as in Scheme 9.1).

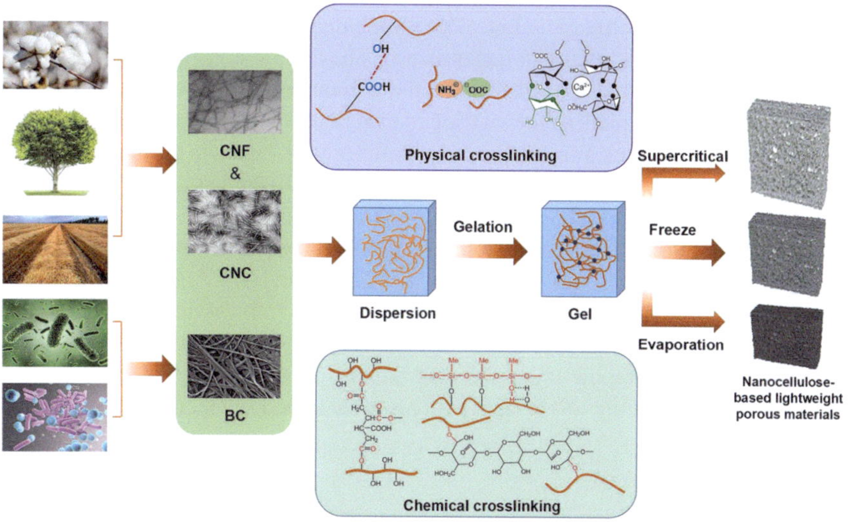

Scheme 9.1 Synthesis pathway of nanocellulose, alongside the fabrication of nanocellulose-based materials. Reprinted with permission from [93]. Copyright, 2020 Elsevier Ltd

Inorganic flame retardants such as graphene oxide (GO) [81, 94], expandable graphite [95], magnesium hydroxide [96] and sodium-montmorillonite (MMT) [97] have been previously integrated with nanocellulose as flame-retardant cellulose nanocomposites. Similarly, nanocellulose has been integrated with organic flame retardants such as deoxyribonucleic acid (DNA), ammonium polyphosphate (APP) and melamine polyphosphate (MPP) [98]. Interestingly, hybrid materials incorporating both inorganic and organic flame retardants into the nanocellulose matrix have been reported to be highly effective as flame retardant nanocomposites [99]. The methods used to fabricate the flame retardant nanocellulose-based materials include the traditional evaporation drying method, supercritical and freeze-drying method (as shown in Scheme 9.1 above), which are discussed in this chapter.

4.1 Synthesis Methods for the Preparation of Cellulose Nanocomposites and Flame Retardant Materials.

4.1.1 Electrospinning Method

Electrospinning is a well exploited process for the synthesis of continuous fibers with diameters ranging from tens of microns to hundreds of nanometres, emancipating from the *in-situ* modification/functionalization of polymers and nanomaterials [100]. Evidently, the use of electrospinning method has been adopted drastically in an effort to produce nanofibers with improved flame retardancy [91, 97, 101]. To this end, Vahabi and co-authors [102] discussed various cellulose nanofibers with incorporated flame retardant materials using the electrospinning method. The authors reported that the flame retardant electrospun fibers possessed enhanced thermal stability, mechanical property, and flame retardant performance. The review suggested that the preparation of core–shell structures in which a flame retardant takes position in the core wrapped in a polymer shell, or preparation of nanofiber mats with integration of flame retardant inside the fibers provided by the of pores present between the nanofibers, can yield improved thermal and flame retardant properties. In their study, Kobayashi and co-authors [103] reported the aerogel with a three-dimensionally ordered nanofiber skeleton of liquid-crystalline nanocellulose obtained by supercritical drying of the hydrogels as a flame retardant, with tough and transparent insulating capabilities. The surface carboxylated cellulose nanofibers were prepared from wood cellulose through an oxidation reaction using 2,2,6,6-tetramethylpiperidine-1-oxyl (TEMPO) as the catalyst. The authors reported that the materials preparation process required only a low energy input, leading to aerogels with properties such as mechanical toughness, transparent and a good heat insulation. These properties are not observed in conventional aerogels, which makes the nanofiber skeleton of liquid-crystalline nanocellulose an extraordinarily high-performance insulator with optical transparency and mechanical toughness.

4.1.2 Hydrothermal Method

The eco-friendly hydrothermal method has been adopted for the synthesis of flame retardant cellulose nanofiber composites. In their study, Fan et al. [104] prepared the cellulose nanofiber/aluminum oxide hydroxide (AlOOH) aerogel nanocomposite as a flame retardant and thermal insulator using the hydrothermal method. Instead of the traditional dispersion of flame retardants in a polymer matrix, the AlOOH flame retardant was wrapped around the CNF or adhered to the CNF, which also favoured the reduction in the agglomeration of AlOOH. The one pot synthesis involving the CNF aqueous solution, aluminium sulphate ($Al_2(SO_4)_3$) and urea ($CO(NH_2)_2$) yielded to the cellulose nanofiber/AlOOH aerogel after heating at 100 °C for 8 h in a Teflon-lined autoclave. The thermal conductivity and diffusivity of the resulting CNFA aerogel were 0.0385 $W{\cdot}m^{-1}\ K^{-1}$ and 0.341 $mm^2\ s^{-1}$, respectively. The authors reported that the combination of the cellulose and AlOOH resulted in a lightweight composite with remarkable flame retardant and thermal insulation performance observed during the flammability tests. Compared to other methods, the hydrothermal method is time consuming as it required the pre-synthesized CNF, which are further treated for 8 h in an autoclave. This is the major disadvantage that leads to researchers working on improving the electrospinning method.

4.1.3 Solution Casting Method

Solution casting has been reported as one of the adopted methods for the synthesis of flame retardant nanocellulose-based nanocomposites [90, 91, 93]. As shown in Scheme 9.2, the previously prepared nanocellulose is dispersed in a solvent and then added to the pre-dispersed polymer that has already been dispersed in the same solvent. The dispersion can be achieved through stirring or sonication, leading to a well dispersed solution. Then, the solvent is completely evaporated from the mixture, leaving nanocellulose composite integrated with a flame retardant component [90]. Similar to the hydrothermal method, this method relies on the pre-electrospinning of CNF and involves the less effective evaporation process either than the freeze drying or the supercritical fluid drying that are more effective for the fabrication of the flame retardant cellulose gel composite.

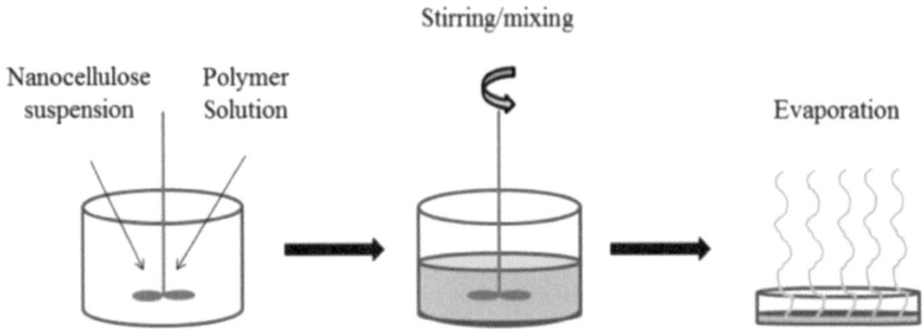

Scheme 9.2 Solvent casting procedure for the preparation of flame retardant nanocellulose-based nanocomposites [90]

4.1.4 Cross Linking Method

Two types of crosslinking techniques have been adopted for the fabrication of cellulose nanocomposites [105–107]. The chemical and physical cross-linking techniques have seen wide applications in an effort to prepare cellulose nanocomposites based on either hydrogen bonding, covalent bonding and electrostatic interactions, amongst other interactions (as shown in Scheme 9.1). For instance, Yue et al. cross linked the already prepared CNFs by melamine formaldehyde (MF), with the resulting cross-linked product further cross-linked and surface decorated with methyltrimethoxysilane (MTMS), resulting in a flame retarded compound aerogels [108]. MF is known as a nitrogen-containing flame retardant that can easily react with the hydroxyl groups on the surface of the cellulose to form a cross-linked structure shown in Scheme 9.2. This arrangement improves the flame retardant properties and the mechanical properties of the cellulose aerogels. The introduction of the MTMS as shown in Scheme 9.3 further improves the flame retardancy as a synergistic flame retardant. The limiting oxygen index (LOI) value of CNF/MF increased from 19.5% to 37.1%, whilst a significant decrease in peak heat release rate (PHRR) and the total heat release (THR) were 50.6% and 64.3%, respectively. Thus, the obtained CNF modified with MF and MTMS composite foam showed improved mechanical performance, hydrophobic property, thermal properties and flame retardancy after tests. The drawback reported for the cross linking technique is the fast decomposition of the cellulose composite due to low mechanical strengths, although recent studies have reported an improved mechanical structure based on the properties of the flame retardant used [90, 91, 106]. In some cases, the absorption of moisture has also been reported, which limits the flame retardancy of the cellulose composite as the fire resistance is altered [109]. This is one of the commonly adopted techniques for the preparation of CNF modified by flame retardants.

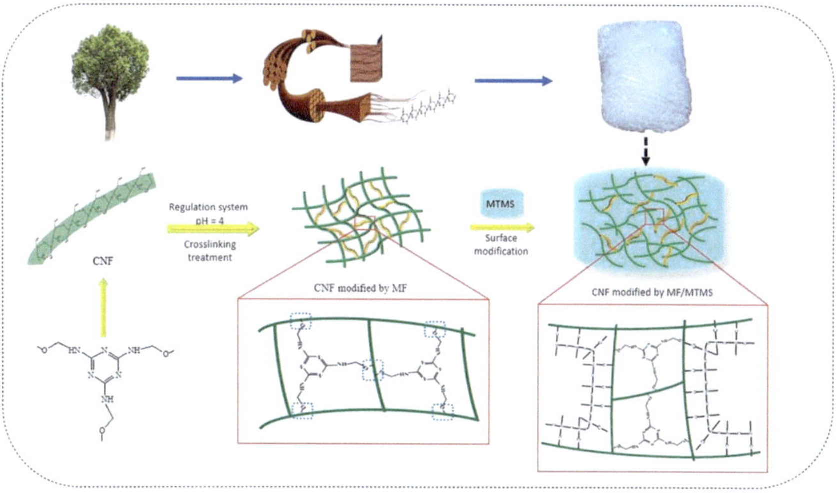

Scheme 9.3 Double crosslinking formation of CNF/MF/MTMS nanocomposite foam [108]

4.2 *Solution Evaporation Methods*

Once the cellulose and the flame retardant materials are prepared, the following methods are critical for the complete fabrication of dry gels that have minimal liquid on their pores to ensure they perform optimally as flame retardants.

4.2.1 Evaporation Method

As a conventional drying method, evaporation is a non-destructive, cost-effective and environmentally friendly process for the removal of the liquid phase of a gel whilst maintaining the highly porous structure of the flame retarded nanocellulose composite [110]. In this method, evaporation drying is carried out at ambient conditions or with the application of a low vacuum in a wide temperature range. Though this process is very economical and allow for the large scale fabrication of cellulose nanocomposites, the shrinkage and the collapse of the highly porous cellulose nanostructure is unavoidable due to capillary forces. In another study, researchers have replaced the use of water with non-polar solvents (such as pentane and hexane) which can reduce the interfacial tension, leading to the improvement on the shrinkage and the porous structure [111]. However, though the low density is achieved, the porosity and surface area are often compromised, which limits the widespread application of the conventional evaporation as a drying method for the cellulose nanocomposite, particularly for flame retardancy applications.

4.2.2 Supercritical Fluid Drying Method

The supercritical fluid drying (SFD) method involves the optimization of the pressure and the temperature such that the solvent reaches a critical point through the drying process. This enables the complete supercritical fluid transition without the gas–liquid interface, as water is not used, instead supercritical fluids such as methanol and ethanol are used [111, 112]. However, since the discovery that methanol and ethanol are not good supercritical fluids due to their explosive nature, supercritical carbon dioxide ($SCCO_2$) is the best option due to its mild temperature (31 °C) and pressure (74 bar) [93]. Thus, SFD can easily dry out the surface and pores liquid in the wet gel with minimal pore structural damage. Nevertheless, the pressure of 74 bar is still high, which requires the use of expensive facilities and consumes more energy, leading to higher costs. Still, if the wet gel is dried at ambient conditions, the porous structure is damages, as the gel shrinks significantly [113, 114].

4.2.3 Freeze-Drying Method

The freeze casting method has been reported to produce green and stable cellulose nanocomposite foams (three-dimensional structure), displaying a unique pore structure and composition, with the nanocomposite foams prepared by freeze-drying a colloidal suspension [115, 116]. The method includes freezing the liquid suspension, followed by the removal of the ice crystals through the process of ice sublimation [117]. The nanocomposite foams prepared from this method are usually comprised of nanocellulose and an excess number of nanomaterials, such as aluminum hydroxide, layered double hydroxide, sepiolite, kaolin and sodium montmorillonite that act as flame retardant additives [101, 118]. For instance, Carosio et al. [116] combined the cellulose nanofibers (CNF) with the 2D montmorillonite (MMT) nanoplatelets flame retardant additive and xyloglucan (XG) hemicellulose binder using a water-based freeze casting approach, forming a CNF/XG/MTM nanocomposite. The nanocomposite flame-retardant foams of 98.5% porosity could withstand the penetration of the flame and the high surface temperature reached during the test, which averaged 800 °C. This structure was thermally stable, with a very low thermal diffusivity, which explains the large temperature drop. Though some significant problems such as adhesion caused by solvent surface tension, and collapse of the internal network structure due to capillary action of the solid matrix have been reported for the freeze-drying method, some studied have overcome these challenges. For instance, Wang and co-workers [119] synthesized CNFs from softwood pulp using a TEMPO-oxidized method. Through the sonication and freeze-drying methods, the authors integrated the formed CNFs with the previously prepared cyclotriphosphazene-bridged periodic mesoporous organosilica (PMOPZ) to form a rigid three-dimensional PMOPZ/CNFs nanocomposite foam shown in Scheme 9.4. In addition to achieving a mechanical stiffness in the freezing direction, an ultra-low thermal conductivity of 27 mW m^{-1} K^{-1}, excellent flame-resistant and self-extinguishing properties with a limiting oxygen index value of 31%, as well as 72.4% and 68.5% reduction in PHRR and

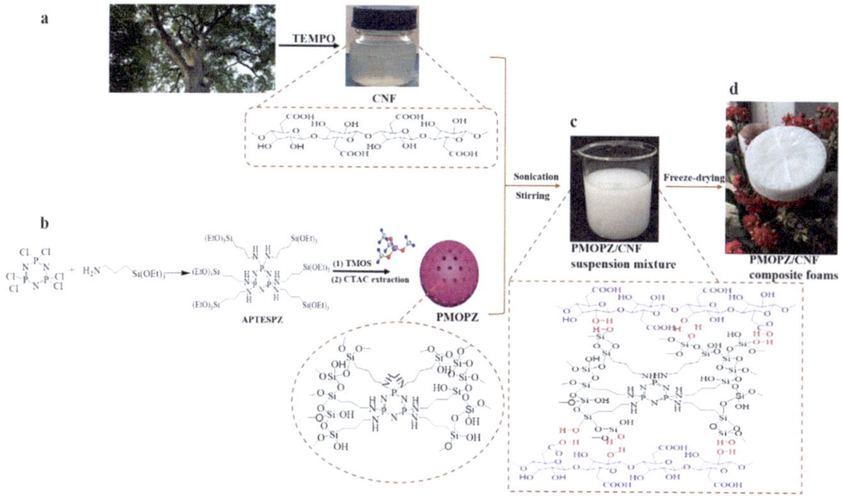

Scheme 9.4 **a** CNFs prepared from softwood pulp via TEMPO-oxidized method, **b** synthesis of PMOPZ hybrid nanospheres using APTESPZ intermediates, **c** PMOPZ/CNFs suspension bonded via hydrogen bonding interactions through sonication, **d** PMOPZ/CNFs composite foam formed after the freeze-drying process [120]

THR, respectively, were achieved by the PMOPZ/CNF nanocomposite foam. The authors suggested that this work provides a novel strategy to achieve nanocellulose-based insulating nanocomposites with high performance as flame retardants, leading to fire safety.

Amongst the three drying methods, the freeze drying method has been used significantly over the supercritical and evaporation methods for the flame retardant cellulose composite fabrication as shown in Table 9.1.

5 Characterization of Flame Retardant Nanocellulose-Based Materials

Flame retardant materials have become a crucial area of research in various industries due to the increasing demand for fire-resistant products. Nanocellulose-based materials, derived from renewable sources, such as wood pulp, have shown promising potential in enhancing flame retardancy properties [123, 124]. This section aims to provide a comprehensive scientific analysis of the characterization techniques employed to evaluate the flame-retardant behaviour of nanocellulose-based materials. The discussion encompasses various aspects, including thermal stability, fire behaviour, gas evolution, and char formation, shedding light on the mechanisms and properties that contribute to their flame retardant efficacy. The findings presented here will serve as a valuable resource for researchers and industrial practitioners in

Table 9.1 Methods used to fabricate flame retardant nanocellulose-based materials

Nanocellulose-based Materials	Flame-retardant	Preparation method	Advantages	Disadvantages	References
MCAPP/polyvinyl alcohol/cellulose nanofibers (MCAPP/PVA/CNFs) hybrid aerogel	0.8 wt% microencapsulated ammonium polyphosphate (MCAPP)	Freeze drying	The aerogel exhibited high thermal insulation ability (thermal conductivity: ~0.04 W/m·K), the PHRR decreased from 222.44 to 107.84 kW/m^2, this resulted in improved anti-combustion performance	MCAPP decomposes itself and dehydrate PVA and CNFs when heated under 35 kW exposure, leading to the formation of char	[117]
DOPO/IA/CNFs composite aerogel	9,10-dihydro-9-oxa-10-phosphaphenantherne-10-oxide/Itaconic Acid (DOPO/IA)	Freeze drying	Significantly enhanced thermal stability, mechanical property, and flame retardant performance compared with the pure cellulose aerogel. Exhibited remarkable reductions in PHRR and THR values which were 67% and 60%, respectively	The cellulose aerogel possesses high flammability due to their inherent chemical structure compositions of C, H and O elements	[121]

(continued)

Table 9.1 (continued)

Nanocellulose-based Materials	Flame-retardant	Preparation method	Advantages	Disadvantages	References
CNFs/graphene oxide (GO)/sepiolite nanorods (SEP) nanocomposite foam	Graphene oxide (GO)/sepiolite nanorods (SEP)	Freeze-casting	Super-insulating, fire-retardant and strong anisotropic foams. excellent combustion resistance and exhibit a thermal conductivity of 15 mW m^{-1} K^{-1}, which is about half that of expanded polystyrene	The typical deformation behaviour of an open honeycomb-like foam, linear elastic behaviour at low strain, followed by a cell collapse-related stress reduction at intermediate strains	[94]
Highly porous silica/ cellulose aerogels	Silica layers	Freeze-drying	The thermal performance was improved and the integration of silica on CNF prevented the collapse of the aerogel during drying	The flame retardant tests were not fully exploited	[122]

the pursuit of developing safer and sustainable flame retardant materials. The flame retardant materials are designed to inhibit or delay the spread of fire, thus reducing the potential risks associated with combustion [125]. Nanocellulose, a nanomaterial derived from cellulose fibres, exhibits excellent mechanical properties, high aspect ratio, low density, and renewability, making it a promising candidate for enhancing flame retardancy in various applications [126]. However, the evaluation and characterization of the flame retardant behaviour of nanocellulose-based materials require advanced analytical techniques and standardized methodologies. This chapter aims to provide a comprehensive overview of the characterization techniques used to assess the flame retardant properties of such materials. To better understand the composition and characteristics of flame-retardant nanocellulose-based materials, a variety of complementary techniques, such as X-ray diffraction (XRD), Fourier Transform Infra-Red (FTIR) spectroscopy, electron microscopy analysis (Scanning Electron Microscope (SEM) and Transmission Electron Microscope (TEM)), Atomic Force Microscopy (AFM), thermogravimetric analysis (TGA), mechanical and microscale combustion calorimeter (MCC) testing are used to name just a few.

5.1 X-Ray Diffraction Technique

The X-ray diffraction (XRD) analysis is a powerful technique used to characterize the crystal structure and crystallinity of materials, including flame-retardant nanocellulose-based materials [127]. In the context of flame retardancy, XRD provide valuable information about changes in the crystal structure and potential modifications induced by flame retardant additives. XRD analysis allows for the determination of the crystal structure of nanocellulose-based materials [127]. By exposing the sample to X-ray radiation, the XRD instrument detects the diffraction pattern resulting from the interaction of X-rays with the crystal lattice. This pattern is unique to the crystal structure and provides valuable information about the arrangement of atoms or molecules within the material. The crystallinity of nanocellulose-based materials plays a crucial role in their physical and mechanical properties, as well as flame retardancy [127, 128]. The XRD analysis can quantify the degree of crystallinity by measuring the intensity of diffraction peaks. Crystalline regions within the sample will produce sharp and intense peaks, while amorphous regions will generate broad and diffuse signals. The crystallinity index can be calculated by comparing the intensity of the crystalline peaks with a reference material or amorphous baseline. The XRD was used to analyse the synergetic integration of thermal conductivity and flame resistance in nacre-like nanocellulose composites as reported by [129].

The characteristic 101 and 002 diffraction peaks are seen in the pure CNFs film at ~15.3° and ~22.7°, respectively, as shown in Fig. 9.1. The CNFs/BNNS composite film clearly exhibits the significant 002 and 004 characteristic diffraction peaks of BNNS at 26.6° and 54.9° after BNNS has been added. Intriguingly, BNNS's 002 and 004 diffraction peaks in CNFs/BNNS-p-APP composites are slowly displaced in comparison to the CNFs/BNNS composite film. This should be attributable to

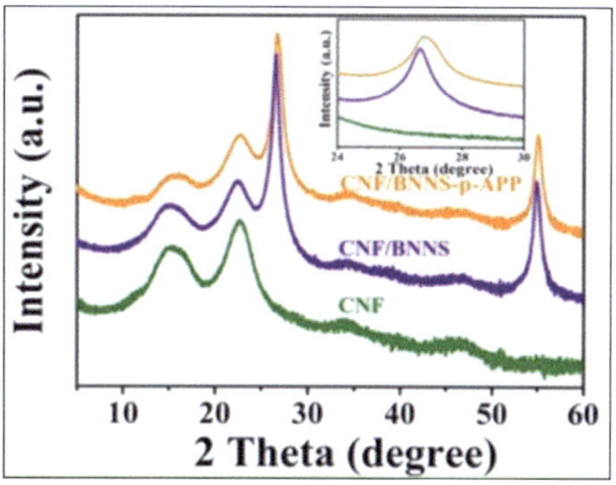

Fig. 9.1 XRD patterns of pure cellulose nanofibers (CNF) film and cellulose nanofibers-based boron nitride nanosheets (CNF/BNNS) composite films [129]

the extensive hydrogen-bonding interaction between BNNS-p-APP and CNFs [125, 129]. Hence, considering the previously mentioned exploration of shapes and surface connections in composites of CNFs/BNNS-p-APP, the well-organized arrangement, robust interfacial bonding, and effective immobilization of APP have the potential to not only minimize the disruption of heat flow but also limit the escape of flammable gases and prevent external oxygen from permeating the internal structure. These combined effects can significantly improve thermal conductivity and flame resistance, ultimately leading to enhanced performance [129]. As depicted from the XRD patterns, it was possible to quantify cellulose's crystallinity index with greater accuracy and it could also work with other cellulose polymorphs [130]. The addition of flame-retardant additives, such as phosphorus-based compounds or intumescent agents, can potentially influence the crystal structure and crystallinity of nanocellulose-based materials. Tong et al. [124] reported CNFs that were highly fibrillated and naturally flame-resistant was used to create transparent fire-protective coatings without the need of mineral fillers. However, Q-CNFs films provide outstanding transparency and flame resistance after post-boronation. XRD analysis can detect any changes or shifts in diffraction peaks that indicate the presence of these additives [131]. This information helps in understanding the interaction between the flame retardant additives and the nanocellulose matrix, shedding light on the mechanisms responsible for flame retardancy. Flame retardant treatments can introduce structural modifications to nanocellulose-based materials [123, 132]. As such, XRD analysis can detect changes in peak positions, peak broadening, or the appearance of new peaks, indicating alterations in the crystal lattice. These modifications may arise from chemical reactions, cross-linking, or the formation of new phases as a result of the flame retardant treatment. During combustion, nanocellulose-based materials undergo charring, resulting in the formation of a protective layer. XRD analysis can provide insights into the structure and composition of the charred residues. By analyzing the XRD patterns of the char layer, it is possible to identify the presence of new crystalline phases or changes in the crystallinity, which may contribute to the formation of a more robust and stable char layer. XRD analysis of flame retardant nanocellulose-based materials enables the characterization of crystal structure, quantification of crystallinity, detection of additives, assessment of structural modifications, and evaluation of char formation. These findings contribute to a comprehensive understanding of the flame retardant mechanisms and provide valuable insights for the development and optimization of flame retardant nanocellulose-based materials.

5.2 Fourier Transform Infrared Spectroscopy Technique

The Fourier Transform Infrared (FTIR) absorption spectroscopy is a powerful analytical technique commonly used for the characterization of flame retardant nanocellulose-based materials [132, 133]. This technique provides valuable information about the chemical structure, functional groups, and interactions present in these materials. In the context of flame retardancy, FTIR analysis helps in understanding the molecular changes that occur during thermal decomposition, flame inhibition, and char formation processes [132, 134]. FTIR spectroscopy allows the identification of specific functional groups present in nanocellulose-based materials. The characteristic absorption peaks in the infrared spectrum correspond to the vibrational modes of chemical bonds [132–135]. By comparing the FTIR spectrum of flame retardant nanocellulose-based materials with reference spectra, it is possible to determine the presence of functional groups introduced by flame retardant additives. Common functional groups associated with flame retardants include phosphorus, nitrogen, halogens, and intumescent compounds [133, 136]. Moreover, FTIR spectroscopy can also be used for quantitative analysis of flame retardant additives in nanocellulose-based materials. As such, by correlating the intensity of specific absorption peaks with known concentrations of flame retardants, it is possible to estimate the loading or concentration of the additives in the material. This information is crucial for optimizing flame retardant performance and ensuring consistent quality control in the production process [136, 137]. During the thermal decomposition of flame-retardant nanocellulose-based materials, FTIR spectroscopy can provide real-time information about the chemical changes occurring [132].

It is possible to track the degradation and release of volatile compounds by monitoring the changes in absorption bands or the appearance of new peaks. This helps in understanding the decomposition pathways, identifying the release of potentially harmful gases, and evaluating the efficiency of flame retardant additives in suppressing the degradation process. Char formation is a vital mechanism in flame retardancy, where a protective carbonaceous layer is formed on the material's surface, preventing further combustion. As such, FTIR spectroscopy can be used to analyze the chemical composition and structure of the charred residue. The FTIR analysis on the facile preparation of flame-retardant cellulose composite with biodegradable and water-resistant properties for electronic device applications was reported by Chandrasekaran et al. [138]. The resulting cellulose, ammonium phosphate (APP)-powder, non-hydrophobised, and hydrophobized cellulose-APP composite with 10 wt% APP ATR-FTIR spectra are displayed in Fig. 9.2.

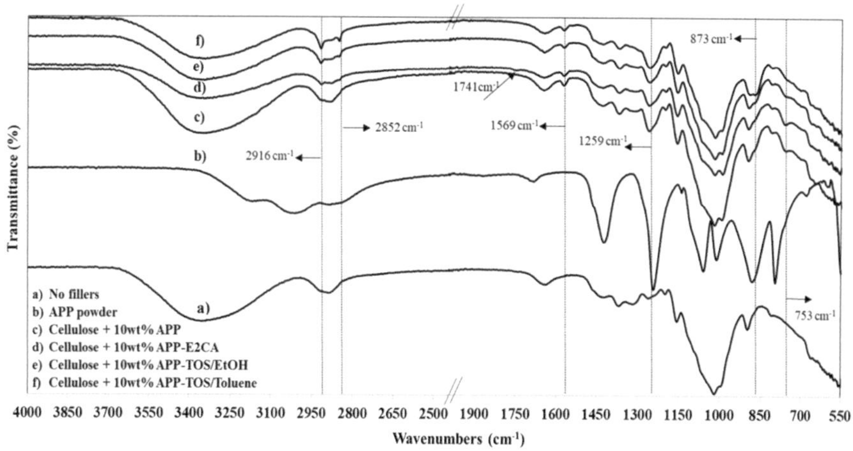

Fig. 9.2 ATR-FTIR of **a** cellulose with no flame-retardants, **b** APP powder, **c** cellulose-APP composite, **d** cellulose-APP composite-E2CA coated, **e** cellulose-APP composite-TOS/EtOH coated and **f** cellulose-APP composite-TOS/Toluene coated [138]

As shown on the figure, the hydroxyl group in cellulose is represented by the band at 3353 cm^{-1}, while the stretching vibration of –CH/–CH_2 groups is indicated by the bands ranging from 2927 to 2879 cm^{-1}. The presence of water in cellulose is associated with the band at 1640 cm^{-1}. The weak band at 1569 cm^{-1} could be a result of the formation of charged –NH^+ or COO^- groups during the dissolution of cellulose in the presence of APP. The bands at 1259 cm^{-1} (–P–O), 873 cm^{-1} (–P–O asymmetrical stretching vibration), and 753 cm^{-1} (O=P–O) were attributed to the presence of APP in the cellulose composites. The ATR-FTIR analysis mentioned earlier has verified that both hydrophobic agents were successfully applied to the surface of the cellulose composites. This deposition of hydrophobic agents has effectively enhanced the water resistance of the cellulose composites [138].

As seen from the figure above, comparing the FTIR spectra of the charred residue with the original material, it is possible to identify changes in functional groups, such as the formation of aromatic compounds and carbon–carbon cross-linking [139, 140]. This analysis provides insights into the char formation process and the effectiveness of flame retardant additives in promoting the formation of stable and protective char layers. FTIR spectroscopy is also useful for investigating the compatibility between nanocellulose and flame retardant additives [140, 141]. By comparing the FTIR spectra of the individual components and the nanocomposite, it is possible to observe any shifts, broadening, or changes in peak intensities, as seen from Fig. 9.2. These spectral changes can indicate the interactions and chemical bonding between nanocellulose and flame retardant additives, which is essential for ensuring good dispersion and uniform distribution of the additives in the nanocellulose matrix [139, 140]. Therefore, FTIR absorption spectroscopy is of great importance in the characterization of flame retardant nanocellulose-based materials. The FTIR also plays a vital role by offering valuable information about various aspects such as the chemical structure, functional groups, thermal decomposition, char formation, and compatibility of these materials. As a result, utilizing FTIR analysis, researchers can acquire crucial insights into the flame retardant mechanisms at work and effectively optimize the design of nanocellulose-based materials. This optimization aims to enhance the fire resistance properties of these materials, leading to improved safety and performance.

5.3 Electron Microscopy Techniques

The electron microscopy techniques, such as scanning electron microscopy (SEM) and transmission electron microscopy (TEM), play a crucial role in characterizing the morphology, structure, and composition of flame-retardant nanocellulose-based materials. These techniques provide high-resolution imaging capabilities that allow for detailed examination of the surface and internal structures of materials at the nanoscale. In the context of flame retardant nanocellulose-based materials, SEM and TEM analysis provide valuable insights into the dispersion of flame retardant additives, interfacial interactions, char formation, and the overall morphology of the materials [142].

5.3.1 Scanning Electron Microscope Technique

SEM is a widely used technique for imaging the surface morphology of materials. It provides information about the size, shape, and distribution of particles or fibres within flame-retardant nanocellulose-based materials. SEM analysis helps in visualizing the dispersion of flame retardant additives within the nanocellulose matrix and their interaction with the cellulose fibres. It also allows for the examination of the surface features of charred residues formed during combustion, providing insights into char morphology and the effectiveness of flame retardancy [18, 142].

5.3.2 Transmission Electron Microscope Technique

TEM is a powerful technique that enables the examination of the internal structure and ultrafine details of materials. It involves transmitting a beam of electrons through a thin section of the material, allowing for high-resolution imaging. TEM analysis of flame-retardant nanocellulose-based materials provides information about the dispersion and localization of flame-retardant nanoparticles at the nanoscale [18, 143]. It can reveal the interfacial interactions between the flame retardant additives and the nanocellulose matrix, providing insights into the mechanisms of flame retardancy. Additionally, TEM analysis can be used to investigate the crystalline structure of nanocellulose fibres and the formation of char layers during combustion [143].

5.3.3 Atomic Force Microscope (AFM) Technique

AFM is a powerful characterization technique used to study the surface topography, morphology, and mechanical properties of materials at the nanoscale. When applied to flame-retardant nanocellulose-based materials, AFM provides valuable insights into the structural features, surface roughness, and interfacial interactions that contribute to their flame-retardant behaviour. This section will discuss the application of AFM in the characterization of flame-retardant nanocellulose-based materials, highlighting its key parameters and findings [144, 145]. The AFM allows for the visualization and measurement of the surface topography and morphology of flame-retardant nanocellulose-based materials with high spatial resolution. It provides three-dimensional images, revealing details of the material's surface features, such as nanocellulose fibres, agglomerates, and any additives or flame retardant coatings. AFM can also determine the nanocellulose's lateral dimensions, including width and length distribution, which are crucial for understanding its dispersion and interaction within the matrix.

The surface roughness plays a significant role in the interfacial properties and flame retardant behaviour of materials. AFM can measure the surface roughness of flame retardant nanocellulose-based materials at the nanoscale level [145]. The root mean square (RMS) roughness parameter obtained from AFM images provides quantitative information about the surface irregularities and can be correlated with the material's flame retardant performance. A smoother surface may indicate better dispersion and interfacial adhesion, leading to improved flame retardancy. AFM can assess the adhesive and cohesive forces between the nanocellulose-based material and other surfaces. Force-distance curves obtained using AFM can quantify the adhesion forces between the material's surface and the AFM tip [145, 146]. This information is crucial for understanding the interaction between flame retardant additives and the nanocellulose matrix, as well as their influence on the overall flame retardant behaviour. Adhesive and cohesive forces can be correlated with the mechanical properties of the material, providing insights into its structural stability during exposure to fire.

AFM-based techniques such as nanoindentation and force spectroscopy can determine the mechanical properties of flame retardant nanocellulose-based materials at the nanoscale [144, 147]. By applying controlled forces, AFM probes can measure the material's stiffness, elasticity, and viscoelastic properties. These mechanical properties are directly related to flame retardant behaviour, as they influence the material's ability to withstand external forces and resist deformation or ignition during fire exposure [147, 148]. The AFM can be coupled with other characterization techniques, such as infrared spectroscopy or Raman spectroscopy, to obtain chemical information about specific regions of the flame-retardant nanocellulose-based material. This enables the identification and mapping of flame retardant additives or other chemical components within the material's matrix [149–153]. The combination of AFM with chemical analysis provides a comprehensive understanding of the chemical composition and spatial distribution of flame retardant components, facilitating the correlation between chemical properties and flame retardant performance.

In summary, AFM is a versatile tool for characterizing flame-retardant nanocellulose-based materials. It provides valuable information on surface topography, morphology, roughness, adhesion, cohesion, mechanical properties, and local chemical composition. By utilizing AFM, researchers can gain crucial insights into these materials' structural and interfacial properties, which contribute to their flame-retardant behaviour. This knowledge aids in the design and optimization of flame-retardant nanocellulose-based materials with enhanced fire resistance properties.

5.4 *Elemental Analysis Technique*

In addition to morphology, SEM and TEM can be coupled with energy-dispersive X-ray spectroscopy (EDS) to perform elemental analysis of flame-retardant nanocellulose-based materials [140, 154]. EDS allows for the identification and mapping of elements present within the material, including flame retardant additives and their distribution. Elemental analysis provides quantitative data on elemental composition, allowing for the characterization of the elemental distribution on the surface and within the bulk of the material [155].

In summary, SEM and TEM analysis, coupled with elemental analysis using EDS, are powerful tools for characterizing flame-retardant nanocellulose-based materials. These techniques enable the visualization and examination of surface morphology, dispersion of flame retardant additives, interfacial interactions, and char formation. The insights gained from SEM and TEM analysis contribute to a deeper understanding of the structure–property relationships of flame-retardant nanocellulose-based materials, facilitating the development of improved fire-resistant materials with enhanced performance and safety.

5.5 *Thermal Stability Analysis Techniques*

Thermal stability is a critical factor in evaluating the flame retardant efficacy of materials. Differential scanning calorimetry (DSC), thermogravimetric analysis (TGA), and thermomechanical analysis (TMA) are commonly employed techniques to investigate the thermal stability and decomposition behaviour of nanocellulose-based materials. These techniques provide valuable information on the onset temperature of degradation, and the weight loss of materials under controlled heating conditions [156, 157].

5.5.1 Fire Behaviour Analysis

The fire behaviour of nanocellulose-based materials can be characterized using several experimental methods. Microscale combustion calorimeter (MCC) is a widely used technique to assess the flammability and heat release rate (HRC) of materials. It provides insights into parameters such as ignition time, PHRR, THR, and smoke production [158]. Flame spread tests, such as the LOI and the horizontal flame spread test, offer further evaluation of the fire propagation characteristics of materials [159, 160]. These tests help in understanding the influence of nanocellulose additives on the flame retardant behaviour.

5.5.2 Gas Evolution Analysis

When exposed to high temperatures, materials undergo thermal decomposition, leading to the evolution of various gases that can affect the combustion process [161]. Gas chromatography-mass spectrometry (GC–MS) is commonly employed to analyze the gaseous products emitted during the thermal degradation of nanocellulose-based materials [161]. This technique enables the identification and quantification of specific gases, such as carbon dioxide, carbon monoxide, and volatile organic compounds (VOCs), providing insights into the flame retardant properties and toxic gas emissions of the materials [162].

5.6 Mechanical Analysis Techniques

Flame-retardant nanocellulose-based materials have gained significant attention due to their potential for enhancing fire resistance. However, ensuring the mechanical integrity and performance of these materials is crucial for their successful application in various industries [163]. This scientific analysis focuses on the mechanical testing techniques employed to evaluate the structural properties, strength, and durability of flame-retardant nanocellulose-based materials. It encompasses an overview of key characterization methods, such as tensile testing, flexural testing, impact testing, and hardness testing, providing valuable insights into the mechanical behaviour and performance of these materials [164, 165]. The findings presented here will serve as a valuable resource for researchers and practitioners working on the development and application of flame-retardant nanocellulose-based materials. Flame-retardant nanocellulose-based materials offer unique opportunities to improve fire resistance in various applications. However, their successful implementation requires a thorough understanding of their mechanical behaviour [166]. This paper focuses on the mechanical testing techniques used to assess the structural integrity, strength, and durability of these materials, shedding light on their mechanical properties and performance.

5.6.1 Tensile Testing

Tensile testing is a fundamental technique employed to evaluate the mechanical behaviour of materials under tension. It provides insights into the tensile strength, Young's modulus, and elongation at the break of flame-retardant nanocellulose-based materials [166, 167]. By subjecting test specimens to controlled forces, tensile testing enables the determination of stress–strain curves, which help characterize the material's deformation and failure mechanisms.

5.6.2 Flexural Testing

Flexural testing, also known as three-point or four-point bending, is utilized to assess the bending strength, flexural modulus, and fracture toughness of materials. This testing method involves subjecting a specimen to a controlled bending moment, enabling the measurement of stress–strain behaviour under flexural loading conditions [167]. By analyzing load–deflection curves, important mechanical properties related to the material's resistance to bending can be obtained.

5.6.3 Impact Testing

Impact testing evaluates a nanocellulose material's ability to withstand sudden dynamic loads or impacts. Charpy and Izod impact tests are commonly employed to measure the impact strength and energy absorption capacity of materials [168]. These tests involve striking a notched or unnotched specimen with a pendulum or falling weight, respectively, and measuring the energy absorbed during fracture. Impact testing provides insights into the material's toughness and resistance to sudden loading conditions.

5.6.4 Hardness Testing

Hardness testing is used to evaluate a material's resistance to indentation or surface deformation. Techniques such as Rockwell, Brinell, and Vickers hardness tests are commonly employed [169]. These tests involve indenting the material's surface with a specific load and measuring the indentation depth or area. Hardness testing provides information on the material's hardness, which is an essential parameter for assessing its resistance to wear and deformation.

5.7 Other Mechanical Testing Techniques

Other mechanical testing techniques, such as compression testing, shear testing, and fatigue testing, can also be applied to assess the mechanical properties of flame retardant nanocellulose-based materials. Compression testing determines the material's compressive strength and deformation behaviour under axial loading. Shear testing evaluates the material's shear strength and its response to shear forces. Fatigue testing assesses the material's resistance to repeated loading, simulating cyclic or dynamic loading conditions to determine its fatigue life [170, 171].

In summary, mechanical testing plays a vital role in assessing the structural integrity, strength, and durability of flame-retardant nanocellulose-based materials. Tensile testing, flexural testing, impact testing, hardness testing, and other techniques provide valuable insights into the mechanical behaviour and performance of these materials. By understanding their mechanical properties, researchers and practitioners can optimize the design and development of flame-retardant nanocellulose-based materials for a wide range of applications, ensuring their successful implementation in fire-resistant products. The characterization of flame retardant nanocellulose-based materials involves a comprehensive analysis of various properties, including thermal stability, fire behaviour, gas evolution, and char formation. Employing advanced techniques such as DSC, TGA, cone calorimetry, GC–MS, SEM–EDS, FTIR, and XRD provides valuable insights into the flame retardant mechanisms and properties of these materials. By enhancing our understanding of characterization techniques, researchers and industrial practitioners can optimize the design and development of flame-retardant nanocellulose-based materials, leading to safer and sustainable products with improved fire resistance properties.

5.8 Char Formation Analysis

Char formation is a crucial mechanism in the flame-retardancy of nanocellulose materials [124]. As discussed above, scanning electron microscopy coupled with energy-dispersive X-ray spectroscopy allows for the examination of the morphological and elemental changes occurring in the charred residues of nanocellulose-based materials after exposure to fire. Additionally, Fourier-transform infrared spectroscopy and X-ray diffraction techniques help in elucidating the chemical structure and crystallinity of the charred residues, providing valuable information on the formation of protective char layers.

6 Applications of Flame Retardant Nanocellulose-Based Materials

Polymeric materials, such as cellulose, are invaluable commodities that have found applications in various fields. However, the flammability of these materials restricted their full potential. Today, the flammability characteristics of many of the everyday products we use are being closely examined and scrutinized. By decreasing the likelihood of serious fires starting from exposure to ignition sources of various intensities up to the design limits of the systems being utilized, flame retardant systems and additives reduce fire losses, fire injuries, and fire deaths. As a consequence, the modification with flame retardant materials will make these type of materials more practical and safer for application. Fire-resistant materials are being used in the manufacture of many industrial components. Flame retardants have a wide range of uses in both commercial and consumer goods, including furniture (e.g., seats foam and covers, even for vehicles, beds, and carpets), appliances (e.g., phones, televisions, computers, and other electrical and electronic equipment), materials for building and construction including spray foam insulation made of polyurethane and styrene, as well as coatings for electrical lines and cables [172, 173]. High thermal stability and less flammable materials are needed for applications in the electrical and transportation industries [5].

Flame retardant nanocellulose-based materials have therefore been described to be employed as valuable additives for boosting thermal stability and decreasing their flammability features in order to overcome this flammability application restriction [174]. The section below will discuss the recent advances in improving the flame retardant properties of nanocelluose materials and their potential applications in a variety of fields.

A potential and fascinating topic of current and future research and development (R&D) is the applicability of nanocellulose-based materials for diverse military applications. There have been some recent advancements in the usage of nanocellulose for military purposes as reported by Norrrahim et al. [175].

Chen et al. fabricated a flame-retardant, anti-dripping, easy-processing and versatile cellulosic material, DOPO-cellulose acrylate (DCA), through the modification of the cellulose chain with acrylate groups and 9,10-dihydro-9-oxa-10-phosphaphenanthrene-10-oxide (DOPO) [176]. The resultant DCA is then utilised as an adaptable and transparent coatings for protecting flammable materials from fire. The flame retardancy results showed that the DCA film self-extinguishes immediately (<1 s) following removal from the flame as compared to the pure cellulose. Moreover, the heat transformation and gas diffusion were blocked off by the formation of the dense, homogeneous and continuous char. The MCC data has indicated a decrease in PHRR and heat release capacity (HRC) following the addition of the DOPO into the cellulose chains. In comparing the pure cellulose to the DCA with a degree of substitution (DS) DOPO of 0.31, both the PHRR and the HRC values decreased from 275.2 W/g to 71.9 W/g and from 274 J/(g·K) to 72 J/(g·K), respectively. Furthermore, the THR of DCAs with DSDOPO values of 0.16–0.58 is noticeably lower than that of pure cellulose. Taking into account the factors affecting flame retardancy, inclusive of self-extinguishing time, char yield, PHRR, HRC and THR, the DCAs with DSDOPO values of 0.31–0.58 have optimum flame-retardant properties, precisely as a result of the formation of a dense and continuous char and the marked decline in released heat.

Melro et al. fabricated a mordenite modified bacterial nanocellulose (BNC_MOR) composite and then assessed its flame retardancy properties for possible applications in textile industries [177]. The flammability tests demonstrated that the BNC took only 3.6 s to thoroughly burn while the BNC_MOR sample did not burn at all. This obviously showed the efficacy of MOR as a flame retardant.

A CNFs-based flame retardant has been produced by Zhang et al. through the phosphorylation of CNFs by a mechanomechanical approach [178]. The resultant flame retardant is then applied to bamboo pulp paper. The flame retardancy data has showed an increased LOI with an increasing loading of the flame retardant: The controlled paper was found to be 17% while the 30 wt% loaded sample achieved 31%, which surpassed the industrial standard for flame retardant materials, 27% [179]. However, the LOI values of 10 and 20 wt % of P2C1NM-10 were nearly the same, which may be because of the size of charred layers not being big enough to slow down the oxygen and heat to the inner material. Interestingly, the charred layers formed from 30 wt% loaded sample were large enough to cover the paper and shield the inner material. As a result, The LOI tendency was the same as that in the literature [180].

In respect of the combustion behaviour, the heat release rate of the P2C1NM-10 loaded paper dropped [181]. Similarly, the PHRR of treated paper (138.7 W/g) declined by about 62.8% as compared to the control paper (379.5 W/g). In addition, the THR of the P2C1NM-10-loaded papers reduced from 12.5 kJ/g of the control paper to 3.4 kJ/g. By and large, the THR of paper is proportional to the carbohydrate amounts. However, phosphorous could accelerate the cellulose dehydration to form charred layers, which could slow down heat and oxygen to the inner material. The paper could not be ignited with the limited oxygen and heat, thus, the ultimate amount of the heat was reduced. The considerable reduction of both the PHRR and THR show that the P2C1NM-10 is an effective flame retardant [181].

The work by Yildirim et al. developed fire-retardant bio-structural panels (BISPs) using CNFs, corn starch, boric acid ($B(OH)_3$), and n-dodecenyl succinic anhydride (n-DDSA) with adhesive-free character [182]. The performance properties were examined and contrasted with other prominent products already on the market. The developed BISP showed 100% fire-retardant character and was found to be statistically better than the other commercially available products. In addition, the smoke development study showed that the BISP generated nearly zero smoke (2.20% maximum smoke density) while the other market products produced 15–20 times more smoke than the BISP.

The nanocrystalline cellulose (NCC) using the acid hydrolysis method and its surface was chemically modified via silylation treatment (MTMS) as reported by Kima et al. [183]. The flammability results have shown that the NCC had a LOI value of 18.4% which suggests that it could burn easily and rapidly in an open atmosphere. However, the materials with different silylation ratios saw an increase in LOI values ranging from 20.2 (S1), 27.2 (S2) and 34.2% (S3). This observation indicates the slow burning of the materials. It's worth mentioning that the LOI value of the S3 was about 60% higher than weight fraction of oxygen in the air (21%).

Farooq et al. used freeze-drying method to prepared sodium bicarbonate (SBC)-modified carbon nanofibrils aerogel flame retardant. The flame retardantcy studies have indicated that the neat carbon nanofribils aerogels ignited and immediately covered in flames in <1 s [184]. Thus, the average combustion velocity of pure carbon nanofibrils aerogel was found to be 5.84 cm s^{-1} while the SBC-carbon nanofribils aerogels showed a distinctly lower burning rate. A 10 wt% of SBC (C-SB10) had a notable flame-retarding effect as compared to the pure CNF aerogel with a burning velocity 0.33 cm s^{-1} along with a self-extinguishing behaviour after 5 s of initial burning. The flame retardant aerogels (C-SB20, C-SB30, and C-SB40) self-extinguished soon after the flame source was removed resulting in a decrease in the combustion velocity. However, after 10 s of burning, the combustion velocity remains unchanged in particular among C-SB20, C-SB30, and C-SB40 samples. Generally, the flammability of carbon nanofribils aerogels was positively impacted by the addition of SBC due to the release of carbon dioxide and water upon heat-induced decomposition.

The work by Yue et al. studied the flame retardancy of cellulose aerogel produced via the formaldehyde (MF)/MTMS double cross-linking approach [185]. It emerged that the LOI value of the pure CNF aerogel was only 19.5%, while the UL-94 grade was NR, confirming the high flammability character of the CNF aerogel. Upon incorporation of the MF and/or MTMS, the flame retardant properties of compound aerogels increased. However, the UL-94 grade of CNF/MF still maintained at NR, and only UL-94 V1 grade could be achieved by higher MF loading (CNF/MF-1). In addition, the LOI value of aerogels modified by MTMS (Si-CNF and Si-CNF-1) reached 30.4% and 33.7%, respectively, and the UL-94 grade of Si-CNF progressed to V1 level even under a relatively low MTMS loading. Furthermore, the addition of both MF and MTMS into CNF matrix resulted in increased flame retardant properties. The LOI value and UL-94 grade of Si-CNF/MF reached 37.1% and V0 level, respectively, which showed that a synergistic effect might have existed between MF and MTMS.

A cellulose-ammonium phosphate (APP) flame retardant was prepared by Chandrasekaran et al. and the resultant product is earmarked for electronic device applications [137]. The flame retardancy results have shown that the LOI value of the cellulose composites increased from 19 to 25.3% upon addition of 10 wt% APP. Furthermore, the horizontal burning test (UL-94 HB) results demonstrated that 10 wt% APP contributed self-extinguishing fire properties to the cellulose. This is as a result of APP producing phosphoric acid capable of stimulating the dehydration of hydroxyl groups of cellulose, thus, preventing the formation of combustible volatiles. The above LOI and UL-94 HB results are a testament that the char created from APP acts as a protective barrier. As a result, the APP has proven to be an excellent intumescent flame-retardant on the cellulose matrix.

The synthesized carbon nanofibril (CNF)/xyloglucan (XG)/montmorillonite clay (MTM) foam using a water-based freeze casting approach indicated the almost immediate self-extinguishing behaviour soon after removal of the igniting blue methane flame [116]. However, a commercially available rigid PU foam exhibited flaming combustion with strong flames and release of dense smoke. As regards the LOI testing, the carbon nanofribils/XG/MTM foam attained a LOI rating of 31.6% which is well above the range for commercially available fire-retardant FR-PU foams [186]. In addition, a comparison of LOI values for fossil-based, fire-retardant FR foams with the best performance: PVC, phenolic resin and polyimide, such a comparison shows the high performance of carbon nanofribil/XG/MTM foams. [187–193]. Furthermore, the carbon nanofribil/XG/MTM foams displayed very low HRR values (below 100 kW m^{-2}), which are comparable to the phenolic resins (among the best foams available) [194]. In addition, the carbon nanofribil/XG/MTM foams were found to be capable of resisting the flame penetration of the torch while maintaining the structural integrity [120].

The flame retardant CNFs/graphene oxide (GO)/sepiolite nanorods (SEP)/boric acid (BA) foam was fabricated by Wicklein et al. using the freeze-casting method [77]. The fire retardancy results showed that an optimized addition of GO (10 wt.%), SEP (10 wt.%) and BA (3 wt.%) demonstrated a very good fire retardancy behaviour. However, foams with suboptimal composition (i.e. with no GO or low clay content) displayed lower fire retardancy properties as compared an optimal composition. In respect of LOI, the nanocomposite foams have considerably higher LOI-value compared to commercial, flame retardant-containing polymer-based foams, which displayed usual LOI-values between 22 and 25 [195].

Berglund et al. fabricated seaweed-derived alginate-CNFs (A-CNFs) aerogel using a three-step approach for insulation applications [196]. First, CNFs were separated from alginate-rich seaweed followed by the assembling of an anisotropic aerogel using ice-templating and freeze-drying methods. Lastly, the flame-retardant behaviour and stability of the resultant products were improved using a crosslinking approach. The flame retardancy results showed that the average combustion velocity of A-CNFs was 1.2 m s^{-1} $\pm$ 0.1, while a lower burning rate (0.75 $\pm$ 0.1 mm s^{-1}) was observed for all cross-linked aerogels. The flameless combustion behaviour slowly consumed the A-CNFs aerogel, leaving behind an uncollapsed, shrunken char residue. In contrast, all crosslinked aerogels were, in part, consumed and exhibited a self-extinguishing behaviour. Fire-retardant behaviour was seen at 1 wt% crosslinking, and the flammability was not impacted at increased concentration. This observation is in agreement with previously studied alginate aerogels, which were found to exhibit non-flammability and only smoldering behavior during combustion [197].

A waterborne polyurethane (WPU)/NCC/magnesium hydroxide (MgOH) flame retardant was prepared by Cheng et al. for wood coating applications [198]. The results have demonstrated that, following combustion as presented in Fig. 9.3a, the residual amount of char for the specimens were different as a function of the coating composition. As regards the coating without NCC (NCC0%–MgOH40%), the sample burnt completely. However, in the presence of 5% of NCC, it was noticed

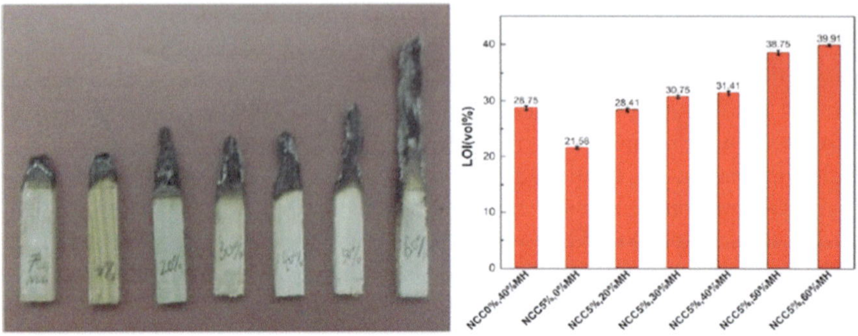

Fig. 9.3 The specimen images of **a** after oxygen index method and **b** LOI bar chart [198]

that the length of the coal residue increased with an increase in the MH in the formulation. At 60.0% MgOH, the coating above the tick mark was not completely blackened, however, a gradual transition from carbon black to brown and then to white was visible. This observation suggests that when the flame burned to the tick mark and went out, the coating above did not burn and decomposed rapidly.

As can be seen from Fig. 9.3b, the oxygen index value of the sample also increased from ~22 vol% to ~40 vol% as the amount of the magnesium hydroxide (MH) increased from 0 wt% to 60 wt%. A comparison of NCC0%–MH40% and NCC5%–MH40% indicated that the addition of cellulose nanowhiskers in the formulation led to an increase in LOI value from 28.75 vol% to 30.75 vol% [198].

7 Conclusions and Perspectives

Nanocellulose is another material that shows promise for usage as a reinforcing filler in composite materials due to its lightweight, high surface area, high mechanical strength, high aspect ratio, and low-density characteristics. The development and characterization of flame-resistant nanocellulose-based materials have drawn a lot of interest in recent years as a result of the rising need for environmentally friendly, high-performance materials that are sustainable and sustainable. Due to its remarkable qualities, including high mechanical strength, low density, excellent biodegradability, and exceptional compatibility with other polymers, nanocellulose, a versatile and renewable natural polymer derived from plant cellulose, has emerged as a promising candidate for a variety of applications. The intrinsic flammability and lack of fire resistance of nanocellulose, however, is a significant disadvantage. As a result, the creation of effective flame-retardant methods for these materials has emerged as a crucial research area. Flame-retardant nanocellulose-based materials have been developed using a variety of fabrication methods. The majority of these involve physically blending or chemically altering the nanocellulose matrix to include flame retardants into it. Surface grafting or covalent attachment of flame-retardant substances to the hydroxyl groups present on the nanocellulose surface is the conventional method of chemical modification. By using this technique, you can be confident that the flame retardants are evenly distributed throughout the material and will hold up even in the worst circumstances. Physical blending, on the other hand, merely entails combining flame retardants with nanocellulose before shaping it into the required shape or structure. This procedure is simpler and less expensive, but it performs flame retardancy less effectively than chemical alteration. The ability of materials based on nanocellulose to resist fire can be improved by using various kinds of flame retardants. Due to their thermal stability, non-toxicity, and environmental friendliness, inorganic compounds such metal hydroxides (for example, aluminium hydroxide), metal oxides (for example, zinc oxide), clay minerals (for example, montmorillonite) and nanomaterials are frequently utilized. By establishing a barrier on the material surface, these substances serve as fire barriers, limiting heat transfer and preventing ignite. They also encourage the production of char during burning, which hinders flame growth even more and lessens smoke release. Although less prevalent than their inorganic equivalents, organic flame retardants have also been studied for their potential to increase the fire resistance of nanocellulose-based products. These include known halogenated substances (such as decabromodiphenyl ether), phosphorus-containing substances (such as ammonium polyphosphate), and boron-containing molecules (such as boric acid and borate salts). Organic flame retardants often work through a variety of processes, such as condensed-phase charring enhancement, gas-phase radical quenching, or both. However, their broad use in nanocellulose applications has been constrained by worries about their environmental impact and potential health risks. Exploiting the synergistic interactions between various flame retardants or combining them with other functional additions like carbon nanotubes or graphene oxide is another interesting method for creating flame-retardant nanocellulose-based materials. This method not only improves the material's overall performance but also aids in getting around various drawbacks related to particular flame retardants. The effectiveness of flame-retardant nanocellulose-based materials must be assessed, and the fundamental mechanisms driving their enhanced fire resistance must be understood, using characterization methodologies. Thermogravimetric analysis (TGA) and differential scanning calorimetry (DSC) are two thermal analysis techniques frequently used to examine the thermal breakdown behaviour, char residue production, and melting properties of these materials. Additionally, flammability tests such as cone calorimetry, limiting oxygen index (LOI), and UL-94 horizontal burning test offer useful data on the timing of ignition, the rate of heat release, the spread of the flame, and the production of smoke during combustion. Using scanning electron microscopy (SEM) or transmission electron microscopy (TEM), morphological characterization can show the distribution of flame retardants within the nanocellulose matrix as well as the structural alterations that take place when exposed to fire.

When creating novel nanocellulose-based materials with improved fire resistance, additional aspects, such as mechanical characteristics, surface hydrophobicity, optical transparency, and biodegradability should also be taken into account. The kind and quantity of flame retardants utilized, the fabrication technique used, and any further processing processes used all have a significant impact on these qualities. Therefore, for these materials to be successfully used in a variety of applications ranging from packaging and construction to the automotive and aerospace sectors, a thorough understanding of the structure–property correlations in these materials is required. Despite the substantial advancements in the fabrication of flame-resistant nanocellulose-based materials to date, a number of issues still need to be resolved in order to fully realize their potential advantages.

Before these materials are widely commercialized, potential environmental dangers related to some flame-retardants must also be carefully evaluated. Furthermore, more sophisticated characterisation methods are needed to acquire a better understanding of the intricate processes that take place during combustion on both a macroscopic and molecular level. Finally, in order to reduce environmental effects while retaining high efficiency, future research approaches may concentrate on investigating new forms of eco-friendly flame-retardants made from waste materials or renewable resources. Their range of applications will also be further expanded by the development of multifunctional nanocellulose composites with enhanced fire-resistant qualities as well as other functional qualities like thermal insulation or electromagnetic interference shielding. In summary, the fabrication and characterization of flame-retardant nanocellulose-based materials represent a rapidly evolving field with tremendous potential for innovation, ultimately contributing to the sustainable development of advanced materials for various industries.

Acknowledgements The authors acknowledge DSI/Mintek Nanotechnology Innovation Centre and Mintek, South Africa, for supporting this work.

References

1. Jefferson Andrew, J., Dhakal, H.N.: Sustainable biobased composites for advanced applications: recent trends and future opportunities—a critical review. Compos. Part C: Open Access **7**, 100220 (2022)
2. Mishra, T., Mandal, P., Rout, A.K., Sahoo, D.: A state-of-the-art review on potential applications of natural fiber-reinforced polymer composite filled with inorganic nanoparticle. Compos. Part C: Open Access **9**, 100298 (2022)
3. Faruk, O., Bledzki, A.K., Fink, H.-P., Sain, M.: Progress report on natural fiber reinforced composites. Macromol. Mater. Eng. **299**, 9–26 (2014). https://doi.org/10.1002/mame.201300008
4. Villamil Watson, D.A., Schiraldi, D.A.: Biomolecules as flame retardant additives for polymers: a review. Polymers **12**(4), 849 (2020)
5. Laoutid, F., Bonnaud, L., Alexandre, M., Lopez-Cuesta, J.-M., Dubois, P.: New prospects in flame retardant polymer materials: from fundamentals to nanocomposites. Mater. Sci. Eng. R: Rep. **63**(3), 100–125 (2009). ISSN 0927-796X. https://doi.org/10.1016/j.mser.2008.09.002
6. Azwa, Z.N., Yousif, B.F., Manalo, A.C., Karunasena, W.: A review on the degradability of polymeric composites based on natural fibres. Mater. Des. **47**, 424–442 (2013). ISSN 0261-3069. https://doi.org/10.1016/j.matdes.2012.11.025

7. Moon, R.J., Martini, A., Nairn, J., Simonsen, J., Youngblood, J.: Cellulose nanomaterials review: structure, properties and nanocomposites. Chem, Soc. Rev. **40**, 3941 (2011)
8. Klemm, D., Heublein, B., Fink, H.-P., Bohn, A.: Cellulose: fascinating biopolymer and sustainable raw material. Angew. Chem. Int. Ed. **44**, 3358–3393 (2005)
9. Gan, P.G., Sam, S.T., Abdullah, M.F., Omar, M.F.: Thermal properties of nanocellulose-reinforced composites: a review. J. Appl. Polym. Sci. **137**, 48544 (2020). https://doi.org/10.1002/app.48544
10. Saba, N., Jawaid, M.: Recent advances in nanocellulose-based polymer nanocomposites. In: Jawaid, M., Boufi, S., Abdul Khalil H.P.S. (eds.) Woodhead Publishing Series in Composites Science and Engineering, Cellulose-Reinforced Nanofibre Composites, pp. 89–112. Woodhead Publishing (2017)
11. Klemm, D., Kramer, F., Moritz, S., Lindström, T., Ankerfors, M., Gray, D., Dorris, A.: Nanocelluloses: a new family of nature-based materials. Angew. Chem. Int. Ed. **50**, 5438–5466 (2011)
12. Fang, Z., Hou, G., Chen, C., Liangbing, H.: Nanocellulose-based films and their emerging applications. Curr. Opin. Solid State Mater. Sci. **23**(4), 100764 (2019)
13. Purser, D.A.: Fire Safety Performance of Flame Retardants Compared with Toxic and Environmental Hazards. Elsevier B.V. (2014). https://doi.org/10.1016/B978-0-444-53808-6.00002-0
14. Speight, J.G.: Sources and Types of Organic Pollutants (2017). https://doi.org/10.1016/b978-0-12-804492-6.00004-6
15. Horrocks, A.R.: Flame retardant challenges for textiles and fibres: new chemistry versus innovatory solutions. Polym. Degrad. Stab. **96**, 377–392 (2011). https://doi.org/10.1016/j.polymdegradstab.2010.03.036
16. Qi, P., Chen, F., Li, Y., Li, H., Gu, X., Sun, J., Zhang, S.: A review of durable flame-retardant fabrics by finishing: fabrication strategies and challenges. Springer Nature Singapore (2023). https://doi.org/10.1007/s42765-023-00255-x
17. Jeong, S.H., Park, C.H., Song, H., Heo, J.H., Lee, J.H.: Biomolecules as green flame retardants: recent progress, challenges, and opportunities. J. Clean. Prod. **368**, 133241 (2022). https://doi.org/10.1016/j.jclepro.2022.133241
18. Trache, D., Tarchoun, A.F., Derradji, M., Hamidon, T.S.: Nanocellulose : from fundamentals to advanced applications. Front. Chem. **8**, 1–33 (2020). https://doi.org/10.3389/fchem.2020.00392
19. Mohamad, N., Mohammad Taib, T., Sherwyn Hamidon, Z.N.G., Trache, D., Uyama, H., Hussin, M.H.: Recent progress in cellulose-based composites towards flame retardancy applications. Polymer (Guildf) **244**, 124677 (2022). https://doi.org/10.1016/j.polymer.2022.124677
20. Norrrahim, M.N.F., Kasim, N.A.M., Knight, V.F., Ujang, F.A., Janudin, N., Razak, M.A.I.A., Shah, N.A.A., Noor, S.A.M., Jamal, S.H.: Nanocellulose : the next super versatile material for the military. Mater. Adv. **2**, 1485–1506 (2021). https://doi.org/10.1039/d0ma01011a
21. Weiqi Leng, Z.C., Li, J.: Synthesis and characterization of cellulose nanofibril-reinforced polyurethane foam. Polymers (Basel) **9**, 1–14 (2017). https://doi.org/10.3390/polym9110597
22. Tan, B.K., Ching, Y.C., Poh, S.C., Abdullah, L.C.: A review of natural fiber reinforced poly (vinyl alcohol) based composites: application and opportunity. Polymers 2205–2222 (2015). https://doi.org/10.3390/polym7111509
23. Brebu, M.: Environmental degradation of plastic composites with natural fillers-a review. Polymers (Basel) **12** (2020). https://doi.org/10.3390/polym12010165
24. Omran, A.A.B., Mohammed, A.A.B.A., Sapuan, S.M., Ilyas, R.A., Asyraf, M.R.M., Koloor, S.S.R., Petrů, M.: Micro-and nanocellulose in polymer composite materials: a review. Polymers (Basel). **13**, 1–30 (2021). https://doi.org/10.3390/polym13020231
25. Isogai, A.: Emerging nanocellulose technologies: recent developments. Adv. Mater. **33**, 1–10 (2021). https://doi.org/10.1002/adma.202000630
26. L.A.S. Costa, de Assis, D.J., Gomes, G.V.P., Da Silva, J.B.A., Fonsêca, A.F., Druzian, J.I.: Extraction and characterization of nanocellulose from corn Stover. Mater. Today Proc. **2**, 287–294 (2015). https://doi.org/10.1016/j.matpr.2015.04.045

27. Bajwa, D.S., Rehovsky, C., Shojaeiarani, J., Stark, N., Bajwa, S., Dietenberger, M.A.: Functionalized cellulose nanocrystals: a potential fire retardant for polymer composites. Polymers (Basel) **11**, 1–7 (2019). https://doi.org/10.3390/polym11081361
28. Fang, Z., Hou, G., Chen, C., Hu, L.: Nanocellulose-based films and their emerging applications. Curr. Opin. Solid State Mater. Sci. **23**, 100764 (2019). https://doi.org/10.1016/j.cossms.2019.07.003
29. Beroual, M., Trache, D., Mehelli, O., Boumaza, L., Tarchoun, A.F., Derradji, M., Khimeche, K.: Effect of the delignification process on the physicochemical properties and thermal stability of microcrystalline cellulose extracted from date palm fronds. Waste Biomass Valorization **12**, 2779–2793 (2021). https://doi.org/10.1007/s12649-020-01198-9
30. Beroual, M., Mehelli, O., Boumaza, L., Trache, D., Tarchoun, A.F.: Synthesis and characterization of microcrystalline cellulose from giant reed using different delignification processes. Mater. Res. Appl. 173–187 (2020). https://doi.org/10.1007/978-981-15-9223-2_7
31. Haldar, D., Purkait, M.K.: Micro and nanocrystalline cellulose derivatives of lignocellulosic biomass: a review on synthesis, applications and advancements. Carbohydr. Polym. **250**, 116937 (2020). https://doi.org/10.1016/j.carbpol.2020.116937
32. Trache, D., Hussin, M.H., Hui Chuin, C.T., Sabar, S., Fazita, M.R.N., Taiwo, O.F.A., Hassan, T.M., Haafiz, M.K.M.: Microcrystalline cellulose: Isolation, characterization and bio-composites application—a review. Int. J. Biol. Macromol. **93**, 789–804 (2016). https://doi.org/10.1016/j.ijbiomac.2016.09.056
33. Tarchoun, A.F., Trache, D., Klapötke, T.M., Khimeche, K.: Tetrazole-functionalized microcrystalline cellulose: a promising biopolymer for advanced energetic materials. Chem. Eng. J. **400**, 125960 (2020). https://doi.org/10.1016/j.cej.2020.125960
34. Bouhoute, M., Taarji, N., de Oliveira Felipe, L., Habibi, Y., Kobayashi, I., Zahar, M., Isoda, H., Nakajima, M., Neves, M.A.: Microfibrillated cellulose from *Argania spinosa* shells as sustainable solid particles for O/W Pickering emulsions. Carbohydr. Polym. **251**, 116990 (2021). https://doi.org/10.1016/j.carbpol.2020.116990
35. Singh, M., Pahal, V., Ahuja, D.: Isolation and characterization of microfibrillated cellulose and nanofibrillated cellulose with "biomechanical hotspots." Carbohydr. Polym. **234**, 115827 (2020). https://doi.org/10.1016/j.carbpol.2020.115827
36. Nechyporchuk, O., Belgacem, M.N., Bras, J.: Production of cellulose nanofibrils: a review of recent advances. Ind. Crops Prod. **93**, 2–25 (2016). https://doi.org/10.1016/j.indcrop.2016.02.016
37. Osong, S.H., Norgren, S., Engstrand, P.: Processing of wood-based microfibrillated cellulose and nanofibrillated cellulose, and applications relating to papermaking: a review. Cellulose **23**, 93–123 (2016). https://doi.org/10.1007/s10570-015-0798-5
38. Ma, T., Hu, X., Lu, S., Liao, X., Song, Y., Hu, X.: Nanocellulose: a promising green treasure from food wastes to available food materials. Crit. Rev. Food Sci. Nutr. **62**, 989–1002 (2022). https://doi.org/10.1080/10408398.2020.1832440
39. Shen, Z., Qin, M., Xiong, F., Zou, R., Zhang, J.: Nanocellulose-based composite phase change materials for thermal energy storage: status and challenges. Energy Environ. Sci. **16**, 830–861 (2023)
40. Kumar, R., Rai, B., Gahlyan, S., Kumar, G.: A comprehensive review on production, surface modification and characterization of nanocellulose derived from biomass and its commercial applications. Express Polym. Lett. **15**, 104–120 (2021). https://doi.org/10.3144/expresspolymlett.2021.11
41. Chen, C., Hu, L.: Nanocellulose toward advanced energy storage devices: structure and electrochemistry. Acc. Chem. Res. **51**, 3154–3165 (2018). https://doi.org/10.1021/acs.accounts.8b00391
42. Naz, S., Ali, J.S., Zia, M.: Nanocellulose isolation characterization and applications: a journey from non-remedial to biomedical claims. Bio-Design Manuf. **2**, 187–212 (2019). https://doi.org/10.1007/s42242-019-00049-4
43. Brinchi, L., Cotana, F., Fortunati, E., Kenny, J.M.: Production of nanocrystalline cellulose from lignocellulosic biomass: technology and applications. Carbohydr. Polym. **94**, 154–169 (2013). https://doi.org/10.1016/j.carbpol.2013.01.033

44. Charreau, H., Foresti, M.L., Vazquez, A.: Nanocellulose patents trends: a comprehensive review on patents on cellulose nanocrystals, microfibrillated and bacterial cellulose. Recent Pat. Nanotechnol. **7**, 56–80 (2012). https://doi.org/10.2174/187221051130106
45. Mariano, M., El Kissi, N., Dufresne, A.: Cellulose nanocrystals and related nanocomposites: review of some properties and challenges. J. Polym. Sci. Part B Polym. Phys. **52**, 791–806 (2014). https://doi.org/10.1002/polb.23490
46. Vazquez, V.P.C.A., Laura Foresti, M., Moran, J.I.: Chapter 7. Extraction and production of cellulose nanofibers. Handb. Polym. Nanocompos. Process. Perform. Appl. Vol. C Polym. Nanocompos. Cellul. Nanoparticles 1–511 (2015). https://doi.org/10.1007/978-3-642-45232-1
47. Carvalho, T., Guedes, G., Sousa, F.L., Freire, C.S.R., Santos, H.A.: Latest advances on bacterial cellulose-based materials for wound healing, delivery systems, and tissue engineering. Biotechnol. J. **14**, 1–19 (2019). https://doi.org/10.1002/biot.201900059
48. Sharma, C., Bhardwaj, N.K.: Bacterial nanocellulose: present status, biomedical applications and future perspectives. Mater. Sci. Eng. C **104**, 109963 (2019). https://doi.org/10.1016/j.msec.2019.109963
49. Trache, D.: Nanocellulose as a promising sustainable material for biomedical applications. AIMS Mater. Sci. **5**, 201–205 (2018). https://doi.org/10.3934/matersci.2018.2.201
50. Liu, W., Du, H., Zhang, M., Liu, K., Liu, H., Xie, H., Zhang, X., Si, C.: Bacterial cellulose-based composite scaffolds for biomedical applications: a review. ACS Sustain. Chem. Eng. **8**, 7536–7562 (2020). https://doi.org/10.1021/acssuschemeng.0c00125
51. Lin, D., Liu, Z., Shen, R., Chen, S., Yang, X.: Bacterial cellulose in food industry: current research and future prospects. Int. J. Biol. Macromol. **158**, 1007–1019 (2020). https://doi.org/10.1016/j.ijbiomac.2020.04.230
52. Sherazi, T.A.: Encyclopedia of membranes. Encycl. Membr. 3–5 (2015). https://doi.org/10.1007/978-3-642-40872-4
53. Carlmark, A., Larsson, E., Malmström, E.: Grafting of cellulose by ring-opening polymerisation—a review. Eur. Polym. J. **48**, 1646–1659 (2012). https://doi.org/10.1016/j.eurpolymj.2012.06.013
54. Karaj-Abad, S.G., Abbasian, M., Jaymand, M.: Grafting of poly[(methyl methacrylate)-block-styrene] onto cellulose via nitroxide-mediated polymerization, and its polymer/clay nanocomposite. Carbohydr. Polym. **152**, 297–305 (2016). https://doi.org/10.1016/j.carbpol.2016.07.017
55. Neves, R.M., Ornaghi, H.L., Zattera, A.J., Amico, S.C.: Recent studies on modified cellulose/nanocellulose epoxy composites: a systematic review. Carbohydr. Polym. **255** (2021). https://doi.org/10.1016/j.carbpol.2020.117366
56. Coseri, S., Biliuta, G., Simionescu, B.C., Stana-Kleinschek, K., Ribitsch, V., Harabagiu, V.: Oxidized cellulose—survey of the most recent achievements. Carbohydr. Polym. **93**, 207–215 (2013). https://doi.org/10.1016/j.carbpol.2012.03.086
57. Tang, Z., Li, W., Lin, X., Xiao, H., Miao, Q., Huang, L., Chen, L., Wu, H.: TEMPO-oxidized cellulose with high degree of oxidation. Polymers (Basel) **9**, 3–4 (2017). https://doi.org/10.3390/polym9090421
58. Zimnitsky, D.S., Yurkshtovich, T.L., Bychkovsky, P.M.: Synthesis and characterization of oxidized cellulose. J. Polym. Sci. Part A Polym. Chem. **42**, 4785–4791 (2004). https://doi.org/10.1002/pola.20302
59. Wang, Y., Wang, X., Xie, Y., Zhang, K.: Functional nanomaterials through esterification of cellulose: a review of chemistry and application. Cellulose **25**, 3703–3731 (2018). https://doi.org/10.1007/s10570-018-1830-3
60. Hansu Kima, Y.S.S., Youna, J.R.: Eco-friendly flame retardant nanocrystalline cellulose prepared via silylation. Nanotechnology **29**, 455702–455723 (2018). https://doi.org/10.1088/1361-6528/aadc87
61. Kim, H., Park, J., Minn, K.S., Pak, S.Y., Lee, D., Youn, J.R., Song, Y.S.: Flame retardant composite foam modified by silylated nanocellulose and tris (2-chloropropyl) phosphate. Fibers Polym. **20**, 2280–2288 (2019). https://doi.org/10.1007/s12221-019-9491-x

62. Salmeia, K.A., Jovic, M., Ragaisiene, A., Rukuiziene, Z., Milasius, R., Mikuioniene, D., Gaan, S.: Flammability of cellulose-based fibers and the effect of structure of phosphorus compounds on their flame retardancy, Polymers (Basel) **8** (2016). https://doi.org/10.3390/polym8080293
63. Gaan, S., Sun, G.: Effect of phosphorus flame retardants on thermo-oxidative decomposition of cotton. Polym. Degrad. Stab. **92**, 968–974 (2007). https://doi.org/10.1016/j.polymdegradstab.2007.03.009
64. Zhang, S., Li, S., Wu, Q., Li, Q., Huang, J., Li, W.: Phosphorus containing group and lignin toward intrinsically flame retardant cellulose nanofibril-based film with enhanced mechanical properties. Compos. Part B. **212**, 108699 (2021). https://doi.org/10.1016/j.compositesb.2021.108699
65. Xie, K., Gao, A., Zhang, Y.: Flame retardant finishing of cotton fabric based on synergistic compounds containing boron and nitrogen. Carbohydr. Polym. **98**, 706–710 (2013). https://doi.org/10.1016/j.carbpol.2013.06.014
66. Cheng, X., Zhu, S., Pan, Y., Deng, Y., Shi, L., Gong, L.: Fire retardancy and thermal behaviors of cellulose nanofiber/zinc borate aerogel. Cellulose **27**, 7463–7474 (2020). https://doi.org/10.1007/s10570-020-03289-1
67. Le, N.D., Trogen, M., Varley, R.J., Hummel, M., Byrne, N.: Effect of boric acid on the stabilisation of cellulose-lignin filaments as precursors for carbon fibres. Cellulose **28**, 729–739 (2021). https://doi.org/10.1007/s10570-020-03584-x
68. Yuan, B., Zhang, J., Yu, J., Song, R., Mi, Q., He, J., Zhang, J.: Transparent and flame retardant cellulose/aluminum hydroxide nanocomposite aerogels. Sci. China Chem. **59**, 1335–1341 (2016). https://doi.org/10.1007/s11426-016-0188-0
69. Han, Y., Zhang, X., Wu, X., Lu, C.: Flame retardant, heat insulating cellulose aerogels from waste cotton fabrics by in situ formation of magnesium hydroxide nanoparticles in cellulose gel nanostructures. ACS Sustain. Chem. Eng. **3**, 1853–1859 (2015). https://doi.org/10.1021/acssuschemeng.5b00438
70. He, C., Huang, J., Li, S., Meng, K., Zhang, L., Chen, Z., Lai, Y.: Mechanically resistant and sustainable cellulose-based composite aerogels with excellent flame retardant, sound-absorption, and superantiwetting ability for advanced engineering materials. ACS Sustain. Chem. Eng. **6**, 927–936 (2018). https://doi.org/10.1021/acssuschemeng.7b03281
71. Luo, J., Wang, H.: Preparation, thermal insulation and flame retardance of cellulose nanocrystal aerogel modified by TiO_2. Int. J. Heat Technol. **36**, 614–618 (2018). https://doi.org/10.18280/ijht.360226
72. Wang, X., Kalali, E.N., Wan, J.T., Wang, D.Y.: Carbon-family materials for flame retardant polymeric materials. Prog. Polym. Sci. **69**, 22–46 (2017). https://doi.org/10.1016/j.progpolymsci.2017.02.001
73. Li, M., Pang, L., Chen, M., Xie, J., Liu, Q.: Effects of aluminum hydroxide and layered double hydroxide on asphalt fire resistance. Materials (Basel) **11** (2018). https://doi.org/10.3390/ma11101939
74. Uddin, M., Alabbad, M., Li, L., Orell, O., Sarlin, E., Haapala, A.: Novel micronized mica modified casein–aluminum hydroxide as fire retardant coatings for wood products. Coatings **12** (2022). https://doi.org/10.3390/coatings12050673
75. Sertsova, A.A., Marakulin, S.I., Yurtov, E.V.: Metal compound nanoparticles: flame retardants for polymer composites. Russ. J. Gen. Chem. **87**, 1395–1402 (2015). https://doi.org/10.1134/S1070363217060421
76. Wang, Z., Han, E., Ke, W.: Effect of nanoparticles on the improvement in fire-resistant and anti-ageing properties of flame-retardant coating. Surf. Coatings Technol. **200**, 5706–5716 (2006). https://doi.org/10.1016/j.surfcoat.2005.08.102
77. Wicklein, B., Kocjan, A., Salazar-Alvarez, G., Carosio, F., Camino, G., Antonietti, M., Bergström, L.: Thermally insulating and fire-retardant lightweight anisotropic foams based on nanocellulose and graphene oxide. Nat. Nanotechnol. **10**(3), 277−283 (2015). https://doi.org/10.1038/nnano.2014.248

78. Pack, S., Kashiwagi, T., Stemp, D., Koo, J., Si, M., Sokolov, J.C., Rafailovich, M.H.: Segregation of carbon nanotubes/organoclays rendering polymer blends self-extinguishing. Macromolecules **42**, 6698–6709 (2009). https://doi.org/10.1021/ma900966k
79. Medina, L., Carosio, F., Berglund, L.A.: Recyclable nanocomposite foams of poly(vinyl alcohol), clay and cellulose nanofibrils—mechanical properties and flame retardancy. Compos. Sci. Technol. **182**, 107762 (2019). https://doi.org/10.1016/j.compscitech.2019.107762
80. Ma, H., Tong, L., Xu, Z., Fang, Z.: Synergistic effect of carbon nanotube and clay for improving the flame retardancy of ABS resin. Nanotechnology **18**, 375602–375610 (2007). https://doi.org/10.1088/0957-4484/18/37/375602
81. Sertsova, A. A., Koroleva, M. Y., Yurtov, E. V., Pravednikova, O. B., Dutikova, O.S., Gal'Braikh, L.S.: Fire-resistant polymer nanocomposites based on metal oxides and hydroxides, Theor. Found. Chem. Eng. **44**, 772–777 (2010). https://doi.org/10.1134/S0040579510050222.
82. DeVree, B.T., Steiner, L.M., Głazowska, S., Ruhnow, F., Herburger, K., Persson, S., Mravec, J.: Current and future advances in fluorescence-based visualization of plant cell wall components and cell wall biosynthetic machineries. Biotechnol. Biofuels **14**, 78 (2021)
83. Zhao, X., Bhagia, S., Gomez-Maldonado, D., Tang, X., Wasti, S., Lu, S., Zhang, S., Parit, M., Rencheck, M.L., Korey, M.: Bioinspired design toward nanocellulose-based materials. Mater. Today. (2023)
84. Godoy, M.G., Amorim, G.M., Barreto, M.S., Freire, D.M.G.: Agricultural residues as animal feed: protein enrichment and detoxification using solid-state fermentation. Curr. Dev. Biotechnol. Bioeng. 235–256 (2018)
85. Huang, Y., Nair, S.S., Chen, H., Fei, B., Yan, N., Feng, Q.: Lignin-rich nanocellulose fibrils isolated from parenchyma cells and fiber cells of western red cedar bark. ACS Sustain. Chem. Eng. **7**, 15607–15616 (2019)
86. Usov, I., Nyström, G., Adamcik, J., Handschin, S., Schütz, C., Fall, A., Bergström, L., Mezzenga, R.: Understanding nanocellulose chirality and structure–properties relationship at the single fibril level. Nat. Commun. **6**, 7564 (2015)
87. Ioelovich, M.: Characterization of various kinds of nanocellulose. Handb. Nanocellulose Cellul. Nanocompos. **1**, 51–100 (2017)
88. Lin, N., Dufresne, A.: Nanocellulose in biomedicine: current status and future prospect. Eur. Polym. J. **59**, 302–325 (2014)
89. Zhao, J., Yuan, X., Wu, X., Liu, L., Guo, H., Xu, K., Zhang, L., Du, G.: Preparation of nanocellulose-based aerogel and its research progress in wastewater treatment. Molecules **28**, 3541 (2023)
90. Aoudi, B., Boluk, Y., El-Din, M.G.: Recent advances and future perspective on nanocellulose-based materials in diverse water treatment applications. Sci. Total Environ. 156903 (2022)
91. Norizan, M.N., Shazleen, S.S., Alias, A.H., Sabaruddin, F.A., Asyraf, M.R.M., Zainudin, E.S., Abdullah, N., Samsudin, M.S., Kamarudin, S.H., Norrrahim, M.N.F.: Nanocellulose-based nanocomposites for sustainable applications: a review. Nanomaterials **12**, 3483 (2022)
92. Teles, F., Martins, G., Antunes, F.: Fire retardancy in nanocomposites by using nanomaterial additives. J. Anal. Appl. Pyrolysis **163**, 105466 (2022)
93. Sun, Y., Chu, Y., Wu, W., Xiao, H.: Nanocellulose-based lightweight porous materials: a review. Carbohydr. Polym. **255**, 117489 (2021)
94. Wang, D., Peng, H., Wu, Y., Zhang, L., Li, M., Liu, M., Fu, S.: Bioinspired lamellar barriers for significantly improving the flame-retardant properties of nanocellulose composites. ACS Sustain. Chem. Eng. (2020). https://doi.org/10.1021/acssuschemeng.9b07745
95. Mazela, B., Batista, A., Grześkowiak, W.: Expandable graphite as a fire retardant for cellulosic materials—a review. Forests **11**, 755 (2020)
96. Li, X., Zhang, K., Shi, R., Ma, X., Tan, L., Ji, Q., Xia, Y.: Enhanced flame-retardant properties of cellulose fibers by incorporation of acid-resistant magnesium-oxide microcapsules. Carbohydr. Polym. **176**, 246–256 (2017)

97. Wang, L., Cui, L., Sánchez-Soto, M., Shou, W., Xia, Z., Liu, Y.: Highly flame retardant melamine-formaldehyde cross-linked cellulose nanofibrils/sodium montmorillonite aerogels with improved mechanical properties. Macromol. Mater. Eng. **303**, 1800379 (2018)
98. Koedel, J., Seibt, S., Callsen, C., Puchtler, F., Weise, M., Weidinger, A., Altstaedt, V., Ruckdäschel, H., Schobert, R., Biersack, B.: DNA as a natural flame retardant for cellulose acetate polymer mixtures. Chem. Select **6**, 3605–3609 (2021)
99. Vasiljević, J., Hadžić, S., Jerman, I., Černe, L., Tomšič, B., Medved, J., Godec, M., Orel, B., Simončič, B.: Study of flame-retardant finishing of cellulose fibres: organic–inorganic hybrid versus conventional organophosphonate. Polym. Degrad. Stab. **98**, 2602–2608 (2013)
100. Wang, H.-S., Fu, G.-D., Li, X.-S.: Functional polymeric nanofibers from electrospinning. Recent Pat. Nanotechnol. **3**, 21–31 (2009)
101. Yang, W.-J., Wei, C.-X., Yuen, A.C.Y., Lin, B., Yeoh, G.H., Lu, H.-D., Yang, W.: Fire-retarded nanocomposite aerogels for multifunctional applications: a review. Compos. Part B Eng. **237**, 109866 (2022)
102. Vahabi, H., Wu, H., Saeb, M.R., Koo, J.H., Ramakrishna, S.: Electrospinning for developing flame retardant polymer materials: current status and future perspectives. Polymer (Guildf) **217**, 123466 (2021)
103. Kobayashi, Y., Saito, T., Isogai, A.: Aerogels with 3D ordered nanofiber skeletons of liquid-crystalline nanocellulose derivatives as tough and transparent insulators. Angew. Chemie Int. Ed. **53**, 10394–10397 (2014)
104. Fan, B., Chen, S., Yao, Q., Sun, Q., Jin, C.: Fabrication of cellulose nanofiber/AlOOH aerogel for flame retardant and thermal insulation. Materials (Basel) **10**, 311 (2017)
105. Yang, D., Peng, X., Zhong, L., Cao, X., Chen, W., Wang, S., Liu, C., Sun, R.: Fabrication of a highly elastic nanocomposite hydrogel by surface modification of cellulose nanocrystals. RSC Adv. **5**, 13878–13885 (2015)
106. Niu, J., Wang, J., Dai, X., Shao, Z., Huang, X.: Dual physically crosslinked healable polyacrylamide/cellulose nanofibers nanocomposite hydrogels with excellent mechanical properties. Carbohydr. Polym. **193**, 73–81 (2018)
107. Darbasizadeh, B., Fatahi, Y., Feyzi-Barnaji, B., Arabi, M., Motasadizadeh, H., Farhadnejad, H., Moraffah, F., Rabiee, N.: Crosslinked-polyvinyl alcohol-carboxymethyl cellulose/ZnO nanocomposite fibrous mats containing erythromycin (PVA-CMC/ZnO-EM): fabrication, characterization and in-vitro release and anti-bacterial properties. Int. J. Biol. Macromol. **141**, 1137–1146 (2019)
108. Yue, X., Zhang, S., He, J., Wang, Z.: Fabrication of flame retarded cellulose aerogel with hydrophobicity via MF/MTMS double cross-linking. J. Nat. Fibers. **20**, 2133053 (2023)
109. Mohammed, L., Ansari, M.N.M., Pua, G., Jawaid, M., Islam, M.S.: A review on natural fiber reinforced polymer composite and its applications. Int. J. Polym. Sci. **2015** (2015)
110. Lavoine, N., Lennart, B.: Nanocellulose-based foams and aerogels: processing, properties, and applications. J. Mater. Chem. **5**, 16105–16117 (2017). https://doi.org/10.1039/c7ta02807e
111. Smith, D.M., Deshpande, R., Jeffrey Brinke, C.: Preparation of low-density aerogels at ambient pressure. MRS Online Proc. Library **271**, 567–572 (1992). https://doi.org/10.1557/PROC-271-567
112. Yamanaka, K., Ohtaki, H.: Supercritical fluids and reductions in environmental pollution. Thermodyn. Solubility Environ. 51–84 (2007)
113. Jessop, P.G.: The utility of carbon dioxide in homogeneously catalyzed organic synthesis. Chem. Inform. **36**, no–no (2005)
114. Şahin, İ, Özbakır, Y., Inönü, Z., Ulker, Z., Erkey, C.: Kinetics of supercritical drying of gels. Gels **4**, 3 (2017)
115. Şahin, İ, Uzunlar, E., Erkey, C.: Investigation of kinetics of supercritical drying of alginate alcogel particles. J. Supercrit. Fluids **146**, 78–88 (2019)
116. Carosio, F., Medina, L., Kochumalayil, J., Berglund, L.A.: Green and fire resistant nanocellulose/hemicellulose/clay foams. Adv. Mater. Interfaces **8**, 2101111 (2021)
117. Prakobna, K., Berthold, F., Medina, L., Berglund, L.A.: Mechanical performance and architecture of biocomposite honeycombs and foams from core–shell holocellulose nanofibers. Compos. Part A Appl. Sci. Manuf. **88**, 116–122 (2016)

118. Min, P., Li, X., Liu, P., Liu, J., Jia, X., Li, X., Yu, Z.: Rational design of soft yet elastic lamellar graphene aerogels via bidirectional freezing for ultrasensitive pressure and bending sensors. Adv. Funct. Mater. **31**, 2103703 (2021)
119. Nomicisio, C., Ruggeri, M., Bianchi, E., Vigani, B., Valentino, C., Aguzzi, C., Viseras, C., Rossi, S., Sandri, G.: Natural and synthetic clay minerals in the pharmaceutical and biomedical fields. Pharmaceutics **15**, 1368 (2023)
120. Wang, D., Feng, X., Zhang, L., Li, M., Liu, M., Tian, A., Fu, S.: Cyclotriphosphazene-bridged periodic mesoporous organosilica-integrated cellulose nanofiber anisotropic foam with highly flame-retardant and thermally insulating properties. Chem. Eng. J. **375**, 121933 (2019)
121. Huang, J., Wang, X., Guo, W., Niu, H., Song, L., Hu, Y.: Eco-friendly thermally insulating cellulose aerogels with exceptional flame retardancy, mechanical property and thermal stability. J. Taiwan Inst. Chem. Eng. **131**, 104159 (2022)
122. Liu, D., Wu, Q., Andersson, R.L., Hedenqvist, M.S., Farris, S., Olsson, R.T.: Cellulose nanofibril core–shell silica coatings and their conversion into thermally stable nanotube aerogels. J. Mater. Chem. A. **3**, 15745–15754 (2015)
123. Tavakoli, M., Ghasemian, A., Dehghani-Firouzabadi, M.R., Mazela, B.: Cellulose and its nano-derivatives as a water-repellent and fire-resistant surface: a review. Materials **15**, 82 (2021). https://doi.org/10.3390/MA15010082
124. Tong, C., Zhang, S., Zhong, T., Fang, Z., Liu, H.: Highly fibrillated and intrinsically flame-retardant nanofibrillated cellulose for transparent mineral filler-free fire-protective coatings. Chem. Eng. J. **419**, 129440 (2021). https://doi.org/10.1016/J.CEJ.2021.129440
125. Zhu, Y., Yu, Z., Zhu, J., Zhang, Y., Ren, X., Jiang, F.: Developing flame-retardant lignocellulosic nanofibrils through reactive deep eutectic solvent treatment for thermal insulation. Chem. Eng. J. **445**, 136748 (2022). https://doi.org/10.1016/J.CEJ.2022.136748
126. Norizan, M.N., Shazleen, S.S., Alias, A.H., Sabaruddin, F.A., Asyraf, M.R.M., Zainudin, E.S., Abdullah, N., Samsudin, M.S., Kamarudin, S.H., Norrrahim, M.N.F.: Nanocellulose-based nanocomposites for sustainable applications: a review. Nanomaterials **12**, 3483 (2022). https://doi.org/10.3390/NANO12193483
127. Daicho, K., Saito, T., Fujisawa, S., Isogai, A.: The crystallinity of nanocellulose: dispersion-induced disordering of the grain boundary in biologically structured cellulose. ACS Appl. Nano Mater. **1**, 5774–5785 (2018). https://doi.org/10.1021/ACSANM.8B01438/ASSET/IMAGES/LARGE/AN-2018-014383_0011.JPEG
128. Thomas, P., Duolikun, T., Rumjit, N.P., Moosavi, S., Lai, C.W., Bin Johan, M.R., Fen, L.B.: Comprehensive review on nanocellulose: recent developments, challenges and future prospects. J. Mech. Behav. Biomed. Mater. **110**, 103884 (2020). https://doi.org/10.1016/J.JMBBM.2020.103884
129. Hu, D., Liu, H., Ding, Y., Ma, W.: Synergetic integration of thermal conductivity and flame resistance in nacre-like nanocellulose composites. Carbohydr. Polym. **264**, 118058 (2021). https://doi.org/10.1016/J.CARBPOL.2021.118058
130. Ju, X., Bowden, M., Brown, E.E., Zhang, X.: An improved X-ray diffraction method for cellulose crystallinity measurement. Carbohydr. Polym. **123**, 476–481 (2015). https://doi.org/10.1016/J.CARBPOL.2014.12.071
131. Wang, C., Wu, Y., Li, Y., Shao, Q., Yan, X., Han, C., Wang, Z., Liu, Z., Guo, Z.: Flame-retardant rigid polyurethane foam with a phosphorus-nitrogen single intumescent flame retardant. Polym. Adv. Technol. **29**, 668–676 (2018). https://doi.org/10.1002/PAT.4105
132. Montoya-Escobar, N., Ospina-Acero, D., Velásquez-Cock, J.A., Gómez-Hoyos, C., Serpa Guerra, A., Gañan Rojo, P.F., Vélez Acosta, L.M., Escobar, J.P., Correa-Hincapié, N., Triana-Chávez, O., Zuluaga Gallego, R., Stefani, P.M.: Use of fourier series in X-ray diffraction (XRD) analysis and fourier-transform infrared spectroscopy (FTIR) for estimation of crystallinity in cellulose from different sources. Polymers (Basel) **14**, 5199 (2022). https://doi.org/10.3390/POLYM14235199/S1
133. Park, S., Baker, J.O., Himmel, M.E., Parilla, P.A., Johnson, D.K.: Cellulose crystallinity index: measurement techniques and their impact on interpreting cellulose performance. Biotechnol Biofuels. **3** (2010). https://doi.org/10.1186/1754-6834-3-10

134. Maponya, T.C., Makgopa, K., Somo, T.R., Modibane, K.D.: Highlighting the importance of characterization techniques employed in adsorption using metal–organic frameworks for water treatment. Polymers **14**, 3613 (2022). https://doi.org/10.3390/POLYM14173613
135. Ahvenainen, P., Kontro, I., Svedström, K.: Comparison of sample crystallinity determination methods by X-ray diffraction for challenging cellulose I materials. Cellulose **23**, 1073–1086 (2016). https://doi.org/10.1007/S10570-016-0881-6
136. Costes, L., Laoutid, F., Khelifa, F., Rose, G., Brohez, S., Delvosalle, C., Dubois, P.: Cellulose/phosphorus combinations for sustainable fire retarded polylactide. Eur. Polym. J. **74**, 218–228 (2016). https://doi.org/10.1016/J.EURPOLYMJ.2015.11.030
137. Tornabene, F., Gorbunova, M., Grunin, L., Morris, R.H., Imamutdinova, A.: Nanocellulose-based thermoplastic polyurethane biocomposites with shape memory effect. J. Compos. Sci. **7**, 168 (2023). https://doi.org/10.3390/JCS7040168
138. Chandrasekaran, S., Cruz-Izquierdo, A., Castaing, R., Kandola, B., Scott, J.L.: Facile preparation of flame-retardant cellulose composite with biodegradable and water resistant properties for electronic device applications. Sci. Rep. **13**(1), 1–12 (2023). https://doi.org/10.1038/s41598-023-30078-0
139. Liu, Y., He, Z., Uchimiya, M.: Comparison of biochar formation from various agricultural by-products using FTIR spectroscopy. Mod. Appl. Sci. **9** (2015). https://doi.org/10.5539/MAS.V9N4P246
140. Yaashikaa, P.R., Kumar, P.S., Varjani, S., Saravanan, A.: A critical review on the biochar production techniques, characterization, stability and applications for circular bioeconomy. Biotechnol. Rep. **28**, e00570 (2020). https://doi.org/10.1016/J.BTRE.2020.E00570
141. Wulandari, W.T., Rochliadi, A., Arcana, I.M.: Nanocellulose prepared by acid hydrolysis of isolated cellulose from sugarcane bagasse. IOP Conf. Ser. Mater. Sci. Eng. **107** (2016). https://doi.org/10.1088/1757-899X/107/1/012045
142. Venkateshaiah, A., Padil, V.V.T., Nagalakshmaiah, M., Waclawek, S., Černík, M., Varma, R.S.: Microscopic techniques for the analysis of micro and nanostructures of biopolymers and their derivatives. Polymers **12**, 512 (2020). https://doi.org/10.3390/POLYM12030512
143. Eichhorn, S.J., Etale, A., Wang, J., Berglund, L.A., Li, Y., Cai, Y., Chen, C., Cranston, E.D., Johns, M.A., Fang, Z., Li, G., Hu, L., Khandelwal, M., Lee, K.Y., Oksman, K., Pinitsoontorn, S., Quero, F., Sebastian, A., Titirici, M.M., Xu, Z., Vignolini, S., Frka-Petesic, B.: Current international research into cellulose as a functional nanomaterial for advanced applications. J. Mater. Sci. **57**, 5697–5767 (2022). https://doi.org/10.1007/S10853-022-06903-8
144. Siedlecki, C.A., Marchant, R.E.: Atomic force microscopy for characterization of the biomaterial interface. Biomaterials **19**, 441–454 (1998). https://doi.org/10.1016/S0142-9612(97)00222-6
145. Wang, M.S., Palmer, L.B., Schwartz, J.D., Razatos, A.: Evaluating protein attraction and adhesion to biomaterials with the atomic force microscope. Langmuir **20**, 7753–7759 (2004). https://doi.org/10.1021/la049849
146. Webb, H.K., Truong, V.K., Hasan, J., Fluke, C., Crawford, R.J., Ivanova, E.P.: Roughness parameters for standard description of surface nanoarchitecture. Scanning **34**, 257–263 (2012). https://doi.org/10.1002/SCA.21002
147. Bałaga, Z., Gnatowski, A., Kulesza, S., Bramowicz, M., Gwoździk, M.: AFM measurements and testing properties of HDPE and PBT composites with fillers in the form of montmorillonite and aluminum hydroxide. Materials **15**, 8738 (2022). https://doi.org/10.3390/MA15248738
148. Cross, S.E., Jin, Y.S., Tondre, J., Wong, R., Rao, J.Y., Gimzewski, J.K.: AFM-based analysis of human metastatic cancer cells. Nanotechnology **19** (2008). https://doi.org/10.1088/0957-4484/19/38/384003
149. Kirmizis, D., Logothetidis, S.: Atomic force microscopy probing in the measurement of cell mechanics. Int. J. Nanomed. **5**, 137 (2010). https://doi.org/10.2147/IJN.S5787
150. Bussetti, G., Menegazzo, M., Mitko, S., Castiglioni, C., Tommasini, M., Lucotti, A., Magagnin, L., Russo, V., Bassi, A.L., Siena, M., Guadagnini, A., Grillo, S., Del Curto, D., Duò, L.: A combined raman spectroscopy and atomic force microscopy system for in situ and real-time measures in electrochemical cells. Materials **16**, 2239 (2023). https://doi.org/10.3390/MA16062239

151. Dazzi, A., Prater, C.B., Hu, Q., Chase, D.B., Rabolt, J.F., Marcott, C.: AFM-IR: combining atomic force microscopy and infrared spectroscopy for nanoscale chemical characterization. Appl. Spectrosc. **66**, 1365–1384 (2012). https://doi.org/10.1366/12-06804
152. Moll, N., Gross, L., Mohn, F., Curioni, A., Meyer, G.: The mechanisms underlying the enhanced resolution of atomic force microscopy with functionalized tips. New J Phys. **12** (2010). https://doi.org/10.1088/1367-2630/12/12/125020
153. Pérez, L.A., Bajales, N., Lacconi, G.I.: Raman spectroscopy coupled with AFM scan head: a versatile combination for tailoring graphene oxide/reduced graphene oxide hybrid materials. Appl. Surf. Sci. **495** (2019). https://doi.org/10.1016/J.APSUSC.2019.143539
154. Yivlialin, R., Ferraro, L., Filoni, C., Majumdar, I., Calloni, A., Goto, F., Finazzi, M., Duò, L., Ciccacci, F., Bussetti, G.: Probing the correlation between morphology and optical anisotropy in ZnTPP films grown at different temperatures. Appl. Surf. Sci. **611** (2023). https://doi.org/10.1016/J.APSUSC.2022.155729
155. Son, D., Cho, S., Nam, J., Lee, H., Kim, M.: X-ray-based spectroscopic techniques for characterization of polymer nanocomposite materials at a molecular level. Polymers **12**, 1053 (2020). https://doi.org/10.3390/POLYM12051053
156. Aisyah, H.A., Paridah, M.T., Sapuan, S.M., Ilyas, R.A., Khalina, A., Nurazzi, N.M., Lee, S.H., Lee, C.H.: A comprehensive review on advanced sustainable woven natural fibre polymer composites. Polymers (Basel). **13**, 1–45 (2021). https://doi.org/10.3390/POLYM13030471
157. Izwan, S.M., Sapuan, S.M., Zuhri, M.Y.M., Mohamed, A.R.: Thermal stability and dynamic mechanical analysis of benzoylation treated sugar palm/Kenaf fiber reinforced polypropylene hybrid composites. Polymers **13**, 2961 (2021). https://doi.org/10.3390/POLYM13172961
158. Sacristán, M., Hull, T.R., Stec, A.A., Ronda, J.C., Galià, M., Cádiz, V.: Cone calorimetry studies of fire retardant soybean-oil-based copolymers containing silicon or boron: Comparison of additive and reactive approaches. Polym. Degrad. Stab. **95**, 1269–1274 (2010). https://doi.org/10.1016/J.POLYMDEGRADSTAB.2010.03.015
159. Brischke, C., Humar, M.: Performance of bio-based building materials. 249–333 (2017). https://doi.org/10.1016/B978-0-08-100982-6.00005-7
160. Nazaré, S., Horrocks, A.R.: Flammability testing of fabrics. Fabric Testing 339–388 (2008). https://doi.org/10.1533/9781845695064.339
161. Bartlett, A.I., Hadden, R.M., Bisby, L.A.: A review of factors affecting the burning behaviour of wood for application to tall timber construction. Fire Technol. **55**, 1–49 (2019). https://doi.org/10.1007/S10694-018-0787-Y/TABLES/3
162. Stasiek, J., Malinowski, R., Ligor, T., Buszewski, B.: GC/MS analysis of gaseous degradation products formed during extrusion blow molding process of PE films. Chem. Pap. **64**, 579–583 (2010). https://doi.org/10.2478/S11696-010-0050-6
163. Takeshita, A., Uemura, Y., Onoe, K.: Quantification of persistent free radicals (PFRs) formed in thermally decomposed cellulose and its correlation with residual carbon amount. J. Anal. Appl. Pyrolysis **150**, 104883 (2020). https://doi.org/10.1016/J.JAAP.2020.104883
164. Liang, S., Neisius, N.M., Gaan, S.: Recent developments in flame retardant polymeric coatings. Prog. Org. Coat. **76**, 1642–1665 (2013). https://doi.org/10.1016/J.PORGCOAT.2013.07.014
165. Ahankari, S., Paliwal, P., Subhedar, A., Kargarzadeh, H.: Recent developments in nanocellulose-based aerogels in thermal applications: a review. ACS Nano **15**, 3849–3874 (2021). https://doi.org/10.1021/ACSNANO.0C09678/ASSET/IMAGES/MEDIUM/NN0C09678_0018.GIF
166. Du, F., Jin, Z., Yang, R., Hao, M., Wang, J., Xu, G., Zuo, W., Geng, Z., Pan, H., Li, T., Zhang, W., She, W.: Thermally insulating and fire-retardant bio-mimic structural composites with a negative Poisson's ratio for battery protection. Carbon Energy e353 (2023). https://doi.org/10.1002/CEY2.353
167. Shesan, O.J., Stephen, A.C., Chioma, A.G., Neerish, R., Rotimi, S.E., Shesan, O.J., Stephen, A.C., Chioma, A.G., Neerish, R., Rotimi, S.E.: Improving the mechanical properties of natural fiber composites for structural and biomedical applications. Renew. Sustain. Compos. (2019). https://doi.org/10.5772/INTECHOPEN.85252

168. Shrivastava, A.: Plastic Properties and Testing. Introduction to Plastics Engineering, pp. 49–110 (2018). https://doi.org/10.1016/B978-0-323-39500-7.00003-4
169. Broitman, E.: Indentation hardness measurements at macro-, micro-, and nanoscale: a critical overview. Tribol. Lett. **65**, 1–18 (2017). https://doi.org/10.1007/S11249-016-0805-5/FIGURES/16
170. Alaoui, A.H., Woignier, T., Scherer, G.W., Phalippou, J.: Comparison between flexural and uniaxial compression tests to measure the elastic modulus of silica aerogel. J. Non Cryst. Solids **354**, 4556–4561 (2008). https://doi.org/10.1016/J.JNONCRYSOL.2008.06.014
171. Brnic, J., Balos, S., Brcic, M., Dramicanin, M., Krscanski, S., Milutinovic, M., Ding, B., Gao, Z.: Testing and analysis of uniaxial mechanical fatigue, charpy impact fracture energy and microhardness of two low-carbon steels. Materials **16**, 884 (2023). https://doi.org/10.3390/MA16020884
172. Mohantya, A.K., Vivekanandhan, S., Tripathi, N., Roy, P., Snowdon, M.R., Drzal, L.T., Misra, M.: Sustainable composites for lightweight and flame retardant parts for electric vehicles to boost climate benefits: a perspective. Compos. Part C: Open Access **12**, 100380 (2023)
173. Zuiderveen, E.A.R., Slootweg, J.C., de Boer, J.: Novel brominated flame retardants—a review of their occurrence in indoor air, dust, consumer goods and food. Chemosphere **255**, 126816 (2020)
174. Wang, X., Kalali, E.N., Wan, J.-T., Wang, D.-Y.: Carbon-family materials for flame retardant polymeric materials. Prog. Polym. Sci. **69**, 22–46 (2017)
175. Norrrahim, M.N.F., Mohd Kasim, N.A., Knight, V.F., Ujang, F.A., Janudin, N., Abdul Razak, M.A.I., Wan Yunus, W.M.Z.: Nanocellulose: the next super versatile material for the military. Mater. Adv. **2**(5), 1485–1506 (2021). https://doi.org/10.1039/d0ma01011a
176. Chen, Z., Xiao, P., Zhanga, J., Tian, W., Jia, R., Nawaz, H., Jina, K., Zhang, J.: A facile strategy to fabricate cellulose-based, flame-retardant, transparent and anti-dripping protective coatings. Chem. Eng. J. **379**, 1222 (2020)
177. Melro, L., Fernandes, R.D.V., Peixoto, J.J., Pinheiro, I., Ribeiro, A., Bouça, V., Dourado, F., Gama, M., Padrão, J., Zille, A.: Antiviral properties of flame retardant bacterial nanocellulose modified with mordenite. AUTEX 2022 Conference Proceedings (2022)
178. Zhang, T., Wu, M., Kuga, S., Ewulonu, C.M., Huang, Y.: Cellulose nanofibril-based flame retardant and its application to paper. ACS Sustain. Chem. Eng. (2020)
179. Li, X., Chen, H., Wang, W., Liu, Y., Zhao, P.: Synthesis of a formaldehyde-free phosphorus−nitrogen flame retardant with multiple reactive groups and its application in cotton fabrics. Polym. Degrad. Stab. **120**, 193–202 (2015)
180. Xu, F., Zhong, L., Zhang, C., Wang, P., Zhang, F., Zhang, G.: Novel high-efficiency casein-based P−N-containing flame retardants with multiple reactive groups for cotton fabrics. ACS Sustain. Chem. Eng. **7**, 13999–14008 (2019)
181. Zhao, H., Sha, L.: Effect of surface modification of ammonium polyphosphate-diatomite composite filler on the flame retardancy and smoke suppression of cellulose paper. J. Bioresour. Bioprod. **2**, 30–35 (2017)
182. Yildirim, N.: Developing fire-retardant and water-repellent bio-structural panels using nanocellulose. MRS Commun. **8**(2), 257–265 (2018)
183. Kim, H., Youn, J.R., Song, Y.S.: Eco-friendly flame retardant nanocrystalline cellulose prepared via silylation. Nanotechnology 4–21 (2018)
184. Farooq, M., Sipponen, M.H., Seppälä, A., Österberg, M.: Eco-friendly flame-retardant cellulose nanofibril aerogels by incorporating sodium bicarbonate. ACS Appl. Mater. Interfaces **10**(32), 27407–27415 (2018)
185. Yue, X., Zhang, S., He, J., Wang, Z.: Fabrication of flame retarded cellulose aerogel with hydrophobicity via MF/MTMS double cross-linking. J. Nat. Fibers **20**(1), 2133053 (2023)
186. Singh, H., Jain, A.: J. Appl. Polym. Sci. **111**, 539 (2009)
187. Wang, Z., Huang, Z., Li, X., Zhou, J.-A.: Compos. A **121**, 180 (2019)
188. Wang, Y., Jiang, H., Ni, J., Chen, J., Zhou, H., Wang, X., Xin, F.: RSC Adv. **9**, 192 (2019)
189. Xu, P., Yu, Y., Chang, M., Chang, J.: Polymers **11**, 1471 (2019)

190. Qiu, J.F., Zhang, M.Q., Rong, M.Z., Wu, S.P., Karger-Kocsis, J., J. Mater. Chem. A **1**, 2533 (2013)
191. Li, M., Luo, J., Huang, Y., Li, X., Yu, T., Ge, M.: J. Appl. Polym. Sci. **131**, 40857 (2014)
192. Wang, C., Wu, Y., Li, Y., Shao, Q., Yan, X., Han, C., Wang, Z., Liu, Z., Guo, Z.: Polym. Adv. Technol. **29**, 668 (2018)
193. Yan, L., Fu, L., Chen, Y., Tian, H., Xiang, A., Rajulu, A.V.: J. Appl. Polym. Sci. **134**, 44828 (2017)
194. Hidalgo, J.P., Torero, J.L., Welch, S.: Fire Mater. **42**, 358 (2018)
195. Singh, H., Jain, A.: Ignition, combustion, toxicity, and fire retardancy of polyurethane foams: a comprehensive review. J. Appl. Polym. Sci. **111**, 1115–1143 (2009)
196. Berglund, L., Nissilä, T., Sivaraman, D., Komulainen, S., Telkki, V.-V., Oksman, K.: Seaweed-derived alginate−cellulose nanofiber aerogel for insulation applications. ACS Appl. Mater. Interfaces **13**, 34899–34909 (2021)
197. Shang, K., Liao, W., Wang, J., Wang, Y.-T., Wang, Y.-Z., Schiraldi, D.A.: Nonflammable alginate nanocomposite aerogels prepared by a simple freeze-drying and post-cross-linking method. ACS Appl. Mater. Interfaces **8**, 643–650 (2016)
198. Cheng, L., Ren, S., Brosse, N., Ju, Z., Li, S., et al.: Synergic flame-retardant effect of nano-cellulose crystalline and Magnesium hydroxide in polyurethane wood coating. J. Wood Chem. Technol. 1–8 (2022)

Eco-Friendly Lignin-Based Flame Retardant from Different Polymeric Materials

Stephen C. Agwuncha, C. G. Anusionwu, and E. R. Sadiku

Abstract The need to save the environment from excessive energy consumption from fossil fuels has seen a growing demand for renewable energy solutions in virtually all man's daily needs. Lignin as the second largest abundant renewable resource from plants, has been grossly underutilized. Only about 5% of the total lignin produced annually is used as raw material for useful products. While the remaining 95% are burnt as fuel. Lignin, with its many functional groups on its molecular structure, has the ability, to interact chemically with many other molecules. Therefore, it can be modified easily to suit many daily chemical needs of man. One such application is in flame retardants, especially for polymer materials. Halogenated flame retardants have been found to produce toxic by-products, making their continuous use a serious health concern issue. Also, the need to use more environmentally sustainable materials requires that alternatives for such products be developed. Lignin based flame retardants are being considered suitable alternatives because of their many advantages, including renewability, availability, eco-friendliness, and non-toxicity. This chapter will discuss the different types of eco-friendly lignin flame retardants used in polymer design, their synthesis, efficiency, and possible application areas.

Keywords Flame-retardants · Lignin · Ammonium polyphosphate · Flame mechanism · Thermal stability · Polymer application · Lignin-nitrogen-phosphorus compound

---

S. C. Agwuncha (✉)
Department of Industrial Chemistry, Federal University Lokoja, Lokoja, Nigeria
e-mail: stephen.agwuncha@fulokoja.edu.ng

Department of Forest Biomaterials, North Carolina State University, Raleigh, NC, USA

C. G. Anusionwu
Department of Chemistry, University of Johannesburg, Johannesburg, South Africa

E. R. Sadiku (✉)
Department of Chemical, Metallurgical and Materials Engineering and The Institute of Nano Engineering Research (INER), Tshwane University of Technology, Pretoria, South Africa
e-mail: sadikur@tut.ac.za

T. C. Mokhena et al. (eds.), *Biomaterials as Green Flame Retardants*, Engineering Materials, https://doi.org/10.1007/978-981-97-6871-4_10

1 Introduction

Lignin, present in lignocellulose plants, is a renewable natural macromolecular polymer material [1, 2]. Its role extends to fortifying plant cell walls, enhancing stiffness, structure, and rigidity. Beyond providing mechanical support, lignin serves as a protective shield against microbial attacks and environmental stress [3, 4]. Functioning as a binder, lignin occupies the space between hemicellulose and cellulose, effectively uniting various plant components and maintaining the integrity of the lignocellulose matrix [5].

The content of lignin in plants varies widely, constituting 15–35% of biomass weight, with fluctuations depending on harvest age. For instance, softwoods typically contain 24–33% lignin, hardwoods range from 19 to 28%, and some other biomass registers levels between 20 and 30% [6, 7]. The average molecular weight (Mw) of isolated lignin varies based on plant type. Studies have reported Mw values of 17,200 g mol^{-1}, 29,600 g mol^{-1}, and 53,850 g mol^{-1} for eucalyptus (*Eucalyptus globulus*), Southern pine (*Pinus palustris*), and Norway spruce (*Picea abies*), respectively. It's worth noting that these variations in Mw may be attributed to different isolation methods [8].

Lignin macromolecules exhibit intricate structures as oxygen-rich aromatic polymers. The chemical composition of lignin comprises phenyl propane units, originating from three aromatic monomer precursors known as basic units (refer to Fig. 10.1). These monolignols—p-coumaryl alcohol (H), coniferyl alcohol (G), and sinapyl alcohol (S)—play a crucial role in the formation of lignin, with their phenolic substructures linked through inter-unit connections (see Fig. 10.2) [9–14]. Lignin is characterized by a diverse array of functional groups, including methoxy, alcohol hydroxyl, phenolic hydroxyl, ethyl group, phenyl, aldehyde, and carbonyl groups. These functional groups offer abundant active sites for potential chemical modifications [15, 16]. It's noteworthy that the H unit lacks methoxy groups, the G unit possesses one, and the S unit contains two [6].

The lignin molecular structure is made up of branching heteropolymers with no specific structure, hence it is regarded as complex [17]. Lignin is composed of many irregular functional groups, as mentioned above. Therefore, there are many interunit

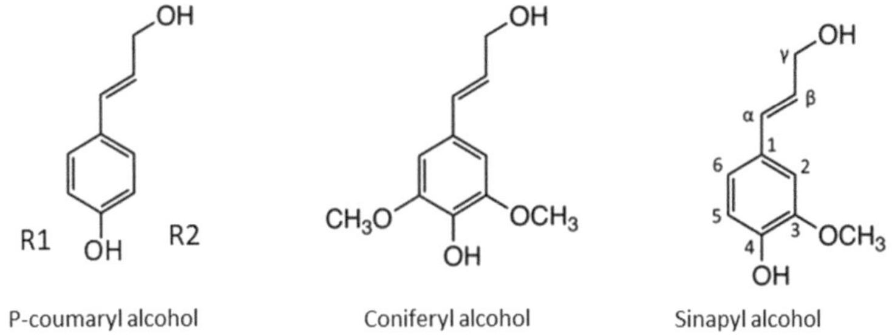

Fig. 10.1 Phenolic primary monolignol structures responsible for the formation of lignin

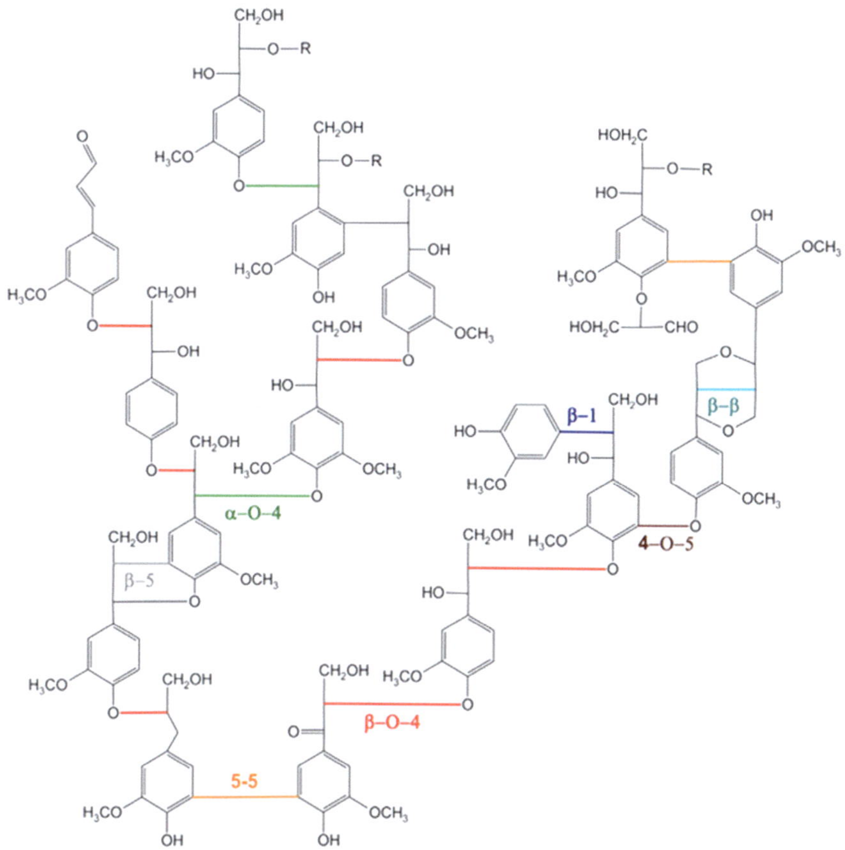

Fig. 10.2 Lignin 3D macromolecule showing the presence of the *p*-cumaryl alcohol, coniferyl alcohol and sinapyl alcohol

linkages. The predominant interunit linkages in lignin are commonly observed in the aryglycerol-β-ether dimer or β-O-4 ether type, constituting approximately 45–50% of the lignin structure. Additional linkages include: (i) aryglycerol-α-ether dimer or α-O-4 type (6–8%), (ii) resinol (β–β′) (9–12%), (ii) phenylcoumaran (β-5) (0–3%), biphenyl/dibenzodioxocin 5–5 × (18–25%), (iv) siaryl ether 4-O-5 moieties (4–8%), and a minor presence of spirodienone [7]. The specific composition of these linkages varies in lignin derived from different plants or sources.

Also, the method of isolation accounts for some observed differences [9].

The determination of functional group contents in lignin can be easily accomplished by gauging the volume of gaseous by-products, such as methane and H_2, produced during the reaction. Additionally, lignin showcases a high density of surface functional groups within the range of 0.8–11 mmol g^{-1}, with a hydroxyl group content of approximately 6.3 mmol g^{-1}. The majority of these functional groups, constituting two-thirds, are aromatic hydroxyls, while the remaining portion is primarily composed of aliphatic hydroxyl groups. The presence of these highly active aromatic hydroxyls imparts excellent antimicrobial and antioxidant activities to lignin, thereby allowing it to effectively stabilize polymer matrices against radical degradation [18–20].

Lignin is commercially generated as a by-product within the paper and pulp industry [21–23]. In industrial settings, the extraction of lignin from pulp waste, known as black liquor, occurs through either the sulfur or non-sulfur method. Lignosulfonate lignin (LL) and Kraft lignin (KL) are commonly obtained through sulfur-based extraction processes [24]. Conversely, organosolv lignin (OL) and soda lignin (SL) are produced through a sulfur-free extraction process involving solvent pulping within the temperature range of 150–200 °C and soda pulping within the range of 150–170 °C at a pH value of 11–13, respectively. The annual production of industrial lignin is estimated to be around 50 million tons [24]. However, a mere fraction, less than 5%, of this industrial lignin is commercially utilized as additives, dispersants, binders, or surfactants in various industries. The remaining approximately 95% is either incinerated for energy recovery or discharged into water systems [11].

With the many merits of lignin which include renewability, abundance, and low cost, widening its utilization has always been attractive but has so far remained a major challenge. Although some industries use lignin as water reducers for cement, dispersants, and fillers [12, 25–29].

The research on utilizing lignin in high-value applications is rapidly expanding. Initially, lignin was commonly employed as a low-cost reinforcing agent to improve mechanical and thermal properties in various composite materials, including thermoplastics, thermosets, rubber, and foam composites [18, 30, 31]. Typically, surface modification of lignin is necessary to enhance its mechanical reinforcement effect. Additionally, lignin has shown to possess compatibilizing properties between flax fibers and thermosetting resin [32]. Numerous other promising technical applications have been proposed in literature, including its use as a dispersant, adsorbent (UV-adsorbent), ion exchanger, and for controlling the release of fertilizers and herbicides in agriculture [33]. Lignin also holds potential for saccharification of lignocelluloses in biofuel production, electrochemical applications, and carbon fiber synthesis [34–36]. Furthermore, lignin can be transformed and modified to generate fine chemicals [37, 38]. Another extensively investigated application for lignin is its potential to enhance the flame retardancy of polymeric materials.

In recent times, numerous materials scientists have redirected their focus toward the development of flame-retardants (FRs) derived from lignin, driven by its notable thermal stability and capacity to form char [39]. This shift is, in part, a response to concerns about the potential toxicity associated with conventional halogenated FRs like decabromodiphenyl ether (DBDPO), tetrabromobisphenol A (TBBA), polybrominated dibenzofuran (PBDF), and hexabromocyclododecane (HBCD), which are gradually facing global bans [40–42]. Another motivating factor is the growing emphasis on sustainable practices in both additives and polymers.

Consequently, lignin has emerged as a promising FR for various polymer matrices, including low-density polyethylene (LDPE) [43], polypropylene (PP) [44–47], polylactic acid (PLA), acrylonitrile–butadiene–styrene copolymer (ABS) [49, 50], polyurethane (PU) [51, 52], and epoxy resins (EPs) [53–55]. This exploration capitalizes on lignin's impressive char-forming capability and its abundance of hydroxyl groups, making it amenable to chemical modification [15, 56].

2 Chemistry of Lignin-Based Flame Retardants

It has been reported that neat lignin samples display certain flame retardancy properties when compounded with polymers, aside from the mechanical improvements observed [2, 24, 57]. Lignin polymer composites were found to display better thermal stability over pristine polymer materials [58]. It is believed that the lignin polymer cross-linked structure which has a number of phenolic groups on it, can decompose and generate a high char yield which slows down the polymer from burning [59, 60]. However, the polymer composites compounded with pristine lignin have been reported to fail the UL-94 test V-0 rating required for it to be applicable in the industry. Therefore, chemical modification of lignin materials to strengthen the flame retardancy property is an active research area in the world today [61–64].

The use of lignin-based flame-retardants as additives for polymers is strongly linked to their high thermal stability and good char-forming capability [65, 66]. The benefits of using bio-based compounds like lignin as flame retardant additives stem from their molecular composition and structure, which give them an inherent ability to form thermally stable charred residues when exposed to fire [67, 68]. The origin, extraction methods, and modification done on the lignin directly impact their thermal decomposition behavior. The primary decomposition temperature of pristine lignin typically ranges from 200 to 500 °C, depending on the source and type [69, 70]. Lignin is roughly classified into several categories with regard to its botanical origin and extraction process [1, 30].

Lignin exhibits moderate stability at elevated temperatures, attributed to its distinctive aromatic superstructure [71]. However, it undergoes decomposition across a wide temperature range due to the presence of oxygen-containing functional groups, as illustrated in Fig. 2. These functional groups possess varying thermal stabilities, resulting in different bond scissions occurring at different temperatures [72]. The thermal degradation process of lignin involves three key steps: (i) cleavage of ether bonds, (ii) cleavage of C–C bonds and dealkylation, and (iii) carbonization demethylation. As temperatures rise, the cleavage of these functional groups and the complete rearrangement of the backbone contribute to the generation of a substantial residue char. [73]. A better understanding of these steps and processes can aid in designing new lignin-based flame retardants, LBFRs or improving existing ones. Furthermore, the flame retardancy behavior of lignin is boosted by the formation of a large char volume as it helps to obstruct the polymer matrix surface from the fire [74, 75]. Costes et al. [76], reported that below 250 °C, lignin hydroxyl groups catalyze the cleavage of acidic functions and the formation of water and CO_2. Between 200 and 500 °C, lignin undergoes depolymerization reactions that lead to the production of various volatile products. The decomposition rate is highest during this temperature range and is directly related to lignin composition. For instance, if lignin has more guaiacyl units, it releases more coniferyl alcohol, coniferyl aldehyde, isoeugenol, and vanillin [75]. Additionally, thermal degradation generates some monomeric radicals that trigger repolymerization reactions, thereby creating thermally stable aromatic networks.

3 Mechanism of Lignin-Based Flame Retardants

As mentioned earlier, polymers are highly flammable materials. Their behavior with fire is an important issue during their design and production [61, 77]. Therefore, their widespread utilization in virtually all sectors of human endeavor makes it very important to fortify with a reliable FR. In polymer materials, FR can either be compounded with the polymer during preparation or be incorporated into the polymer chain as part of the polymer primary chain. Whichever method is used, once the FR is incorporated into the polymer, two possible action paths can be taken to decrease or control the spread of fire. That is in the gas phase or/and the condensed phase (Fig. 10.3) [78].

Yang et al. [17] categorized the combustion behavior of polymeric materials into distinct zones to enhance comprehension of the mechanism. These zones include the combustion zone, degradation zone, char layer, and the underlying matrix. The combustion reaction in the gas phase is influenced by three primary factors: (i) fuel (flammable volatiles), (ii) heat, and (iii) oxygen. Oxygen from the air reacts with flammable volatiles produced from the degradation of the polymer matrix,

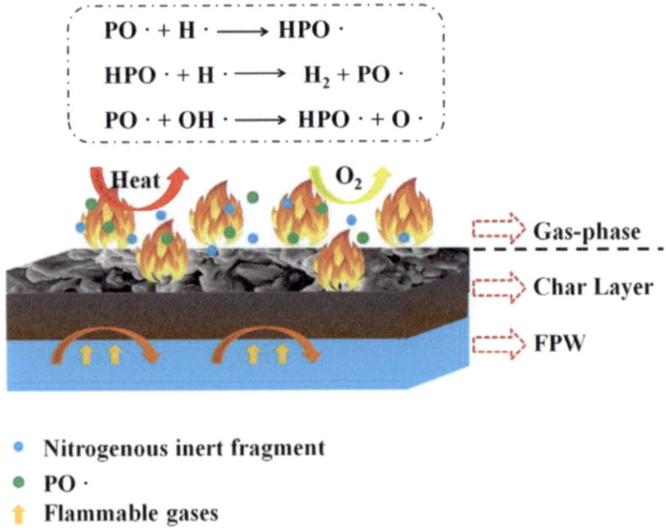

Fig. 10.3 Schematic representation of lignin flame retardant mechanism of functionalized lignin/polymer material [79]

sustaining the combustion reaction and releasing heat along with free radicals such as H˙ and OH˙. The combustion zone, where these processes occur, is considered the gas phase. Flame retardants (FRs) that impede combustion processes by capturing free radicals operate in the gas phase. Conversely, FRs functioning in the condensed phase facilitate the formation of a continuous and compact char layer on the surface. This char layer acts as a barrier, hindering the rapid release of volatiles from the degradation zone and limiting heat transfer to the underlying polymer during a fire incident. [31].

In the condensed phase mechanism, FR operates by producing a thermally insulating char layer in the condensed phase to protect the underlying polymer matrix from fire. This char layer acts as a barrier, inhibiting the transfer of heat to the polymer matrix and slowing down the release of combustible gases that can fuel the combustion zone [71]. The production of a larger number of residues during degradation is a key factor for achieving highly efficient flame retardancy in the condensed phase, as it implies that less fuel is released into the combustion zone. To effectively prevent the transfer of heat and mass between the gas and condensed phases, it is essential to have a char layer that is strong, compact, continuous, and provides thermal insulation. Interestingly, char formation may release non-flammable gas molecules such as water, NH_3, and CO_2. These gases help to dilute the volatile gaseous products from decomposition [80, 81].

The mechanism in the gas phase is more straightforward. The flame is retarded by radical species produced by the FR by interrupting the combustion reaction of flammable volatile. These radical species first dilute the flammable volatiles, thereby slowing down the reaction, then it reacts with the volatile species to stop the flame of fire [42, 77]. Halogen-containing compounds and some phosphorus compounds have been found to generate active species that can interact with the flammable volatiles, thereby interrupting the combustion reaction [61, 82–84].

4 Different Types of Lignin-Based Flame Retardants and Their Preparation

Recently, concern for the environment has been a major issue, hence the attraction towards the use of lignin-based FR. Moreover, halogenated FR have caused more harm than good, as they have been implicated in many health-related diseases due to their use as FR [40–42, 81]. Lignin is not only renewable but also degradable and readily available in large quantities [31]. Only 5% of the total quantity generated is utilized for industrial applications [85]. Hence the interest in lignin. In a review by Yang and co. [17], LBFR can be classified into four main groups, namely single component, (ii) lignin composite, (iii) chemically modified lignin FR, and (iv) nanolignin FR (see Fig. 10.4).

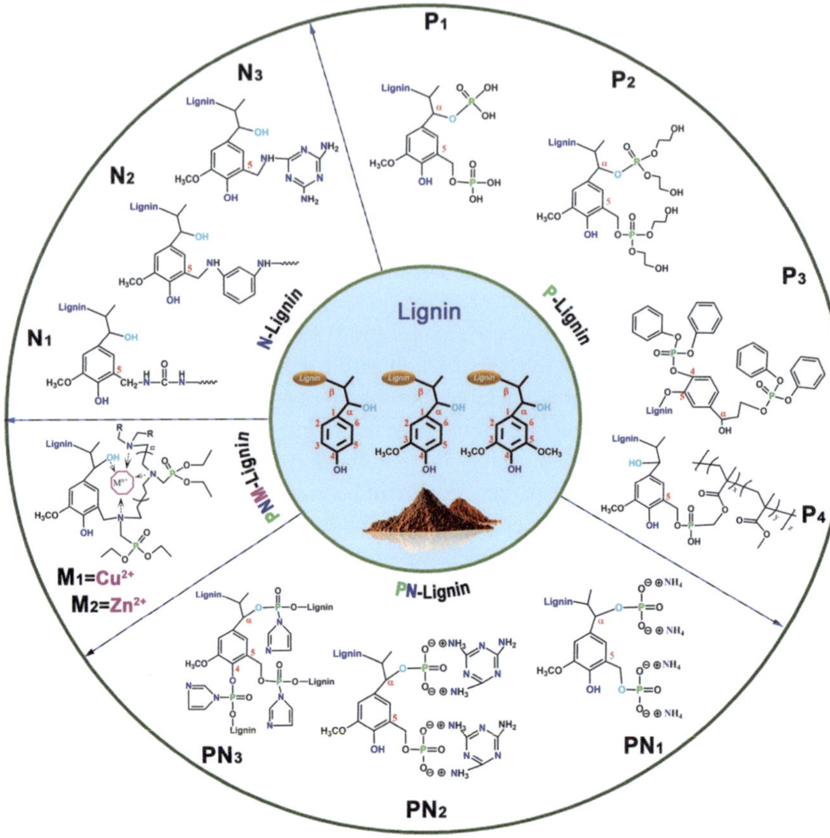

Fig. 10.4 Summary of the different types of functionalized LBFR commonly in use [58]

4.1 Single Component Lignin-Based Flame Retardants

Unmodified lignin used in compounding a polymer matrix is referred to as single component lignin FR. The use of lignin as FR started in the early 1998 by the Audisio group when they compounded PP with lignin as FR [68]. The group's findings showed that lignin was able to play a dual role as a FR as well as a toughening agent For the PP matrix used. The group also observed that the addition of 20 weight% of unmodified lignin led to a significant enhancement of thermal stability.

In the early days of polymer/lignin composites research, it was more about the influence on mechanical and morphological properties [86–89]. Lignin as reinforcement led to reasonable improvement in mechanical and thermal properties. This observation was linked to the extent of interaction between the polymer matrix and the lignin reinforcement. Song and co-workers et al. [50] prepared acrylonitrile butadiene styrene (ABS) composites reinforced with alkaline lignin (AL) through melt blending, and the flame retardancy of lignin was their focus. The thermal analysis of the composites with unmodified AL showed decreased thermal stability, which was attributed to poor compatibility between the polymer matrix and the AL. However, the presence of the AL did increase the formation of char at elevated temperatures. A cone calorimeter test demonstrated that lignin significantly reduces the heat release rate and slow down the combustion process. Sallem-Idrissi and co-researchers [85] prepared ecofriendly bio-composites of polyamide 11 (PA11) using lignin as reinforcement. They aimed to evaluate the mechanical and thermal improvement impacted by unmodified commercial, technical grade lignin and all the impurities present. The composites were found to be miscible and morphological characterization showed that the composites exhibit homogeneity behaviour of a blend. The presence of the lignin was reported to significantly delay the kinetics of PA11 crystallization. Their work showed that lignin may interact differently as a reinforcement with different polymers.

On their part, Costes and co-researchers tried to design a simple and efficient route for enhancing the thermal and flame retardancy of polylactide (PLA) by using lignin and phytic acid [76]. The composites were prepared using melt bending, and their thermal, mechanical, and fire properties were evaluated. They concluded that adding phytic acid improved the dispersion of lignin particles in the PLA matrix, thereby limiting the thermal degradation of PLA induced by Lignin. However, combining lignin and phytic acid led to about 40% reduction of heat release rate during the cone calorimeter test. Lignin alone was unable to achieve the industrial test classification required (V-2 classification) in (UL-94 test) to be used as FR. In another work, Kaewtatip and Thongmee, [90] successfully used kraft lignin (KL) and esterified lignin (EL) to compound thermoplastic starch (TPS) polymer. The composites showed better tensile strength when compared to the pristine TPS. Also, thermal stability was significantly enhanced. The degradation of TPS/lignin was delayed by 28% for KL and 22% for EL. They reported improved compatibility between TPS and lignin. In all the reported works on the use of unmodified lignin, their influence on FR properties of the composites depended on the type of polymer matrix, the molecular weight distribution of the Lignin, the preparation method, and the condition of preparation. These factors will define the extent of compatibility or otherwise. Hence, the performance of lignin as FR. However, the modification of this lignin showed remarkable improvement, even with polymer of poor compatibility.

4.2 *Combined Lignin-Based Flame Retardants*

This group of lignin-based FR is a combination of unmodified lignin and other FR. Researchers have been evaluating the synergistic behaviour of combining raw lignin with traditional FR, such as ammonium dihydrogen polyphosphate (ADP), magnesium hydroxide, ammonium polyphosphate (APP), and boric acid [91, 92]. These FR have been reported to promote crosslinking and char formation by catalysing the cleavage of C–O bonds in the lignin through the formation of highly conjugated systems. In the research reported by Verdolotti and co. [93], AL and APP were combined to develop a sustainable composite using thermo-plasticized zein, (TPZ). The physio-chemical interaction which occurred between the polymer matrix and the reinforcements was evaluated to understand their influence on the mechanical, thermal, and fire behaviours. To simulate different fire scenarios, both cone calorimetry and self-sustained vertical combustion tests were used. The results obtained showed that unmodified lignin was able to intercalate with the X-helix and B-sheet structures of the polymer chains, thereby improving the fire retardancy of the resulting composites. However, the addition of APP enhanced the fire retardancy of the polymer/lignin composites by enhancing the char formation capacity, preventing further combustion. Composites containing 3 wt% AL and 10 wt% APP were observed to exhibit excellent fire retardancy behaviour in different fire scenarios and gave satisfying mechanical properties that are 30% higher than that of the pristine TPZ. This was similar to the results obtained by Costes and co. [76] that combined Lignin and phytic acid in PLA polymer. They reported an increase in mechanical behaviour from 3.1 to 12.6% for elongation at break. This increase was attributed to the combination of lignin and phytic acid, which was able to give a V-2 classification of UL-94 test for flame retardants. In combining lignin and polyphosphate, phosphate acted as the acid source and, also the gas source [94, 95]. A reaction between lignin and the phosphorus acid released initiated crosslinking between the components. This often leads to expansion and charring in the composite due to the non-combustible gas produced in the system. Also, the dehydration and carbonization of the lignin present cause further expansion and foaming. Therefore, the lignin and APP combination are referred to as intumescent FR.

4.3 Chemically Modified Lignin-Based Flame Retardants

Lignin is a structurally complex compound that contains a diverse range of functional groups, making it amenable to chemical modifications [11]. These modifications can be classified into three categories: depolymerization or fragmentation, the introduction of new functional groups, and functionalization of hydroxyl groups. Lignin depolymerization can be achieved through pyrolysis or oxidation. The introduction of new functional groups can be accomplished via halogenation, sulfomethylation, amination, hydroxyalkylation, alkylation/dealkylation, and nitration. Hydroxyl group modification has been the focus of much research, as lignins contain many reactive phenolic and aliphatic hydroxyl groups. Esterification, etherification, isocyanate reaction, silylation, phenolation, and oxidation/reduction are the primary reactions used to functionalize lignin hydroxyl groups. The goal of lignin modification is to create high-value chemicals (e.g., vanillin, aromatics, quinones) or to broaden their applications by enhancing their reactivity, solubility in organic solvents, and ease of processing. These reactive sites can be used to modify the lignin with a variety of molecules to generate new compounds of interest. A variety of modified lignin for

FR applications have been reported by researchers, and research is still ongoing for different fields of application. These various modified lignins can be classified into the following groups as described below:

4.3.1 Phosphorus Modified Lignin-Based Flame Retardants

There is a growing interest in the phosphorylation of lignin [96], primarily to enhance its FR properties. Phosphorylated lignin, with its thermally stable aromatic structure and phosphorous groups, is known to have a higher potential for forming char [49]. This is due to the dehydrating effect of the phosphoric acid compounds that are produced during thermal degradation. The presence of phosphorus-containing compounds facilitates the breakdown of hydrocarbon molecules into char during combustion. Lignin modified with phosphorus moiety can generate volatile phosphorus -containing species which can act as flame inhibitors through the recombination process of catalytic radicals [73, 77, 96, 97]. Phosphorus modified lignin is prepared by the phosphorylation of lignin using APP. APP has been reported to function as an acid and as a gas source [96]. It has high content of phosphorus and nitrogen, good thermal stability, low smoke, and it is non-toxic [98, 99]. It has been widely reported as an inorganic flame retardant [36, 49, 71, 100]. When compounded with lignin, it is an excellent intumescent with good carbon source, acid source, and gas source. The phosphorylation of APP and lignin is believed to occur at the phenolic OH and carbonyl groups of the lignin molecules. Some researchers have reported the use of phosphorus pentoxide, diphenyl phosphoryl chloride (DPC), triphenyl phosphate (TPP); diethyl (2-(triethoxysilyl)ethyl) phosphate [49, 81, 95, 101]. Dai and co-workers [102] prepared novel lignin-based flame retardants with concurrent excellent smoke-suppression properties by modifying lignin with functional groups containing nitrogen and phosphorus. In their work, two components of known excellent epoxy resin flame retardancy were used, namely: piperazine (PA) and 9, 10 dihydro-9-oxa-10-phosphaphenanthrene-10-oxide (DOPO) for the preparation of intermediates (PA-DOPO) and grafted onto lignin. This was then chemically bonded to epoxy resin to form composites with varying amounts of lignin content. Their results show that the best flame retardancy was achieved with the functionalized lignin epoxy composites with an elevated phosphorus content. These composites were able to achieve a V-0 rating of the UL-94 tests and exhibited excellent smoke suppression, substantially reducing the total heat release and smoke produced by 46.6% and 53% respectively. These improvements observed were explained by several mechanisms which worked coherently in both the condensed phase and the gas phase. In the condensed phase, the decomposition of DOPO modified lignin produced substances that played very important roles including phosphoric acid. These acids were believed to have promoted the dehydration reaction of lignin and EP to produce carbonaceous materials; that formed a char layer which reduced the transfer heat; insulated oxygen gas and prevented further burning of degraded polymer materials. Their results also allocated the excellent charring property of lignin. In the gas phase, they believed that the non-flammable gases produced by the decomposition of piperazine during combustion diluted the oxygen concentration in the combustion zone, delaying the process and reducing heat release. In addition, the phosphorus-based radicals from the decomposition of DOPO reacted with the active radicals to quench them and inhibit combustion. The synergistic effects of the two phases lead to excellent FR behaviour. This work emphasized the importance of high phosphorus content in the modified LBFR. However, the unmodified LBTR remains a good charring material for carbonization in polymer composite fire prevention. Similarly, Hu and co-researchers [79] also prepared modified LBFR using DOPO for wood/PLA biocomposite uses. However, the modified LBFR was synthesized using 4,4-diphenylmethane diisocyanate (MDI) in DMF and triethylamine.

During composites preparation, 1,3,5-triglycidylisocynate (TGIC) was added to the mixture. The results showed that composites samples with higher contents of modified LBFR gave better thermal stability flame retardancy due to the formation of higher char contents. The presence of TGIC and modified LBFR produced a synergistic effect. The phosphenanthrene group in the modified LBFR decomposed into PO free radicals that helped to terminate the chain reaction the active free radicals were propagating in the gas phase. The limiting oxygen index value for the sample with 25% modified LBFR and 4% TGIC reached 28.6%, while the residue at 700 °C was up to 31.3% all these helped to increase the retardancy. In all, the percentages of lignin and phosphorus in the FR are very important, as well as the presence of a catalyst that can speed up the formation of $PO^{\cdot}$ acidic radicals [58].

4.3.2 Nitrogen Modified Lignin-Based Flame Retardants

Researchers have explored the use of nitrogen-containing compounds to functionalize lignin and enhance its fire-retardant properties [103, 104]. The resulting nitrogen-containing lignin can act as an intumescent flame-retardant system in polymeric materials by releasing non-combustible gases upon exposure to fire. This is due to the nitrogen compound that is grafted onto lignin, which causes the char to expand. Mannich reaction between lignin and urea was used to prepare nitrogen-containing lignin [103]. This method resulted in a covalently grafted nitrogen group onto the lignin, which amounted to 14.45 wt% nitrogen. When incorporated into PLA at a concentration of 23 wt% along with APP/urea modified-lignin at a 4:1 ratio, the resulting composite displayed remarkable fire-retardant properties, including a LOI value of 34.5 and a UL 94 V-0 classification. Furthermore, in cone calorimeter tests, this composite showed the best fire behavior with more stable carbonaceous char. However, the resulting composite showed a reduction in the mechanical properties, which can be attributed to the poor interfacial interaction between the urea modified-LBFR and the PLA matrix. In general, most nitrogen-modified LBFR (Nm-LBFR) have been reported to perform maximally in combination/synergistically with phosphorus moiety, which will be discussed in greater detail next.

4.3.3 Phosphorus and Nitrogen (PN)LBFR

Researchers employ two primary strategies to enhance flame retardancy in lignin: augmenting the quantity of hydroxyl groups prior to phosphorylation and grafting compounds containing phosphorus and nitrogen [94]. The incorporation of nitrogen-containing lignin and phosphorous compounds can provide a beneficial intumescent flame-retardant system for polymeric materials. The phosphorous compound serves as the acid source, with lignin acting as the carbon source to encourage the formation of char. Furthermore, the nitrogen-containing compound grafted onto lignin aids in the release of non-combustible gases, expanding the char layer. As mentioned in the earlier research report by Zhang and colleagues [103], urea-modified lignin proves to be a more effective charring agent when combined with APP compared to unmodified lignin. This effectiveness is attributed to their synergistic approach in safeguarding the polymer material.

In their report, Costes and co-workers [48] prepared a phosphorus-nitrogen LBFR (PN-LBFR) of ammonium phosphate-modified lignin containing 7 wt% nitrogen and 9 wt% phosphorus, using $POCl_3$ and ammonium hydroxide. Incorporating 20 wt% of this PN-LBFR into PLA resulted in UL 94 V-0 classification being achieved, with a 40% reduction in pHRR, without affecting the time to ignition when compared to neat PLA (Fig. 10.5).

The ability of lignin to form char is highly related to the amount of grafted phosphorous. Therefore, to increase the percentage content of phosphorus in the final FR, Yu et al. [100] first increased the hydroxyl groups of the lignin by using formaldehyde before grafting phosphorus/nitrogen compounds. The phosphorus-nitrogen intermediate was synthesized by reacting imidazole with $POCl_3$, which was then reacted

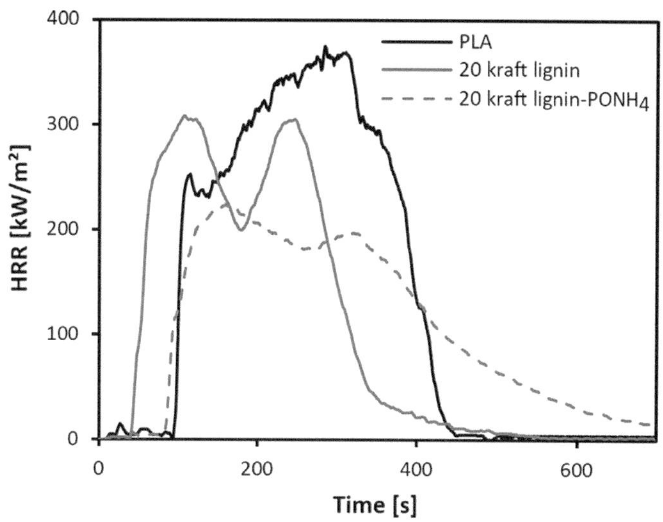

Fig. 10.5 Reported 40% reduction of pHRR of PLA after addition of PN-LBFR [48]

with lignin-OH to produce a PN-lignin. This resulted in a PN-LBFR containing 8.1 wt% phosphorus and 7.2 wt% nitrogen, which was then incorporated into polypropylene. The flame resistance of the composite containing PN-LBFR was examined and compared to untreated lignin. PN-LBFR demonstrated enhanced thermal and fire behavior, attributed to the generation of a continuous and highly compact charred residue. Introducing 20 wt% PN-LBFR into polypropylene resulted in a 72% reduction in pHRR (peak heat release rate) and a 15% reduction in THR (total heat release) compared to neat polypropylene. This reduction was even greater than that obtained with the polypropylene filled with 30 wt% unmodified lignin.

A report by Costes et al. [48]. These results were attributed to the dehydration action of phosphoric acid derivatives, which is also contribute to the high char residue of PN-lignin. The high char content has the potential to enhance the thermal stability and flame retardancy of polymers [105]. In a separate study, Zhou et al. [106] observed that incorporating 2 wt% PN-lignin resulted in an enhancement of the LOI (limiting oxygen index) value for the epoxy composite, increasing it from 22.3 to 28.3%. Simultaneously, a V-1 grade was attained. Furthermore, with the addition of 7 wt% PN-lignin, the epoxy composite achieved a V-0 grade in the vertical burning test.

Zhong et al. [27] demonstrated the development of a halogen-free red phosphorus FR by microencapsulating piperazine-modified lignin and aluminum phosphate into red phosphorus FRs, resulting in multi-level FR properties. The integration of a fire-resistant (FR) grade into an ABS composite, comprising 25% FR, enables the attainment of a V-0 performance level, accompanied by an LOI value of 26.1%. In comparison to ABS resin lacking FR, the peak heat release rate (pHRR) and total smoke generated during ABS combustion decrease by 25.8% and 63.1%, respectively. Achieving both halogen-free environmental protection and effective flame retardancy necessitates the formation of a dense carbon layer catalyzed by the phosphorus element and the release of ammonia gas facilitated by the nitrogen element.

In general, studies have shown that the addition of PN-lignin to composites can decrease the time to ignition (TTI), peak heat release rate (pHRR), and total heat release (THR). As an example, a polypropylene (PP) composite with 30 wt% PN-lignin demonstrates elevated TTI along with decreased pHRR and THR compared to a PP/lignin composite. Additionally, the average mass loss rate (AMLR) for the PP/PN-lignin composite is lower than that observed in the PP/lignin composite [105]. Similarly, by incorporating 7 wt% PN-lignin into epoxy has been found to reduce the composite's TTI, pHRR, and THR and significantly decrease the total smoke production rate (TSP) [106].

4.4 *PN-Metal-LBFR*

Weng et al. [107] attempted to develop LBFR with exceptional flame retardancy and mechanical performance properties for phenolic foam. In their work, lignin, doped with nitrogen and phosphorus was prepared by modifying reactive lignin with

DOPO and triethylenetramine (TETA). These reagents were utilized to partially replace phenol in the phenolic foam to serve as FR. Their results showed that the nitrogen-phosphorus doped LBFR enhanced the compressive strength of the foam as well as the flame retardancy. The limiting oxygen index (LOI) and compressive strength of the prepared foams were found to have increased by 33.3% and 43.7%, respectively. The peak heat release rate (pHRR) and peak smoke release rate (rSRR) decreased by 48.0 and 53.2%. Their work showed that nitrogen and phosphorus present work together synergistically to promote polymer carbonization and generate rich char residue layer of nitrogen and phosphorus that insulates heat and oxygen as an effective flame retardant in in phenolic foam.

Zhou et al. [106] functionalized lignin with nitrogen and phosphorus to produce a PN-LBFR with 5 wt% N_2 and 12 weight% P. The addition of 7 wt% of the PN-LBFR to epoxy resin composite gave a UL-94 grade of V–O and LOI of 31.4%, indicating excellent flame retardancy by the PN-LBFR even though the thermal stability was slightly influenced negatively. The higher the amount of modified lignin used, the thicker the char produced, releasing ammonia gas facilitated by the presence of nitrogen. Char formation is promoted by phosphorus, as reported by many other researchers. From the mechanism studied, they found that lignin not only reduced the combustibles in the gas phase but also promoted carbon fixation in the condensed phase. The amino groups on the modified lignin surface reacted with the epoxy resin substrate to form a strong interaction force, thereby leading to increased nitrogen compounds in the gas phase products that can reduce the oxygen concentration around the substrate. These products diluted the combustible gases and thus prevented the spread of fire. Therefore, the combination of nitrogen and phosphorus to modify lignin can improve the efficiency of lignin to function as FR. The lignin origin and chemistry can affect the modification process. PN-lignin conferred superior flame retardancy on polymer composite when compared to that of unmodified lignin. PN-LBFR has been reported to have excellent FR behavior. The amount required before it could exhibit this behavior has also been reported to be detrimental to the mechanical and thermal properties of the polymers. Therefore, one solution preferred was the addition of metal. The incorporation of metal to the mix, has been reported to reduce smoke production and catalyze the process [61, 77], thereby, reducing the amount of PN-LBFR loading required to give effective flame retardancy. Metal–organic frameworks (MOF's) are functional porous materials. Most of the MOF's flame-retardants reported have been prepared using their acetate salts.

5 Application of Lignin-Based Flame Retardants in Polymers

There have been numerous publications on the performances of LBFRs polymer composites. On these bases, the performance of LBFRs in polymer composites has been very remarkable and effective in controlling or reducing the rate at which these polymer composite materials burn in flame when compared to their neat polymers. These properties have been reported based on their peak heat release rate (pHRR), total heat release (THR), and time to ignition (TTI). These measurable properties have been found to correlate well with their combustion behavior. Some selected LBFRs polymer composites will be discussed based on their possible applications as they relate to different published reports. Yang et al. [17] published a more comprehensive list of LBFRs polymer systems, tabulating these properties in this review based on char yield, effective heat of combustion, total mass loss, residue, and total heat evolved.

5.1 Polymer Films and Sheets

Polymers are widely used in packaging industries because of their versatility in properties and application, coupled with ease of processing. Their application cut across food, beverages, pharmaceutics, cosmetics, etc. According to Drobny [108], polymer films are thin continuous materials with a thickness of approximately 200 μm. However, if the thickness is more than 200 μm, it is referred to, as a polymer sheet. Polymer films are made from thermoplastics such as PLA, PP, PE, PBSA, and many more. Their properties can vary widely depending on the polymer used and the intended application. Most applications of films and sheets include but are not limited to packaging, bags, labels, automobile, aerospace, electrical and electronic, in the chemical industry, military, building, construction, and landscaping. Therefore, flame retardance requirements for these materials will be based on the polymer used and intended application.

5.1.1 PLA

Polylactic acid is a biodegradable polymer with good mechanical strength and processability [109] It is widely applied as a single polymer, blended with other polymers, or as composites. It has been used as films in the food industry, casing for cellphones and computers in electrical and electronics, interior decoration materials in the automobile and aerospace industries. However, the flammability of PLA has seriously limited its wider application in these and other fields. Hence, the use of flame retardants. To sustain its biodegradability and environmental friendliness. Lignin flame retardants are gaining more ground in developing PLA materials [2, 30, 58, 89, 107].

Spiridon et al. [110] conducted an evaluation of PLA-lignin bioplastic composites both before and after undergoing accelerated weathering. They documented that the inclusion of lignin led to enhancements in PLA's thermal stability and mechanical attributes. Similarly, Gordobil et al. [111] explored the utilization of kraft lignin as an additive in PLA to enhance the polymer's flexibility and thermal characteristics. Their research demonstrated that the elongation at break experienced an increase with the addition of small amounts (0.5 and 1%) of acetylated kraft lignin. In a separate study, Park et al. [112] delved into incorporating lignin fractions into UV-resistant lignin-PLA biocomposites through lignin-lactide grafting. The resulting composites displayed favorable tensile properties and the capability to block UV light, suggesting potential applications in packaging and UV-protective scenarios. PLA/lignin composites have been widely investigated because of their potential to replace polymers for fossil fuel sources [113, 114]. Sun et al. [114] synthesized a group of lignin-derived toughening agents by chemically co-polymerizing two biobased acrylate monomers (lauryl methacrylate and tetrahydrofurfuryl methacrylate) to enhance the toughness of PLA. This graft modification technique notably improved the compatibility between PLA and lignin at the interface. Despite being a extensively researched subject, there remain unresolved challenges within the realm of PLA-lignin composites.

More recently, Sun et al. [115] prepared PLA/lignin composites with improved toughness and reduced flammability by melt blending the mixture with the addition of APP and PEG. The resultant composites showed high strain at break of 280% and impact strength of 14.9 kJ m^{-2}. There were increases of 38% for strain at break and 2% for impact strength. Their results showed that apart from acting as a flame retardant, modified lignin can still act as a compatibilizer. Podkosciena et al. [63], in their work, reported that the use of modified lignin as a flame retardant is more effective and beneficial than unmodified lignin, which mainly produces char to slow down the flame. However, this had been reported long ago by Costes and his co-researchers [48] when they reported the improvement of lignin-based flame retardance in PLA composites after modification using phosphorus and nitrogen derivatization. It is obvious that the functionalization of lignin led to improved flame retardance. The properties of the polymers were now of interest as they would dictate their applications. The use of functionalized LBFR has led to improved mechanical and flame-retardant properties.

In another work by Costes et al. [76], PLA/lignin composites were prepared in combination with phytic acid through melt blending. The presence of phytic acid led to improved dispersity of lignin in the PLA matrix, leading to improved thermal stability. More importantly, the combined presence of lignin and phytic acid led to a significant reduction in the pHRR, of up to 44%. This gave a class of the PLA composite a V-2 classification in UL-94 test, a very important improvement. This is aside from the mechanical properties improvements also reported. With all the improvements, Hu and Co. [79] reported the preparation of wood/PLA biocomposites by using lignin-based flame retardants mixed with triglicidyl isocyanate. The biocomposites showed improved mechanical properties and flame retardancy. The wood/PLA bicomposite is believed to be applicable in industries such as building construction, automobile, and aerospace because of its improved mechanical and flame retardancy properties.

5.1.2 Polyamide

Polyamide 11 (PA11), also known as Nylon II, derived from castor oil, is a bio-based thermoplastic polymer with excellent mechanical and chemical properties. However, its high flammability has limited its wider applications because it drips during burning. Mandlekar and co-researcher [35] demonstrated LBFR can reduce the flammability and HRR of PA11. In their work, lignins from different sources were used to compound PA II by melt bending. The results showed that lignin with sulfonate group had better capability to reduce the pHRR and THR by 51% and 22%, respectively. In addition, sulfonated lignin gave a V-I classification of self-extinction and promoted char formation in the condensed phase.

In another work by the same group [36], PA11/lignin composites with good flame retardancy were developed by the combination of zinc phosphinate (ZnP) and low sulfonate lignin in PA11 matrix. They studied the chemical structure of the condensed phase using FTIR, while the flame retardancy was evaluated using UL-94 and cone calorimetry. Their investigation revealed that ZnP helps inhibit the flame by radical trapping and partially diluting the fuel with phosphinate ions and phosphonic acid released in the gas phase. On the other hand, the lignin promoted char formation, which helped in slowing down the flame.

Sallem-Idrissi and co-workers [85] also used PA11 to prepare 100% biocomposites that are eco-friendly using raw lignin. This was done by directly extruding PA11 and lignin without any chemical pre- or *in-situ* modification. The impurities inherent in the technical grade lignin were allowed deliberately so that the effect of the whole lignin material on the structure evolution of PA11 will be evaluated. From their investigation, the PA11 and lignin were found to be miscible up to 25% lignin by wt. The mechanical properties were found to be outstanding up to 12.5% lignin wt. With the charring properties of lignin, this might be one way to improve the sustainability of the PA11 and its wider applications. Polymer/lignin composites have also been prepared using polyacrylonitrile [82], acrylonitrile–butadiene–styrene copolymer (ABS) [49, 50] and polybutylene succinate [98]. However, all the reports showed that the lignin materials used were phosphorylated to improve their flame retardancy.

5.2 *Foams*

Polymer foams are common in the packaging industries and structural applications. Especially for lightweight applications [2, 51, 116]. However, their flammability has always been a source of concern. Incorporating lignin into foam matrices like tannin furanic and polyurethane helps improve their environmental sustainability and acceptability, aside from other benefits, such as improved mechanical strength, density, specific surface area, pore volume, and pore size [98]. Furthermore, lignin has been considered to replace petroleum-based polyols and react with isocyanate groups, thereby providing a new choice of readily available raw material [117].

The reaction of lignin with tolylene 2, 4-diisocyanate by Li and co-workers [118] provided the impetus in this regard. They obtained elastomers with excellent mechanical properties and toughness and, also with high char-forming ability that helped to slow down the burning rate to a reasonable extent. To achieve excellent interaction of lignin with isocyanate, the lignin material is mostly modified [116, 119, 120], which impacts their cost and degradation properties. However, Wang and co-workers [121] reported the preparation of lignin-based flame retardants containing phosphorus which was used as pre-polymer material to prepare lignin/PU foams. The resulting foam had a limiting oxygen index (LOI) of 29.9% maximum with a lignin pre-polymer content of 30 wt%. These foam materials exhibited excellent flame-retardant properties, flexibility of 2 mm, hardness of 2H, adhesion of O, and an impact resistance greater than 100 cm/1000 g. All this with little impact on the overall cost.

Weng and co-researchers developed LBFR for phenolic foam [107]. The LBFR was doped with nitrogen and phosphorus by using 9, 10-dihydro-9-oxa-10-phosphaphenanthrene-10-oxide (DOPO) and Triethylenetetramine (TETA). These reagents were used to partially replace phenol, while preparing the flame-retardant phenolic foam. Analysis of the foam properties showed that the nitrogen-phosphorus-doped LBFR enhanced the compressive strength and flame retardancy of the foam material developed. The pHRR and peak smoke release rate (pSRR) decreased by 48% and 53.2%, respectively. Further analysis showed that the nitrogen and phosphorus in the LBFR worked together to promote carbonization and generate a rich phosphorus and nitrogen char residue layer that insulates heat and oxygen to achieve its flame retardance.

5.3 *Resin/Lignin Composites for Structural Application*

Epoxy resins have found wide applications as electric laminates, carbon foams, molding compounds, circuit boards, coating, and adhesives [61]. These wide applications of phenolic resins are attributed to their extensive cross-linked structure and aromaticity [122, 123]. Therefore, there is a great interest in identifying alternative polyphenol to serve as precursors, and lignin fit in well [124, 125]. Lu and co-researchers [61] prepared lignin-based epoxy resins using phosphorus/nitrogen-treated cobalt metal–organic frameworks. They considered using a metal–organic framework because it has an excellent LOI. Figure 10.6 shows the production of carbon residue when functionalized lignin was used as flame retardant at varying wt% loading. According to the report, the amount of carbon residue produced increases as the wt% of LBFR added increases. Furthermore, Liang and his group [74] showed that the use of ammonium polyphosphate (APP), melamine (MEL), and alkaline lignin prepared in a one-step reaction can significantly improve the flame retardancy of epoxy resin. In their work, the carbon residue increased by 15.3 wt% while LOI value increased from 26.7 to 36.1%. The new material showed an excellent smoke suppression effect and a low heat release rate.

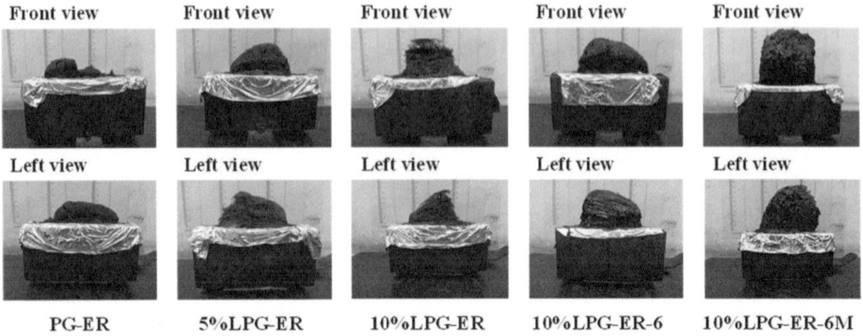

Fig. 10.6 Experimental report of carbon residue produced by lignin/phenol-glyoxal epoxy resin materials as reported by [61]

In addition, a two-step fabrication of LBFR for epoxy resin was reported by Zhou et al. [106], by using polyethylenimine (PEI) and polyphosphoric acid. They found that the nitrogen/phosphorus-modified lignin exhibited lower char-forming temperature and enhanced thermal stability. The UL-94 grade and LOI were up to V-0 grade and 31.4%, respectively, while the maximum decomposition temperature was slightly reduced from 387 to 364 °C. This method was described as a gentle and efficient multifunctional process that enhances the application of lignin-based FR in epoxy resin composites.

6 Future of Lignin-Based Flame Retardants

Lignin remains a promising green material for the future. Its abundance, renewability, and eco-friendliness are advantages over other materials in the same category (inorganic and organic flame retardants). The ability to function as a flame retardant either in its raw or modified state makes lignin a perfect material for the future [6, 30, 31, 74, 97, 107, 126]. Scientists and engineers are working round the clock, trying to improve the existing research outputs. The need to develop LBFR which can act as flame retardant as well as compatibilizer and filler at the same time is very much on the front burner. The advantage of having multiple functional groups present on it makes the application of lignin more attractive, making lignin a very reactive additive. A lot still has to be done. This includes developing specific LBFR products for specific polymers of specific applications. Just as it is obtainable now in the manufacturing community, where every application has its own material requirements and should be produced to specification. Knowing what to add and the quality requirements. So also, it is necessary to develop specific LBFR for specific applications. Results from recent research are encouraging but need to be specific in terms of application for real progress to be achieved. Also, in order not to lose its advantages as it relates to environmental sustainability, researchers must focus equal energy on the biodegradability of the LBFR and its fate after use. Many of the environmental pollution problems in today's world started just like this.

7 Conclusion

Lignin, a renewable biomaterial, is the second largest bioresource from lignocellulose biomass after cellulose. It is largely generated from paper industries as a by-product. Its aromatic superstructure helps to generate rich carbon residue that can act as a barrier to flame in a burning material, thereby conferring lignin as a flame retardant. To improve the flame retardancy of lignin, factors such as type of lignin, lignin percentage loading, and type of modification carried out were considered to maintain its environmental sustainability. Even though pristine lignin decomposes thermally over a range of temperatures, it only produces char which acts as a barrier to prevent the continuation of a flame. This was considered inadequate for polymer materials required in some high-risk fire zones. To achieve its fire barrier properties, lignin was modified with different reagents that helped introduce phosphorus and nitrogen atoms onto the lignin molecules. So that the lignin flame-retardants can function both in the condensed and gas phases of the flame. Therefore, eco-friendly LBFRs for polymer materials can be designed to be single component LBFR, combined components, chemically modified LBFRs (P-LBFR, N-LBFR, or PN-LBFR), or metal-PN LBFR. The use of PN-modified LBFR has been shown to combine well with many of the polymers used in most industries, and it produces non-toxic gas. Modified lignin is considered a better charring agent than unmodified. The modification also helps to improve compatibilization with the polymer matrix, thereby leading to improved mechanical properties. Modified LBFR have been reported to perform well as flame retardants in PLA, PA11, epoxy resins, polyurethane, and polypropylene, and the resulting composites have been reportedly applied in industries such as food packaging, automobile interiors, aerospace, electrical and electronics, building, and structural applications. However, more research is still needed in this area to develop specialized LBFR for specific polymers and for specific applications.

Acknowledgements The authors acknowledge and appreciate following: North Carolina State University, NCSU for the use of resources, especially the time provision of journal papers not readily available online. Federal University Lokoja, FUL for their continuous support and encouragement during this time.

References

1. Abdelaziz, O.Y., Brink, D.P., Prothmann, J., Ravi, K., Sun, M., García-Hidalgo, J., Sandahl, M., Hulteberg, C.P., Turner, C., Lidén, G., Gorwa-Grauslund, M.F.: Biological valorization of low molecular weight lignin. Biotechnol. Adv. (2016)

2. Fagbemigun, T.K., Mai, C.: Production and characterisation of self-blowing lignin-based foams. Eur. J. Wood Wood Prod. **81**, 579–590 (2023). https://doi.org/10.1007/s00107-022-01908-1
3. Voelker, S.L., Lachenbruch, B., Meinzer, F.C., Strauss, S.H.: Reduced wood stiffness and strength, and altered stem form, in young antisense 4CL transgenic poplars with reduced lignin contents. New Phytol. **189**, 1096–1109 (2011). https://doi.org/10.1111/j.1469-8137.2010.03572.x
4. Youssefian, S., Rahbar, N.: Molecular origin of strength and stiffness in bamboo fibrils. Sci. Rep. **5** (2015). https://doi.org/10.1038/srep11116
5. Zeng, Y., Himmel, M.E., Ding, S.Y.: Visualizing Chemical Functionality in Plant Cell Walls. Mike Himmel (2017)
6. Vásquez-Garay, F., Carrillo-Varela, I., Vidal, C., Reyes-Contreras, P., Faccini, M., Mendonça, R.T.: A review on the lignin biopolymer and its integration in the elaboration of sustainable materials. Sustainability (2021)
7. Lu, Y., Lu, Y.C., Hu, H.Q., Xie, F.J., Wei, X.Y., Fan, X.: Structural characterization of lignin and its degradation products with spectroscopic methods. J. Spectrosc. (2017)
8. Tolbert, A., Akinosho, H., Khunsupat, R., Naskar, A.K., Ragauskas, A.J.: Characterization and analysis of the molecular weight of lignin for biorefining studies. Biofuels Bioprod. Biorefin (2014)
9. Watkins, D., Nuruddin, M., Hosur, M., Tcherbi-Narteh, A., Jeelani, S.: Extraction and characterization of lignin from different biomass resources. J. Market. Res. **4**, 26–32 (2015). https://doi.org/10.1016/j.jmrt.2014.10.009
10. Zhao, S., Abu-Omar, M.M.: Renewable thermoplastics based on lignin-derived polyphenols. Macromolecules **50**, 3573–3581 (2017). https://doi.org/10.1021/acs.macromol.7b00064
11. Laurichesse, S., Avérous, L.: Chemical modification of lignins: towards biobased polymers. Prog. Polym. Sci. (2014)
12. Upton, B.M., Kasko, A.M.: Strategies for the conversion of lignin to high-value polymeric materials: review and perspective. Chem. Rev. (2016)
13. Harman-Ware, A.E., Happs, R.M., Davison, B.H., Davis, M.F.: The effect of coumaryl alcohol incorporation on the structure and composition of lignin dehydrogenation polymers Mike Himmel. Biotechnol. Biofuels **10** (2017). https://doi.org/10.1186/s13068-017-0962-2
14. Garedew, M., Lin, F., Song, B., DeWinter, T.M., Jackson, J.E., Saffron, C.M., Lam, C.H., Anastas, P.T.: Greener routes to biomass waste valorization: lignin transformation through electrocatalysis for renewable chemicals and fuels production. Chem. Sus. Chem. (2020)
15. Sen, S., Patil, S., Argyropoulos, D.S.: Thermal properties of lignin in copolymers, blends, and composites: a review. Green. Chem. (2015)
16. Chung, H., Washburn, N.R.: Improved lignin polyurethane properties with Lewis acid treatment. ACS Appl. Mater. Interfaces **4**, 2840–2846 (2012). https://doi.org/10.1021/am300425x
17. Yang, H., Yu, B., Xu, X., Bourbigot, S., Wang, H., Song, P.: Lignin-derived bio-based flame retardants toward high-performance sustainable polymeric materials. Green Chem. (2020)
18. Thakur, V.K., Thakur, M.K., Raghavan, P., Kessler, M.R.: Progress in green polymer composites from lignin for multifunctional applications: a review. ACS Sustain. Chem. Eng. **2**, 1072–1092 (2014). https://doi.org/10.1021/sc500087z
19. Pucciariello, R., Villani, V., Bonini, C., D'Auria, M., Vetere, T.: Physical properties of straw lignin-based polymer blends. Polymer (Guildf) **45**, 4159–4169 (2004). https://doi.org/10.1016/j.polymer.2004.03.098
20. Yang, W., Fortunati, E., Dominici, F., Giovanale, G., Mazzaglia, A., Balestra, G.M., Kenny, J.M., Puglia, D.: Effect of cellulose and lignin on disintegration, antimicrobial and antioxidant properties of PLA active films. Int. J. Biol. Macromol. **89**, 360–368 (2016). https://doi.org/10.1016/j.ijbiomac.2016.04.068
21. Calvo-Flores, F.G., Dobado, J.A.: Lignin as renewable raw material. Chem. Sus. Chem. (2010)
22. Zong, E., Huang, G., Liu, X., Lei, W., Jiang, S., Ma, Z., Wang, J., Song, P.: A lignin-based nano-adsorbent for superfast and highly selective removal of phosphate. J. Mater. Chem. A Mater. **6**, 9971–9983 (2018). https://doi.org/10.1039/c8ta01449c

23. Li, Y., Wu, M., Wang, B., Wu, Y., Ma, M., Zhang, X.: Synthesis of magnetic lignin-based hollow microspheres: a highly adsorptive and reusable adsorbent derived from renewable resources. ACS Sustain. Chem. Eng. **4**, 5523–5532 (2016). https://doi.org/10.1021/acssuschemeng.6b01244
24. Solihat, N.N., Sari, F.P., Falah, F., Ismayati, M., Lubis, M.A.R., Fatriasari, W., Santoso, E.B., Syafii, W.: Lignin as an active biomaterial: a review. Jurnal Sylva Lestari. **9**, 1 (2021). https://doi.org/10.23960/jsl191-22
25. Nge, T.T., Tobimatsu, Y., Takahashi, S., Takata, E., Yamamura, M., Miyagawa, Y., Ikeda, T., Umezawa, T., Yamada, T.: Isolation and characterization of polyethylene glycol (PEG)-modified glycol lignin via PEG solvolysis of softwood biomass in a large-scale batch reactor. ACS Sustain. Chem. Eng. **6**, 7841–7848 (2018). https://doi.org/10.1021/acssuschemeng.8b00965
26. Xia, Z., Kiratitanavit, W., Facendola, P., Thota, S., Yu, S., Kumar, J., Mosurkal, R., Nagarajan, R.: Fire resistant polyphenols based on chemical modification of bio-derived tannic acid. Polym. Degrad. Stab. **153**, 227–243 (2018). https://doi.org/10.1016/j.polymdegradstab.2018.04.020
27. Song, Y., Zong, X., Wang, N., Yan, N., Shan, X., Li, J.: Preparation of γ -divinyl-3-aminopropyltriethoxysilane modified lignin and its application in flame retardant poly(lactic acid). Materials. **11** (2018). https://doi.org/10.3390/ma11091505
28. Liu, Y., Zhang, Y., Fang, Z.: Design of novel fire retardant (2012)
29. Kai, D., Tan, M.J., Chee, P.L., Chua, Y.K., Yap, Y.L., Loh, X.J.: Towards lignin-based functional materials in a sustainable world. Green Chem. (2016)
30. Andrade-Guel, M., Cabello-Alvarado, C., Avila-Orta, C.A., Pérez-Alvarez, M., Cadenas-Pliego, G., Reyes-Rodríguez, P.Y., Rios-González, L.: Green flame-retardant composites based on PP/TiO_2/Lignin obtained by melt-mixing extrusion. Polymers (Basel) **14** (2022). https://doi.org/10.3390/polym14071300
31. Liang, Y., Jian, H., Deng, C., Xu, J., Liu, Y., Park, H., Wen, M., Sun, Y.: Research and application of biomass-based wood flame retardants: a review. Polymers (2023)
32. Thielemans, W., Wool, R.P.: Butyrated kraft lignin as compatibilizing agent for natural fiber reinforced thermoset composites. Compos. Part A: Appl. Sci. Manuf. 327–338 (2004)
33. Łukawski, D., Grześkowiak, W., Lekawa-Raus, A., Widelicka, M., Lisiecki, F., Dudkowiak, A.: Flame retardant effect of lignin/carbon nanotubes/potassium carbonate composite coatings on cotton roving. Cellulose **27**, 7271–7281 (2020). https://doi.org/10.1007/s10570-020-03270-y
34. Norgren, M., Edlund, H.: Lignin: recent advances and emerging applications. Curr. Opin. Colloid Interface Sci. (2014)
35. Mandlekar, N., Cayla, A., Rault, F., Giraud, S., Salaün, F., Malucelli, G., Guan, J.: Thermal stability and fire retardant properties of polyamide 11 microcomposites containing different lignins. Ind. Eng. Chem. Res. **56**, 13704–13714 (2017). https://doi.org/10.1021/acs.iecr.7b03085
36. Mandlekar, N., Malucelli, G., Cayla, A., Rault, F., Giraud, S., Salaün, F., Guan, J.: Fire retardant action of zinc phosphinate and polyamide 11 blend containing lignin as a carbon source. Polym. Degrad. Stab. **153**, 63–74 (2018). https://doi.org/10.1016/j.polymdegradstab.2018.04.019
37. Maity, S.K.: Opportunities, recent trends and challenges of integrated biorefinery: Part II. Renew. Sustain. Energy Rev. (2015)
38. Jiang, L.Y., Zhu, J.M.: Separation technologies for current and future biorefineries-status and potential of membrane-based separation. Adv. Bioenergy Sustain. Challenge (2013)
39. Rodríguez Correa, C., Stollovsky, M., Hehr, T., Rauscher, Y., Rolli, B., Kruse, A.: Influence of the carbonization process on activated carbon properties from lignin and lignin-rich biomasses. ACS Sustain. Chem. Eng. **5**, 8222–8233 (2017). https://doi.org/10.1021/acssuschemeng.7b01895

40. Law, R.J., Allchin, C.R., de Boer, J., Covaci, A., Herzke, D., Lepom, P., Morris, S., Tronczynski, J., de Wit, C.A.: Levels and trends of brominated flame retardants in the European environment. Chemosphere **64**, 187–208 (2006). https://doi.org/10.1016/j.chemosphere.2005.12.007
41. Darnerud, P.O.: Toxic effects of brominated flame retardants in man and in wildlife. Environ. Int. **29**, 841–853 (2003). https://doi.org/10.1016/S0160-4120(03)00107-7
42. Zhang, M., Buekens, A., Li, X.: Brominated flame retardants and the formation of dioxins and furans in fires and combustion. J. Haz. Mater. (2016)
43. Iyer, K.A., Torkelson, J.M.: Sustainable green hybrids of polyolefins and lignin yield major improvements in mechanical properties when prepared via solid-state shear pulverization. ACS Sustain. Chem. Eng. **3**, 959–968 (2015). https://doi.org/10.1021/acssuschemeng.5b00099
44. Tanjung, F.A., Husseinsyah, S., Hussin, K., Hassan, A.: Mechanical and thermal properties of organosolv lignin/sodium dodecyl sulphate binary agent-treated polypropylene/chitosan composites. Polym. Bull. **73**, 1427–1445 (2016). https://doi.org/10.1007/s00289-015-1556-4
45. De Chirico, A., Armanini, M., Chini, P., Cioccolo, G., Provasoli, F., Audisio, G.: Flame retardants for polypropylene based on lignin. Polym. Degrad. Stab. (2003)
46. Yeo, J.S., Seong, D.W., Hwang, S.H.: Chemical surface modification of lignin particle and its application as filler in the polypropylene composites. J. Ind. Eng. Chem. **31**, 80–85 (2015). https://doi.org/10.1016/j.jiec.2015.06.010
47. Maldhure, A.V., Ekhe, J.D.: Effect of modifications of lignin on thermal, structural, and mechanical properties of polypropylene/modified lignin blends. J. Thermoplast. Compos. Mater. **30**, 625–645 (2017). https://doi.org/10.1177/0892705715610402
48. Costes, L., Laoutid, F., Aguedo, M., Richel, A., Brohez, S., Delvosalle, C., Dubois, P.: Phosphorus and nitrogen derivatization as efficient route for improvement of lignin flame retardant action in PLA. Eur. Polym. J. **84**, 652–667 (2016). https://doi.org/10.1016/j.eurpolymj.2016.10.003
49. Prieur, B., Meub, M., Wittemann, M., Klein, R., Bellayer, S., Fontaine, G., Bourbigot, S.: Phosphorylation of lignin to flame retard acrylonitrile butadiene styrene (ABS). Polym. Degrad. Stab. **127**, 32–43 (2016). https://doi.org/10.1016/j.polymdegradstab.2016.01.015
50. Song, P., Cao, Z., Fu, S., Fang, Z., Wu, Q., Ye, J.: Thermal degradation and flame retardancy properties of ABS/lignin: effects of lignin content and reactive compatibilization. Thermochim. Acta **518**, 59–65 (2011). https://doi.org/10.1016/j.tca.2011.02.007
51. Yang, L., Wang, X., Cui, Y., Tian, Y., Chen, H., Wang, Z.: Modification of renewable resources-lignin-by three chemical methods and its applications to polyurethane foams. Polym. Adv. Technol. **25**, 1089–1098 (2014). https://doi.org/10.1002/pat.3356
52. Avelino, F., Almeida, S.L., Duarte, E.B., Sousa, J.R., Mazzetto, S.E., de Souza Filho, M. de S.M.: Thermal and mechanical properties of coconut shell lignin-based polyurethanes synthesized by solvent-free polymerization. J. Mater. Sci. **53**, 1470–1486 (2018). https://doi.org/10.1007/s10853-017-1562-z
53. Kumar, S., Krishnan, S., Mohanty, S., Nayak, S.K.: Synthesis and characterization of petroleum and biobased epoxy resins: a review. Polym. Int. (2018)
54. Ferdosian, F., Yuan, Z., Anderson, M., Xu, C.C.: Thermal performance and thermal decomposition kinetics of lignin-based epoxy resins. J. Anal. Appl. Pyrolysis **119**, 124–132 (2016). https://doi.org/10.1016/j.jaap.2016.03.009
55. Nikafshar, S., Zabihi, O., Moradi, Y., Ahmadi, M., Amiri, S., Naebe, M.: Catalyzed synthesis and characterization of a novel lignin-based curing agent for the curing of high-performance epoxy resin. Polymers (Basel) **9** (2017). https://doi.org/10.3390/polym9070266
56. Zhao, S., Abu-Omar, M.M.: Synthesis of renewable thermoset polymers through successive lignin modification using lignin-derived phenols. ACS Sustain. Chem. Eng. **5**, 5059–5066 (2017). https://doi.org/10.1021/acssuschemeng.7b00440
57. Chollet, B., Lopez-Cuesta, J.M., Laoutid, F., Ferry, L.: Lignin nanoparticles as a promising way for enhancing lignin flame retardant effect in polylactide. Materials **12** (2019). https://doi.org/10.3390/ma12132132

58. Solihat, N.N., Hidayat, A.F., Taib, M.N.A.M., Hussin, M.H., Lee, S.H., Ghani, M.A.A., Edrus, S.S.A.O.A., Vahabi, H., Fatriasari, W.: Recent developments in flame-retardant lignin-based biocomposite: manufacturing, and characterization. J. Polym. Environ. (2022)
59. Goliszek, M., Podkościelna, B., Klepka, T., Sevastyanova, O.: Preparation, thermal, and mechanical characterization of UV-cured polymer biocomposites with lignin. Polymers (Basel) **12** (2020). https://doi.org/10.3390/POLYM12051159
60. Goliszek, M., Podkościelna, B., Sevastyanova, O., Fila, K., Chabros, A., Pączkowski, P.: Investigation of accelerated aging of lignin-containing polymer materials. Int. J. Biol. Macromol. **123**, 910–922 (2019). https://doi.org/10.1016/j.ijbiomac.2018.11.141
61. Lu, X., Lee, A.F., Gu, X.: Improving the flame retardancy of sustainable lignin-based epoxy resins using phosphorus/nitrogen treated cobalt metal-organic frameworks. Mater. Today Chem. **26** (2022). https://doi.org/10.1016/j.mtchem.2022.101184
62. Podkościelna, B., Wnuczek, K., Goliszek, M., Klepka, T., Dziuba, K.: Flammability tests and investigations of properties of lignin-containing polymer composites based on acrylates. Molecules **25** (2020). https://doi.org/10.3390/MOLECULES25245947
63. Podkościelna, B., Gargol, M., Goliszek, M., Klepka, T., Sevastyanova, O.: Degradation and flammability of bioplastics based on PLA and lignin. Polym Test. **111** (2022). https://doi.org/10.1016/j.polymertesting.2022.107622
64. Matsushita, Y., Hirano, D., Aoki, D., Yagami, S., Takagi, Y., Fukushima, K.: A biobased flame-retardant resin based on lignin. Adv. Sustain. Syst. **1** (2017). https://doi.org/10.1002/adsu.201700073
65. Kim, J.Y., Oh, S., Hwang, H., Kim, U.J., Choi, J.W.: Structural features and thermal degradation properties of various lignin macromolecules obtained from poplar wood (*Populus albaglandulosa*). Polym. Degrad. Stab. **98**, 1671–1678 (2013). https://doi.org/10.1016/j.polymdegradstab.2013.06.008
66. Brebu, M., Tamminen, T., Spiridon, I.: Thermal degradation of various lignins by TG-MS/FTIR and Py-GC-MS. J. Anal. Appl. Pyrolysis **104**, 531–539 (2013). https://doi.org/10.1016/j.jaap.2013.05.016
67. Costes, L., Laoutid, F., Brohez, S., Dubois, P.: Bio-based flame retardants: when nature meets fire protection. Mater. Sci. Eng. R. Rep. (2017)
68. Widsten, P., Tamminen, T., Paajanen, A., Hakkarainen, T., Liitiä, T.: Modified and unmodified technical lignins as flame retardants for polypropylene. Holzforschung **75**, 584–590 (2021). https://doi.org/10.1515/hf-2020-0147
69. Klapiszewski, Ł., Bula, K., Sobczak, M., Jesionowski, T.: Influence of processing conditions on the thermal stability and mechanical properties of PP/silica-lignin composites. Int. J. Polym. Sci. **2016** (2016). https://doi.org/10.1155/2016/1627258
70. Tao, J., Hosseinaei, O., Delbeck, L., Kim, P., Harper, D.P., Bozell, J.J., Rials, T.G., Labbé, N.: Effects of organosolv fractionation time on thermal and chemical properties of lignins. RSC Adv. **6**, 79228–79235 (2016). https://doi.org/10.1039/c6ra16296g
71. Li, B., Zhang, X., Su, R.: An investigation of thermal degradation and charring of larch lignin in the condensed phase: the effects of boric acid, guanyl urea phosphate, ammonium dihydrogen phosphate and ammonium polyphosphate. Polym. Degrad. Stab. **75**, 35–44 (2002)
72. Brebu, M., Vasile, C.: Thermal degradation of lignin-a review. Cellul. Chem. Technol. (2010)
73. Hu, J., Shen, D., Wu, S., Zhang, H., Xiao, R.: Effect of temperature on structure evolution in char from hydrothermal degradation of lignin. J. Anal. Appl. Pyrolysis **106**, 118–124 (2014). https://doi.org/10.1016/j.jaap.2014.01.008
74. Liang, D., Zhu, X., Dai, P., Lu, X., Guo, H., Que, H., Wang, D., He, T., Xu, C., Robin, H.M., Luo, Z., Gu, X.: Preparation of a novel lignin-based flame retardant for epoxy resin. Mater. Chem. Phys. **259** (2021). https://doi.org/10.1016/j.matchemphys.2020.124101
75. Brodin, I., Sjöholm, E., Gellerstedt, G.: The behavior of kraft lignin during thermal treatment. J. Anal. Appl. Pyrol. **87**, 70–77 (2010). https://doi.org/10.1016/j.jaap.2009.10.005
76. Costes, L., Laoutid, F., Brohez, S., Delvosalle, C., Dubois, P.: Phytic acid–lignin combination: a simple and efficient route for enhancing thermal and flame retardant properties of polylactide. Eur. Polym. J. **94**, 270–285 (2017). https://doi.org/10.1016/j.eurpolymj.2017.07.018

77. Braun, U., Bahr, H., Sturm, H., Schartel, B.: Flame retardancy mechanisms of metal phosphinates and metal phosphinates in combination with melamine cyanurate in glass-fiber reinforced poly(1,4-butylene terephthalate): the influence of metal cation. Polym. Adv. Technol. **19**, 680–692 (2008). https://doi.org/10.1002/pat.1147
78. Liu, L., Qian, M., Song, P., Huang, G., Yu, Y., Fu, S.: Fabrication of green lignin-based flame retardants for enhancing the thermal and fire retardancy properties of polypropylene/wood composites. ACS Sustain. Chem. Eng. **4**, 2422–2431 (2016). https://doi.org/10.1021/acssuschemeng.6b00112
79. Hu, W., Zhang, Y., Qi, Y., Wang, H., Liu, B., Zhao, Q., Zhang, J., Duan, J., Zhang, L., Sun, Z., Liu, B.: Improved mechanical properties and flame retardancy of wood/PLA all-degradable biocomposites with novel lignin-based flame retardant and TGIC. Macromol. Mater. Eng. **305** (2020). https://doi.org/10.1002/mame.201900840
80. Zhang, Y.M., Zhao, Q., Li, L., Yan, R., Zhang, J., Duan, J.C., Liu, B.J., Sun, Z.Y., Zhang, M.Y., Hu, W., Zhang, N.N.: Synthesis of a lignin-based phosphorus-containing flame retardant and its application in polyurethane. RSC Adv. **8**, 32252–32261 (2018). https://doi.org/10.1039/c8ra05598j
81. Mendis, G.P., Dietenberger, M., Youngblood, J.P., Weiss, S.G., Howarter, J.A., Korey, M., Boardman, C.R.: Phosphorylated lignin as a halogen-free flame retardant additive for epoxy composites. Green Compos. **4**, 150–159 (2016)
82. Guo, Y., Cheng, C., Huo, T., Ren, Y., Liu, X.: Highly effective flame retardant lignin/polyacrylonitrile composite prepared via solution blending and phosphorylation. Polym. Degrad. Stab. **181** (2020). https://doi.org/10.1016/j.polymdegradstab.2020.109362
83. Schartel, B., Hull, T.R.: Development of fire-retarded materials—interpretation of cone calorimeter data. Fire Mater. **31**, 327–354 (2007). https://doi.org/10.1002/fam.949
84. Levchik, S.V., Weil, E.D.: A review of recent progress in phosphorus-based flame retardants. J. Fire Sci. **24**, 345–364 (2006). https://doi.org/10.1177/0734904106068426
85. Sallem-Idrissi, N., Van Velthem, P., Sclavons, M.: Fully bio-sourced Nylon 11/raw lignin composites: thermal and mechanical performances. J. Polym. Environ. **26**, 4405–4414 (2018). https://doi.org/10.1007/s10924-018-1311-7
86. Gallina, G., Bravin, E., Badalucco, C., Audisio, G., Armanini, M., De Chirico, A., Provasoli, F.: Application of cone calorimeter for the assessment of class of flame retardants for polypropylene. Fire Mater. **22**, 15–18 (1998). https://doi.org/10.1002/(SICI)1099-1018(199801/02)22:1%3c15::AID-FAM626%3e3.0.CO;2-3
87. Nie, S., Hu, Y., Song, L., He, Q., Yang, D., Chen, H.: Synergistic effect between a char forming agent (CFA) and microencapsulated ammonium polyphosphate on the thermal and flame retardant properties of polypropylene. Polym. Adv. Technol. **19**, 1077–1083 (2008). https://doi.org/10.1002/pat.1082
88. Chien, W.P., Seader, J.D.: Prediction of specific optical density for smoke obscuration in an NBS smoke density chamber. Fire Technol. **11**, 206–218 (1975)
89. Shi, K., Liu, G., Sun, H., Weng, Y.: Polylactic acid/lignin composites: a review. Polymers (Basel) **15**, 2807 (2023). https://doi.org/10.3390/polym15132807
90. Kaewtatip, K., Thongmee, J.: Effect of Kraft lignin and esterified lignin on the properties of thermoplastic starch. Mater. Des. **49**, 701–704 (2013). https://doi.org/10.1016/j.matdes.2013.02.010
91. Maqsood, M., Langensiepen, F., Seide, G.: The efficiency of biobased carbonization agent and intumescent flame retardant on flame retardancy of biopolymer composites and investigation of their melt-spinnability. Molecules **24** (2019). https://doi.org/10.3390/molecules24081513
92. Laoutid, F., Duriez, V., Brison, L., Aouadi, S., Vahabi, H., Dubois, P.: Synergistic flame-retardant effect between lignin and magnesium hydroxide in poly(ethylene-co-vinyl acetate). Flame Retardancy Therm. Stab. Mater. **2**, 9–18 (2019). https://doi.org/10.1515/flret-2019-0002
93. Verdolotti, L., Oliviero, M., Lavorgna, M., Iannace, S., Camino, G., Vollaro, P., Frache, A.: On revealing the effect of alkaline lignin and ammonium polyphosphate additives on fire retardant properties of sustainable zein-based composites. Polym. Degrad. Stab. **134**, 115–125 (2016). https://doi.org/10.1016/j.polymdegradstab.2016.10.001

94. Ye, D., Kong, J., Gu, S., Zhou, Y., Huang, C., Xu, W., Zhang, X.: Selective aminolysis of acetylated lignin: toward simultaneously improving thermal-oxidative stability and maintaining mechanical properties of polypropylene. Int. J. Biol. Macromol. **108**, 775–781 (2018). https://doi.org/10.1016/j.ijbiomac.2017.10.168
95. Bourbigot, S., Fontaine, G.: Modified Lignin as Flame Retardant for Polymeric Materials. PhD Thesis (2016)
96. Illy, N., Fache, M., Ménard, R., Negrell, C., Caillol, S., David, G.: Phosphorylation of bio-based compounds: The state of the art. Polym. Chem. (2015)
97. Xiong, L., Huang, S., Zhong, R., Tang, W., Liu, C., Jin, Y.: Preparation, characterization, and performance of lignin-based microencapsulated red phosphorus flame retardant for ABS. J. Wuhan Univ. Technol. Mater. Sci. Ed. **37**, 292–299 (2022). https://doi.org/10.1007/s11595-022-2530-2
98. Chen, S., Lin, S., Hu, Y., Ma, M., Shi, Y., Liu, J., Zhu, F., Wang, X.: A lignin-based flame retardant for improving fire behavior and biodegradation performance of polybutylene succinate. Polym. Adv. Technol. **29**, 3142–3150 (2018). https://doi.org/10.1002/pat.4436
99. Shi, Y., Yu, B., Duan, L., Gui, Z., Wang, B., Hu, Y., Yuen, R.K.K.: Graphitic carbon nitride/phosphorus-rich aluminum phosphinates hybrids as smoke suppressants and flame retardants for polystyrene. J. Hazard. Mater. **332**, 87–96 (2017). https://doi.org/10.1016/j.jhazmat.2017.03.006
100. Yu, Y., Fu, S., Song, P., Luo, X., Jin, Y., Lu, F., Wu, Q., Ye, J.: Functionalized lignin by grafting phosphorus-nitrogen improves the thermal stability and flame retardancy of polypropylene. Polym. Degrad. Stab. **97**, 541–546 (2012). https://doi.org/10.1016/j.polymdegradstab.2012.01.020
101. Bourbigot, S., Duquesne, S.: Fire retardant polymers: recent developments and opportunities. J. Mater. Chem. **17**, 2283–2300 (2007). https://doi.org/10.1039/b702511d
102. Dai, P., Liang, M., Ma, X., Luo, Y., He, M., Gu, X., Gu, Q., Hussain, I., Luo, Z.: Highly efficient, environmentally friendly lignin-based flame retardant used in epoxy resin. ACS Omega **5**, 32084–32093 (2020). https://doi.org/10.1021/acsomega.0c05146
103. Zhang, R., Xiao, X., Tai, Q., Huang, H., Hu, Y.: Modification of lignin and its application as char agent in intumescent flame-retardant poly(lactic acid). Polym. Eng. Sci. **52**, 2620–2626 (2012). https://doi.org/10.1002/pen.23214
104. Cheng-Han, H., Yung-Chan, L., En-Nan, L.: Flame retardants agents and flame retardant materials comprising the same. (2013)
105. Yu, Y., Song, P., Jin, C., Fu, S., Zhao, L., Wu, Q., Ye, J.: Catalytic effects of nickel (cobalt or zinc) acetates on thermal and flammability properties of polypropylene-modified lignin composites. Ind. Eng. Chem. Res. **51**, 12367–12374 (2012). https://doi.org/10.1021/ie301953x
106. Zhou, S., Tao, R., Dai, P., Luo, Z., He, M.: Two-step fabrication of lignin-based flame retardant for enhancing the thermal and fire retardancy properties of epoxy resin composites. Polym. Compos. **41**, 2025–2035 (2020). https://doi.org/10.1002/pc.25517
107. Weng, S., Li, Z., Bo, C., Song, F., Xu, Y., Hu, L., Zhou, Y., Jia, P.: Design lignin doped with nitrogen and phosphorus for flame retardant phenolic foam materials. React. Funct. Polym. **185** (2023). https://doi.org/10.1016/j.reactfunctpolym.2023.105535
108. Drobny, J.G.: Introduction. In: Applications of Fluoropolymer Films, pp. 3–38. Elsevier (2020)
109. Agwuncha, S.C., Ray, S.S., Jayaramudu, J., Khoathane, C., Sadiku, R.: Influence of boehmite nanoparticle loading on the mechanical, thermal, and rheological properties of biodegradable polylactide/poly(ε-Caprolactone) blends. Macromol. Mater Eng. **300** (2015). https://doi.org/10.1002/mame.201400212
110. Spiridon, I., Leluk, K., Resmerita, A.M., Darie, R.N.: Evaluation of PLA-lignin bioplastics properties before and after accelerated weathering. Compos. B Eng. **69**, 342–349 (2015). https://doi.org/10.1016/j.compositesb.2014.10.006
111. Gordobil, O., Delucis, R., Egüés, I., Labidi, J.: Kraft lignin as filler in PLA to improve ductility and thermal properties. Ind. Crops Prod. **72**, 46–53 (2015). https://doi.org/10.1016/j.indcrop.2015.01.055

112. Park, S.Y., Kim, J.Y., Youn, H.J., Choi, J.W.: Utilization of lignin fractions in UV resistant lignin-PLA biocomposites via lignin-lactide grafting. Int. J. Biol. Macromol. **138**, 1029–1034 (2019). https://doi.org/10.1016/j.ijbiomac.2019.07.157
113. Park, C.W., Youe, W.J., Kim, S.J., Han, S.Y., Park, J.S., Lee, E.A., Kwon, G.J., Kim, Y.S., Kim, N.H., Lee, S.H.: Effect of lignin plasticization on physico-mechanical properties of lignin/poly(lactic acid) composites. Polymers (Basel) **11** (2019). https://doi.org/10.3390/polym11122089
114. Sun, Y., Ma, Z., Xu, X., Liu, X., Liu, L., Huang, G., Liu, L., Wang, H., Song, P.: Grafting lignin with bioderived polyacrylates for low-cost, ductile, and fully biobased poly(lactic acid) composites. ACS Sustain. Chem. Eng. **8**, 2267–2276 (2020). https://doi.org/10.1021/acssuschemeng.9b06593
115. Sun, Y., Sun, S., Chen, L., Liu, L., Song, P., Li, W., Yu, Y., Fengzhu, L., Qian, J., Wang, H.: Flame retardant and mechanically tough poly(lactic acid) biocomposites via combining ammonia polyphosphate and polyethylene glycol. Compos. Commun. **6**, 1–5 (2017). https://doi.org/10.1016/j.coco.2017.07.005
116. Gómez-Fernández, S., Ugarte, L., Calvo-Correas, T., Peña-Rodríguez, C., Corcuera, M.A., Eceiza, A.: Properties of flexible polyurethane foams containing isocyanate functionalized kraft lignin. Ind. Crops Prod. **100**, 51–64 (2017). https://doi.org/10.1016/j.indcrop.2017.02.005
117. Yan, R., Yang, D., Zhang, N., Zhao, Q., Liu, B., Xiang, W., Sun, Z., Xu, R., Zhang, M., Hu, W.: Performance of UV curable lignin based epoxy acrylate coatings. Prog. Org. Coat. **116**, 83–89 (2018). https://doi.org/10.1016/j.porgcoat.2017.11.011
118. Li, H., Sun, J.T., Wang, C., Liu, S., Yuan, D., Zhou, X., Tan, J., Stubbs, L., He, C.: High modulus, strength, and toughness polyurethane elastomer based on unmodified lignin. ACS Sustain. Chem. Eng. **5**, 7942–7949 (2017). https://doi.org/10.1021/acssuschemeng.7b01481
119. Li, H., Liang, Y., Li, P., He, C.: Conversion of biomass lignin to high-value polyurethane: a review. J. Bioresourc. Bioprod. (2020)
120. Antonino, L.D., Sumerskii, I., Potthast, A., Rosenau, T., Felisberti, M.I., dos Santos, D.J.: Lignin-based polyurethanes from the blocked isocyanate approach: synthesis and characterization. ACS Omega (2023). https://doi.org/10.1021/acsomega.3c03422
121. Wang, Y., Zhang, Y., Liu, B., Zhao, Q., Qi, Y., Wang, Y., Sun, Z., Liu, B., Zhang, N., Hu, W., Xie, H.: A novel phosphorus-containing lignin-based flame retardant and its application in polyurethane. Compos. Commun. **21** (2020). https://doi.org/10.1016/j.coco.2020.100382
122. Pilato, L.: Phenolic resins: 100 years and still going strong. In: Reactive and Functional Polymers, pp. 270–277 (2013)
123. García, D.E., Glasser, W.G., Pizzi, A., Lacoste, C., Laborie, M.P.: Polyphenolic resins prepared with maritime pine bark tannin and bulky-aldehydes. Ind. Crops Prod. **62**, 84–93 (2014). https://doi.org/10.1016/j.indcrop.2014.08.010
124. Naseem, A., Tabasum, S., Zia, K.M., Zuber, M., Ali, M., Noreen, A.: Lignin-derivatives based polymers, blends and composites: a review. Int. J. Biol. Macromol. **93**, 296–313 (2016). https://doi.org/10.1016/j.ijbiomac.2016.08.030
125. Zhao, M., Jing, J., Zhu, Y., Yang, X., Wang, X., Wang, Z.: Preparation and performance of lignin-phenol-formaldehyde adhesives. Int. J. Adhes. Adhes. **64**, 163–167 (2016). https://doi.org/10.1016/j.ijadhadh.2015.10.010
126. Wang, Y., Zhang, S., Wu, X., Lu, C., Cai, Y., Ma, L., Shi, G., Yang, L.: Effect of montmorillonite on the flame-resistant and mechanical properties of intumescent flame-retardant poly(butylene succinate) composites. J. Therm. Anal. Calorim. **128**, 1417–1427 (2017). https://doi.org/10.1007/s10973-017-6092-z

Preparation and Characterization of Nanomaterials from Biomaterials for Flame Retardancy

M. P. Motaung, W. Moloto, B. Ntsendwana, T. C. Mokhena, and M. J. Mochane

Abstract This chapter focuses on the synthesis and characterization of nanomaterials derived from biomaterials, which signifies a viable path for improving flame retardancy for various applications. This chapter begins outline the major issues in conventional flame retardant materials, as well as the requirement for sustainable alternatives; and investigates the methods used to produce nanomaterials from biomaterials in order to improve flame retardant properties. Furthermore, chapter discusses the preparation procedures as well as the review of data released on the various mechanical, structural, morphological, and thermal properties of the synthesized nanomaterials. Future perspectives are highlighted to identify new research avenues in terms of tuning properties of bio-based nanomaterials and improve their efficiency for flame retardancy application.

Keywords Bio-nanomaterials · Thermal stability · Non-halogenated flame-retardants · Smoke production rate

1 Introduction

Intumescent flame retardant technologies centered on halogen-containing flame retardants such [1]. Halogenated-based FRs are the most commonly employed flame retardants and have been continuously employed since the initial development of flame retardants. Bromine and chlorine are the most effective components in halogen groups in inhibiting the combustion process by continually producing free radicals. These halogens decompose at a higher temperature (280–290 °C) [2]. Fluorine

M. P. Motaung (✉) · W. Moloto · B. Ntsendwana · T. C. Mokhena
DSI/National Innovation Centre, Advanced Materials Division, Mintek, 200 Malibongwe Drive, Private Bag X 3015, Johannesburg, Gauteng, South Africa
e-mail: MathatoM@mintek.co.za

M. J. Mochane
Department of Life Sciences, Central University of Technology, Bloemfontein, Free State, South Africa

T. C. Mokhena et al. (eds.), *Biomaterials as Green Flame Retardants*, Engineering Materials, https://doi.org/10.1007/978-981-97-6871-4_11

compounds, on the other hand, are not typically employed in flame retardant applications because fluorine radicals are generated at temperatures considerably above the point of degradation of conventional polymers [2, 3]. Free radicals from halogenated flame retardant terminated the ˙OH and ˙H radicals generated during the combustion phase. In the gas phase, most halogen compounds function as free radical scavengers, reducing the combustion process through radical quenching while producing a crystalline protective coating as char [4–6]. Different types of flame-retardants have been established and are basically categorized into two types according to their component composition, such as halogenated and non-halogenated. Non-halogenated flame retardants are classified into phosphorus-based, nitrogen-based, silicon-based, boron-based, and metal hydroxide-based varieties [7–10]. These non-halogenated flame-retardants were found to work more effectively in combination than they do separately due to a synergistic effect. Presently, a number of halogen-free flame retardants have been explored for improved flame retardancy, such phosphorous-based compounds, intumescent flame retardants, and metal hydroxides [11, 12]. Despite the fact that halogenated flame-retardants excel over other types of flame-retardants in their vapour phase by scavenging active radicals, they are prohibited because they release components that are poisonous. Even though the halogenated-based compound is a common flame retardant, it has a shortcoming of producing dense, hazardous fumes as it burns, which have a negative impact to human health [2, 13, 14]. Hence, innovative bio-based flame retardants are required to be developed as alternatives for petroleum-based flame retardant additives in order to generate sustainable, eco-friendly, and highly effective flame retardant materials [15, 16].

2 Sources of Biomaterials

Biomaterials from renewable resources have recently attracted enormous attention owing to increasing environmental consciousness, waste accumulation, and destruction, as well as the inevitable depletion of fossil resources [17]. Biomaterials are potentially marketed materials that have been widely used in biomedical devices, textiles, and agricultural applications due to their outstanding biocompatibility, biodegradability, non-toxicity, relatively simple production process, and high mechanical durability [17–19]. Bio-based FRs originate from two sources: biomass (cellulose, starch, lignin, tannins, phytic acid, proteins and oil) and animals (DNA and chitosan) as illustrated in Fig. 11.1.

Biodegradable materials are biodegradable and compostable. Table 11.1 shows the properties of currently used bio-based materials for flame retardant.

Bio-based materials must be adjusted to reduce this negatively impacting property, which is usually very expensive. It must fulfil a number of criteria, the most essential of which is that it must not have a negative impact on the qualities of the biomaterials. Fire retardant treatment, also known as flammability reduction, modifies the combustion process to change the ability of the material to absorb heat, degenerate, or

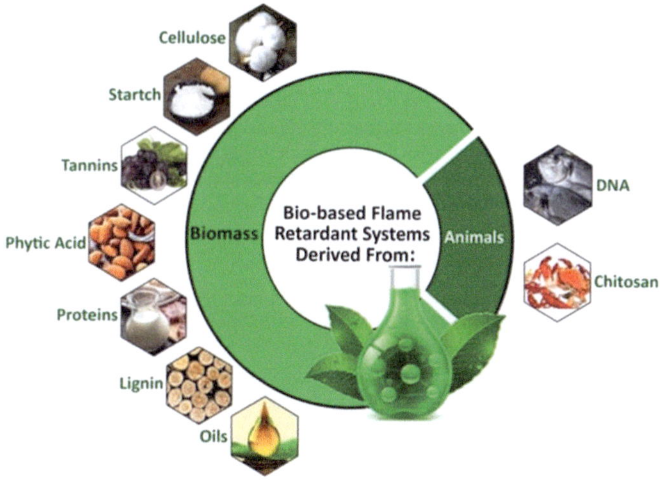

Fig. 11.1 The biomass and animals as a source of bio-based flame retardant [20]

the stage of combustion. The strategies for improving flame retardancy performance listed below [37–39].

- Insulating coatings that are flame retardant.
- Decomposition into flammable products is stabilized.
- Additives that are decomposed by heat to form non-combustible coatings.
- Production of a carbon residue that prevents fire from spreading through the material.
- Antioxidant additives, i.e. their disintegration forms an inert layer that extinguishes a flame.
- Materials that bind free radicals inhibit the chain reaction mechanism.

Flame retarders catalyse chemical equilibrium by altering the rate of reaction, but not the intensity of thermal flux, lack or excess of oxygen affecting burning conditions. However, it may affect or impact the fuel generation process as well as its ignitability. Flame retardation options are already limited during later stages of the burning process, particularly at large heat fluxes [40, 41]. The processes that occur at the start of each burning stage, namely initiation stage, propagation stage and termination stage, are essential for successful flame retardation. Flame retardant property is primarily dependent on materials' properties [15, 42, 43].

Table 11.1 Properties of the most commonly used bio-based materials

Bio-based material	Description of biomaterials	References
Cellulose	Cellulose is a linear homopolymer molecule composed of D-glucose units bonded by 1–4 glycosidic linkages. Aliphatic groups decrease during cellulose decomposition, while benzene rings and furan compounds begin to grow dramatically in the condensed phase. A char layer forms at high temperatures, such as 800 °C. The decomposition of cellulose can be influenced by the presence of other components and the heating conditions	[20–24]
Starch	Starch has a chemical composition formula that is comparable to cellulose since it contains D-glucose units. Starch polymer has two types of macromolecules: amylose, a linear polymer with glucose units joined through 1–4 linkage, and amylopectin, a branching polymer with glucose units connected through 1–6 linkage	[25–29]
Phytic acid	Phytic acid, also known as inositol hexakisphosphate, is a naturally occurring compound found in many plant-based foods, particularly grains, seeds, and nuts. It is not typically used for flame retardant purposes because it does not possess the characteristics necessary to effectively reduce flammability in materials	[14, 15, 30]
Lignin	Lignin is a complex organic polymer found in plant cell walls that is produced as a byproduct of the pulp and paper industry. The primary constituents of lignin are phenolic chemicals, specifically coniferyl alcohol, sinapyl alcohol, and p-coumaryl alcohol. To create lignin, these monolignols undergo oxidative polymerization. Various links are formed during the polymerization process, including ether bonds (aryl-ether linkages), carbon–carbon bonds (aryl–alkyl linkages), and carbon–oxygen bonds (aryl-ester linkages)	[31–35]
Chitosan	Chitosan is a copolymer composed of D-glucosamine and *N*-acetyl-D-glucosamine units linked by 1–4 linkages. Chitosan is produced via the enzymatic deacetylation of chitin, which is present in shrimp shells. Chitosan decomposes thermally in three stages. The first weight loss happens around 130–140 °C, which corresponds to the dehydration of weakly linked molecules. The second degradation process, which happens between 260 and 360 °C, is chitosan depolymerization and deacetylation. In the third phase, the residual decomposition reaction causes a very low weight loss rate above 400 °C, and char residue up to 30% wt. was formed at 500 °C	[15, 17, 36]

3 Nanomaterials

Nanomaterials are frequently categorized based on their composition, dimensions, and shape. The composition of nanomaterials varies greatly depending on their intended purpose and the source material [44, 45]. The dimensional properties of the nanomaterials refer to the number of external dimensions (height, width, and depth) that lie in the nanoscale. The morphology of nanomaterials is subdivided into the properties of size, shape, and aspect ratio [46]. The properties of nanomaterial can have significant effects on their efficacy in flame retardant applications. In addition, nanomaterials have shown potential in enhancing the flame retardant properties of various materials, due to their unique properties at the nanoscale. Depending on their use, nanomaterials can be categorized based on a variety of chemical or physical characteristics, such as composition, electrical, optical, or magnetic capabilities [44, 47–49]. Furthermore, nanomaterials produced from biomaterials for flame retardancy have shown an increased fire safety while having a smaller negative environmental impact compared to classic flame-retardants, which could contain dangerous chemicals. However, considerable investigation and evaluation of these materials is required to verify that they are safe and effective in specific applications. However, using such nanomaterials in commercial products, regulatory compliance and requirements should be taken into account [33, 50, 51].

Nanomaterials derived from biomaterials have showed potential in a number of applications, including flame retardancy. Biomaterial-based nanomaterials provide an environmentally friendly and potentially safer flame retardant alternative [3, 45, 52]. Flame retardant materials have been reported for coating textile fabrics, however after treatment a number of issues have been raised, such tensile strength failure, hazardous gas emissions from fire and poor stability. The addition of nanoscale components provides novel features for efficiently attaining the objectives of the necessary textile fabric (as shown in Fig. 11.2) [53, 54].

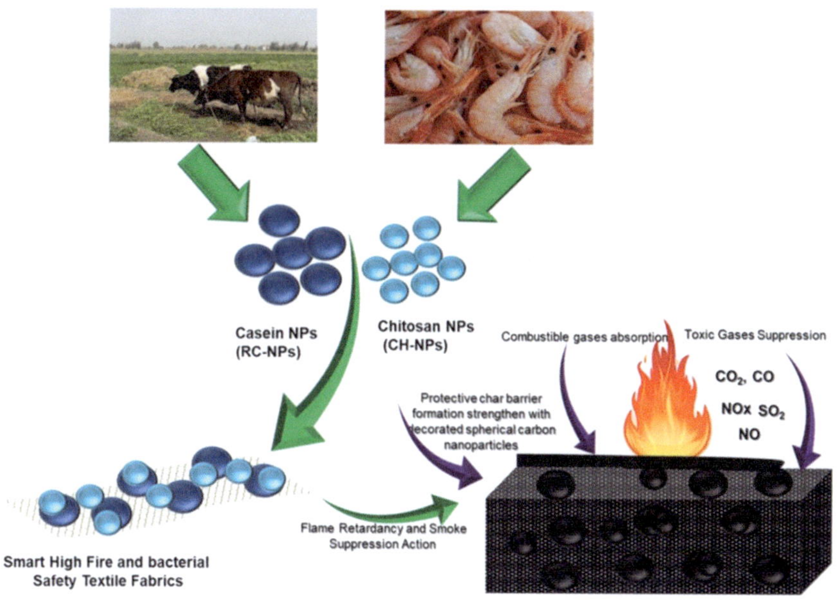

Fig. 11.2 Schematic diagram representing green coating process of textile fabrics nanocomposites via layer-by-layer approach; flame retardancy and toxic gases suppression actions of newly developed textile fabrics. Reprinted with permission from [53]. Copyrights 2022 Elsevier

Therefore, there is a significant demand for the design and green synthesis of nanomaterials with tunable size from renewable bio-precursors employing cost effective techniques in flame retardant. Additionally, this may yield efficient and stable nanomaterials for a number of applications in flame retardant [6, 55, 56]. Bio-based nanomaterials can be made of organic, inorganic, carbon-based, or composite-based materials in a variety of size and form morphologies. The composition of nanobiomaterials includes the core of the material, shell, and any surface modified. In general, the chemical composition of nanobiomaterials and their properties should be improved for stability and low toxicity of the material. Carbon-containing organic compounds formed from man-made or organic polymers make up organic nanobiomaterials [25, 57, 58]. The majority of the properties of biomaterials enables the synthesis of nanocrystals, nanofiller, nanofibers, nanofilms, and nanocoatings. Synthesizing nanobiomaterials for diverse purposes such as flame retardant applications, has been made possible due to the outstanding characteristics and plasticity of natural polymers like chitosan and lignin as well as synthetic polymers [26, 59, 60].

These biomaterials are appealing due to low toxicity, controllable decomposition, stability, and physical properties [61, 62]. Carbon-based nanobiomaterials include fullerenes, carbon nanotubes (CNTs), graphene and its derivatives (graphene oxide), nanodiamonds, and carbon-based quantum dots. Their unique structural dimensions and excellent physical, electrical, thermal, optical, and chemical properties have attracted tremendous interest in flame retardancy applications. Lignocellulosic materials have the potential to reduce fire hazards a study reported by Song et al. [63]. Graphene nanoplatelets and carbon nanotubes were pre-adsorbed with alkali lignin and then thin films of these materials were formed utilizing the Meyer rod method as illustrated in Fig. 11.3. The materials exhibited enhanced thermal stability, self-extinguishing behaviour and; decreased weight loss and burning area. These materials work synergistically to improve physical barrier properties and thermal management, opening doors for the creation of effective, affordable, and environmentally friendly flame retardants.

Lukawski et al. [31] have reported a new composite flame retardant coating for cotton roving by fusing organic lignin (L) with non-toxic potassium carbonate and carbon nanotubes (CNT). Adding L/CNT/K_2CO_3 to cotton fibre enhanced the limiting oxygen index, lowered temperature and mass loss during vertical burning, and decreased the peak heat release rate and total heat released, according to experiments. Their results were supported by scanning electron microscopy and Raman spectroscopy allowed also the explanation of the complementary mechanisms responsible for the overall fire retardant effect [31].

3.1 *Classification of Bio-based Nanomaterial*

Bio-based nanomaterials can be classified into several categories are based on their structure chemical composition, source, and properties of the biomaterial.

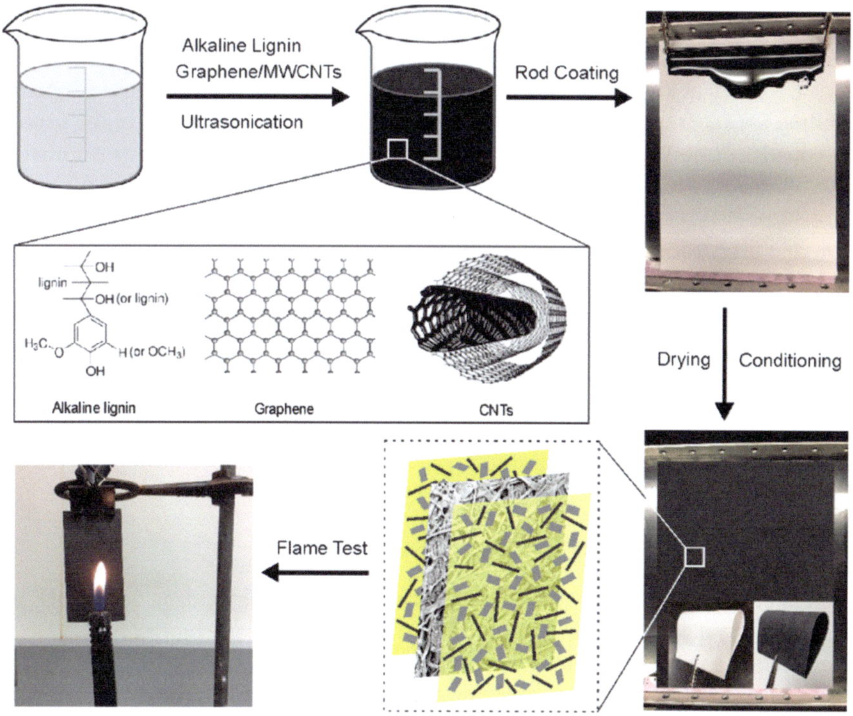

Fig. 11.3 The schematic showing the fabrication and flame test of the GnP/CNT/lignin-coated papers [63]

3.1.1 Bio-Nanofillers

Bio-nanofillers are nanoscale components that are obtained from biomaterials and used to improve the properties of a variety of materials, such as nanoclays, carbon nanotubes (CNT), nanoparticles or nanofibers, into polymers, composites, and other materials.

These materials were reported to exhibit enhance physical, thermal, electrical, and chemical properties of the host polymeric materials. Bio-nanofillers have various advantages over regular nanofillers, including the fact that they are frequently biodegradable, renewable, and can be produced with less environmental effect. Bio-based nano-fillers, improve flame retardancy by enhancing carbon formation, slowing heat release, prolonging igniting time, and reducing volatile flammable pollutants in fire [55, 59, 64]. These materials have drawn an enormous amount of attention in recent years due to their biocompatibility, renewable nature, and potential to improve the efficiency of flame retardancy. Nanocomposites have stronger thermostability than the original polymer because nano-fillers reduce polymer mobility and free radical capture, and their flame retandancy is governed by the geometrical shape and chemical structure of the nanofillers.

The particular surface area of nanofillers is large due to their high specific surface area/volume ratio, the interaction between the polymer and the nanofillers can be improved even at low loading. In addition to their low cost, nanofillers are a superior solution for ensuring that overloading does not degrade the properties of the natural polymer [65–68]. Wu et al. [69] reported chemical modification and co-precipitation to synthesize a green functional lignin-based nanofiller (lignin-diethylenetriamine/red phosphorus nanoparticles, Lignin-N-P NPs). The physical, chemical, and flame retardant properties of the blends after blending with commercial acrylonitrile butadiene styrene copolymers (ABS) reveal that the Lignin-N-P NPs/ABS blend has suitable process ability, mechanical properties, and significantly improved thermal stability and fire performance. They observed that the peak heat release rate and total heat released per unit area values were reduced by 67.8% and 77.5%, respectively. This suggest that there is a potential to open pathways for novel designs for both flame retardants and lignin-based materials.

3.1.2 Bio-Nanocrystal

Bio-nanocrystals, also known as biological nanocrystals or biogenic nanocrystals, are nanoscale crystalline structures produced or synthesized from biomaterials. These nanomaterials are derived from biomaterials, a natural polymer found in plant cell walls Fig. 11.4 shows a typical example of nanocrystals derived from cellulose [18]. They possess properties like high mechanical strength, biodegradability, and biocompatibility, making them suitable for applications in various fields such as biomedical engineering, packaging, and electronics [55, 70].

Bio-nanocrystals can be introduced into a variety of materials to improve their flame retardant qualities, including polymers, textiles, and coatings. When exposed to high temperatures or flames, these nanocrystals can aid in hindering or delaying burning through formation of a barrier that prevents ignition, releasing flame-extinguishing gases, or encouraging char formation on the surface of the material [59, 60, 71]. Three separate precursor materials (China cotton, South African cotton, and waste tissue paper) were used for preparing nanocellulose via acid hydrolysis. Shape and size of the nanocellulose were determined, as shown in Fig. 11.5a–c. In comparison to other precursors, nanocellulose was found to have greater thermal stability and crystallinity (TGA/DSC study) according to Fig. 11.5d [39].

Lignin can be used in the production of cellulose-nanocrystals shows the importance of understanding its structure and impact on other lignocellulose components in the plant cell wall. Lignin is a highly cross-linked phenolic polymer that acts as an adhesive to hold the cell wall components together. It has greater surface polarity/hydrophobicity than cellulose and can inhibit hydrogen bonding in cellulose. When utilized to reinforce general purpose polymers, such as polyethylene, polypropylene, or polystyrene, the incorporation of lignin in cellulose nanoparticles helps overcome the issues of poor compatibility presented by cellulose-nanocrystals [63, 72]. Zhang et al. [72] suggested a sustainable approach for producing lignin-cellulose nanocrystals (L-CNCs) from hemp hurd. The nanomaterials are homogenous and

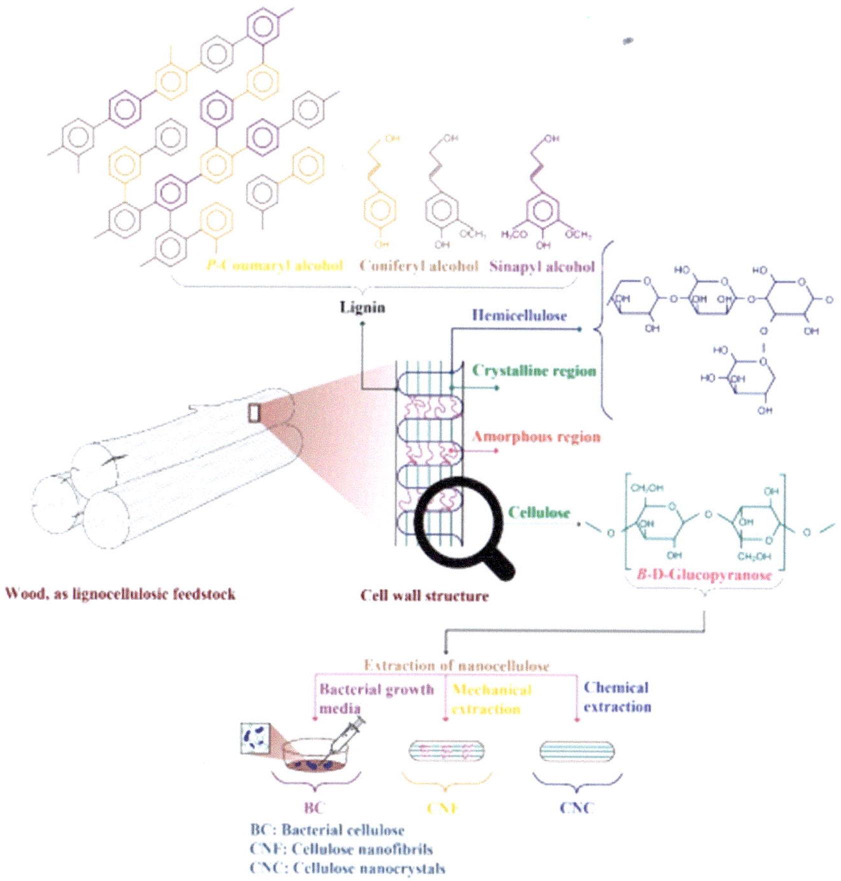

Fig. 11.4 A schematic of cellulose biopolymer derivation from wood, which is converted to CNF, CNC, or BCs, as the three major nanoscale forms of cellulose [18]

were prepared by mild acid hydrolysis. These L-CNCs are employed as fillers in polyvinyl alcohol (PVA) film, resulting in a light-coloured nanocomposite film with strong UV-protection and enhanced tensile qualities. The absorbance of UV at 400 nm and permeability in the visible-light range at 550 nm were 116 times and 70% higher than those of pure PVA. In addition to these advantages, the nanocomposite film demonstrated water vapour transmission properties equivalent to commercial food package film, indicating future applications, as shown in Fig. 11.6. Water vapour transmission is also a property of the film [72].

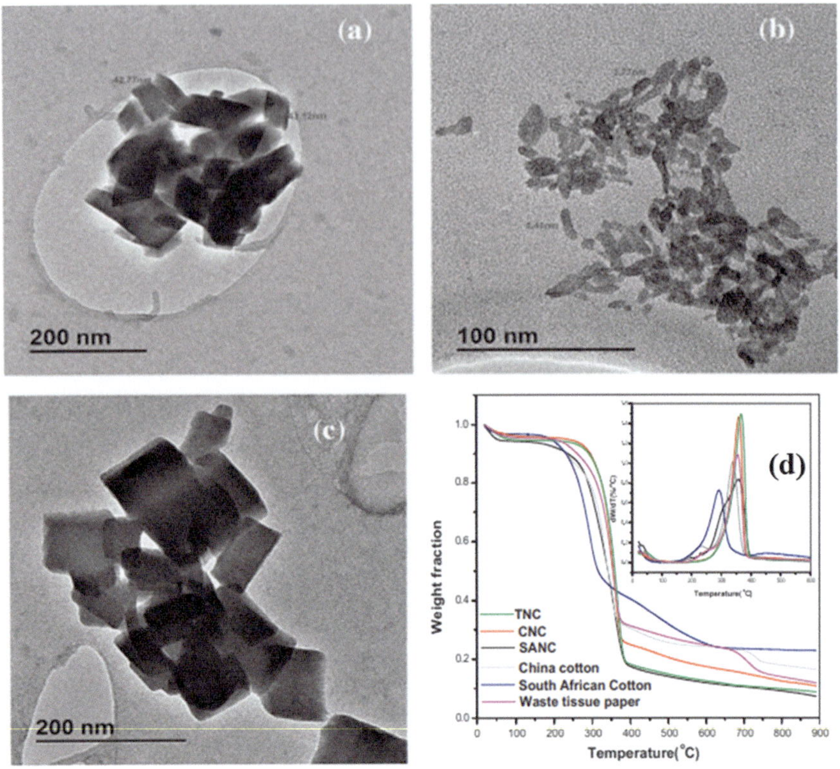

Fig. 11.5 TEM image of **a** CNC, **b** SANC and **c** TNC. **d** Comparative TGA and DTGA of nanocellulose samples with their corresponding precursor [39]

3.1.3 Bio-Nanoparticles

Bio-based nanoparticles have gained significant attention in recent years as potential flame retardant additives for various materials. Flame-retardants are substances added to materials to reduce their flammability and slow down the spread of flames in the event of a fire. Bio-based nanoparticles offer several advantages as flame-retardants, including their renewable and sustainable nature, as well as their low toxicity compared to some traditional flame-retardants.

Bio-based nanoparticles can be derived from a variety of renewable sources, such as cellulose, lignin, starch, chitosan, and proteins. These nanoparticles can be in the form of nanocellulose, nanolignin, nanostarch, or other biopolymers. Bio-based nanoparticles can function as flame retardants through several mechanisms, including forming a protective char layer when exposed to heat, diluting the combustible gases released during combustion, and absorbing heat energy. Fire-resistant foams with biobased components that can replace petroleum-based foams were studied by Carosio et al. [73]. They used a water-based freeze casting method,

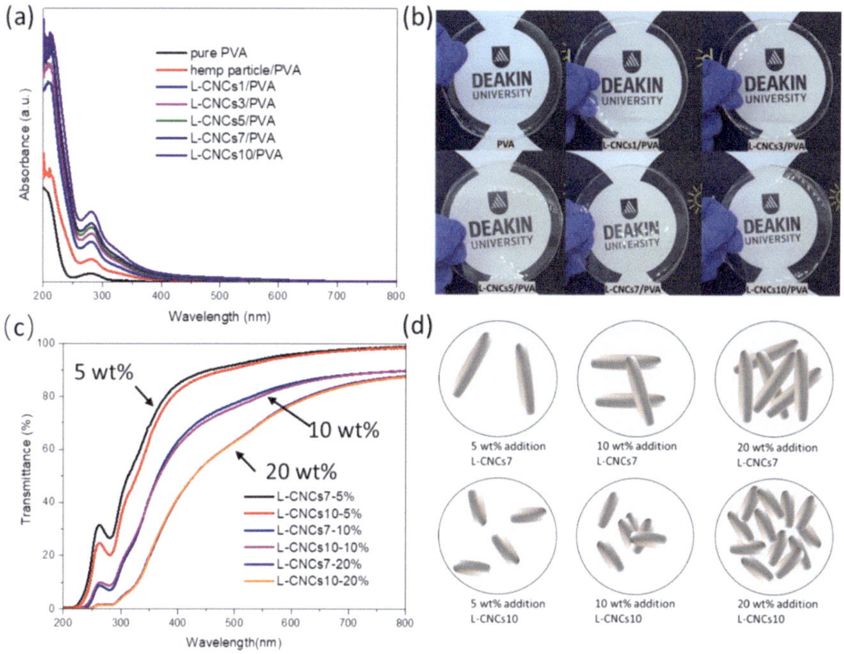

Fig. 11.6 **a** Absorption spectra of pure PVA, hemp particle/PVA, and L-CNCs/PVA films (5 wt.%), **b** digital images of pure PVA and L-CNCs/PVA films (5 wt.%), **c** transmission spectra of L-CNCs7(10)/PVA at 5 wt.%, 10 wt.%, and 20 wt.% ratio and **d** schematic diagram of L-CNCs' distribution in PVA film [72]

cellulose nanofibers are mixed with 2D montmorillonite nanoplatelets and a native xyloglucan hemicellulose binder. These lightweight foams self-extinguish the flame during flammability tests due to the silicate nanoplatelets. The limiting oxygen index can reach 31.5%, which is comparable to the best fire-retardant synthetic foams on the market. Furthermore, the foams can endure the penetration of a flame torch concentrated on one side of the specimen (T on surface 800 °C) while maintaining structural integrity. Figure 11.7 shows, the exposed side was insulated, as evidenced by a through-thickness temperature reduction of 680 °C cm^{-1}. The findings provide a tremendous opportunity for the creation of fire-safe foams that combine good sustainability with multifunctional performance [73].

Yue et al. [74] developed flame retardant compound aerogels through cross-linking cellulose nanofibrils with melamine formaldehyde (MF) and methyltrimethoxysilane (MTMS). Cross-linking between the MF/MTMS and CNF matrices was detected by FTIR spectra, leading in better mechanical stability. Figure 11.8 shows possible flame-retardant mechanism of Si-CNF/MF material, the compound aerogels were also exceptionally hydrophobic and flame-retardant when compared to pure CNF aerogel, the limited oxygen index increased from 19.5 to 37.1%. However, the peak

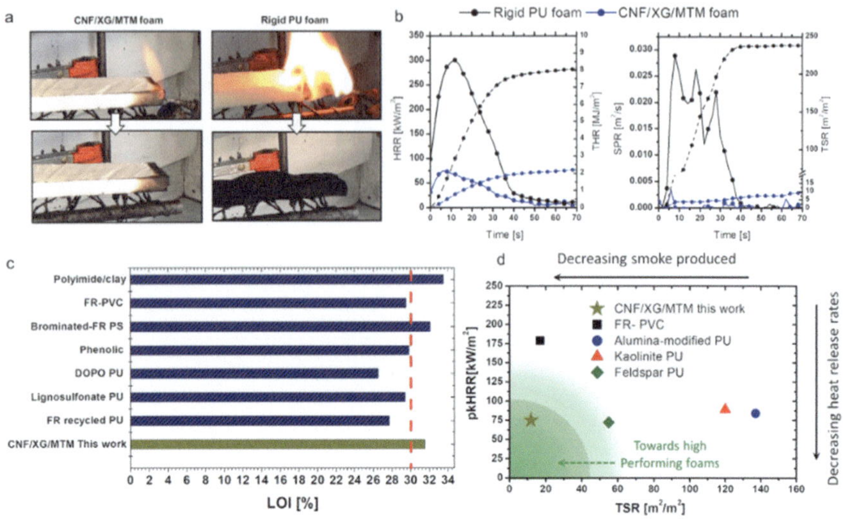

Fig. 11.7 Flame retardant characterization: **a** snapshots of the foams during horizontal flammability tests, **b** cone calorimetry plots for heat and smoke release rates HRR and SPR (solid lines) and integral parameters total heat release rate THR and total smoke release TSR (dotted line), **c** LOI values of tested foams in comparison with the literature, and **d** comparison of cone calorimetry data (peak heat release rate) pkHRR and (total smoke release) TSR with literature data. The foam porosity is 98.5% and the composition is CNF/XG/MTM 40/27/33 wt% [73]

heat release rate and total heat release decreased by 50.6% and 64.3%, respectively. These aerogels are useful for heat retention to their low thermal conductivity [74].

Kim et al. [75] proposed a flame retardant polymer composite foam, which enables the improved performance, good formability, and reduced environmental burden while burning. Their strategy was to incorporate sylilated nanocellulose into

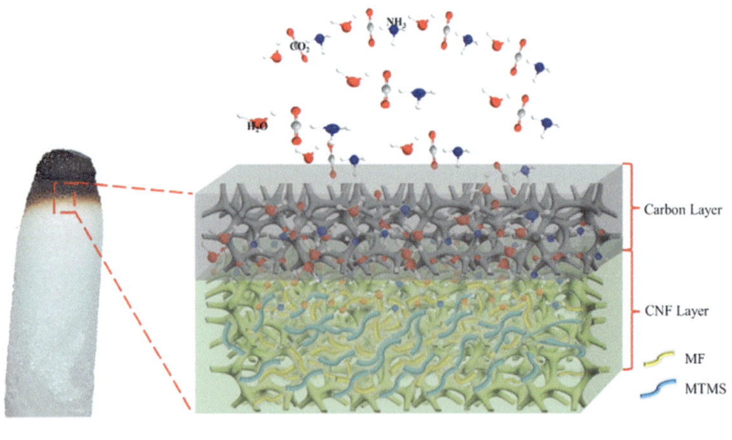

Fig. 11.8 Possible flame retardant mechanism of Si-CNF/MF [74]

a polyurethane matrix containing a conventional flame retardant, tris(2-chloroethyl) phosphate (TCEP). This strategy leads to the generation of char layer faster during combustion, resulting in a delayed flame propagation. The limiting oxygen index (LOI) of the samples increased by 28%, and the production rate of toxic gas emission was considerably reduced as shown in Fig. 11.9. The chemical, thermal, mechanical, and morphological analyses were carried out to understand the underlying physics. In the cone calorimetry test, the PUF/Si-CNFs/TCEP sample showed low HRR and SPR values, suggesting a comparatively small environmental burden in addition to high flame retardancy. The technique described in this study will reduce the environmental impact of insulating materials while improving flame retardant efficiency.

Chollet et al. [76] reported the effect of employing nanoscale lignin as a novel flame-retardant addition for polylactide (PLA). A dissolution–precipitation procedure was used to create lignin nanoparticles (LNP) from Kraft lignin microparticles (LMP) to improve their flame-retardant properties in PLA, both micro and nano lignins were functionalized with diethyl chlorophosphate and diethyl (2-(triethoxysilyl)ethyl) phosphonate. The use of inductively coupled plasma (ICP) spectrometry indicates that a significant amount of phosphorus was embedded onto the nanoparticles. Phosphorylated lignin nanoparticles, on the other hand, appear

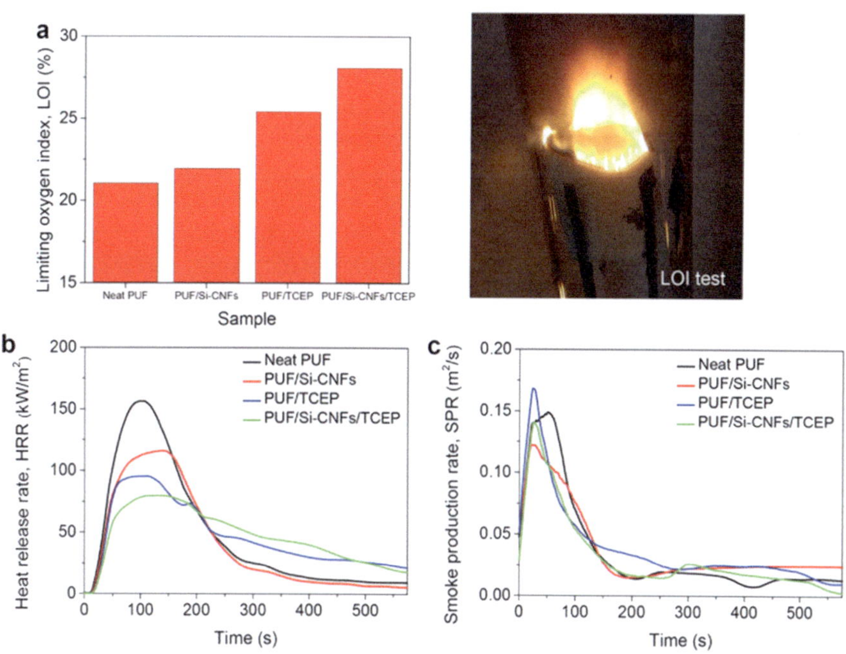

Fig. 11.9 Flame retardancy of the synergetic composite foams. **a** Limiting oxygen index (LOI) results of the fabricated foams (left) and the photograph of the LOI test (right), **b** time-dependent heat release rate (HRR), and **c** smoke production rate (SPR) of the samples while burning in the cone calorimetry. Reprinted with permission from [75]. Copyrights 2022 Elsevier

to inhibit PLA degradation during melt processing, and the nanocomposites have been demonstrated to be relatively thermally stable. Cone calorimeter measurements demonstrated that the addition of untreated lignin, regardless of particle size, resulted in an increase in pHRR. This adverse effect can be reduced by using phosphorylated lignin nanoparticles, especially those that have been treated with diethyl (2-(triethoxysilyl)ethyl) phosphonate. Although when using only 5 weight percent LNP-SiP, the pHRR is also considerably decreased. Biobased nanoparticles are considered to be ecologically friendly and sustainable compared to some conventional flame retardants, which may contain halogenated or toxic compounds [33, 73]. Biobased nanoparticles for flame retardancy can be used in a wide range of applications, including textiles, plastics, coatings, and construction materials. They are especially valuable in industries where there is a growing emphasis on sustainability and reduced environmental impact. While biobased nanoparticles show promise as flame retardants, there are challenges to overcome, such as ensuring consistent performance across different materials and optimizing their dispersion and compatibility within the target materials [42, 76, 77].

4 Synthesis of Bio-based Nanomaterials

Biomaterials at the nanoscale can also be produced employing a variety of top-down and bottom-up approaches. The first approach produces nanometric blocks by fracturing bulk cellulose, whereas the second method produces nonmetric biomaterial by using individual atoms or molecules [3, 27]. Nanostructured biomaterials, such as nanoparticles, nanofibers, nanowires, and nanocomposites have attracted interest in flame retardant applications [5, 27, 78]. Nanobiomaterials have a variety of shapes, with spheres being the most prevalent, including rods, needles, platelets, and polygons. The physical, chemical, electrical, mechanical, and optical properties of nanobiomaterials are influenced by their size and shape, in contrast to typical bulk materials making it possible to improve modification and efficiency [6, 25, 52, 79]. Bio-based nanomaterials have been synthesized using a variety of techniques through chemical method (which is primarily carried out by acid hydrolysis, enzyme-assisted hydrolysis, and mechanical treatments) and physical method (such as high-pressure homogenizing, grinding, and ultrasonication) as shown in Fig. 11.10.

These methods can also be combined to produce new nanomaterials, acid hydrolysis has emerged as a viable technique for modifying the chemical and physical properties of biomaterials. Acid hydrolysis produces nanocrystalline biomaterials by hydrolyzing the hydroxyl functional group of biomaterial under specific temperature, time, and acid conditions [72, 80, 81].

Matebie et al. [76] isolation cellulose nanocrystals (CNCs) using a brewery's spent grain (BSG) by acid hydrolysis while optimizing the hydrolysis parameters (hydrolysis time, temperature, liquid–solid ratio, and acid concentration). BSG was treated with alkali and bleaching to remove hemicellulose and lignin. The central composite design (CCD) optimization approach was used to attain the best value of cellulose

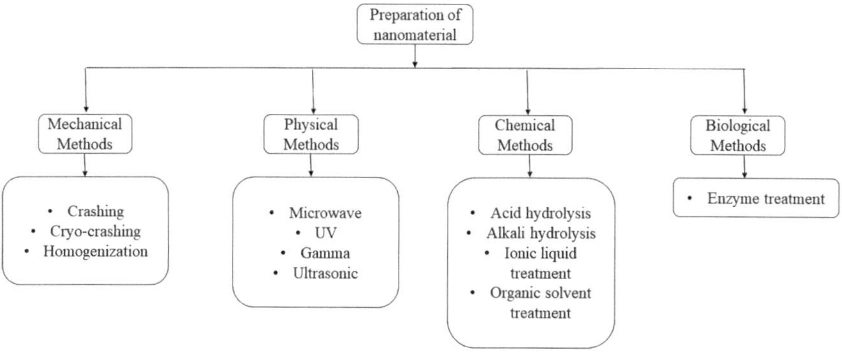

Fig. 11.10 Synthesis of nanomaterials derived from biomaterials

nanocrystal (CNC) yield at optimum hydrolysis conditions of 50 °C, 51 wt% acid concentrations, 41 min, and liquid–solid ratio of 19 mL g^{-1}, the maximum cellulose nanocrystal (CNC) yield of 43.24% was attained. SEM, XRD, particle analyzer, FTIR, and differential scanning calorimeter (DSC) were used to characterize the raw brewery discarded grain, alkali-treated fiber, bleached fiber, and obtained CNC. The results showed that the synthesized cellulose nanocrystal (CNC) yielded a rod-like whisker shape, a crystallinity of 76.3%, and an average particle size of 309.4 nm. Yu et al. [81] prepare cellulose nanocrystals (CNCs) from Pennisetum hydridum fertilized with municipal sewage sludge (MSS) using sulfuric acid hydrolysis at various acid concentrations (40–65%), temperatures (room temperature 55 °C), and reaction times (50–120 min). The results demonstrated that the produced CNC had steady dispersion in water. Under the conditions of room temperature, 65% acid concentration, and 120 min reaction time, the length of CNCs reached 272.5 nm, and the diameter was within 10 nm. Furthermore, Fourier transform infrared (FTIR) analysis revealed that the CNC retained its cellulose type I structure. When compared to delignified Pennisetum hydridum fibers, the crystallinity of CNCs increased to a maximum of 18.34%. Thermogravimetry (TG) revealed that CNCs have a reduced thermal stability compared the cellulose of delignified Pennisetum hydridum fibers this was ascribed to the addition of sulfate groups into the cellulose. Their study showed that Pennisetum hydridum fertilized with MSS could be a good raw material for CNCs. This implies that MSS and Pennisetum hydridum are being used effectively.

The approach used to isolate bio-based nanomaterial is determined by the type of bio-based material, nanomaterial characteristics preferred, and proposed application. During the preparation of bio-based nanomaterials, it is critical to examine their environmental sustainability and biocompatibility. Furthermore, characterisation techniques such as electron morphology and particle size analysis are required to confirm the properties of the yield nanomaterials as summarized in Table 11.2.

Table 11.2 The general overview of bio-based nanomaterials preparation methods

Biomaterial	Method	Obtained nanomaterial	Particle size (nm)	References
Cellulose	Acid hydrolysis	Cellulose nanocrystals	272.5–459.2	[81]
Cellulose	Acid hydrolysis	Cellulose nanocrystals	309.4	[80]
Lignin-cellulose	Acid hydrolysis	Lignin–cellulose nanocrystals	550	[72]
Starch	Alkali hydrolysis	Starch nanoparticles	200–500	[82]
Starch	Ultra-sonication	Starch nanoparticles	200–500	[82]
Lignin	Co-precipitation	Lignin-based nanofiller	–	[69]

5 Characterization of Nanomaterials

Characterization of bio-based nanomaterials flame retardants various researchers employed a wide range of complementary techniques, which include electron microscopy analysis (SEM and TEM), atomic force microscopy (AFM), Fourier transform infrared (FTIR) spectroscopy, thermogravimetric analysis (TGA), Cone Calorimeter and testing limiting oxygen index (LOI) in analyzing the structure and properties of flame retardants [55, 72]. SEM are frequently used to examine sample surface morphology and create 3D structural micrographs. While TEM, enhances morphology imaging and size distribution while providing improved resolution and better scanning of large images due to its higher cost and longer processing time, SEM is favoured over TEM because it can handle larger samples at higher magnification and better resolution [61, 83, 84]. AFM is a technique that is frequently used to describe the characteristics of flame retardant materials, it measures mechanical parameters like stiffness and adhesive force and provides high resolution, 3D topography. The ambient environment of AFM makes it advantageous as it prevents sample modification, however its limited utilization is due to the lack of chemical data [85]. XRD characterization is used to identify crystalline solids and study more about their chemical form and crystallographic structure. The height ratio between the intensities of the crystalline and non-crystalline peak is used to determine the crystallinity index using Bragg diffraction peaks. Crystallite sizes are calculated using the Scherrer equation sizes as shown in Eq. 1 [86]:

$$L = \frac{k\lambda}{\beta \cos \theta} \tag{1}$$

where L is the crystallite size in nm, λ signifies the X-ray wavelength, k denotes the Scherrer constant, β is the full width half maximum of lattice plane reflection in radian, and θ is attributed to the resultant Bragg angle (reflection angle). The two most reliable techniques commonly employed to investigate degradation and thermal properties of the nanomaterials are thermogravimetric analysis (TGA) and differential scanning calorimetry (DSC). Commonly, the TGA curves for most of the flame-retardants show both endothermic and exothermic peaks. The cone calorimeter calculates variables including heat release rate, smoke suppression effect, igniting time, and mass loss to estimate real-time fire behavior. Calculating the amount of heat released requires knowledge about the oxygen consumption during burning. The peak of heat release rate (pHRR) is one of the most important parameters that indicate the flashover conditions in a real fire situation. This is vital when modelling heat release rate, finite element analysis is advantageous, particularly when taking into account dangerous chemicals and processes [19, 21, 87]. Table 11.3 show summary of nanomaterials generated data obtained from literature for various characterization approaches.

The development of carbon-based (graphene, carbon nanotubes) and ceramic (boron nitride, Al_2O_3, AlN, and SiC) nanomaterials for the formation of thermally conductive fillers has been greatly investigated. Carbon-based materials have strong thermal conductivities, but they also have very high electrical conductivities, which can lead to uncontrolled leakage currents that malfunction devices [1]. An efficient way to increase the thermal conductivity and flame retardance of the polymer matrix is to incorporate thermal conductivity fillers, such as graphene, carbon nanotubes, Al_2O_3, and hexagonal boron nitride (h-BN), into the matrix. h-BN is a new type of 2D thermally conductive material with great dielectric breakdown strength (794 mV mm^{-1}), thermal stability, and high in-plane thermal conductivity (400 W mK^{-1}). These characteristics give h-BN the ability to enhance PTCC materials' thermal conductivity and flame retardance [96, 97]. Xu et al. [96] reporter a thermal conductive filler which was added to composites based on cellulose nanofibrils (CNF) to increase their thermal conductivity, but doing so invariably results in the mechanical characteristics of composite being lost (as illustrated in Fig. 11.11). This study showed an increased toughness and thermal conductivity of the CNF/boron nitride nanosheets (BNNS) composite from the standpoint of the addition of thermally conductive filler and the modification of the CNF crystal. Before being added to CNF, the hydrophilic-modified BNNSs were effectively produced through ball-milling with xylose assistance. When loading 30% BNNS, the in-plane TC of the CNF/BNNS composite (12.68 W/m·K)) increased dramatically by 846% in comparison to that of the CNF film (1.34 W/m·K)). Subsequently, mercerization with 12.5% NaOH solution greatly improved the toughness (8.0 MJ·m^{-3}, increased ~250%) and TC (14.7 W/m·K), increased ~16%) of the CNF/BNNS composite. In related investigations, toughness and TC have improved together, which is exceptional and helps to prepare thermal management materials effectively.

Two-dimensional nanomaterials in particular, with their special structure and capabilities, have garnered a lot of interest in the field of smoke suppression and flame retardant. Due to their layered structures, which have good thermal stability

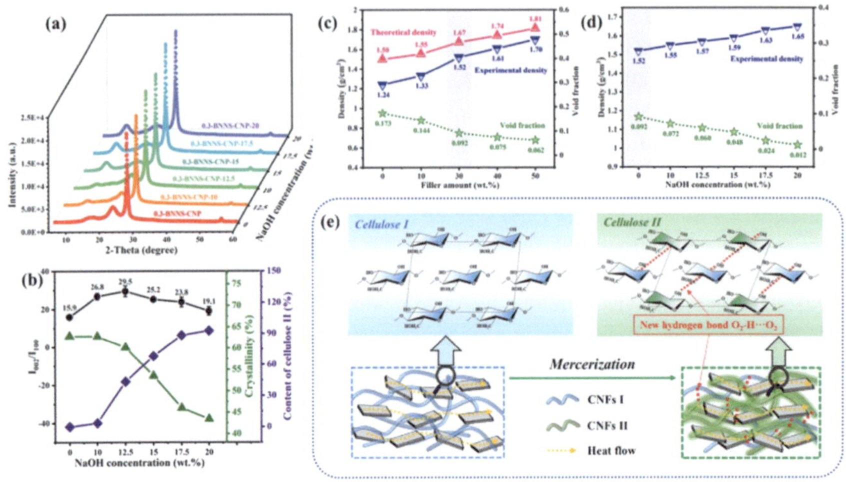

Fig. 11.11 Effect of mercerization on the CNF/BNNS films: **a** the XRD characteristic peak region of mercerized films; **b** effects of mercerization on the I(002)/I(100) ratio, cellulose crystallinity, and content of cellulose II; **c** curves of theoretical density, experimental density, and void fraction of films with different filler amounts; **d** curves of theoretical density, experimental density, and void fraction with different mercerization degrees; and **e** schematic diagram of how mercerization enhanced the hydrogen bond structure of the film. Reprinted with permission from [96]. Copyrights 2022 Elsevier

and can create a physical barrier in a polymer matrix, two-dimensional nanomaterials have emerged as a superior flame retardant choice when compared to zero- and one-dimensional nanomaterials. In addition, the mechanical characteristics of polymers can be effectively enhanced by the homogenous dispersion of a modest amount of two-dimensional nanoparticles in polymers. The high cost and low yield of peeled sheet structures are thought to be the main barriers to the deployment of two-dimensional nanomaterials as flame retardant additives. Flame-retardant applications have been made possible by several investigations on the large-scale manufacturing of two-dimensional nanosheets, as evidenced by the sharp rise in publications pertaining to these newly developed two-dimensional nanomaterials. The high cost and low yield of peeled sheet structures are thought to be the main barriers to the deployment of two-dimensional nanomaterials as flame retardant additives. Flame-retardant applications have been made possible by several investigations on the large-scale manufacturing of two-dimensional nanosheets, as evidenced by the sharp rise in publications pertaining to these newly developed two-dimensional nanomaterials [96, 98]. Figure 11.12 presents the suggested flame-retardant mechanism based on the previously discussed structural analysis and related results. Through hydrogen bonding and interaction, the presence of phosphorus-containing groups and lignin can create a homogeneous and extremely dense polymer structure resembling nacre in this BHL-PCNF film. Because phosphates, a char promoter, are present during combustion, cellulose chains containing lignin rapidly carbonize to become

Table 11.3 Summary of characterization technique

Method	Techniques	Data obtained	References
Microscopic techniques	SEM	Surface morphology, structure particle, size, shape, and distribution of nanomaterials	[16, 35]
	TEM	Morphology, size, and dispersion of nanomaterials	[80, 88]
	AFM	Stiffness, elasticity, adhesion, and deformation behaviour of the nanomaterial	[6, 65]
Spectroscopic techniques	FTIR	Chemical composition and interactions of nanomaterials	[89, 90]
	Raman	Vibrations of molecules in nanomaterials, chemical compounds and chemical modification	[31]
	UV–visible	Formulations Functional groups, unsaturation, and chain structure	[72, 91]
X-ray techniques	XRD	Crystalline structure and phase purity of nanomaterial	[35, 88]
Thermal analysis	TGA	Thermal stability and decomposition profiles	[92, 93]
	DSC	Heat flow and phase transitions	
Flame retardancy	Cone calorimeter	Flammability and fire behaviour of materials	[30, 76]
	LOI	Flammability of the material and its ability to resist combustion in the presence of oxygen	[32, 94, 95]

carbonated polymers and further build carbon layer. Concurrently, the phosphorus-containing group undergoes a quick transformation into the PxOy complex through thermal breakdown and dehydration, which also releases certain gases like CO/CO_2 and H_2O. Additionally, the protective layer that is composed of carbon and PxOy-based layers will redeposit onto the broken surface of the film, effectively preventing heat and oxygen from reaching the interior polymer network. The underlying combustion of CNF is thereby repressed and heat- and oxygen-exchanged, leading to a clearly enhanced flame resistance for the BHL-PCNF film [99–101].

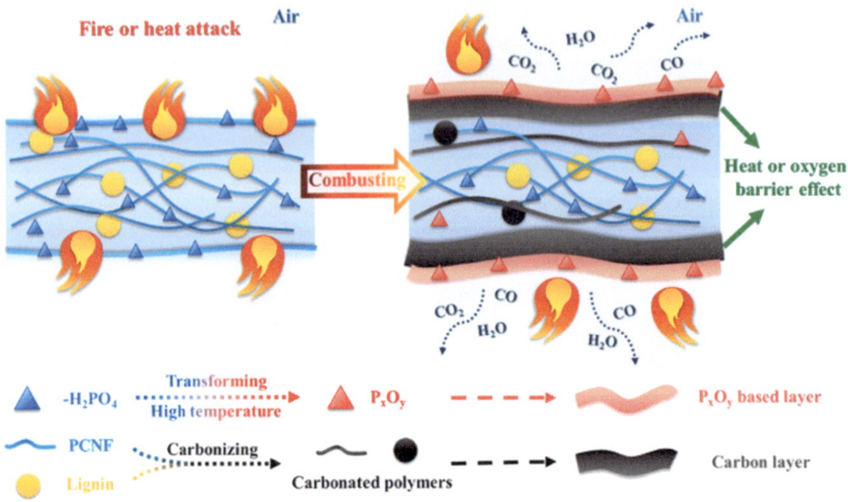

Fig. 11.12 The proposed flame mechanism of BHL-PCNF films during the combusting process. Reprinted with permission from [99]. Copyrights 2021 Elsevier

6 Conclusion and Future Perspective

The demand for environmentally friendly flame retardants has increased due to concerns about toxicity and environmental have increasing demand for conventional synthetic alternatives. Biomaterials offer a promising source for nanomaterial synthesis due to their abundance, renewability, and biodegradability. The preparation and characterization of nanomaterials derived from biomaterials have a significant potential for nontoxic, stable and more environmentally friendly fire-resistant materials. Advanced techniques, such as TEM, SEM, XRD, and FTIR are used to understand the morphology, structure, and composition of these nanomaterials. Through various methods, nanomaterials produced from biomaterials exhibit flame retardant characteristics. This includes the development of a protective char layer, a slower rate of heat release, and gas-phase flame reduction. Identifying such routes is essential to developing efficient flame retardant materials. The incorporation of these nanoparticles into polymers, fabrics, and coatings can improve their flame retardancy. These nanomaterials derived from biomaterial have the potential to increase the safety of a wide range of applications. One of the primary benefits of using biomaterial-derived nanomaterials is that they have a lower environmental impact than conventional flame retardants because these materials are often biodegradable, they may aid in the reduction of dangerous chemical accumulation in the environment. However, challenges such as scalability, cost efficiency, and long-term durability require further investigation on mass production of these bio-based flame retardance materials and development of techno-economic frameworks to ensure sustainability in the market place.

References

1. Malucelli, G.: Flame-retardant systems based on chitosan and its derivatives: state of the art and perspectives. Molecules **25**(18) (2020)
2. Nurul, M., Mohammad, A., Antov, P., Savov, V., Fatriasari, W., Seng, L., et al.: Current progress of biopolymer-based flame retardant. Osvaldov a. **205** (2022)
3. Lokhande, P.E., Chavan, U.S., Nurul, M., Mohammad, A., Antov, P., Savov, V., et al.: Efficient fabrication of flame-retarding silicone rubber/hydroxylated boron nitride nanocomposites based on volumetric extensional rheology. Chem. Eng. J. [Internet]. **118**, 108184 (2022). Available from: https://doi.org/10.1016/j.compositesa.2016.06.017
4. Ji, W., Wang, D., Guo, J., Fei, B., Gu, X., Li, H., et al.: The preparation of starch derivatives reacted with urea-phosphoric acid and effects on fire performance of expandable polystyrene foams. Carbohydr. Polym. [Internet]. **233**, 115841 (2020). Available from: https://doi.org/10.1016/j.carbpol.2020.115841
5. Vahabi, H., Laoutid, F., Mehrpouya, M., Reza, M., Dubois, P.: Materials Science & Engineering R Flame retardant polymer materials: an update and the future for 3D printing developments. Mater. Sci. Eng. R [Internet] **144**, 100604 (2021). Available from: https://doi.org/10.1016/j.mser.2020.100604
6. Nurul, M., Mohammad, A., Sherwyn, T., Garba, Z.N., Trache, D., Uyama, H., et al.: Recent progress in cellulose-based composites towards flame retardancy applications. Polymer (Guildf) [Internet] **244**, 124677 (2022). Available from: https://doi.org/10.1016/j.polymer.2022.124677
7. Sykam, K., Hussain, S.S., Sivanandan, S., Narayan, R., Basak, P.: Progress in organic coatings non-halogenated UV-curable flame retardants for wood coating applications: review. Prog. Org. Coatings [Internet]. **179**, 107549 (2023). Available from: https://doi.org/10.1016/j.porgcoat.2023.107549
8. Afriyie, R., Shanmugam, V., Narayanan, S., Sally, J., Babu, K., Esmaeely, R., et al.: A review of sustainable and environment-friendly flame retardants used in plastics. **108** (2022)
9. Cheng, Y., Liu, K., Liu, Y., Ding, D., Li, J., Zhang, F.: Industrial crops & products A durable flame retardant with N−P=O (O−NH^{4+}) 2 based on amino acid for cotton fabrics. Ind. Crop. Prod. [Internet]. **204**(PB), 117327 (2023). Available from: https://doi.org/10.1016/j.indcrop.2023.117327
10. Vahabi, H., Laoutid, F., Mehrpouya, M., Reza, M., Dubois, P.: Materials science & engineering R flame retardant polymer materials: an update and the future for 3D printing developments. Mater. Sci. Eng. R [Internet]. **144**, 100604 (2021). Available from: https://doi.org/10.1016/j.mser.2020.100604
11. Yi, O., Zahirasri, M., Tohir, M., Choong, T., Yeaw, S., Abdul, M., et al.: Progress in organic coatings fire-resistant and flame-retardant surface finishing of polymers and textiles: a state-of-the-art review. Prog. Org. Coatings [Internet]. **175**, 107330 (2023). Available from: https://doi.org/10.1016/j.porgcoat.2022.107330
12. Costes, L., Laoutid, F., Brohez, S., Dubois, P.: Bio-based flame retardants: when nature meets fire protection. Mater. Sci. Eng. R [Internet] **117**, 1–25 (2017). Available from: https://doi.org/10.1016/j.mser.2017.04.001
13. Li, M., Zhao, H., Cheng, J., Wang, T., Fu, T., Zhang, A., et al.: An effective green porous structural adhesive for thermal insulating, flame-retardant, and smoke-suppressant expandable polystyrene foam. Engineering [Internet] **17**, 151–160 (2022). Available from: https://doi.org/10.1016/j.eng.2020.08.032
14. Sas, G., Das, O., Sykam, K.: Industrial crops & products phytic acid : a bio-based flame retardant for cotton and wool fabrics. Indus. Crops Prod. **164** (2021)
15. Fang, Y., Sun, W., Li, J., Liu, H., Liu, X.: Eco-friendly flame retardant and dripping-resistant of polyester/cotton blend fabrics through layer-by-layer assembly fully bio-based chitosan/phytic acid coating. Int. J. Biol. Macromol. **175**, 140–146 (2021)

16. Fang, Y., Liu, X., Tao, X.: Intumescent flame retardant and anti-dripping of PET fabrics through layer-by-layer assembly of chitosan and ammonium polyphosphate. Prog. Org. Coatings **134**, 162–168 (2019)
17. Ma, X., Wu, N., Liu, P., Cui, H.: Fabrication of highly efficient phenylphosphorylated chitosan bio-based flame retardants for flammable PLA biomaterial. Carbohydr. Polym. [Internet] **287**, 119317 (2022). Available from: https://doi.org/10.1016/j.carbpol.2022.119317
18. Article, R., Lu, Y., He, Q., Fan, G., Cheng, Q., Song, G.: Extraction and modification of hemicellulose from lignocellulosic biomass: a review. Green Process. Synthesis 779–804 (2021)
19. Liu, Y., Zhang, A., Cheng, Y., Li, M., Cui, Y., Li, Z.: Recent advances in biomass phytic acid flame retardants. Polym. Test [Internet]. **124**, 108100 (2023). Available from: https://doi.org/10.1016/j.polymertesting.2023.108100
20. Vahabi, H., Jouyandeh, M., Parpaite, T., Saeb, M.R., Ramakrishna, S.: Coffee wastes as sustainable flame retardants for polymer materials. Coatings **11**(9), 1–14 (2021)
21. Lu, Y., Liu, C., Mei, C., Sun, J., Lee, J., Wu, Q., et al.: Recent advances in metal organic framework and cellulose nanomaterial composites. Coord. Chem. Rev. [Internet] **461**, 214496 (2022). Available from: https://doi.org/10.1016/j.ccr.2022.214496
22. Khodayari, A., Thielemans, W., Hirn, U., Vuure, A.W., Van Seveno, D.: Cellulose-hemicellulose interactions—a nanoscale view. Carbohydr. Polym. [Internet]. **270**, 118364 (2021). Available from: https://doi.org/10.1016/j.carbpol.2021.118364
23. Soman, S., Chacko, A.S., Prasad, V.S., Anju, P., Surya, B.S., Vandana, K.: Self-assembly of oleylamine modified nano-fibrillated cellulose from areca husk fibers into giant vesicles. Carbohydr. Polym. **2018**(182), 69–74 (2017)
24. Tavakoli, M., Ghasemian, A., Dehghani-firouzabadi, M.R., Mazela, B.: Cellulose and its nano-derivatives as a water-repellent and fire-resistant surface: a review. Materials (2022)
25. Muzammil, M., Ejaz, M., Shah, R., Afaq, S.K., Song, J.: A bio-based approach to simultaneously improve flame retardancy, thermal stability and mechanical properties of nano-silica filled jute/thermoplastic starch composite. Mater. Chem. Phys. [Internet]. **289**, 126485 (2022). Available from: https://doi.org/10.1016/j.matchemphys.2022.126485
26. Prabhakar, M.N., Shah, R., Song, J.: Improved flame-retardant and tensile properties of thermoplastic starch/flax fabric green composites. Carbohydr. Polym. [Internet]. **168**, 201–211 (2017). Available from: https://doi.org/10.1016/j.carbpol.2017.03.036
27. Calderas, F., Sanchez-olivares, G., Rabe, S., Ricardo, P., Schartel, B.: Industrial-waste agave fibres in flame-retarded thermoplastic starch biocomposites. Compos. Part B: Eng. **177** (2019)
28. Xia, Y., Chai, W., Liu, Y., Su, X., Liao, C., Gao, M., et al.: Facile fabrication of starch-based, synergistic intumescent and halogen-free flame retardant strategy with expandable graphite in enhancing the fire safety of polypropylene. Ind. Crop. Prod. [Internet] **184**, 115002 (2022). Available from: https://doi.org/10.1016/j.indcrop.2022.115002
29. Pigłowska, M., Kurc, B., Rymaniak, Ł: Starch as the flame retardant for electrolytes. Materials (Basel) **15**, 523 (2022)
30. Wang, D., Wang, Y., Li, T., Zhang, S., Ma, P., Shi, D., et al.: A bio-based flame-retardant starch based on phytic acid. ACS Sustain. Chem. Eng. **8**(27), 10265–10274 (2020)
31. Łukawski, D., Grześkowiak, W., Lekawa-Raus, A., Widelicka, M., Lisiecki, F., Dudkowiak, A.: Flame retardant effect of lignin/carbon nanotubes/potassium carbonate composite coatings on cotton roving. Cellulose **27**(12), 7271–7281 (2020)
32. Wang, Y., Zhang, Y., Liu, B., Zhao, Q., Qi, Y., Wang, Y., et al.: A novel phosphorus-containing lignin-based flame retardant and its application in polyurethane. Compos Commun. **21**(May), 100382 (2020)
33. Liu, L., Qian, M., Song, P., Huang, G., Yu, Y., Fu, S.: Fabrication of green lignin-based flame retardants for enhancing the thermal and fire retardancy properties of polypropylene/wood composites. ACS Sustain. Chem. Eng. **4**(4), 2422–2431 (2016)
34. Yue, X., Li, C., Li, Y.: Using colloidal lignin intercalated montmorillonite nanosheets as synergistic and reinforced agent for flame-retardant poly(butylene succinate) composites. Polym. Adv. Technol. **32**(6), 2552–2565 (2021)

35. Andrade-guel, M., Cabello-alvarado, C., Avila-orta, C.A., Marissa, P., Cadenas-pliego, G., Yahaira, P.: Green flame-retardant composites based on PP/TiO_2/lignin obtained by melt-mixing extrusion. Polymers (Basel) **14**, 1–13 (2022)
36. Li, P., Liu, C., Xu, Y.J., Jiang, Z.M., Liu, Y., Zhu, P.: Novel and eco-friendly flame-retardant cotton fabrics with lignosulfonate and chitosan through LbL: flame retardancy, smoke suppression and flame-retardant mechanism. Polym. Degrad. Stab. [Internet] **181**, 109302 (2020). Available from: https://doi.org/10.1016/j.polymdegradstab.2020.109302
37. Flejszar, M., Chmielarz, P., Hoch, A.: Advances and opportunities in synthesis of flame retardant polymers via reversible deactivation radical polymerization. Polym. Degrad. Stab. **214** (April)
38. Hussin, M.H., Nelson, J., Eng, N., Hanis, N., Latif, A., Brosse, N., et al.: A recent advancement on preparation , characterization and application of nanolignin. Int. J. Biol. Macromol. [Internet] **200**, 303–326 (2022). Available from: https://doi.org/10.1016/j.ijbiomac.2022.01.007
39. Maiti, S., Jayaramudu, J., Das, K., Reddy, S.M., Sadiku, R., Ray, S.S., et al.: Preparation and characterization of nano-cellulose with new shape from different precursor. Carbohydr. Polym. **98**(1), 562–567 (2013)
40. Lv, L.Y., Cao, C.F., Qu, Y.X., Zhang, G.D., Zhao, L., Cao, K., et al.: Smart fire-warning materials and sensors: design principle, performances, and applications. Mater. Sci. Eng. R [Internet] **150**, 100690 (2022). Available from: https://doi.org/10.1016/j.mser.2022.100690
41. Pielichowska, K., Nowicka, K.: Analysis of nanomaterials and nanocomposites by thermoanalytical methods. Thermochim. Acta [Internet] **675**, 140–163 (2019). Available from: https://doi.org/10.1016/j.tca.2019.03.014
42. Daniel, Y.G., Howell, B.A.: Flame retardant properties of isosorbide bis-phosphorus esters. Polym. Degrad. Stab. [Internet] **140**, 25–31 (2017). Available from: https://doi.org/10.1016/j.polymdegradstab.2017.04.005
43. Gupta, P., Verma, C., Maji, P.K.: Flame retardant and thermally insulating clay based aerogel facilitated by cellulose nanofibers. J. Supercrit. Fluids [Internet] **152**, 104537 (2019). Available from: https://doi.org/10.1016/j.supflu.2019.05.005
44. Wang, M., Yin, G.Z., Yang, Y., Fu, W.: Bio-based flame retardants to polymers: a review. Adv. Indus. Eng. Polym. Res. **6** (2023)
45. Phanthong, P., Reubroycharoen, P., Hao, X., Xu, G., Abudula, A., Guan, G.: Nanocellulose: extraction and application. Carbon Resour. Convers. [Internet] **1**(1), 32–43 (2018). Available from: https://doi.org/10.1016/j.crcon.2018.05.004
46. Attia, N.F., Mohamed, A., Hussein, A., El-demerdash, A.G.M., Kandil. H.: Bio-inspired one-dimensional based textile fabric coating for integrating high flame retardancy, antibacterial, toxic gases suppression, antiviral and reinforcement properties. Polym. Degrad. Stab. [Internet]. **205**, 110152 (2022). Available from: https://doi.org/10.1016/j.polymdegradstab.2022.110152
47. Pan, M., Gu, Y., Yun, Y., Li, M., Jin, X., Wang, S.: Nanomaterials for electrochemical immunosensing. Sensors (Switzerland) **17**(5) (2017)
48. Ghorai, S.: Chemical, physical and mechanical properties of nanomaterials and its applications (2013)
49. Zhang, Y., Jing, J., Liu, T., Xi, L., Sai, T., Ran, S.: A molecularly engineered bioderived polyphosphate for enhanced flame retardant, UV-blocking and mechanical properties of poly (lactic acid). Chem. Eng. J. [Internet] **411**, 128493 (2021). Available from: https://doi.org/10.1016/j.cej.2021.128493
50. Wang, M., Yin, G.Z., Yang, Y., Fu, W., Díaz Palencia, J.L., Zhao, J., et al.: Bio-based flame retardants to polymers: a review. Adv. Ind. Eng. Polym. Res. **6**(2), 132–155 (2023)
51. Nurfajrin, N., Alif, S., Hidayat, F., Nurul, M., Mohammad, A.: Recent developments in flame-retardant lignin-based biocomposite: manufacturing, and characterization. J. Polym. Environ. 4517–4537 (2022)
52. Noremylia, M.B., Zaki, M., Ismail, Z.: Recent advancement in isolation, processing, characterization and applications of emerging nanocellulose: a review. Int. J. Biol. Macromol. [Internet] **206**, 954–976 (2022). Available from: https://doi.org/10.1016/j.ijbiomac.2022.03.064

53. Attia, N.F., Mohamed, A., Hussein, A., El-Demerdash, A.G.M., Kandil, S.H.: Greener bio-based spherical nanoparticles for efficient multilayer textile fabrics nanocoating with outstanding fire retardancy, toxic gases suppression, reinforcement and antibacterial properties. Surf. Interfaces **36**, 102595 (2023)
54. Cheedarala, R.K., Prabhakar, M.N., Cho, B.G., Park, Y.B., Song, J.I.: Novel 3D-networked melamine-naphthalene-polyamic acid nanofillers doped in vinyl ester resin for higher flame retardancy. Mater. Adv. **2**(13), 4339–4351 (2021)
55. Marturano, V., Marotta, A., Agustin, S., Ambrogi, V., Cerruti, P.: Recent advances in bio-based functional additives for polymers. Prog. Mater. Sci. [Internet] **139**, 101186 (2023). Available from: https://doi.org/10.1016/j.pmatsci.2023.101186
56. Li, M., Prabhakar, M.N., Song, J.I.: Synthesis of novel Si–P–N complex coating for the fabrication of flame-retardant and hydrophobic cotton textile potentially suitable for rail seat cover. Prog. Org. Coatings [Internet] **172**, 107144 (2022). Available from: https://doi.org/10.1016/j.porgcoat.2022.107144
57. Crapina, L.C., Dzene, L., Brendlé, J., Fourcade, F., Amrane, A., Limousy, L.: Review: clay-modified electrodes in heterogeneous electro-fenton process for degradation of organic compounds: the potential of structural Fe(iii) as catalytic sites. Materials **14** (2021)
58. Prabhakar, M.N,. Song, J.I.: Fabrication and characterisation of starch/chitosan/flax fabric green flame-retardant composites. Int. J. Biol. Macromol. [Internet] **119**, 1335–1343 (2018). Available from: https://doi.org/10.1016/j.ijbiomac.2018.07.006
59. Yang, X.M., Qiu, S., Yusuf, A., Sun, J., Zhai, Z., Zhao, J., et al.: Recent advances in flame retardant and mechanical properties of polylactic acid: a review. Int. J. Biol. Macromol. [Internet] **243**, 125050 (2023). Available from: https://doi.org/10.1016/j.ijbiomac.2023.125050
60. Wang, S., Zhang, P., Li, Y., Li, J., Li, X., Yang, J., et al.: Recent advances and future challenges of the starch-based bio-composites for engineering applications. Carbohydr. Polym. [Internet] **307**, 120627 (2023). Available from: https://doi.org/10.1016/j.carbpol.2023.120627
61. Hou, B., Wang, Y., Li, B., Li, N., Yang, D., Liu, H., et al.: Bio-based flame-retardant plasticizers derived from l-lactic acid: effects of valence states of phosphorus structure on fire safety, flexibility and transparency of poly (lactic acid). Chem. Eng. J. [Internet] **474**, 145563 (2023). Available from: https://doi.org/10.1016/j.cej.2023.145563
62. Chemistry, O., Dutta, S., Wu, K.C.: Enzymatic breakdown of biomass: enzyme active sites, immobilization, and biofuel production. Green. Chem. [Internet] **16**, 53–66 (2014). Available from: https://doi.org/10.1039/c4gc01405g
63. Song, K., Ganguly, I., Eastin, I., Dichiara, A.B.: Lignin-modified carbon nanotube/graphene hybrid coating as efficient flame retardant. Int. J. Mol. Sci. **18**(11) (2017)
64. Xu, Y., Zhang, X., Wang, G., Zhang, X., Luo, J., Li, J., et al.: Preparation of a strong soy protein adhesive with mildew proof, flame-retardant, and electromagnetic shielding properties via constructing nanophase-reinforced organic—inorganic hybrid structure. Chem. Eng. J. **447** (2022)
65. Xu, Y., Han, Y., Li, Y., Li, J., Li, J., Gao, Q.: Preparation of a strong, mildew-resistant , and flame-retardant biomimetic multifunctional soy protein adhesive via the construction of an organic-inorganic hybrid multiple-bonding structure. Chem. Eng. J. [Internet] **437**(P2), 135437 (2022). Available from: https://doi.org/10.1016/j.cej.2022.135437
66. Xiao, D., Lv, J.X., Wu, F.J., Wang, Z.B., Harre, K., Chen, J.H., et al.: Development of multifunctional highly-efficient bio-based fire-retardant poly (lactic acid) composites for simultaneously improving thermal, crystallization and fire safety properties. Int. J. Biol. Macromol. [Internet] **215**, 646–656 (2022). Available from: https://doi.org/10.1016/j.ijbiomac.2022.06.158
67. Nurul, M., Mohammad, A., Antov, P., Savov, V., Fatriasari, W., Seng, L., Osvaldová, L.M., et al.: Current progress of biopolymer-based flame retardant. Polym. Degrad. Stab. **205** (2022)
68. Hou, Y., Xu, Z., Chu, F., Gui, Z., Song, L., Hu, Y., et al.: A review on metal-organic hybrids as flame retardants for enhancing fire safety of polymer composites. Compos. Part B [Internet] **221**, 109014 (2021). Available from: https://doi.org/10.1016/j.compositesb.2021.109014

69. Wu, Q., Ran, F., Dai, L., Li, C., Li, R., Si, C.: A functional lignin-based nanofiller for flame-retardant blend. Int. J. Biol. Macromol. **190**, 390–395 (2021)
70. Almeida, J.A., Oliveira, A.S., Rigoti, E., Neto, J.C.D., De Alcântara, A.C.S.: Design of solid foams for flame retardant based on bionanocomposites systems. Appl. Clay Sci. [Internet] **180**, 105173 (2019). Available from: https://doi.org/10.1016/j.clay.2019.105173
71. Ning, Y., Liu, R., Chi, W., Liu, W., Zhu, Q., Xu, S., et al.: Food Hydrocolloids Simultaneously flame-retardant and antibacterial films from κ-carrageenan/carboxylated cellulose nanofibril/ phytic acid for the preservation of cooked pork. Food Hydrocoll. [Internet] **143**, 108915 (2023). Available from: https://doi.org/10.1016/j.foodhyd.2023.108915
72. Zhang, Y., Haque, A.N.M.A., Naebe, M.: Lignin–cellulose nanocrystals from hemp hurd as light-coloured ultraviolet (Uv) functional filler for enhanced performance of polyvinyl alcohol nanocomposite films. Nanomaterials **11**(12) (2021)
73. Carosio, F., Medina, L., Kochumalayil, J., Berglund, L.A.: Green and fire resistant nanocellulose/hemicellulose/clay foams. Adv. Mater Interfaces **8**(18) (2021)
74. Yue, X., Zhang, S., He, J., Wang, Z.: Fabrication of flame retarded cellulose aerogel with hydrophobicity via MF/MTMS double cross-linking. J. Nat. Fibers [Internet] **20**(1) (2023). Available from: https://doi.org/10.1080/15440478.2022.2133053
75. Kim, H., Park, J., Minn, K.S., Youn, J.R., Song, Y.S.: Eco-friendly nanocellulose embedded polymer composite foam for flame retardancy improvement. Macromol. Res. **28**(2), 165–171 (2020)
76. Chollet, B., Lopez-Cuesta, J.M., Laoutid, F., Ferry, L.: Lignin nanoparticles as a promising way for enhancing lignin flame retardant effect in polylactide. Materials (Basel) **12**(13) (2019)
77. Qiu, S., Sun, J., Li, H., Gu, X., Fei, B., Zhang, S.: A green way to simultaneously enhance the mechanical, flame retardant and anti-ultraviolet aging properties of polylactide composites by the incorporation of tannic acid derivatives. Polym. Degrad. Stab. [Internet] **196**, 109831 (2022). Available from: https://doi.org/10.1016/j.polymdegradstab.2022.109831
78. Qiu, S., Wang, X., Yu, B., Feng, X., Mu, X., Yuen, R.K.K., et al.: Flame-retardant-wrapped polyphosphazene nanotubes: a novel strategy for enhancing the flame retardancy and smoke toxicity suppression of epoxy resins. J. Hazard. Mater. [Internet] **325**, 327–339 (2017). Available from: https://doi.org/10.1016/j.jhazmat.2016.11.057
79. Xu, Y.J., Qu, L.Y., Liu, Y., Zhu, P.: An overview of alginates as flame-retardant materials: pyrolysis behaviors, flame retardancy, and applications. Carbohydr. Polym. [Internet] **260**, 117827 (2021). Available from: https://doi.org/10.1016/j.carbpol.2021.117827
80. Matebie, B.Y., Tizazu, B.Z., Kadhem, A.A., Venkatesa Prabhu, S.: Synthesis of cellulose nanocrystals (CNCs) from Brewer's spent grain using acid hydrolysis: characterization and optimization. J. Nanomater. **2021** (2021)
81. Yu, X., Jiang, Y., Wu, Q., Wei, Z., Lin, X., Chen, Y.: Preparation and characterization of cellulose nanocrystal extraction from Pennisetum hydridum fertilized by municipal sewage sludge via sulfuric acid hydrolysis. Front. Energy Res. **9**, 1–10 (2021)
82. Ahmad, M., Gani, A., Hassan, I., Huang, Q., Shabbir, H.: Production and characterization of starch nanoparticles by mild alkali hydrolysis and ultra-sonication process. Sci. Rep. 1–11 (2020)
83. Liu, Y., Du, L., Gu, K., Zhang, M.: Effect of Tm dopant on luminescence, photoelectric properties and electronic structure of In2S3 quantum dots. J. Lumin. [Internet]. **217**, 116775 (2020). Available from: https://doi.org/10.1016/j.jlumin.2019.116775
84. Ewulonu, C.M., Liu, X., Wu, M., Yong, H.: Lignin-containing cellulose nanomaterials: a promising new nanomaterial for numerous applications. J. Bioresour. Bioprod. [Internet] **4**(1), 3–10 (2019). Available from: https://doi.org/10.21967/jbb.v4i1.186
85. Daraio, C., Jin, S.: Synthesis and patterning methods for nanostructures useful for biological applications. In: Nanotechnology for Biology and Medicine: At the Building Block Level, pp. 27–45 (2012)
86. Albulescu, D., Ursu, D., Rusnac, L.M., Nitu, S., Miclau, M., Vajda, M.: Investigation of UV dye-sensitized solar cells based on water electrolyte: a new insight for wavelength-selective greenhouse. Crystals 12(1) (2022)

87. Zhang, J., Wang, H., Yuan, X., Zeng, G.: Tailored indium sulfide-based materials for solar-energy conversion and utilization. J. Photochem. Photobiol. C Photochem. Rev. [Internet]. **38**, 1–26 (2019). Available from: https://doi.org/10.1016/j.jphotochemrev.2018.11.001
88. Gebresas, G.A., Szabó, T., Marossy, K.: A comparative study of carboxylic acids on the cross-linking potential of corn starch films. J. Mol. Struct. 1277 (2023)
89. Gao, Y.Y., Deng, C., Du, Y.Y., Huang, S.C.: A novel bio-based flame retardant for polypropylene from phytic acid. Polym. Degrad. Stab. [Internet] **161**, 298–308 (2019). Available from: https://doi.org/10.1016/j.polymdegradstab.2019.02.005
90. Cui, Z., Chen, Y., Meng, D., Wang, S., Sun, T., Sun, J., et al.: Reactive flame-retardants prepared by transesterification between erythritol and dimethyl methyl phosphonate for rigid polyurethane foams. Polym. Degrad. Stab. [Internet] **208**, 110238 (2022). Available from: https://doi.org/10.1016/j.polymdegradstab.2022.110238
91. Zhan, Y., Wu, X., Wang, S., Yuan, B., Fang, Q., Shang, S., et al.: Synthesis of a bio-based flame retardant via a facile strategy and its synergistic effect with ammonium polyphosphate on the flame retardancy of polylactic acid. Polym. Degrad. Stab. [Internet] **191**, 109684 (2021). Available from: https://doi.org/10.1016/j.polymdegradstab.2021.109684
92. Zhang, S., Liu, X., Jin, X., Li, H., Sun, J., Gu, X.: The novel application of chitosan: effects of cross-linked chitosan on the fire performance of thermoplastic polyurethane. Carbohydr. Polym. **2018**(189), 313–321 (2017)
93. Passauer, L.: Thermal characterization of ammonium starch phosphate carbamates for potential applications as bio-based flame-retardants. Carbohydr. Polym. [Internet] **211**, 69–74 (2019). Available from: https://doi.org/10.1016/j.carbpol.2019.01.100
94. Battegazzore, D., Frache, A., Carosio, F.: Layer-by-Layer nanostructured interphase produces mechanically strong and flame retardant bio-composites. Compos. Part B Eng. **200**, 108310 (2020)
95. Prabhakar, M.N., Song, J.: Fabrication and characterisation of starch/chitosan/flax fabric green flame-retardant composites. Int. J. Biol. Macromol. **119**, 1335–1343 (2018)
96. Xu, Y., Chen, X., Zhang, C., Ragauskas, A.J., Wen, J.L., Zhao, P., et al.: Enhancing thermal conductivity and toughness of cellulose nanofibril/boron nitride nanosheet composites. Carbohydr. Polym. **296**, 1–23 (2022)
97. Su, K.H., Su, C.Y., Shih, W.L., Lee, F.T.: Improvement of the thermal conductivity and mechanical properties of 3D-printed polyurethane composites by incorporating hydroxylated boron nitride functional fillers. Materials (Basel) **16**(1) (2023)
98. Lu, S., Hong, W., Chen, X.: Nanoreinforcements of two-dimensional nanomaterials for flame retardant polymeric composites: an overview. Adv. Polym. Technol. **2019** (2019)
99. Zhang, S., Li, S.N., Wu, Q., Li, Q., Huang, J., Li, W., et al.: Phosphorus containing group and lignin toward intrinsically flame retardant cellulose nanofibril-based film with enhanced mechanical properties. Compos. Part B Eng. [Internet] **212**, 108699 (2021). Available from: https://doi.org/10.1016/j.compositesb.2021.108699
100. Xue, Y., Feng, J., Huo, S., Song, P., Yu, B., Liu, L., et al.: Polyphosphoramide-intercalated MXene for simultaneously enhancing thermal stability, flame retardancy and mechanical properties of polylactide. Chem. Eng. J. [Internet] 397, 125336 (2020). Available from: https://doi.org/10.1016/j.cej.2020.125336
101. Velencoso, M.M., Battig, A., Markwart, J.C., Schartel, B., Wurm, F.R.: Molecular firefighting—how modern phosphorus chemistry can help solve the challenge of flame retardancy. Angew. Chemie. Int. Ed. **57**(33), 10450–10467 (2018)

Recent Advances in Protein-Based Flame-Retardant Materials

T. Ntuli, M. S. Xaba, B. Ntsendwana, T. C. Mokhena, and M. J. Mochane

Abstract The development of sustainable and effective flame-retardant materials is important in addressing fire safety concerns in various industries. This chapter presents a critical analysis of recent advances in the use of proteins as a renewable and eco-friendly source for flame-retardant materials. Proteins, derived from natural sources such as soy, whey, keratin, casein, and cystine, have gained significant attention due to their inherent flame-resistant properties and biodegradability. The chapter further discusses innovative strategies for incorporating proteins into flame-retardant materials, including their application in textiles, transportation, plastics, construction and coatings. Recent research has focused on enhancing the flame-retardant performance of proteins through chemical modification, nanocomposite and nanostructures, and synergistic approaches with other flame-retardant additives. Additionally, the mechanisms underlying the flame-retardant action of proteins, such as the formation of char layers, burning rates and heat release rates are explored. Furthermore, this chapter highlights the environmental benefits of protein-based flame-retardant materials, emphasizing their potential to reduce the environmental impact associated with traditional flame-retardants.

Keywords Proteins · Flame-retardant · Biodegradable · Char formation · Textiles

---

T. Ntuli (✉) · M. S. Xaba · B. Ntsendwana · T. C. Mokhena (✉)
Nanotechnology Innovation Centre (NIC), Advanced Materials Division, Mintek, Private Bag X3015, Randburg 2125, South Africa
e-mail: ThembaN@mintek.co.za

T. C. Mokhena
e-mail: mokhenateboho@gmail.com

M. J. Mochane
Department of Life Sciences, Central University of Technology, Bloemfontein, Free State, South Africa

T. C. Mokhena et al. (eds.), *Biomaterials as Green Flame Retardants*, Engineering Materials, https://doi.org/10.1007/978-981-97-6871-4_12

1 Introduction

Throughout the years, fire outbreaks have been one of the major concerns for human existence, wildlife, and different industries. This is due to the devastating effect that a fire outbreak could have on industrial property, buildings, the environment, and human lives. Efforts to mitigate the risk of fire include but are not limited to, the introduction of flame-retardant (FR) materials. These materials include halogenated compounds, inorganic acids, nitrogen-containing compounds, phosphorous-containing compounds, and heavy metals. They are used to protect combustible materials by mitigating the risk of a fire that is usually started by heat sources such as candles, electrical faults, and cigarettes [1]. FR materials have gained a huge amount of attention due to their potential to save lives and protect property and the environment by preventing or slowing down fires. The aforementioned FR materials come with restrictions such as compatibility with other materials, preventing ignition, extinguishing fires, the release of toxic fumes, and environmental accumulation. Thus, it is imperative to develop materials that are not only resistant to fire and environmentally friendly but also compatible with different materials for wider application in fire safety efforts. Such materials include, but are not limited to, protein-based flame retardants (PFRs). They function through different mechanisms, such as (i) releasing water or non-flammable gases to weaken or dilute the fuel and suppress the fire. (ii) The formation of a char layer that insulates the material, and (iii) flame interaction to inhibit its propagation. Several advantages come with PFRs compared to traditional flame-retardants include biodegradability, non-toxicity, and tunable properties, making them environmentally friendly, safer to handle, and compatible with other materials [2].

PFRs that are often employed as include casein, wheat gluten, soy, gelatin, whey, and cysteines. Casein is found in cow's milk and forms a durable char that is water-resistant when subjected to heat or increased temperatures [3]. Wheat gluten is a plant-based protein that acts as an insulator when heated by creating a protective foam [4]. Soy is also plant-based and forms an oxygen-limiting char, preventing the flame from gaining momentum [5]. Gelatin, a protein derived from collagen (an animal body part), forms a carbon layer to prevent fire [6]. Whey is produced during the manufacturing of cheese and casein; it is obtainable after milk has been curdled and strained. Which works by forming a protective carbon layer in the event of a fire [7]. PFRs have gained momentum in the field of flame retardants in efforts to advance the fire safety of a wide range of materials. These advances include protein modification, such as the formation of composites and coatings, to improve the effectiveness of the materials. One of the common themes in fire safety is materials with a low ignition threshold; as a result, the application of FRs becomes necessary [8].

Many commercially available flame retardants are comprised of organic compounds derived from oil sources. Being a component of oil-based products, these flame retardants share common challenges, including geopolitical issues and contributions to global warming [9]. Additionally, there are concerns about their potential to pose specific risks to both health and the environment. The flammability of polymeric materials is contingent upon thermal degradation processes initiated when they are subjected to heat. Recent awareness of previously overlooked or underestimated hazards necessitates novel approaches to traditional fire retardant methods. Successfully addressing this demanding task is likely only if we gain access to more comprehensive insights into the mechanistic aspects involved. Researchers have investigated how parameters, such as oxygen concentration, pressure, and heating rate impact the thermal behavior of polymeric materials concerning combustion conditions [10, 11]. Researchers are being propelled towards the development of valuable alternatives as they strive to discover highly effective environmentally friendly flame retardant products. These alternatives are proteins; which can be harnessed through straightforward and eco-friendly methods such as cross-linking and layer-by-layer, chemical modification, and composite formation techniques. The produced PFRs have potential applications in the textile, electronic, transport and construction industries. The type of protein, their production methods and applications have been presented in Fig. 12.1. This chapter aims to provide insights into the current state of research and development in the field of flame retardant proteins, highlighting their preparation, characterization, mechanisms, applications, and future prospects.

1.1 Overview of Protein-Based Flame Retardants

Proteins are crucial constituents of living organisms, consisting of one or more extended sequences of amino acids, forming large macromolecules, and playing vital roles in various cellular processes [12]. Due to their distinct compositions and chemical structures, proteins have garnered significant attention for their potential use as flame retardants. The chemical compositions of proteins, notably casein, hydrophobins, and whey proteins, confer them with inherent flame retardant capabilities. Casein, the predominant milk protein, can be acquired as a by-product of skim milk manufacturing, and it is a molecule that contains phosphorus [13]. Hydrophobins, an extensive group of proteins abundant in cysteine, consist of sulfur and nitrogen atoms, and they are synthesized by filamentous fungi [14]. Initially explored for creating edible and environmentally friendly films for food packaging, whey proteins have found a new role as flame-retardant coatings due to their effective oxygen barrier characteristics and impressive water vapor absorption capabilities [15]. The growing attention towards minimizing the ecological impact of materials has spurred the advancement of novel biobased additives derived from renewable resources [16]. These additives are aimed at creating sustainable flame-retardant solutions. Caseins can be regarded as biomacromolecules that exhibit intumescent properties when exposed to specific levels of radiant heat in a cone calorimeter or when subjected to a propane flame [17]. In both cases, they can produce an expanded carbon structure filled with bubbles, to prevent or limit the spread of a fire.

Around 100 °C, proteins undergo decomposition, leading to the liberation of carboxylic-acid-functionalized amino-acid derivatives and oligopeptides. This

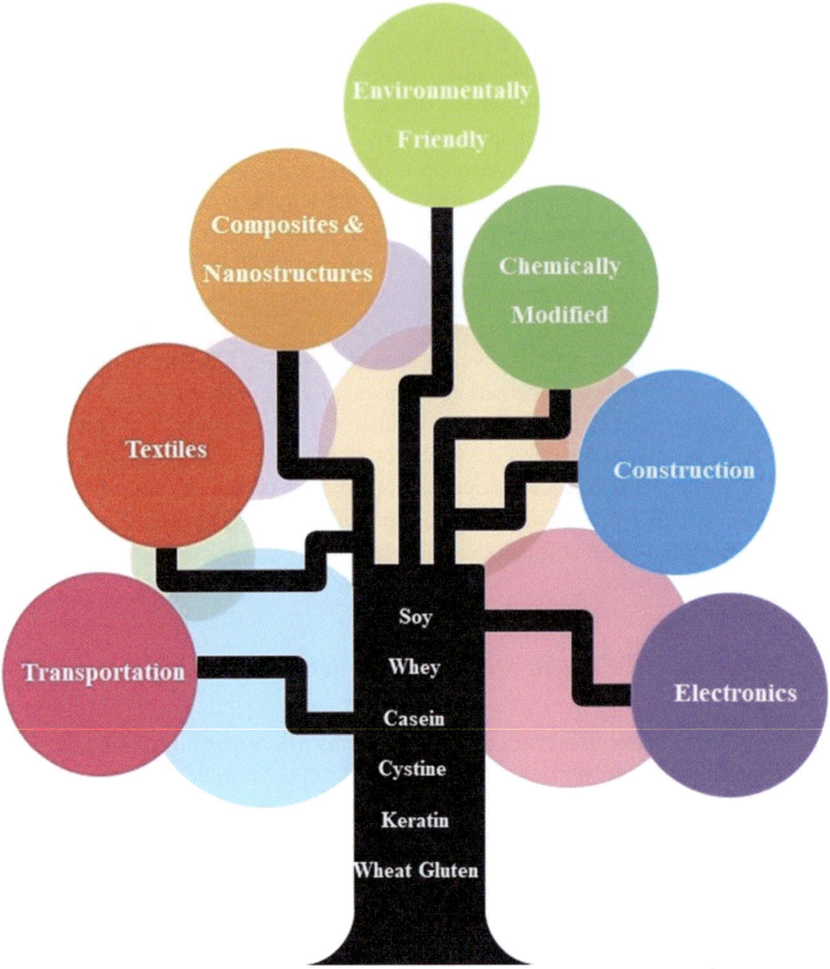

Fig. 12.1 Protein-based flame-retardants produced via different methods and applied in multiple industries

process further facilitates the degradation of materials like cellulose fibers. When exposed to high temperatures, typically around 750 °C, abundant nitrogen, acting as a blowing agent, contributes to the formation of char or carbon layer [18]. Proteins, as natural polymers, have a unique ability to form a protective char layer when exposed to heat and flames [5]. This char layer acts as a shield, insulating the underlying material and preventing the spread of fire. Protein-based flame-retardants (PFRs) are gaining increasing attraction in the field of fire safety and materials engineering.

These innovative compounds play a crucial role in reducing the flammability of various materials, ranging from textiles and plastics to foams and coatings [19]. What makes PFRs particularly valuable is their eco-friendly and biodegradable nature [5,

20]. Unlike traditional flame-retardants that often contain hazardous chemicals, these protein-based alternatives are more sustainable and pose fewer health risks to both humans and the environment. Moreover, PFRs can be tailored to specific applications, making them versatile and adaptable for various industries. In a world where fire safety and environmental concerns are paramount, PFRs offer a promising solution that aligns with the principles of sustainability and safety. Their development and integration into everyday products contribute to safer environments and a greener future.

Whey proteins exhibit globular structures containing a significant proportion of α-helix patterns, where the distribution of acidic/basic and hydrophobic/hydrophilic amino acids along the polypeptide chains is relatively even [21, 22]. The β-lactoglobulin, α-lactalbumin, bovine serum albumin, and immunoglobulin stand as the primary constituents of whey proteins. Making up roughly 20% of the total protein content in milk (with caseins constituting around 80%), whey proteins possess notable nutritional significance due to their rich reservoir of essential amino acids, notably those containing sulfur such as methionine and cysteine [23, 24]. Constituting the predominant portion of milk proteins, caseins have garnered extensive research attention and stand as some of the most thoroughly examined food proteins. Whey proteins demonstrate outstanding gas barrier characteristics, particularly concerning oxygen, due to the presence of hydrophilic groups and the formation of intermolecular and intramolecular interactions through hydrogen bonding [25]. Within bovine milk, casein represents the primary protein component, characterized by its significant phosphorylation with 8 or 9 phosphate groups per molecule. Notably abundant in glutamines bearing $-NH_2$ groups, caseins possess a distinctive feature of a solitary predominant phosphorylation site located close to the N-terminus [22].

In addition to their role in cheese production, caseins have primarily been employed as a food ingredient aimed at improving various physical attributes. These include augmenting properties such as foaming and whipping capabilities, thickening and water retention, emulsification, and texture enhancement, catalysis and drug delivery [26–28]. Furthermore, caseins have been utilized to enhance the nutritional value of food products [11, 29]. Additionally, caseins possess unique coating properties that broaden their scope of application to include papermaking, printing, leather finishing, and the manufacture of synthetic fibers [13, 30]. Traditional flame-retardants (TFRs) are harmful to the environment and humans are not spared from the detrimental effects of TFRs with overexposure leading to health complications such as cancer and developmental disorders [31]. This has prompted researchers to explore alternative, sustainable solutions. Thus, the significance of producing alternative FRs that are more environmentally friendly such as protein-based flame-retardants (PFRs). One of the attractive prospects in the field of flame safety is the development of protein-based flame-retardants (PFRs), which come with a unique combination of fire safety and less harmful to the environment. Since they are non-toxic, biodegradable and easy to functionalize. They have the advantage of high combustion temperatures making them suitable for use in fire prevention or limiting the spread of a fire in multiple industries. Other advantages include sustainability since they are derived from natural sources such as plants and animals alike. Their protective mechanism usually involves the formation of a protective char or carbon layer, which promotes the limiting of a fire or combustion. The subsequent sections explore the significance of protein-based flame-retardants, by providing different methods of production, application advantages and the feasible prospect of addressing the challenges presented by the use of TFRs.

2 Preparation of Protein-Based Flame-Retardant Materials

Creating soy protein adhesives that are both versatile and possess strong bonding capabilities presents a significant undertaking due to the need to replace hazardous formaldehyde-based adhesives commonly employed in the production of wood-based panels. Taking cues from arthropod cuticles, an organic–inorganic hybrid adhesive structure was engineered using epoxide cross-linked soy protein isolate (SPI) to establish a covalently bonded network, combined with dopamine-functionalized waste paper fiber [32]. The flame-retardant capabilities of soy protein adhesives were assessed using micro-combustion calorimetry. Upon removal from the flame, the adhesive promptly extinguished, indicating a substantial enhancement in its flame-retardant properties. Another study presented the initial findings on the preparation, and characterization of bio-based composites composed of jute fabric, and soy protein concentrate that has been modified with glycerol and/or halloysite nanotubes, with a particular focus on assessing their flammability [33]. The results indicated that soy protein concentrate exhibited a lower level of flammability, as measured by its heat release capacity, in comparison to petrochemical-based resins like epoxies and vinyl esters. Furthermore, the inclusion of a 5% mass fraction of halloysite nanotubes was observed to reduce the flammability of the composite without adversely affecting its mechanical properties. Another research study detailed a successful manufacturing method for a soy protein-based adhesive that is free of formaldehyde, featuring enhanced bonding strength, water resistance, and flame retardant properties through the integration of custom-made silica hybrids [34]. The tannic acid, serving as a versatile reactive platform, was employed to catalyze the hydrolysis and condensation of siloxane in an alkaline environment, resulting in the formation of a silica hybrid structure [34, 35]. Furthermore, the soy protein-based adhesive was intentionally formulated for potential use in the production of furniture particleboards. Its flame retardant properties were achieved through a synergistic interaction between the borax denaturant and the flame retardant in the hybrids. As a result, the soy protein-based adhesive exhibited a limiting oxygen index (LOI) value of 28.7% and achieved a V-1 rating in the UL-94 test (vertical flammability test). Consequently, such research offered a viable approach for manufacturing soy protein-based adhesives that are both formaldehyde-free and multifunctional.

The selection of suitable proteins for flame retardancy is a critical consideration in various industries, including textiles [11, 23], construction [36] and electronics [37]. Proteins, derived from natural sources such as plants and animals, have gained prominence as eco-friendly alternatives to conventional flame-retardants containing hazardous chemicals. The choice of proteins for flame retardancy hinges on their inherent properties and compatibility with the intended application. One of the primary factors influencing the selection of proteins is their inherent flame-retardant properties. Certain proteins, like casein and gelatin, possess inherent flame retardancy due to their molecular structure and composition [38]. These proteins release non-combustible gases and form protective char layers when exposed to heat or flames. Casein, for instance, forms a stable, insulating char layer that acts as a barrier against further combustion, making it a popular choice for fire-resistant coatings and paints [39]. Another vital consideration in protein selection is compatibility with the target material or substrate. Proteins must be compatible with the material they are meant to protect, ensuring they adhere effectively and do not compromise the material's mechanical or aesthetic properties. For textiles, proteins like keratin, derived from sources like wool and feathers, offer excellent flame-retardant properties while maintaining the fabric's softness and comfort. In construction materials, soy protein-based adhesives have been used to enhance fire resistance in engineered wood products like plywood and particleboard [8].

The technique of layer-by-layer self-assembly, using different combinations of egg white protein and phytic acid (PA), has been investigated for its potential to provide double coatings on cotton fabrics with flame-retardant properties [40]. Similarly, the treated cellulosic textiles demonstrate remarkable flame-retardant capabilities. This is substantiated by observing reduced values for peak heat release rate (PHRR) and total heat release (THR), along with prolonged time to ignition and increased residues post-combustion, when compared to the untreated fabrics. Sequentially treating the fabric with protein and PA revealed an optimal enhancement in flame retardancy efficiency, highlighting a notable synergistic effect derived from phosphorus and nitrogen. This method achieved a substantial weight gain rate of approximately 40%. The strong electrostatic attraction was verified through the observable stretching vibrations of P=O, amide I, and amide II, along with corresponding elemental shifts, as observed through FTIR and surface element composition analyses. These comparisons were made through assessments using cone calorimeter analyses and vertical flammability tests. Py-GC/MS analyses indicate the generation of non-combustible pyrolysis by-products such as CO_2, H_2O, and NH_3 throughout the pyrolysis process. These results provide compelling evidence for the fire-resistant effects of the treatment in both the gaseous and solid phases when the treated fabrics undergo combustion. In the future, alternative adhesives will be created using natural materials that are non-toxic, cost-effective, and sustainable. By utilizing borate chemistry, strong adhesives by crosslinking soy protein (SP) and soy polysaccharides (SPSs) can be formed, which constitute the fundamental structural components found in higher plants [41]. To enhance toughness, they introduced hyperbranched polyester into the adhesive formulation. The resulting cross-linked protein adhesive exhibited exceptional bonding strength with wood, sometimes surpassing the wood's inherent strength. Additionally, the borate cross-linked protein adhesive exhibited antimicrobial and flame-retardant properties. This straightforward borate chemistry offered a pathway to produce economical, renewable, and high-performance materials.

Using two distinct synthesis approaches, a range of self-curable waterborne dispersions of casein waterborne polyurethane were created, all of which were based on polycaprolactone diol (PCL) and included (3-aminopropyl)triethoxysilane as a crosslinking agent [42]. The addition of casein after waterborne polyurethane synthesis did not alter the particle size, but employing the in-situ method resulted in an increase in particle size, indicating the bonding of casein to the prepolymer during synthesis. Micro combustion calorimetry results demonstrated that an increase in casein content led to delayed combustion of the material. The UL-94 results highlighted the effectiveness of casein as a flame retardant, as it reduced the after-flame time and enhanced the thermal properties of the sample when it was covalently bonded to the waterborne polyurethane [42, 43].

Ortelli and co-workers reported the immobilization of metal oxide nanoparticles on the surface of cotton and further investigated the fire behaviour [44]. They harnessed the capability of amino acid-based biomacromolecules to permanently envelope nanoparticles (NPs), alongside the attraction between metal oxide NPs and inherently hydrophilic fibers. Egg white protein, a widely available and cost-effective protein in our daily lives, is rich in amino acids, calcium, iron, and phosphorus [45]. Furthermore, its diverse amino acid composition can lead to strong electrostatic interactions when combined with eco-friendly bio-based compounds like phytic acid (PA), which contain a unique cyclic structure with six phosphate groups [40]. This combination results in the formation of a green phosphorus-nitrogen flame retardant system, as illustrated in Fig. 12.2. Therefore, egg white protein shows great potential as an affordable alternative to more expensive flame-retardants.

The optimization of protein concentration plays a pivotal role in the development of effective fire retardant applications. In this context, achieving the ideal protein concentration is essential for enhancing the flame-resistant properties of materials. Through careful experimentation and analysis, researchers aim to strike a balance between the amount of protein used and its ability to form a protective barrier [46]. Too little protein may result in inadequate fire resistance, while an excess can lead to reduced flexibility and functionality of the treated materials [5]. Achieving the optimal protein concentration involves a delicate interplay of factors, including the type of protein, the substrate material, and the specific fire retardant mechanisms involved. This pursuit of the perfect protein concentration represents a critical step in advancing fire safety across various industries, from textiles to construction materials. A sustainable mineral-organic approach has been documented for the creation of a film based on soy protein. With this method, nitrogen-coordinating boronic diester was derived from the combination of boronic acid and diols, leading to the formation of strengthened cross-linking networks via dynamic adhesive connections [47]. Hydroxyapatite, an inorganic material, and caffeic acid, a naturally occurring phenolic compound, were subjected to a reaction with boronic diester at varying

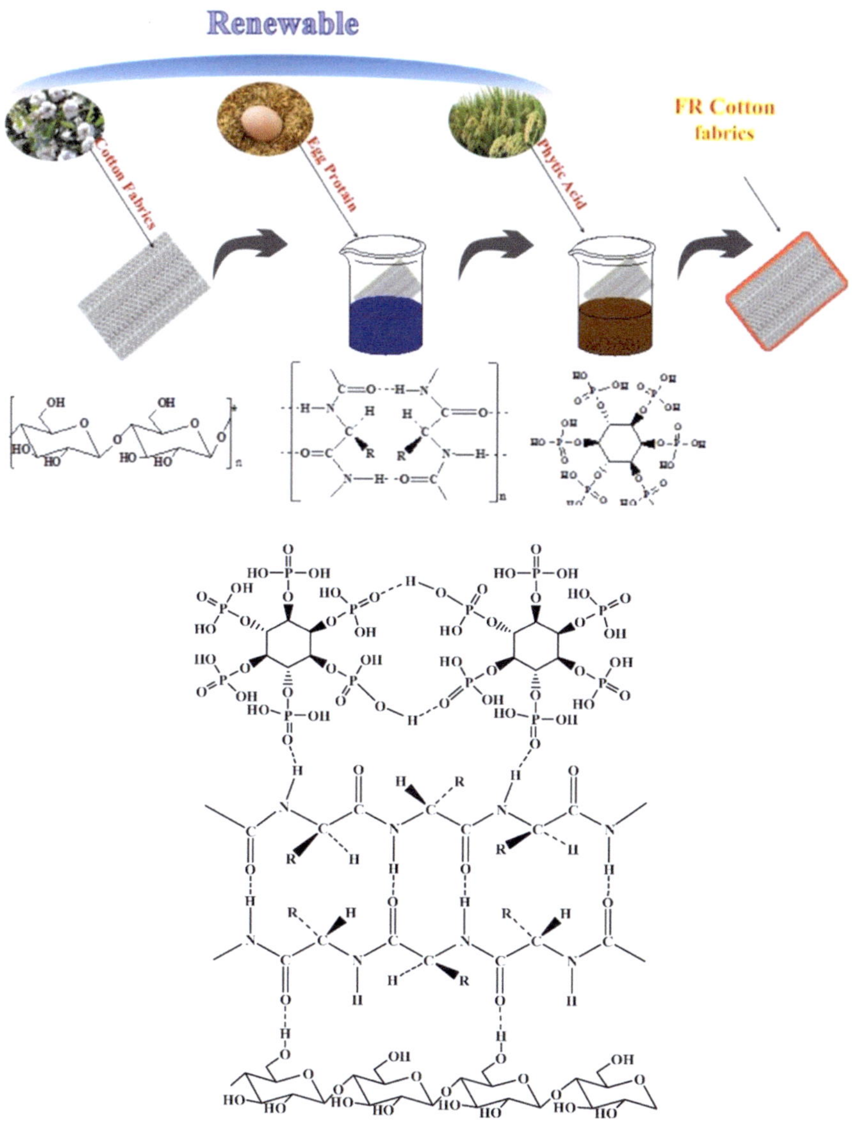

Fig. 12.2 A generic method for sequentially applying double coatings on cellulosic fabrics through layer-by-layer self-assembly, involving egg white proteins and phytic acid. Reproduced with permission from Reference [40]

concentrations to produce the core@shell hybrids. Due to the process of bio mineralization and the presence of phenolic components, the core@shell hybrids integrated into the film provide it with favorable characteristics such as water resistance, flame retardancy, UV protection capabilities, and antibacterial activity. Innovation in the production approach of PFRs involves different preparation methods such as layer-by-layer assembly, chemical modification, bio-composites and incorporation of nanostructured materials.

2.1 The Layer-By-Layer Assembly Approach

The technique of layer-by-layer self-assembly, using different combinations of egg white protein and PA, has been investigated for its potential to provide double coatings on cotton fabrics with flame-retardant properties, where a protein layer is layered with another material [40]. Fang et al. reported that chemical retreatment of the polymer layer influences the flame retardancy of the material. Where the polymer was pre-treated with 1 M NaOH at 90 °C for 30 min [48]. A schematic representation of the layer-by-layer assembly approach has been presented in Fig. 12.3. In the textile industry, cotton, a cellulosic material, plays a significant role in the business, and fire has proven detrimental to life and properties. Cotton readily catches fire due to its flammable nature; thus, there is a need to protect it from fire by improving its flame retardancy. The layer-by-layer assembly method was used, where the cotton material was double-coated with a combination of egg white protein and PA through electrostatic attraction. PA and egg white protein are both biodegradable, and phytic acid contains phosphorus as part of its backbone chemical structure, making it a flame-retardant material. An in-depth preparation of the egg white and phytic acid coating using the layer-by-layer assembly. Where, the cotton fabric or material was soaked in water at 50 °C, to remove debris, then, placed under vacuum and dried at 30 °C. This was followed by a two-step soaking process where the dry material was soaked in the egg white solution to form a cotton-egg material and a rotary drum was used to remove excess egg white (step 1). The cotton-egg material was soaked in a phytic acid solution to form a cotton-egg-acid material, and a rotary drum was used to remove unreacted acid (step 2). The cotton-egg-acid material was dried at 80 °C until the mass of the material did not change. To obtain a double coating, the two-step process was repeated [40]. The mechanism for the layer-by-layer double coating is presented in Fig. 12.3.

Where the hydroxyl groups from the cellulosic cotton form a hydrogen bond with the amino acid group from the egg white protein, resulting in a cellulose-amino acid complex. Then the phytic acid binds to the complex, forming a protein-based flame retardant material. This is because the egg white and phytic acid layers form a protective or preventative layer around the cotton, thus limiting the reach of heat and oxygen to the cotton. Polyelectrolyte molecules such as phytic acid, lactic acid, chitosan, casein, and polyethyleneimine have found prominence in protein-based FR, replacing the toxic halogenated FR. This is due to their ability to ionize in

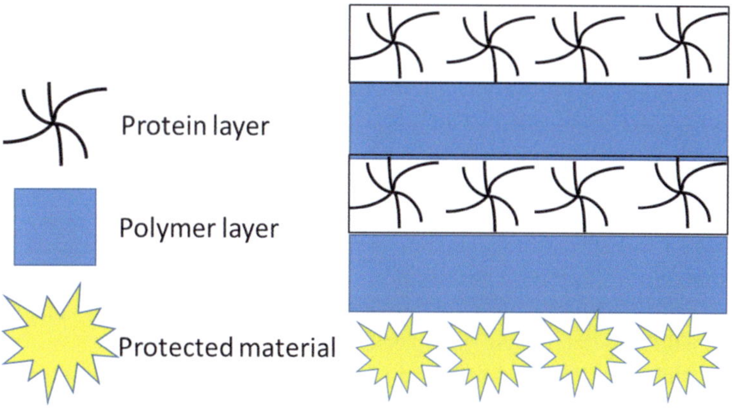

Fig. 12.3 Illustration of a layer-by-layer assembly of double coating

solution and bond with ionic compounds, they have phosphorus or nitrogen as part of their long-chain polymeric structure; and they are biodegradable [49]. The layer-by-layer approach for preparing protein-based flame-retardants offers the flexibility of assembling materials with opposite surface charges in a layer on top of another layer for improved application in fire safety. It provides the option to select more environmentally friendly materials such as biodegradable polymer materials. The method also comes with the advantage of selectivity as each layer could be designed in such a way that is compatible with different or specific substrates.

2.2 Chemical Modification of Flame Retardant Protein-Based Materials

The chemical modification of protein-based FR materials has presented the opportunity to enhance flame retardant properties, and thermal stability and increase fire safety in different applications. This is because chemical modification allows targeted tailoring of the material and comes with the possibility of ensuring compatibility and sustainability. Cautious contemplations of the category or choice of chemical modification to be employed in the production of PFR materials could ensure that the resultant PFR material is environmentally friendly due to biodegradability. Such might be made probable by introducing substances that are not harmful to the environment and are biodegradable. When employing chemical modification, the core production method of the PFR material is retained.

The chemical modification can flawlessly be introduced to coatings, polymers and foams during or post-production, sustaining the integrity of the production method. Generally, PFR materials have a structural composition that is inherently rich in nitrogen and thus their application in fire safety. The chemical modification with sulphur or phosphorous-containing compounds respectively or separately has the potential to enhance the fire safety properties of PFR materials. Dong and co-workers reported on the facile production of phosphorus nitrogen-modified soy-bean PFR for poly(lactic acid) using the Mannich reaction [46]. This reaction centered on the creation of aminomethylated compounds, involving three fundamental elements: the ketone or aldehyde (acting as the electrophile), the amine (functioning as the nucleophile), and the enolizable compound (either an enol or an enolate). The soy-bean protein was stirred in water then, dimethyl phosphite was introduced into the protein mixture followed by the addition of formaldehyde and heated at 85 °C for an hour and dried in an oven for characterization and application. The product together with poly(lactic acid) was heated at 60 °C for 12 h to remove moisture. The melt blending technique was used to produce the PFR composite at 180 °C for 8 min. Some of the characterization techniques include scanning electron microscopy (SEM) for morphological evaluation, for chemical and structural determination, X-Photoelectron spectroscopy (XPS) and Fourier transform infrared (FTIR) spectra were used for functional group determination. The thermal properties of the composites were determined using TGA. The morphological characteristic of the material before the Mannich reaction showed spherical particles with a relatively smooth surface. The surface exhibited rougher edges after the Mannich reaction. The TGA and DTG curves of the polymer and the composites showed that after chemical modification with the polymer the composite exhibited improved charring ability from 2.57 to 7.16% (char residue) compared to the 0.75% of the polymer at 600 °C. The XPS and FTIR spectra showed that the composites material consisted of nitrogen moieties which are essential for the application as flame retardants. This results indicate the possibility of successful chemical modification of soy bean protein with poly(lactic acid). The addition of the soy bean protein increased the limited oxygen index (LOI) value of the poly(lactic acid) from 19.4 to 26.3% and achieved V0 rating in the UL-94 vertical burning tests [46]. The chemical modification approach involves surface and structural modification of proteins to enhance their flame-retardant properties. This process can be achieved through various techniques, such as grafting with other proteins, cross-linking to stabilize the protein structure, or blending with other compounds for synergistic effects. Just like the layer-by-layer approach this method comes with flexibility in the choice of material used and has the potential to be more effective and environmentally friendly materials.

2.3 Bio-Polymer Composites

Another method of producing effective protein-based flame retardant materials is the formation of bio-polymer composites. It is a method that incorporates the protein with a suitable bio-polymer to improve the thermal properties and enhance the overall flame retardancy of the material. All the components of the composite are derived from biological compounds and thus, are an environmentally friendly and sustainable alternative compared to synthetic materials because they have the potential to be biodegradable. Some of the compounds to construct the composite materials are extracted from plant-based materials. For example, wheat gluten and wool fibre were combined to produce a flame-retardant material in a two-step process that involved mixing and compression moulding. The wool fibre and the wheat gluten were dried for 48 h. The dry wheat was added to glycerol and blended for a few minutes. During this process the wool was introduced into the wheat-glycerol blend, resulting in a wheat-glycerol-wool mixture. In a hydraulic press, the mixture was granulated and heated for 4 min at 70 °C, such conditions were selected to avoid premature polymerization and crosslinking. Thereafter, the temperature was increased to 120 °C to allow gluten flow and prevent the formation of air bubbles. This was followed by increasing the press temperature to 140 and 160 °C for the formation of the bio-polymer composite. The composite was cooled to room temperature and stored for use. The wool content (ca. 78%) and glycerol content (20 and 30 wt%) proved to be significant factors in both fire retardancy and mechanical performance of the composites. The aforementioned contents were suitable for was achieving a self-extinguishing flame during the vertical burn test (V-1) and the total heat release of 34%. The mechanical performance results showed values of 67 and 48% for the tensile modulus and strength, respectively. The FTIR spectra of the composite showed the presence of C–O stretching around 1027 cm^{-1} attributed to glycerol, amide group at around 1612–1640 cm^{-1} and overlapping peaks of –OH(–NH_2) between 3600 and 3000 cm^{-1} indicating the presence of glycerol in all the composites peaks [50]. For the most effective outcomes, the ratio balance between the wool-glycerol-wheat needs to be optimized. A similar production method of mixing followed by compression moulding to produce a protein-based composite to protect cotton fabric. The components of the composite were chicken feathers, sodium pyrophosphate, glyoxal-melamine, to enhance the flame retardancy of the composite borax and boric acid were incorporated [51]. The approach of producing flame retardants by combining bio-polymers such as glycerol and glyoxal-melamine with proteins extracted from natural sources like soy, casein, or gelatin results in bio-polymer-protein composites offers a viable approach method that has the potential to be sustainable and environmentally friendly approach. This is due to the options of harnessing the inherent properties of bio-polymers and proteins, similar to the previous two methods of production (layer-by-layer and chemical modification) which also has the option to select less toxic chemicals that are more environmentally friendly materials and materials that have a wider application due to compatibility with different materials.

2.4 *Nanotechnology-Enhanced Protein-Based Flame Retardant Materials*

The wide application of nanotechnology has seen application in coatings to improve flame retardancy of different materials [52]. It is with this motivation that nanoparticles in combination with protein-based flame retardants are explored. Traditional halogenated flame-retardants have been applied successfully in suppressing and/or delaying combustion. The need arises for more environmentally friendly alternative flame retardants as technology continues to evolve. Thus, the combining of protein-based flame retardants with nanoparticles presents an exciting opportunity to produce advanced materials to improve fire safety. Such a combination prevents structural damage of the material even at elevated temperatures. The protein part of the material serves as a protective layer to cover the metallic or nanoparticle part of the material, preventing it from or limiting the combustion process by limiting reaction with oxygen. The nanoparticles serve as heat absorbers resulting in reduced temperatures and therefore, limiting the spread of the fire.

Casein and whey proteins were used as coats for TiO_2 nanoparticles to improve the fire safety of cotton fabrics. The initial steps in the production of the protein-based-TiO_2 coatings was the formation of the suspensions. Where whey protein was dissolved in water at ambient conditions for 1 h. The casein was dissolved in a NaOH solution for 15 h at 80 °C this was because this type of protein is hydrophobic. TiO_2 nanoparticles were added to the solutions of anionic risen and sodium citrate, respectively, resulting in two suspensions of TiO_2, TiO_2-risen and TiO_2-citrate. The whey protein was soaked in the TiO_2-risen suspension whilst the casein was soaked in the TiO_2-citrate suspension resulting in TiO_2-whey and TiO_2-casein respectively. Using the dip-pad-cure method the produced protein-based-TiO_2 flame retardants were incorporated into the cotton fabrics. The fabrics were soaked in the TiO_2-whey and TiO_2-casein respectively for 3 min, followed by padding at 100 °C. The materials were then, cured for 10 min at 130 °C and cooled to room temperature for characterization and application in fire testing. Figure 12.4 is the schematic representation of the production of the protein-based TiO_2 flame retardant materials. The Zeta potential for the nanoparticles was negative whilst for the whey protein it was positive conditions suitable for electrostatic attraction and resulting coating of the TiO_2 with the whey protein. For the casein it was not the case, because it had a negative Eta potential, indicating repulsion between the faces of the nanoparticles and the protein. However, surface coating may have taken place through weak intermolecular forces. The results indicated that the TiO_2/casein coatings outperformed the TiO_2/whey coatings due to the higher residue percentage of the casein coating (22.4 and 41.3%) compared to (5.0 and 7.9%) for the whey coatings. The uncoated cotton fabric burned completely as it did not show any residue left, the results showed that the cotton fabric benefited from being coated with the nanocoating. The morphological representation of the cotton fabric showed a smooth surface that was changed to a rough surface when coated with the respective TiO_2/casein and TiO_2/whey coatings [44]. The nanoparticle-enhanced protein-based method of producing flame-retardants is not exclusive to the textile industry. But inclusive to diverse industries such as construction, transport, electronic industries and in the mitigation of wildfires. This is possible through the incorporation of nanoparticle-enhanced protein-based flame-retardants to improve fire safety. It should me be mentioned that though the long term effects of using nanoparticles is relatively unknown, caution should prevail and efforts given to research the long term environmental effects of using such materials.

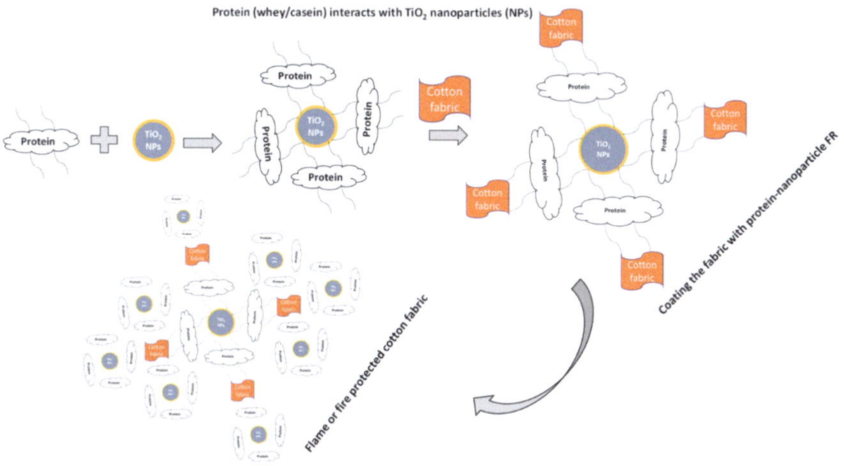

Fig. 12.4 The production of the protein-based TiO_2 flame retardant materials. Reproduced with permission from [44]

It should me be mentioned that though the long term effects of using nanoparticles is relatively unknown, caution should prevail and efforts given to research the long term environmental effects of using such materials. This includes production time, scalability, long-term stability and the possible cost of production. In mitigating such issues the toxicity of nanoparticles and the production time should be prioritized to improve the production method to ensure growth and stability. A collaborative effort between all stakeholders such as regulatory bodies, academia, industries and other affected communities should be encouraged to decrease or prevent environmental harm from the use of the proposed materials.

3 Characterization and Performance of Protein-Based Flame Retardant Materials

The choice of TG-FTIR as the technique was primarily driven by its capability to provide simultaneous information on both thermal degradation (via thermogravimetric analysis) and the evolving gas-phase composition (via Fourier-transform infrared spectroscopy), making it ideal for monitoring the reaction's progress and product formation. The TG-FTIR technique was employed to investigate the thermal properties of casein for the very first time [53]. Thermogravimetric analyses conducted in an air environment revealed a notable enhancement in char residues, reaching 32.9 wt.% at 600 °C and 12.0 wt.% at 800 °C. These outcomes are clearly more impressive compared to findings documented in existing literature involving alternative protein sources, due to their improved thermal stability [40]. The recorded

TG, DTG, and DTA curves reveal a multifaceted thermal degradation process, characterized by two distinct phases. The first phase is endothermic and occurs between 45.6 and 248 °C, followed by a highly exothermic phase spanning from 248 to 770.6 °C. It's noteworthy that within the temperature range of 30–248 °C, the degradation mechanism in an air environment (as depicted in Fig. 12.5a) appears to be identical to that in a nitrogen atmosphere (as shown in Fig. 12.5b), as evidenced by the matching thermograms [53]. As illustrated in Fig. 12.5a, b, the weight loss percentage of casein due to degradation in an air environment between 172.4 and 248 °C was equivalent to that observed under a nitrogen atmosphere. This stage was evidently characterized by an endothermic nature and occurs across the identical temperature ranges. This finding reinforces the earlier conclusion of a consistent degradation mechanism operating within this specific temperature range [53]. The examination of casein using TG, DTG, and DTA under both air and nitrogen atmospheres revealed a distinctive and intricate degradation mechanism. This mechanism can be divided into two distinct domains when considering degradation in air. The first domain, which is endothermic, mirrors the degradation process under nitrogen atmosphere within the same temperature range. The second domain, an exothermic phase, was unique to the degradation process in an air environment and was not observed in the nitrogen atmosphere.

In the quest for novel green flame retardants for cotton fabrics, certain proteins sourced from animals or microbes have been explored, despite the utilization of harmful and environmentally unfriendly chemicals [3, 14, 51, 54, 55]. Upon examining the residue morphologies following the tests, it becomes apparent that two casein proteins exhibit different protective mechanisms, as evidenced by the comparison between the morphologies of cotton caseins depicted in Fig. 12.6a–d and cotton hydrophobins illustrated in Fig. 12.6e–h. Caseins, for instance, are capable of forming globular structures resembling phosphorus-rich bubbles that rupture during combustion as shown in Fig. 12.6a–d. Conversely, when cotton is treated with hydrophobins, numerous unruptured pearl-like bubbles persist and are observable after the flammability tests as demonstrated in Fig. 12.6e–h. This phenomenon was ascribed to the breaking of disulfide bonds, followed by the subsequent formation of cross-links between amide groups [56, 57]. SEM examination confirms that the protein-inner and PA-outer layer coating preserves the fabric's surface structure [40]. Moreover, this arrangement promoted the generation of increased carbonaceous residues during the combustion process.

Ammonium polyphosphate and soy protein were employed in the creation of biodegradable intumescent flame-retardants. These were subsequently blended with halloysite nanotubes to produce environmentally friendly intumescent flame-retardants. This was carried out to improve the flame-retardant properties of poly(butylene succinate) [5]. With an LOI value of 22.4%, poly(butylene succinate) experiences significant melt dripping when it undergoes combustion. UL-94, a flammability standard published by Underwriters Laboratories (USA), categorizes plastics based on their burning behavior under different orientations and part thicknesses. It divides them into six distinct classifications, ranging from the least flame-retardant to the most flame-retardant. While, the limiting oxygen index, expressed

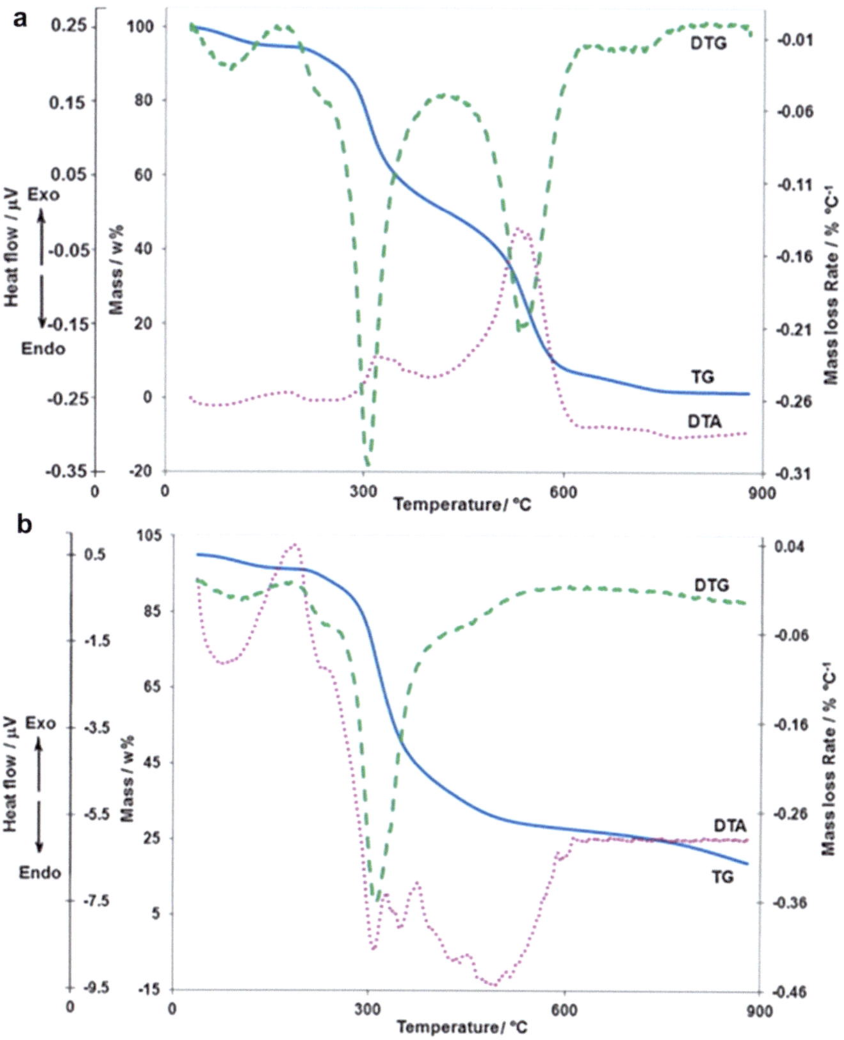

Fig. 12.5 TG, DTG, DTA thermograms of casein in air (**a**) and under nitrogen atmosphere (**b**). Reproduced from Reference [53]

as a percentage, represents the lowest oxygen concentration required to sustain the combustion of a polymer. This measurement involves passing a blend of oxygen and nitrogen over a burning specimen and gradually decreasing the oxygen content until a critical threshold is attained. Table 12.1 provides data on the results of LOI, and combustion velocities for specific flame-retardants derived from proteins.

The introduction of 20 wt.% APP enhances the LOI of PBS to 28.5% and encourages the development of char [5]. Nevertheless, even with this improvement, PBS/APP20 still exhibits dripping behaviour and attains only a V-2 rating. On the other

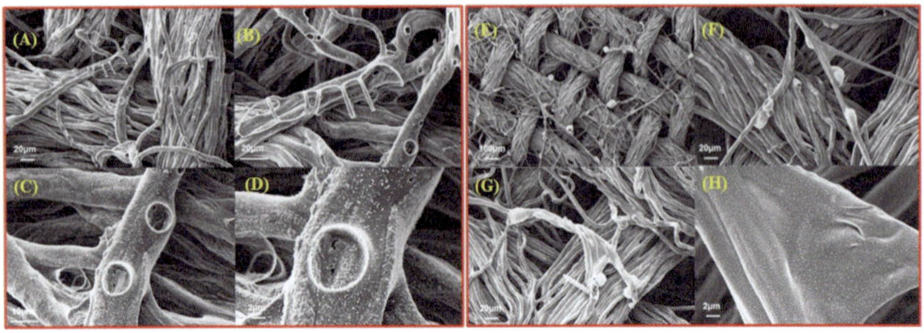

Fig. 12.6 SEM magnification of the residues left by cotton caseins (**a–d**) and cotton hydrophobins (**e–h**) after flammability tests. Reproduced from Reference [14]

Table 12.1 The LOI test and burning rate of some protein based flame-retardants

Protein type	LOI (%)	Burning rate (mms^{-1})	References
Soy	22.4%	–	[5]
Casein	39.5%	–	[3]
Keratin	30%	–	[58]
19% DNA + cotton	28%	3.0	[59]
10% DNA + cotton	25%	1.9	[59]
5% DNA + cotton	23%	1.6	[59]
Cotton + 20 DNA-BLs	24%	1.5	[60]
Cotton + 10 DNA-BLs	23%	1.9	[60]
Cotton + 5 DNA-BLs	21%	2.2	[60]
Cotton + Casein	24%	0.4	[13]
PET + 20% caseins	26%	0.6	[13]
PET–Ctton + 20% caseins	21%	0.7	[13]
PA66 + 3% SPI	23.4 ± 0.2%	–	[61]
PA66 + 4% SPI	24.0 ± 0.3%	–	[61]
PA66 + 5% SPI	25.5 ± 0.2%	–	[61]
PA66 + 6% SPI	25.1 ± 0.3%	–	[61]
PA66 + 7% SPI	25.0 ± 0.4%	–	[61]
10 g/L CPP + Fe	27.8%	–	[62]
10 g/L CPP + Ti	28.0%	–	[62]
HDPP + Silk	30%	–	[63]
L/C/PAN	18.8 ± 0.2	–	[64]
FR/PAN-5	26.3 ± 0.2	–	[64]
FR/PAN-8	29.8 ± 0.2	–	[64]
FR/PAN-10	32.5 ± 0.3	–	[64]
FR/PAN-12	33.2 ± 0.3	–	[64]
3% CES	31.5%	–	[65]

hand, the incorporation of 20 wt.% SP has an adverse impact on PBS, leading to a reduced LOI of 20.5%, with PBS/SP20 failing to qualify in the UL-94 test. This may be attributed to SP promoting char formation during combustion, enhancing PBS's resistance to dripping, and thereby mitigating the influence of melt dripping on the LOI assessment.

To evaluate how various concentrations of PA affect the flame retardant properties of polyacrylonitrile (PAN), Guo and co-workers conducted vertical burning and LOI tests on both untreated and modified PAN samples, as illustrated in Fig. 12.7. For ease of reference, different flame-retardant PAN samples treated with 5%, 8%, 10%, and 12% weight/volume PA solution were designated as FR-PAN-5, FR-PAN-8, FR-PAN-10, and FR-PAN-12, respectively. As shown in Fig. 12.7a, untreated PAN underwent rapid combustion with the production of numerous molten droplets, which quickly ignited the absorbent cotton, leading to the rapid spread of fire. Furthermore, the LOI for untreated PAN was a mere 17.0%, indicating its high flammability with no vertical burning rating. In Fig. 12.7b, even with the inclusion of lignin and casein, designated as L/C/PAN, the molten droplets were somewhat restrained during combustion. However, absorbent cotton still ignited, failing to meet the UL-94 V-0 grade; although, the ignition time was delayed. Additionally, the LOI for L/C/PAN exhibited a slight increase. In contrast, as depicted in Fig. 12.7c–f, following treatment with various concentrations of PA solution, the flame retardancy of FR-PAN significantly improved, achieving a UL-94 V-0 rating. Figure 12.7c–e shows that the char residues for FR-PAN-5 and FR-PAN-8 were longer than those for FR-PAN-10. Moreover, the total burning time for FR-PAN-5 and FR-PAN-8 was 7 and 5 s, respectively, which was longer than that for FR-PAN-10 (1 s). Furthermore, the LOI for FR-PAN-5, FR-PAN-8, and FR-PAN-10 increased to 23.6%, 29.8%, and 32.5%, respectively, significantly surpassing that of untreated PAN (17%). This indicated that the intumescent char layer had a substantial positive impact on the flame retardancy of FR-PAN (Guo et al. 2023). In comparison to FR-PAN-10, there were only marginal changes in the total burning time, length of char residue, and LOI for FR-PAN-12. This minimal variation could be attributed to the gradual saturation of the reaction between PA and lignin/casein. Consequently, treating the PAN composite with a 10 wt% PA solution appeared to provide a favorable charring effect. It is worth noting that when FR-PAN-10 came into contact with an open flame, it promptly developed a noticeable swelling char layer in the ignition zone. Upon removal of the ignition source, this char layer self-extinguished. This char layer acted as a physical barrier, effectively delaying the spread of flames and reducing the rate of combustion. This phenomenon bestowed excellent flame retardant properties upon FR-PAN-10. In simpler terms, the fully bio-based IFR had the ability to rapidly generate an intumescent char residue in the ignition zone when exposed to flames.

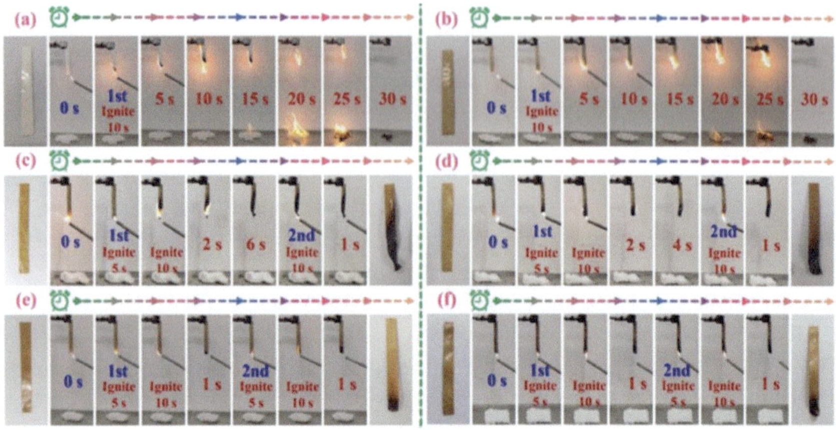

Fig. 12.7 UL-94 test images for various materials: PAN (**a**), L/C/PAN (**b**), FR-PAN-5 (**c**), FR-PAN-8 (**d**), FR-PAN-10 (**e**), and FR-PAN-12 (**f**). Reproduced from Reference [64]

3.1 Flame Retardant Efficiency

Protein-based coatings applied to cotton fabrics accelerate the early thermal breakdown of cellulose, promoting the formation of char instead of depolymerization and the release of volatile compounds [16]. This effect was attributed to the liberation of acidic substances like phosphoric or sulfuric acid. With their effective oxygen barrier attributes and notable water vapor absorption capacity, whey proteins prove to be excellent candidates for flame retardant coatings [15, 16]. A cone calorimeter test was conducted to further explore the impact of a fully bio-based intumescent flame retardant on the flame resistance of FR-PAN. The test results included heat release rate (HRR), total heat release (THR), smoke production rate (SPR), and total smoke production (TSP) curves for both the control and modified PAN materials. Figure 12.8a reveals that PAN easily ignited and rapidly reached its peak heat release rate (PHRR) at 678.3 kW/m^2, indicating poor flame resistance. However, with the inclusion of lignin and casein, the PHRR of L/C/PAN decreased slightly to 559.3 kW/m^2. In contrast, the introduction of PA led to a significant reduction in the PHRR of FR-PAN, down to 194.0 kW/m^2. The total heat release (THR) for pristine PAN, L/C/PAN, and FR-PAN were measured at 11.2, 10.5, and 3.2 MJ/m^2, respectively. Notably, compared to PAN, L/C/PAN showed a slight decrease in THR, while FR-PAN exhibited a substantial reduction. In general, lower PHRR and THR values indicate a lower fire hazard. As depicted in Fig. 12.8a, b, both the PHRR and THR of FR-PAN were reduced by 71.4%, indicating a high level of flame resistance. This significant reduction in PHRR and THR was attributed to the fully bio-based IFR, which promoted the formation of extensive char residue in FR-PAN, effectively blocking thermal transmission.

Figure 12.8c shows that the trend of the SPR curve paralleled that of the HRR curve. The peak of smoke production rate (PSPR) for PAN, L/C/PAN, and FR-PAN

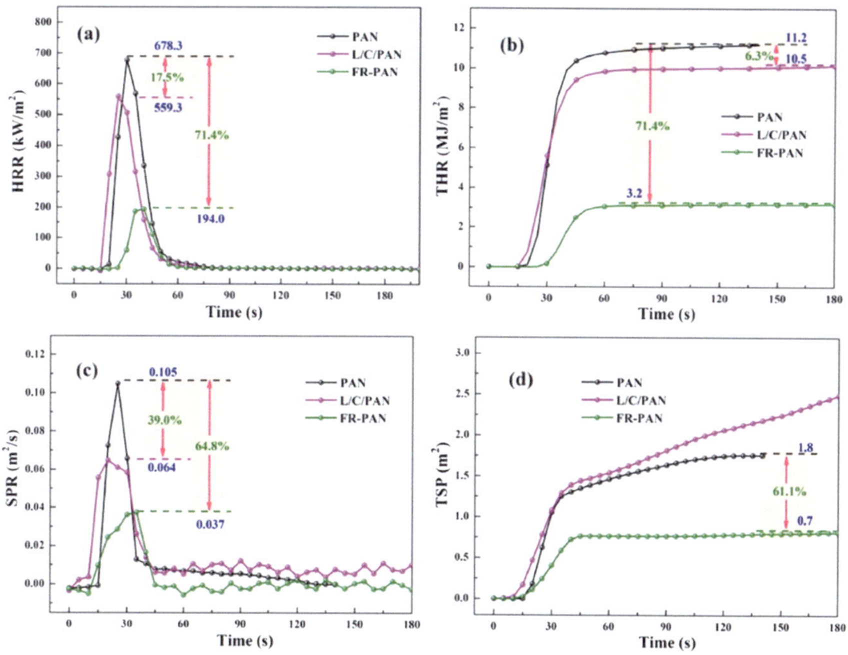

Fig. 12.8 HRR (**a**), THR (**b**), SPR (**c**) and TSP (**d**) curves of PAN, L/C/PAN, and FR-PAN. Reproduced from Reference [64]

were measured at 0.105, 0.064, and 0.037 m^2/s, respectively. Compared to PAN, both L/C/PAN and FR-PAN exhibited reductions of 39.0% and 64.8%, respectively, in PSPR [64]. Figure 12.8d displays the total smoke production (TSP) values for PAN, L/C/PAN, and FR-PAN, which were 1.8, 3.2, and 0.7 m^2, respectively. When compared to PAN, both PSPR and TSP for FR-PAN decreased by 64.8% and 61.1%, respectively, showcasing excellent smoke suppression capabilities. This effect was due to FR-PAN's ability to generate more char residue, effectively inhibiting the release of smoke. These results demonstrate that FR-PAN not only significantly reduced the HRR but also notably suppressed smoke emission, indicating its outstanding fire safety performance.

3.2 *Thermal Stability Enhancement*

One of the challenges facing protein-based flame-retardants is the issue of their thermal stability since they are temperature-sensitive. Under high temperatures, PFRs tend to undergo degradation which could result in the loss of the flame-retardant characteristic. For example, the thermal stability of casein can be attributed to its structure, which is thought to lack stable secondary and tertiary configurations due to the substantial presence of propyl residues [66]. In addition, the phosphoric acid content of casein directly affects its solubility and thermal stability. On the flip side, whey proteins have garnered attention as a flame retardant due to their capacity to act as a barrier against a range of gases, including oxygen, as well as their capability to enhance water absorption through protein hydrolysates [67]. It can be inferred that the effective water-retention capacity of whey proteins plays a significant role in hindering the fire's propagation.

Thus, the need for thermal stability enhancement, different strategies have been incorporated for this purpose. They include chemical modification, combining with nanostructured materials, intumescent coatings and combining PFRs with other materials that have the properties of flame retardancy [50]. The aforementioned methods of enhancing the thermal stability of the PFRs are the same as the different methods of producing PFRs. This indicates that the thermal stability is improved during the production stage. The chemical modification of soybean protein via grafting of surface groups that are heat-resistant was reported by [46]. The protein was reacted with poly(lactic acid) using the Mannich reaction, followed by grafting with dimethyl phosphite using the short reaction process, to produce a phosphate-nitrogen PFR material with improved flame retardant efficiency. This was attributed to the enhanced thermal stability of the phosphate-nitrogen PFR material. The thermal analysis showed that the material started undergoing degradation at temperatures above 206 °C. The fire-safety mechanism included the formation of phosphate-containing compounds at lower temperatures (below 200 °C). These compounds cross-linked on the surface of the PFR material. At higher temperatures (above 200 °C) un-flammable NH_3 was produced on the surface of the protein. This was followed by the formation of that carbon or charred layer at temperatures around 600 °C. The un-flammable NH_3 was responsible for decreasing the temperature of the material and thus, the grafting of the protein showed potential to be used as a flame retardant.

The application of nanostructured PFRs is relatively a recent field and therefore, literature regarding their use is rather limited at the time of writing this book chapter. Regardless, (Ortelli et al. 2018) reported on the combination of TiO_2 nanoparticles with casein and whey protein respectively. The thermal degradation of the proteins was significantly reduced by the TiO_2 nanoparticles as they act as a heat barrier for the proteins. Though the thermal degradation temperatures were not presented, an indication of the improved thermal stability of the nanoparticle-protein materials could be estimated by using the percentage residue and the burning rate. The nanoparticle-protein material was applied to cotton fabric via coating. The cotton without the coating burnt completely and upon coating the residue increased to 5 and 7.9% for whey and 22.4 and 41.3 for the casein. The burning rate decreased by 10–13% for the coated materials compared to the bare cotton. Indicating that the nanoparticle coating does improve the fire safety of the cotton [44]. The fire-safety mechanism involved the formation of a carbon or charred layer. The ability to enhance the thermal stability of protein-based flame-retardants has increased the attraction of such materials. Moreover, the field is still wide open when it comes to material compatibility with other compounds. This includes synthetic polymers such as polyacrylic, polyamide and polyester come with limited functional groups which might limit the application of PFRs. However, prospects could involve functionalizing the aforementioned polymer materials for improved compatibility.

3.3 Environmental Considerations

Environmental and health considerations are also paramount in the choice of proteins for flame retardancy, as the demand for sustainable and non-toxic flame-retardants grows, proteins offer a compelling solution [31, 68]. They are biodegradable and do not introduce harmful chemicals into the environment, making them an attractive option for industries striving to reduce their ecological footprint. This aligns with the increasing focus on eco-friendly alternatives in modern manufacturing and construction practices. The selection of proteins for flame retardancy is a multifaceted decision influenced by their inherent properties, compatibility with materials, and environmental considerations. Proteins offer a promising avenue for developing safer and more sustainable flame-retardant solutions across various industries. As research and innovation continue to advance, the diversity of proteins and their applications in flame retardancy will likely expand, contributing to safer and more environmentally friendly products and processes.

4 Protein-Based Flame Retardants Used in Different Industries

Different industries have prioritized fire safety and have employed PFRs in efforts to mitigate the risk associated with fire outbreaks. The PFRs could be applied to clothing, electrical components or appliances, and building materials. The innovation around the use of PFRs includes not losing the quality of the material that has been incorporated with the PFRs and lesser negative effects on the ecosystem and human health.

4.1 Protein-Based Flame Retardants Used in the Textile Industry

Over the years, accidental fire has had catastrophic consequences on property, wildlife and other life forms. Clothes can at times have the potential to be the ignition point of the fire. A fire that has clothes as an ignition point is close to the victim and thus presents a category 1 (Most dangerous fire to the victim). This is due to the proximity of the fire to the body of the victim, increasing the chances of fatality [69]. Thus, the textile industry has made significant strides in prioritizing fire-safe textiles, especially for firefighters [70]. This has seen the rise in the development of protein-based flame retardants not only as a fire safety measure but also as an environmentally friendly alternative. In the textile industry, PFRs have the potential to be applied in different areas such as apparel and home textiles such as curtains to reduce the risk of a fire. The PFR materials could be incorporated into the fabrics without having to compromise on style and comfort, offering an additional layer of protection.

An assembly of egg white protein and phytic acid was used to enhance the fire safety of cotton fabric that where derived from cellulose by using the double-layer coating method. It was reported that after the fabrics were treated with the protein-based flame retardants they demonstrated that the fabrics took longer to undergo ignition. The heat release rate showed lower values and improved thermal stability, all indicating that the treatment with the PFRs significantly enhanced the fire safety of the cotton fabrics [40]. In another study, a combination of casein and hydrophobins was applied for the enhancement of cotton fabric fire safety by using the homogeneous deposition technique. After treatment, the fabric exhibited higher thermal stability and the ignition time improved by 40%, with the total burning rate decreasing by 35% [14]. The aforementioned studies show that to improve the fire safety of a fabric one should consider thermal properties, time taken to ignition and the burning rate. While the PFRs have shown effective in enhancing the fire safety of fabrics, it is imperative to consider production cost, stability, long-term durability, scalability and environmental impact for a much more sustainable solution to providing fire-safe clothing.

4.2 Protein-Based Flame Retardants in the Construction Industry

The global economy in the past decades has been heavily dependent on urbanisation and industrialization for economic growth [71]. At the heart of urbanisation and industrialisation is the construction industry. Contributing immensely to the development of infrastructure such as homes and buildings that are critical for livelihoods and for economic endeavours [72]. However, ensuring the fire safety of these structures remains vitally important to preserve lives and protect the buildings. It is not a bone of contention that the construction industry utilizes numerous resources, resulting in the generation of large-scale industrial waste [73]. Such sites have the potential to be a breathing ground for fire outbreaks. This has seen protein-based flame retardant (PFR) materials gaining attraction in the construction industry and could be incorporated into parts of the building structure such as insulation, flooring, roofing etc., to mitigate fire outbreaks. Such innovation has the potential to result in fire-smart construction, where even in the case of abandonment of construction sites or buildings the risk of an uncontrollable fire would be mitigated.

In recent years adhesive proteins combined with flame-retardant proteins have been developed to produce durable material for wood coatings. The aforementioned PFR material was produced by the fusion of mussel foot protein or cellulose-binding domain to serve as an adhesive domain. With casein protein serving as the flame-retardant protein material. The fused protein assembly was then coated on different wood materials and tested for flammability. It was reported that the wood showed 40–50% improvement in flame retardancy. The moisture content of the coated and non-coated wood was between 12 and 13.5% an indication that the increase in flame retardancy was attributed to the protein coating rather than different moisture content. The confocal imaging indicated that the protein coating was effective even after rinsing or soaking with water the coated wood materials due to the adhesion characteristics [2]. The combination of adhesive proteins and flame retardant proteins shows potential to be used in the construction industry. What is unique in this system is the adhesive property that enables the coating to adhere to the wood material even after several runs with water and still be flame retardant. This is promising since most of the time constructions or buildings usually are bare and not protected from rain. [36] reported on the use of tannin and soybean protein foam for flame retardancy and insulation in construction materials. To promote synergy and enhance activity the foams were combined with acids such as boric and phosphoric and clay material (montmorillonite). The foams were thermally stable undergoing combustion at temperatures above 150 °C. The LOI was used to determine the flame resistance of the foams. The foams showed an LOI percentage between 35 to 50%, this values indicated that the foams were flame-resistant since they have LOI values greater than 22% for combustible materials and greater than 27% for flame-resistant materials [74].

The hot plate method was used to determine the thermal conductivity of the foams. The foams displayed higher thermal conductivity than the control material (0.0391 W/m K). This was attributed to the foams having higher densities than the control material. Overall the foams exhibited significant flame retardancy properties attributed to the successful crosslinking between the tannin and the soybean proteins and the low emission of formaldehyde is an indication of an environmentally friendly material. The attributes of retardancy and environmentally friendliness of the foams derived from protein-based flame-retardants are promising for the construction industry. However, questions of production cost and scalability remain for such materials to be applied at an industrial scale in efforts to provide fire-safe and environmentally friendly infrastructure.

4.3 *Protein-Based Flame Retardants Used in the Electronics Industry*

Throughout the years the electronic industry has grown exponentially, with devices becoming more powerful than before and forming an inseparable part of human existence. That does not exempt electronic components from the risk of fire and thus the need to improve fire safety in the electronic industry. This is because electrical components or devices have been identified as ignition points for fire or as fuel for a fire [75]. At most the electronic industry is dependent on different types of plastics or polymers for their components and devices, these materials are flammable [76]. Thus, the development of protein-based flame retardants to overcome the fire risk that comes with electronic devices or components. PFRs could be incorporated into the plastic or polymer used to manufacture electrical components and have fire safety properties. Guna et al. reported on the fabrication of novel materials that were used as printed circuit boards. These materials were derived from banana fibres and wheat gluten protein. The banana fibres were carded into a mat and the wheat gluten was sprayed onto the fibres at different fibre: protein ratios to form composites. The resulting composites showed considerable flame-resistant properties, though the composites underwent combustion the fire self-extinguished.

After combustion, the composites did not melt but rather were converted to ash. This was said to be desirable as melting could promote burning and damage to other components. Not only did the composites exhibit good fire safety properties but also good electronic properties [77]. The composites were biodegradable indicating that they are an alternative since they are more environmentally friendly. A similar method of producing printed circuit boards (PCBs) from a combination of protein materials and other polymers was used by [78]. It was reported that a combination of soybean, chicken feather fibre, E-glass fibre, melamine polyphosphate and diethylphosphinic salt were used to fabricate bio-composites at different ratios. The composites were fabricated with a copper foil to enable electronic application. The composites showed improved thermal stability this was demonstrated by thermal analysis as the composites underwent combustion at temperatures above 300 °C. The composites exhibited comparable mechanical, flammable and electronic properties. They, however, demonstrated undesirable peel strength as was shown by the peel strength values around 700 (N/m) against displacement for the materials without flame retardant materials. Whilst the composites exhibited peel strength of close to 600 (N/m) against displacement.

This presented an opportunity to significantly improve the properties of the protein-based flame-retardants to enable large-scale application. The PFRs show considerable compatibility with electronic components without hindering the optimum functioning of the components. Thus, efforts should be in the development of PFRs powered electronics for an environmentally friendly, sustainable, fire-safety and efficient electronic device.

5 Conclusions

Protein-based flame-retardants have shown potential application in industries such as textile, construction, electronics and transportation. They can be produced via different methods such as chemical modification and combining with nanostructured materials. Such production methods enhance their physical, chemical and mechanical properties which enable them to be effectively used as flame retardants. Some of the techniques used for characterization and testing include but are not limited to SEM, FTIR, TGA, and LOI.

They have an advantage over traditional flame-retardants due to their biodegradability, release of less smoke, and abundance in nature. Making them effective in fire safety and less harmful to the ecosystem. Their challenges include compatibility with materials that have limited functional groups. The production of protein-based flame-retardants has not fully extended to large-scale industrial production. This is mainly because of issues relating to infrastructure and production costs. Testing of adhesion is currently lacking as the current method includes soaking the materials in water for several minutes or dipping them in water several times. What happens when the material is in contact with water for an extended period?

The release of formaldehyde compounds and the leaching of nanoparticles into the environment may result in ecological harm, especially with large-scale production. Be that as it may, the production and application of protein-based flame retardants have opened up the field of fire safety. Future prospects may move in the direction of improving the material's durability, resistance, stability, sustainability and toxicity. Scientists are also working on developing smart protein-based coatings that can be applied to various surfaces. These coatings can remain dormant until exposed to heat or flames, at which point they activate and form a protective barrier. This innovation ensures that flame retardancy is only triggered when needed, reducing the overall environmental impact and extending the lifespan of the material. The future of protein-based flame-retardants holds great promise, with ongoing innovations aimed at improving their performance, versatility, and sustainability. These developments align with the growing demand for safer and more eco-friendly flame retardant solutions in industries ranging from textiles and construction to electronics and transportation. As scientific knowledge and technology continue to advance, we can expect protein-based flame-retardants to play a pivotal role in enhancing fire safety while minimizing environmental impact.

References

1. Mazela, B., Batista, A., Grześkowiak.: Expandable graphite as a fire retardant for cellulosic materials—a review. Forests **11**, 1–10 (2020). https://doi.org/10.3390/f11070755
2. Leong, W.I., Lo, O.L.I., Cheng, F.T., Cheong, W.M., Seak, L.C.U.: Using recombinant adhesive proteins as durable and green flame-retardant coatings. Synthetic Syst. Biotech. **6**, 369–376 (2021). https://doi.org/10.1016/j.synbio.2021.10.005
3. Xu, F., Zhong, L., Zhang, C., Wang, P., Zhang, F., Zhang, G.: Novel high-efficiency casein-based P–N-containing flame retardants with multiple reactive groups for cotton fabrics. ACS Sustain. Chem. Eng. **7**, 13999–14008 (2019). https://doi.org/10.1021/acssuschemeng.9b02474
4. Wu, Q., Andersson, R. L., Holgate, T., Johansson, E., Gedde, U.W., Olsson, R.T., Hedenqvist, M.S.: Highly porous flame-retardant and sustainable biofoams based on wheat gluten and in situ polymerized silica. J. Mater. Chem. A **2**, 20996–21009 (2014). https://doi.org/10.1039/c4ta04787g

5. Wang, Y., Liu, C., Lai, J., Lu, C., Wu, X., Cai, Y., Gu, L., Yang, L., Zhang, G., Shi, G.: Soy protein and halloysite nanotubes-assisted preparation of environmentally friendly intumescent flame retardant for poly(butylene succinate). Polym. Testing **81**, 106174 (2020). https://doi.org/10.1016/j.polymertesting.2019.106174
6. Zhu, F., Chen, L., Feng, Q.: Waste gelatin based layer by layer assembly for sustainable solution to cotton fabrics flame retardancy. Prog. Organ. Coatings **163**, 106688 (2022). https://doi.org/10.1016/j.porgcoat.2021.106688
7. Malucelli, G.: Biomacromolecules and bio-sourced products for the design of flame retarded fabrics: current state of the art and future perspectives. Molecules **24**, 1–27 (2019). https://doi.org/10.3390/molecules24203774
8. Xinsheng, H., Yang, G., Zhang, Q.: Application status and development prospects of bio-based flame retardants in packaging materials. Eur. J. Wood Wood Prod. (2023). https://doi.org/10.1007/s00107-023-01977-w
9. Sonnier, R., Taguet, A., Ferry, L., Lopez-Cuesta, J.-M.: Biobased flame retardants. 33–72 (2018)
10. Camino, G., Costa, L., Luda di Cortemiglia, M.P.: Overview of fire retardant mechanisms. Polym. Degrad. Stab. **33**, 131–154 (1991). https://doi.org/10.1016/0141-3910(91)90014-I
11. Malucelli, G., Bosco, F., Alongi, J., Carosio, F., Di Blasio, A., Mollea, C., Cuttica, F., Casale, A.: Biomacromolecules as novel green flame retardant systems for textiles: an overview. RSC Adv. 4, 46024–46039 (2014). https://doi.org/10.1039/c4ra06771a
12. Blanco, A., Blanco, G.: Proteins. 21–71 (2017)
13. Carosio, F., Di Blasio, A., Cuttica, F., Alongi, J., Malucelli, G.: Flame retardancy of polyester and polyester-cotton blends treated with caseins. Indus. Eng. Chem. Res. **53**, 3917–3923 (2014). https://doi.org/10.1021/ie404089t
14. Alongi, J., Carletto, R.A., Bosco, F., Carosio, F., Di Blasio, A., Cuttica, F., Antonucci, V., Giordano, M., Malucelli, G.: Caseins and hydrophobins as novel green flame retardants for cotton fabrics. Polym. Degrad. Stab. **99**(1), 111–117 (2014). https://doi.org/10.1016/j.polymdegradstab.2013.11.016
15. Bosco, F., Carletto, R.A., Alongi, J., Marmo, L., Di Blasio, A., Malucelli, G.: Thermal stability and flame resistance of cotton fabrics treated with whey proteins. Carbohyd. Polyme. **94**, 372–377 (2013). https://doi.org/10.1016/j.carbpol.2012.12.075
16. Costes, L., Laoutid, F., Brohez, S., Dubois, P.: Bio-based flame retardants: When nature meets fire protection. Mater. Sci. Eng. R Rep. **117**, 1–25 (2017). https://doi.org/10.1016/j.mser.2017.04.001
17. Alongi, J., Cuttica, F., Blasio, A. Di, Carosio, F., Malucelli, G.: Intumescent features of nucleic acids and proteins. Thermochimica Acta **591**, 31–39 (2014). https://doi.org/10.1016/j.tca.2014.06.020
18. Alongi, J., Carosio, F., Malucelli, G.: Current emerging techniques to impart flame retardancy to fabrics: an overview. Polym. Degrad. Stab. **106**, 138–149 (2014). https://doi.org/10.1016/j.polymdegradstab.2013.07.012
19. Kim, N.K., Dutta, S., Bhattacharyya, D.: A review of flammability of natural fibre reinforced polymeric composites. Compos. Sci. Technol. **162**, 64–78 (2018). https://doi.org/10.1016/j.compscitech.2018.04.016
20. Song, F., Tang, D.-L., Wang, X.-L., Wang, Y.-Z.: Biodegradable soy protein isolate-based materials: a review. Biomacromolecules **12**, 3369–3380 (2011). https://doi.org/10.1021/bm200904x
21. Baba, W.N., McClements, D.J., Maqsood, S.: Whey protein–polyphenol conjugates and complexes: production, characterization, and applications. Food Chem. **365**, 130455 (2021). https://doi.org/10.1016/j.foodchem.2021.130455
22. Etzel, M.R.: Manufacture and use of dairy protein fractions. J. Nutr. **134**, 996S (2004). https://doi.org/10.1093/jn/134.4.996s
23. Basak, S., Ali, S.W.: Sustainable fire retardancy of textiles using bio-macromolecules. Polym. Degrad. Stab. **133**, 47–64 (2016). https://doi.org/10.1016/j.polymdegradstab.2016.07.019

24. De Wit, J.N.: Nutritional and functional characteristics of whey proteins in food products. J. Dairy Sci. **81**, 597–608 (1998). https://doi.org/10.3168/jds.S0022-0302(98)75613-9
25. Chen, H.: Formation and properties of casein films and coatings, pp. 181–209 (2002)
26. Elzoghby, A.O., Helmy, M.W., Samy, W.M., Elgindy, N.A.: Spray-dried casein-based micelles as a vehicle for solubilization and controlled delivery of flutamide: formulation, characterization, and in vivo pharmacokinetics. Eur. J. Pharm. Biopharm. **84**, 487–496 (2013). https://doi.org/10.1016/j.ejpb.2013.01.005
27. Liu, Y., Liu, L., Yuan, M., Guo, R.: Preparation and characterization of casein-stabilized gold nanoparticles for catalytic applications. Colloids Surf. A: Physicochem. Eng. Aspects **417**, 18–25 (2013). https://doi.org/10.1016/j.colsurfa.2012.08.050
28. Razmi, M., Divsalar, A., Saboury, A.A., Izadi, Z., Haertlé, T., Mansuri-Torshizi, H.: Beta-casein and its complexes with chitosan as nanovehicles for delivery of a platinum anticancer drug. Colloids Surf. A Biointerfaces **112**, 362–367 (2013). https://doi.org/10.1016/j.colsurfb.2013.08.022
29. Sinha, S., Mishra, S., Sen, G.: Microwave initiated synthesis of polyacrylamide grafted Casein (CAS-g-PAM)-its application as a flocculant. Int. J. Biol. Macromol. **60**, 141–147 (2013). https://doi.org/10.1016/j.ijbiomac.2013.05.012
30. Ma, J., Xu, Q., Zhou, J., Gao, D., Zhang, J., Chen, L.: Nano-scale core-shell structural casein based coating latex: synthesis, characterization and its biodegradability. Prog. Organ. Coatings. **76**, 1346–1355 (2013). https://doi.org/10.1016/j.porgcoat.2013.04.006
31. Mensah, R.A., Shanmugam, V., Narayanan, S., Renner, J.S., Babu, K., Neisiany, R.E., Försth, M., Sas, G., Das, O.: A review of sustainable and environment-friendly flame retardants used in plastics. Polym. Testing **108**, 107511 (2022). https://doi.org/10.1016/j.polymertesting.2022.107511
32. Xu, Y., Han, Y., Li, Y., Li, J., Li, J., Gao, Q.: Preparation of a strong, mildew-resistant, and flame-retardant biomimetic multifunctional soy protein adhesive via the construction of an organic-inorganic hybrid multiple-bonding structure. Chem. Eng. J. **437**, 135437 (2022). https://doi.org/10.1016/j.cej.2022.135437
33. Nakamura, R., Netravali, A.N., Morgan, A.B., Nyden, M.R., Gilman, J.W.: Effect of halloysite nanotubes on mechanical properties and flammability of soy protein based green composites. Fire and Mater. **37**, 75–90 (2013). https://doi.org/10.1002/fam.2113
34. Xu, Y., Zhang, X., Liu, Z., Zhang, X., Luo, J., Li, J., Shi, S.Q., Li, J., Gao, Q.: Constructing SiO_2 nanohybrid to develop a strong soy protein adhesive with excellent flame-retardant and coating ability. Chem. Eng. J. **446**, 137065 (2022). https://doi.org/10.1016/j.cej.2022.137065
35. Zeng, Y., Yang, W., Xu, P., Cai, X., Dong, W., Chen, M., Du, M., Liu, T., Jan Lemstra, P., Ma, P.: The bonding strength, water resistance and flame retardancy of soy protein-based adhesive by incorporating tailor-made core–shell nanohybrid compounds. Chem. Eng. J. **428**, 132390 (2022). https://doi.org/10.1016/j.cej.2021.132390
36. Chen, X., Li, J., Essawy, H., Pizzi, A., Fredon, E., Gerardin, C., Du, G., Zhou, X.: Flame-retardant and thermally-insulating tannin and soybean protein isolate (SPI) based foams for potential applications in building materials. Constr. Build. Mater. **315**, 125711 (2022). https://doi.org/10.1016/j.conbuildmat.2021.125711
37. Xu, Y., Zhang, X., Wang, G., Zhang, X., Luo, J., Li, J., Shi, S.Q., Li, J., Gao, Q.: Preparation of a strong soy protein adhesive with mildew proof, flame-retardant, and electromagnetic shielding properties via constructing nanophase-reinforced organic–inorganic hybrid structure. Chem. Eng. J. **447**, 137536 (2022). https://doi.org/10.1016/j.cej.2022.137536
38. Sanchez-Salvador, J.L., Balea, A., Monte, M.C., Negro, C., Blanco, A.: Chitosan grafted/cross-linked with biodegradable polymers: a review. Int. J. Biol. Macromol. **178**, 325–343 (2021). https://doi.org/10.1016/j.ijbiomac.2021.02.200
39. Wen, O.Y., Tohir, M.Z.M., Yeaw, T.C.S., Razak, M.A., Zainuddin, H.S., Hamid, M.R.A.: Fire-resistant and flame-retardant surface finishing of polymers and textiles: a state-of-the-art review. Prog. Organ. Coatings **175**, 107330 (2023). https://doi.org/10.1016/j.porgcoat.2022.107330
40. Liu, X., Zhang, Q., Peng, B., Ren, Y., Cheng, B., Ding, C., Su, X., He, J., Lin, S.: Flame retardant cellulosic fabrics via layer-by-layer self-assembly double coating with egg white protein and phytic acid. J. Clean. Prod. **243**, 118641 (2020). https://doi.org/10.1016/j.jclepro.2019.118641

41. Gu, W., Li, F., Liu, X., Gao, Q., Gong, S., Li, J., Shi, S.Q.: Borate chemistry inspired by cell walls converts soy protein into high-strength, antibacterial, flame-retardant adhesive. Green Chem. **22**, 1319–1328 (2020). https://doi.org/10.1039/C9GC03875B
42. Cobos, M., Pagalday, E., Puyadena, M., Cabido, X., Martin, L., Múgica, A., Irusta, L., González, A.: Waterborne hybrid polyurethane coatings containing Casein as sustainable green flame retardant through different synthesis approaches. Prog. Organ. Coatings **174**, 107278 (2023). https://doi.org/10.1016/j.porgcoat.2022.107278
43. Zhou, J., Dong, F., Liu, C., Lu, K.: P, N, Si synergistic flame-retarding water polyurethane coating with superior flame retardancy and hydrophobicity. J. Polym. Res. **30**, 260 (2023). https://doi.org/10.1007/s10965-023-03635-4
44. Ortelli, S., Malucelli, G., Cuttica, F., Blosi, M., Zanoni, I., Costa, A.L.: Coatings made of proteins adsorbed on TiO_2 nanoparticles: a new flame retardant approach for cotton fabrics. Cellulose **25**, 2755–2765 (2018). https://doi.org/10.1007/s10570-018-1745-z
45. Zhu, Y., Vanga, S.K., Wang, J., Raghavan, V.: Effects of ultrasonic and microwave processing on avidin assay and secondary structures of egg white protein. Food Bioprocess Technol. **11**, 1974–1984 (2018). https://doi.org/10.1007/s11947-018-2158-6
46. Dong, L., Xue, Y., Huang, H., Shen, D., Gao, W., Xu, F., Weng, Y., Zhang, Y.: Facile synthesis of soybean protein-based phosphorus-nitrogen flame retardant for poly(lactic acid). Polym. Degrad. Stab. **214**, 110412 (2023). https://doi.org/10.1016/j.polymdegradstab.2023.110412
47. Li, K., Jin, S., Jiang, S., Li, X., Li, J., Shi, S.Q., Li, J.: Bioinspired mineral–organic strategy for fabricating a high-strength, antibacterial, flame-retardant soy protein bioplastic via internal boron–nitrogen coordination. Chem. Eng. J. **428**, 132616 (2022). https://doi.org/10.1016/j.cej.2021.132616
48. Fang, Y., Liu, X., Wang, C.: Layer-by-layer assembly flame-retardant and anti-dripping treatment of polyethylene terephthalate fabrics. J. Eng. Fibers Fabr. **14**, 1–8 (2019). https://doi.org/10.1177/1558925019870301
49. Wang, X., Wang, W., Wang, S., Yang, Y., Li, H., Sun, J., Gu, X., Zhang, S.: Self-intumescent polyelectrolyte for flame retardant poly (lactic acid) nonwovens. J. Clean. Prod. **282**, 124497 (2021). https://doi.org/10.1016/j.jclepro.2020.124497
50. Kim, N. K., Bruna, F.G., Das, O., Hedenqvist, M. S., Bhattacharyya, D.: Fire-retardancy and mechanical performance of protein-based natural fibre-biopolymer composites. Compos. Part C: Open Access **1**, 100011 (2020). https://doi.org/10.1016/j.jcomc.2020.100011
51. Wang, X., Lu, C., Chen, C.: Effect of chicken-feather protein-based flame retardant on flame retarding performance of cotton fabric. J. Appl. Polym. Sci. **131**, 40584 (2014). https://doi.org/10.1002/app.40584
52. Vo, D. K., Do, T. D., Nguyen, B. T., Tran, C. K., Nguyen, T. A., Nguyen, D. M., Pham, L. H., Nguyen, T. D., Nguyen, T. D., Hoang, D. Q.: Effect of metal oxide nanoparticles and aluminum hydroxide on the physicochemical properties and flame-retardant behavior of rigid polyurethane foam. Constr. Build. Mater. **356**, 129268 (2022). https://doi.org/10.1016/j.conbuildmat.2022.129268
53. Mocanu, A.M., Moldoveanu, C., Odochian, L., Paius, C.M., Apostolescu, N., Neculau, R.: Study on the thermal behavior of casein under nitrogen and air atmosphere by means of the TG-FTIR technique. Thermochimica Acta **546**, 120–126 (2012). https://doi.org/10.1016/j.tca.2012.07.031
54. Song, F., Zhao, Q., Zhu, T., Bo, C., Zhang, M., Hu, L., Zhu, X., Jia, P., Zhou, Y.: Biobased coating derived from fish scale protein and phytic acid for flame-retardant cotton fabrics. Mater. Des. **221**, 110925 (2022). https://doi.org/10.1016/j.matdes.2022.110925
55. Yew, M.C., Ramli Sulong, N.H., Yew, M.K., Amalina, M.A., Johan, M.R.: Eggshells: a novel bio-filler for intumescent flame-retardant coatings. Prog. Organ. Coatings **81**, 116–124 (2015). https://doi.org/10.1016/j.porgcoat.2015.01.003
56. Ling, C., Guo, L., Wang, Z.: A review on the state of flame-retardant cotton fabric: mechanisms and applications. Indus. Crops Prod. **194**, 116264 (2023). https://doi.org/10.1016/j.indcrop.2023.116264

57. Tian, C.M., Li, Z., Guo, H.Z., Xu, J.Z.: Study on the thermal degradation of flame retardant wools. J. Fire Sci. **21**, 155–162 (2003). https://doi.org/10.1177/0734904103021002004
58. Patankar, K.C., Maiti, S., Singh, G.P., Shahid, M., More, S., Adivarekar, R. V: Chemically modified wool waste keratin for flame retardant cotton finishing. Clean. Eng. Technol. **5**, 100319 (2021). https://doi.org/10.1016/j.clet.2021.100319
59. Alongi, J., Carletto, R.A., Di Blasio, A., Cuttica, F., Carosio, F., Bosco, F., Malucelli, G.: Intrinsic intumescent-like flame retardant properties of DNA-treated cotton fabrics. Carbohydr. Polym. **96**, 296–304 (2013). https://doi.org/10.1016/j.carbpol.2013.03.066
60. Carosio, F., Di Blasio, A., Alongi, J., Malucelli, G.: Green DNA-based flame retardant coatings assembled through Layer by Layer. Polymer (Guildf) **54**, 5148–5153 (2013). https://doi.org/10.1016/j.polymer.2013.07.029
61. Meng, D., Guo, J., Wang, A., Gu, X., Wang, Z., Jiang, S., Zhang, S.: The fire performance of polyamide66 fabric coated with soybean protein isolation. Prog. Organ. Coatings **148**, 105835 (2020). https://doi.org/10.1016/j.porgcoat.2020.105835
62. Zhang, Y., Wang, J., Sun, J., Zhang, Y., Chen, Y.: Protein-based flame retardants for cotton fabrics: a review. **43**, 1–18 (2019)
63. Guan, J.-P., Chen, G.-Q.: Flame retardancy finish with an organophosphorus retardant on silk fabrics. Fire Mater. Int. J. **30**, 415–424 (2006). https://doi.org/10.1002/fam.920
64. Guo, Y., Zuo, C., Liu, Y., Chen, X., Ren, Y., Liu, X.: Construction of a fully bio-based intumescent flame retardant for improving the flame retardancy of polyacrylonitrile. Polym. Degrad. Stab. **214**, 110385 (2023). https://doi.org/10.1016/j.polymdegradstab.2023.110385
65. Xu, Z., Chu, Z., Yan, L., Chen, H., Jia, H., Tang, W.: Effect of chicken eggshell on the flame-retardant and smoke suppression properties of an epoxy-based traditional APP-PER-MEL system. Polym. Compos. **40**, 2712–2723 (2019). https://doi.org/10.1002/pc.25077
66. Bhat, M.Y., Dar, T.A., Singh, L.R.: Casein proteins: structural and functional aspects. 1–17 (2016). https://doi.org/10.5772/64187
67. Ito, K., Yamaguchi, M., Noma, T., Yamaji, T., Itoh, H., Oda, M.: Whey protein hydrolysates enhance water absorption in the perfused small intestine of anesthetized rats. Biosci. Biotechnol. Biochem. **80**, 1587–1593 (2016). https://doi.org/10.1080/09168451.2016.1166931
68. Watson, D.A.V., Schiraldi, D.A.: Biomolecules as flame retardant additives for polymers: a review. Polymers (Basel) **12**, 849 (2020). https://doi.org/10.3390/POLYM12040849
69. Runefors, M., Jonsson, A., Bonander, C.: Factors contributing to survival and evacuation in residential fires involving older adults in Sweden. Fire. Saf. J. **122**, 103354 (2021). https://doi.org/10.1016/j.firesaf.2021.103354
70. Mandal, S., Song, G.: Thermal sensors for performance evaluation of protective clothing against heat and fire: a review. Textile Res. J. **85**, 101–112 (2015). https://doi.org/10.1177/0040517514542864
71. Khan, S., Majeed, M. T.: Toward economic growth without emissions growth: the role of urbanization & industrialization in Pakistan. J. Environ. Stud. Sci. **13**, 43–58 (2023). https://doi.org/10.1007/s13412-022-00797-3
72. Norouzi, M., Chàfer, M., Cabeza, L. F., Jiménez, L., Boer, D.: Circular economy in the building and construction sector: a scientific evolution analysis. J. Build. Eng. **44**, 102704 (2021). https://doi.org/10.1016/j.jobe.2021.102704
73. Dumo, M.B.: Construction project abandonment: An appraisal of causes, effects and remedies. World J. Innov. Modern Technol. **1**, 1–10 (2017)
74. Chen, X., Li, J., Pizzi, A., Fredon, E., Gerardin, C., Zhou, X., Du, G.: Tannin-furanic foams modified by soybean protein isolate (SPI) and industrial lignin substituting formaldehyde addition. Indus. Crops Prod. **168**, 113607 (2021). https://doi.org/10.1016/j.indcrop.2021.113607
75. Salvi, U., Lattimer, B.Y., Alhadhrami, S., Wang, J., Sahin, E., Duarte, J.P.: Analysis of historic fires to determine most frequent challenging events. Prog. Nucl. Energy **146**, 104146 (2022). https://doi.org/10.1016/j.pnucene.2022.104146
76. Vouvoudi, E.C., Rousi, A.T., Achilias, D.S.: Thermal degradation characteristics and products obtained after pyrolysis of specific polymers found in waste electrical and electronic equipment. Front. Environ. Sci. Eng. **11**, 1–10 (2017). https://doi.org/10.1007/s11783-017-0996-5

77. Guna, V.K., Murugesan, G., Basavarajaiah, B.H., Ilangovan, M., Olivera, S., Krishna, V., Reddy, N.: Plant-based completely biodegradable printed circuit. Boards **63**, 4893–4898 (2016). https://doi.org/10.1109/TED.2016.2619983
78. Zhan, M., Wool, R.P.: Design and evaluation of bio-based composites for printed circuit board application. Compos. Part A: Appl. Sci. Manuf. **47**, 22–30 (2013). https://doi.org/10.1016/j.compositesa.2012.11.014

DNA: Future Bio-Macromolecule Flame Retardant

Giulio Malucelli

Abstract The search for novel efficient flame retardants, showing low environmental impact, and exploiting bio-derived sources, has pushed academia towards the exploration and exploitation of new performing systems. In the last decade, biomacromolecules and bio-derived products have been thoroughly investigated for their potential fireproof properties. In this quest therefore, deoxyribonucleic acid (DNA) has clearly demonstrated its suitability for protecting different polymer substrates from exposure to a direct flame or an irradiative heat source. Undoubtedly, the structure of DNA and nuclei acids mimics that of an intumescent system, as it consists of nitrogen-containing bases (i.e., adenine, cytosine, guanine, and thymine), which can release ammonia, (poly)saccharide deoxyribose units that behave as carbon sources and blowing systems (upon heating they undergo dehydration, promoting the release of water and favoring charring), and phosphate groups that can give rise to phosphoric acid, acting as char former. This chapter reviews the current state of the art about the use of DNA as an effective flame retardant for different types of polymers (namely, bulk thermoplastics and thermosets, textiles, and foams). Finally, some of the present limitations of the use of these biomacromolecules will be discussed, as well as some viewpoints for the forthcoming years.

Keywords Deoxyribonucleic acid (DNA) · Flame retardance · Flammability tests · Forced-combustion tests · Bulk polymers · Textiles · Foams

1 Introduction

Polymeric materials, if not intrinsically flame-retarded, can easily burn, when exposed to an irradiative heat source or a flame. This characteristic makes them unsuitable for all the applications, in which fireproofing is strongly demanded, not only for avoiding fire spread, but also for limiting the emission of hazardous and toxic

G. Malucelli (✉)
Department of Applied Science and Technology, Politecnico Di Torino, Viale Teresa, Michel 5, 15121 Alessandria, Italy
e-mail: giulio.malucelli@polito.it

T. C. Mokhena et al. (eds.), *Biomaterials as Green Flame Retardants*, Engineering Materials, https://doi.org/10.1007/978-981-97-6871-4_13

smokes. The focus on this issue started in the 1950s, with the design, synthesis, and application of flame-retardants (FRs). These are products that are added to a polymer system and can lower its ease of flammability by decreasing the susceptibility to ignition or reducing the flame spread rate when needed [1].

From a practical point of view, flaming combustion involves oxidative reactions occurring in the gas phase, and requires the presence of oxygen/air in the neighboring atmosphere. Flaming combustion is always anticipated by the degradation of the polymer system, which originates combustible gaseous products that combine with the atmospheric oxygen. The exothermicity of the flame may promote a combustion cycle (Fig. 1), which is self-sustaining when enough heat is transferred to the surface of the polymer.

The progress in the development of effective flame retardants for polymer systems (suitable for stopping the combustion cycle of Fig. 1), has been very impressive and continuously growing. Besides, it involves, in different research periods, the synthesis "on purpose" of specific products, based on halogens, nitrogen, boron, metals, phosphorus, sulfur, either selected as single specific elements or in combination, aiming, in the latter case, at assessing possible joint or even synergistic effects [2, 3].

Different specific characteristics make the flame retardants effective in stopping the self-sustaining combustion cycle, addressing their action towards flame extinction or, at least, determining a lowered burning rate to the polymer. Firstly, the flame retardant should be active in decreasing the heat evolved below the amount needed for supporting the combustion process. In this view, the flame retardant should also reduce the heat flow that is redirected to the surface of the underlying polymer, hence restricting the pyrolysis reactions that occur in the polymer system and allowing

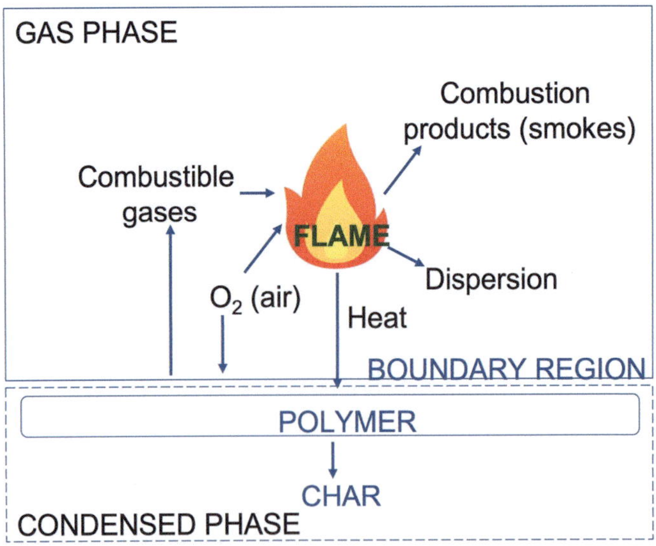

Fig. 1 Self-sustaining combustion cycle for polymers

for the formation of combustible volatile decomposition products. Then, the flame retardant should lead the polymer pyrolysis toward forming a stable, protective, and coherent char, rather than developing combustible gaseous products fueling the flame. Specifically referring to the gas phase action, the flame retardant should also favor the release of flame inhibitors, able to entrap the H and OH radicals that are responsible for the propagation of the combustion reactions.

From a historical point of view, the design and development of flame retardants started with the synthesis of halogenated systems (mainly chlorinated or brominated), phosphorus-, and organo-phosphorus- based systems (from 1950 to 1980s). Then, from 2000s onwards, it evolved toward the design of intumescent flame retardants, often coupled with char-forming nanomaterials (such as nanoparticles and nanoclays), exploiting the presence of such elements as phosphorus and/or nitrogen (and sulfur). The severe issues related to the toxicity (and even mutagenic character) of some halogen-based flame retardants (namely, polychlorinated biphenyls, pentabromodiphenyl and decabromodiphenyl ethers, among others) have further stimulated the academic community toward the seeking for new products, exhibiting high effectiveness, low environmental impact, and low toxicity/hazardousness as well [4–8].

In this context, several efforts have been and are still being made in the field of biomacromolecules and bio-sourced products/extracts (such as, whey proteins, caseins, DNA and nucleic acids, banana pseudo-stem sap, phytic acid, lignin, and pomegranate rind extract, among a few to mention) to be employed as effective flame retardants for bulk polymers, textiles, or even foamed materials. In fact, the "chemistry" of these "green" products accounts for the improved resistance to an irradiative heat flux or to flame propagation. In addition, the usage of these biomacromolecules or bio-derived products well matches the current circular economy concept [9], as some of them are wastes or by-products deriving from the agri-food sector: therefore, their potential use for flame retardant purposes undoubtedly represents a re-valorization at the end of life of these products. After the first pioneering papers dealing with the flame retardance provided by some proteins (namely whey proteins [10] and caseins [11], which are recovered from milk), and hydrophobins (i.e., cysteine-rich proteins extracted from filamentous fungi) [12], the attention was moved to deoxyribonucleic acid (DNA), the hereditary material that is present in humans and almost all living organisms and carries specific genetic information. The latter is stored as a code consisting of four chemical bases, namely adenine, guanine, cytosine, and thymine. The order (also called "sequence") of the bases (paired up with each other, i.e., adenine with thymine and cytosine with guanine), defines the information available for the buildup and support of an organism. Each base is also linked to a sugar structure and a phosphate group as well. Nucleotides, i.e., the results of the combination of a base, sugar, and phosphate are organized in two extended strands, forming a double helix, as shown in Fig. 2.

Up to now, the propensity of DNA to give rise to double-stranded structures has already been employed for obtaining many DNA-based nanomaterials, which can be utilized for different applications, ranging from life science, computing [14, 15], genomics [16], to environmental monitoring and bio-based sensors [17].

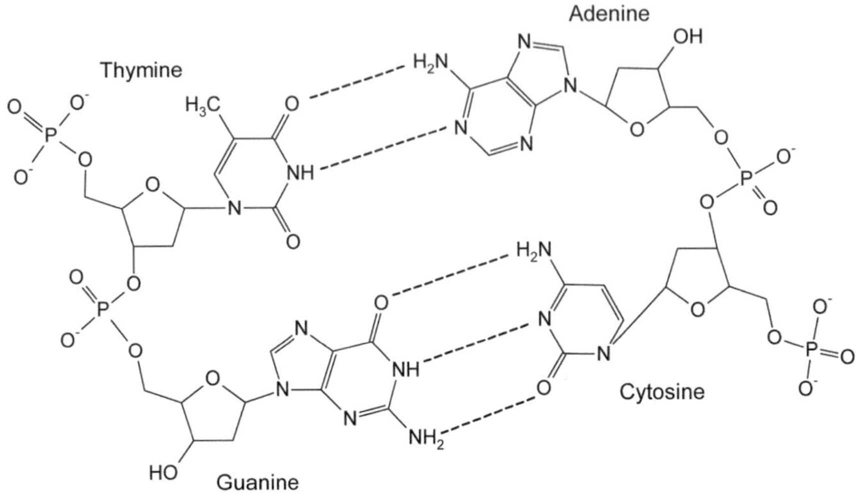

Fig. 2 Molecular structure of deoxyribonucleic acid (DNA). Reproduced with permission from [13]. Copyright Elsevier, 2017

Apart from these "conventional" uses, the chemical structure of DNA is intriguing as it mimics that of an intumescent flame retardant formulation: in fact, the nitrogen-containing bases are responsible for ammonia release (and therefore for the swelling phenomena occurring during the application of a direct flame or an irradiative heat source), the (poly)saccharide deoxyribose units perform both as carbon source (by means of auto-crosslinking reactions that originate stable aromatic structures) and blowing agents (upon heating they undergo dehydration, hence releasing water and making char); finally, the phosphate groups decompose to phosphoric acid (that is the acid source of the intumescent system), acting as charring agents.

The use of DNA as an efficient flame retardant has been the focus of a significant number of scientific publications, as witnessed by the quite intense research activities that have been carried out in the last ten years (Fig. 3). Therefore, this chapter is aimed at describing the general characteristics of DNA as a flame retardant (i.e., its thermal and thermo-oxidative stability, and combustion behavior), providing the reader with some successful examples of its use as an efficient fireproof additive (alone or in combination with other products) for bulk polymers, textiles, or foams. Finally, the current and challenging limitations of the use of the biomacromolecule will be discussed, together with some perspectives for the next future.

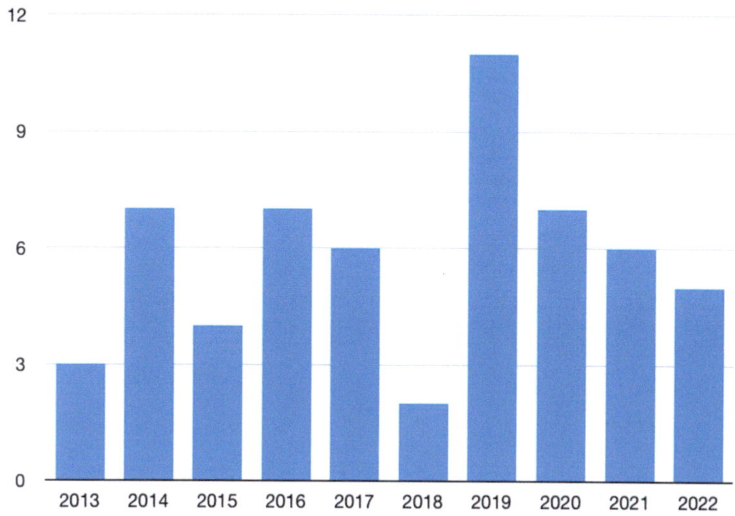

Fig. 3 Number of publications (from 2013 to 2022) in peer-reviewed journals, dealing with flame retardant DNA (data collected from Web of Science™ database, accessed on April 5, 2023)

2 Thermal and Thermo-Oxidative Behavior of DNA

Before describing how DNA affects the reaction to a fire of a polymer, it is worth summarizing how the biomacromolecule decomposes both in an inert and oxidative atmosphere. To this aim, Alongi and co-workers [18] performed thermogravimetry coupled with infrared spectroscopy or combined with differential thermal analysis. Figure 4 shows the results from thermogravimetric analyses, carried out in nitrogen and air atmosphere, by using thermogravimetry and high-resolution thermogravimetry (suitable for assessing very small weight losses, below 5 wt.%). First, it is worth noticing that the thermal degradation of DNA is somehow independent of the selected atmosphere, notwithstanding a faintly larger weight loss (within 250 and 350 °C) observed in nitrogen with respect to air, which clearly indicates that the degradation of DNA is not very sensitive to the presence of oxygen. Besides, the degradation phenomena take place according to three subsequent steps, namely: absorbed water removal (between 100 and 150 °C, which corresponds to about 8% weight loss), physical modification of the biomacromolecule (between 150 and 200 °C, which is accompanied by about a weight loss below 10%) with the formation of a closed cell-based structure that is further disrupted into an open cell-based structure at higher temperatures (between 200 and 400 °C, which corresponds to about 20% weight loss) as shown in Fig. 5.

Concurrently with the formation of this open cell-based structure, DNA gives rise to the formation of a significant amount of volatile products. Finally, at higher temperatures (i.e., from 400 to 700 °C), the resulting residue further loses around 20 wt.%, hence demonstrating that it is evolving into a progressively more stable aromatic

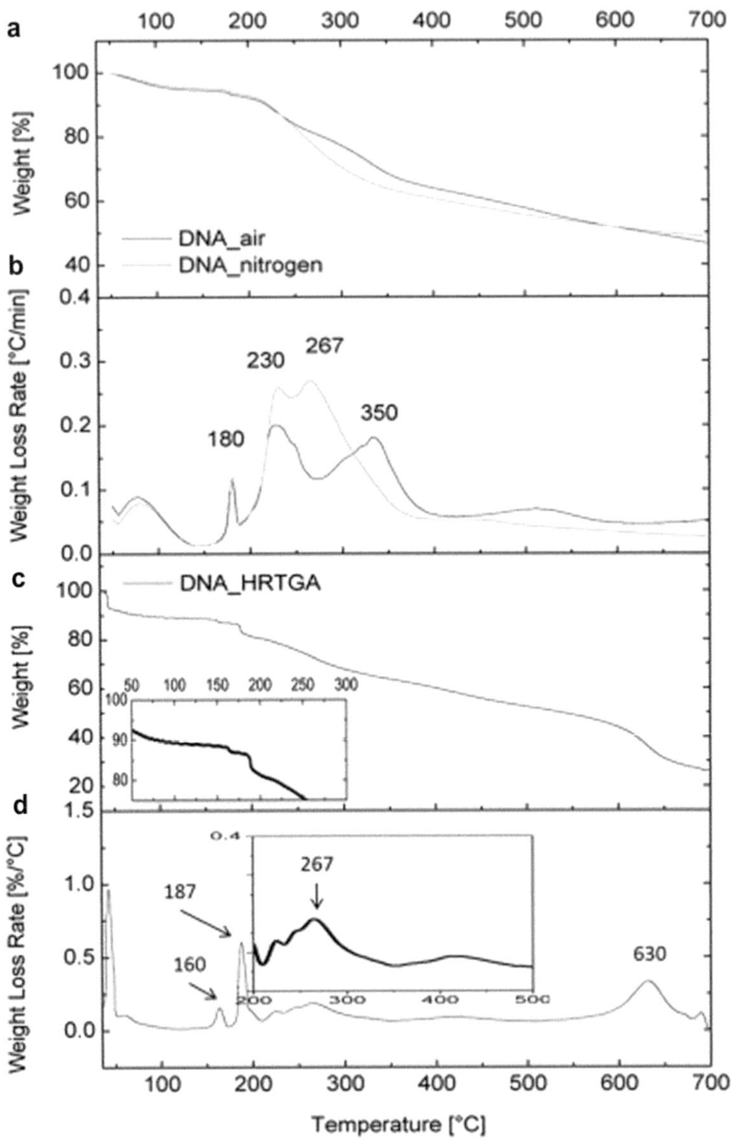

Fig. 4 Results from standard thermogravimetric analyses and high-resolution thermogravimetric analyses (with the corresponding derivative curves) performed on DNA, in nitrogen and air atmosphere. Reprinted with permission from [18]. Copyright 2015, Elsevier

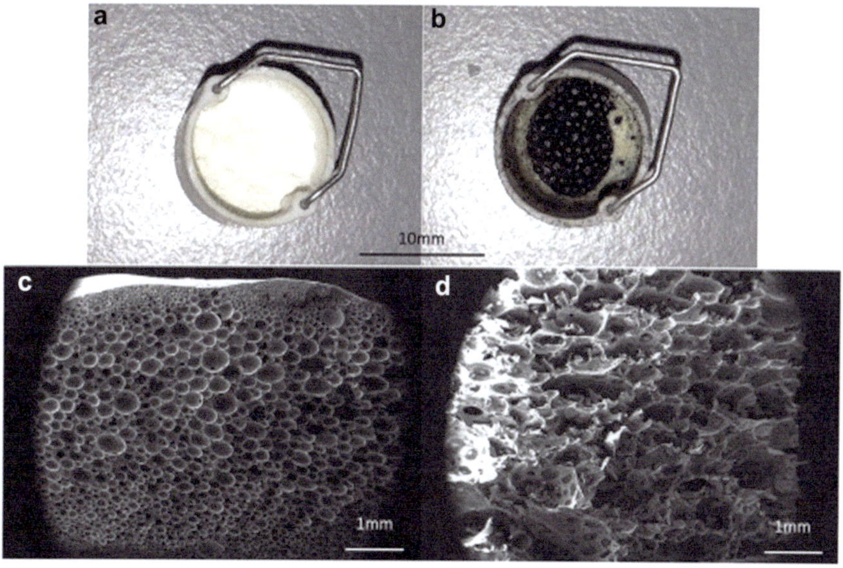

Fig. 5 Images of DNA powder (**a**), DNA residue at 180 °C (**b**), SEM magnification of the DNA residue at 180 °C (**c**) and 230 °C (**d**). Reprinted with permission from [18]. Copyright 2015, Elsevier

structure due to the occurrence of either cyclization, crosslinking, or aromatization reactions. The final residue at 700 °C is about 48 wt.%.

3 Intumescent Properties of DNA as Assessed by Forced Combustion Tests

To evaluate the intumescent feature of DNA, forced-combustion tests at various irradiative heat fluxes (namely: 25, 35, 50, and 75 kW/m^2) have been performed [19]. In particular, prior to the cone calorimetry tests, DNA powder (3 g) from herring sperm was deposited on an aluminum foil and compressed under 5 MPa pressure, operating at room temperature.

Figure 6 shows the heat release rate (HRR) vs. time curves obtained at different heat fluxes. The residues obtained at the end of the measurements are presented in Fig. 7.

It is worthy to note that the selected irradiative heat flux significantly, affects the intumescence of the biomacromolecule. Specifically, the lower the irradiative heat flux, the higher the expansion of the char. Besides, when irradiated at 25 kW/m^2, DNA does not ignite and pyrolyzes until the completion of the test (i.e., after 100 s), giving rise to a heat-resistant expanded cellular carbon structure. The maximum volumetric expansion and the final residue are ~1600% and ~66 wt.%, respectively. Conversely, an immediate ignition (TTI, i.e., Time to Ignition is equal to 1 s) is

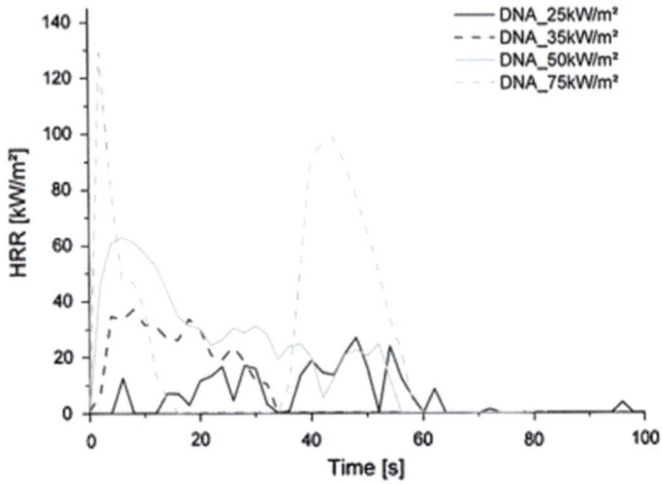

Fig. 6 Heat Release Rate (HRR) versus time curves for DNA tested at various irradiative heat fluxes. The samples are coded as DNA_XX kW/m^2, where XX represents the selected irradiative heat flux. Reprinted with permission from [19]. Copyright Elsevier, 2014

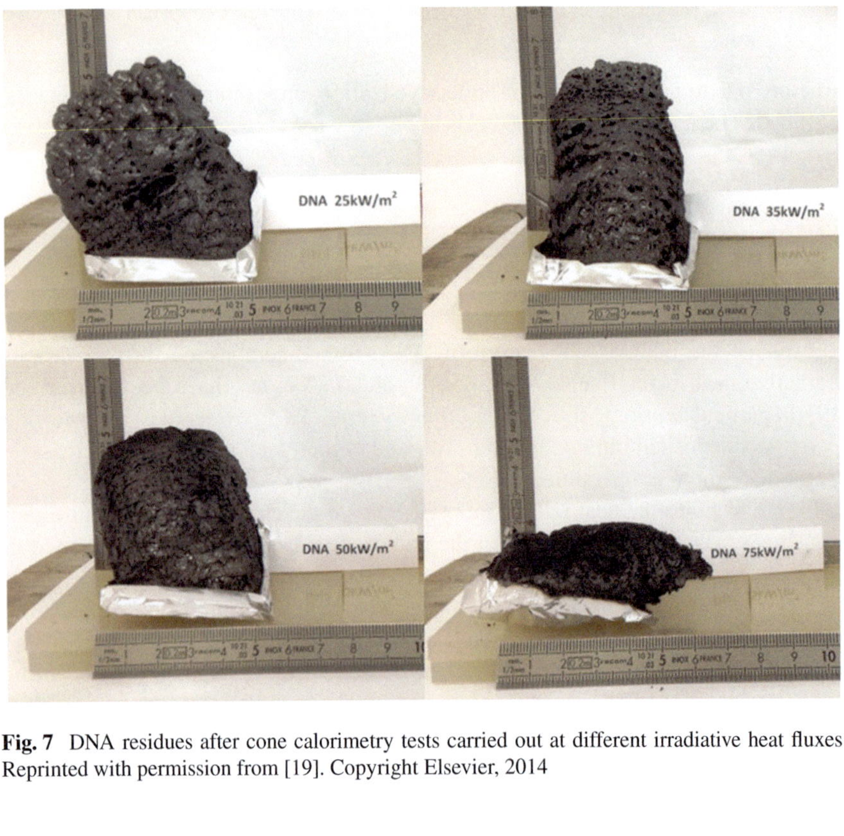

Fig. 7 DNA residues after cone calorimetry tests carried out at different irradiative heat fluxes. Reprinted with permission from [19]. Copyright Elsevier, 2014

observed when DNA is tested at 35 kW/m^2: the peak of Heat Release Rate (pkHRR, 37 kW/m^2) and the flame out are reached in 8 and 12 s, respectively. In addition, it is interesting to highlight that the intumescence continues even after the flame out, achieving a final volumetric expansion of about 1700% (Fig. 7). An additional increase of the irradiative heat flux to 50 kW/m^2 accounts for a combustion behavior like that observed at 50 kW/m^2 accounts, apart from a lower volumetric expansion (around 1200%): this finding can be attributed to the high irradiative heat flux that, in turn, determines the explosion of the bubbles, responsible for the swelling of the DNA char. Finally, the forced-combustion measurements, carried out at the highest irradiative heat flux (i.e., under 75 kW/m^2), clearly indicate a very quick ignition (after 1 s) and achievement of the pkHRR (129 kW/m^2, time to peak of 2 s); the swollen char that originates can resist up to 34 s. Then, the biomacromolecule ignites again, achieving a second pkHRR (close to 99 kW/m^2) after 44 s because of the formation of some cracks placed at the bottom of the expanded char. Besides, the latter shows important growth during the combustion process, but then it collapses because of its weak compactness (the final volumetric expansion does not exceed 750%). The appearance of several bubbles in the swollen char clearly proves the intumescence of DNA, as observed in scanning electron microscopy (SEM) analyses (Fig. 8).

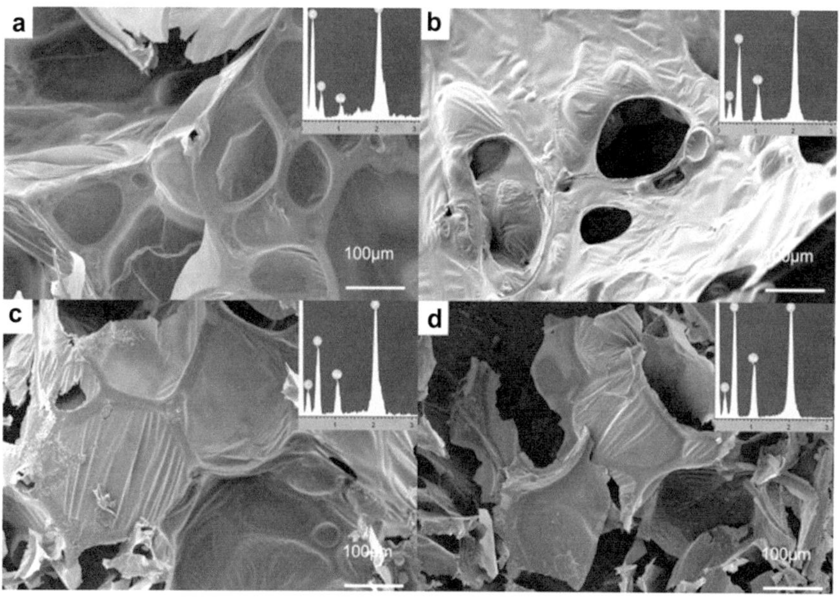

Fig. 8 Typical SEM micrographs (magnification: 500X) of DNA exposed to 25 (**a**), 35 (**b**), 50 (**c**), and 75 kW/m^2 (**d**). Reprinted with permission from [19]. Copyright Elsevier, 2014

4 Applications of DNA as Flame Retardant for Bulk Polymers

One of the pioneering works that dealt with the use of DNA as an effective flame retardant for bulk polymers, dates to 2014 and compares the fire resistance provided by the biomacromolecule, either incorporated into ethylene vinyl acetate (EVA) copolymer (by melt compounding in an internal mixer, at 15 wt.% loading), or as a hot-compressed coating (keeping constant the biomacromolecule loading) deposited on the copolymer surface [20]. As shown by forced-combustion tests (35 kW/m^2), the bulk incorporation of the biomacromolecule accounted for significant anticipation of ignition (TTI of 28 vs. 62 s, for the compound containing 15 wt.% of DNA and the unfilled copolymer, respectively), as well as for a significant lowering of pkHRR (by about 36%), as presented in Fig. 9. Conversely, the DNA coating deposited on the copolymer was more effective with respect to the incorporation in bulk, as the ignition of the tested specimens occurred after about 300 s; meanwhile, their Total Heat Release (THR) was lowered by 38% with respect to the unfilled copolymer.

These differences between the bulk and surface approaches were further highlighted when the irradiative heat flow was raised to 50 kW/m^2 (Fig. 10). In fact, unlike the compounds embedding the biomacromolecule, which immediately started to burn upon irradiation, the DNA-coated counterparts ignited at much higher times (i.e., at about 116 s), and significantly lowered the combustion kinetics.

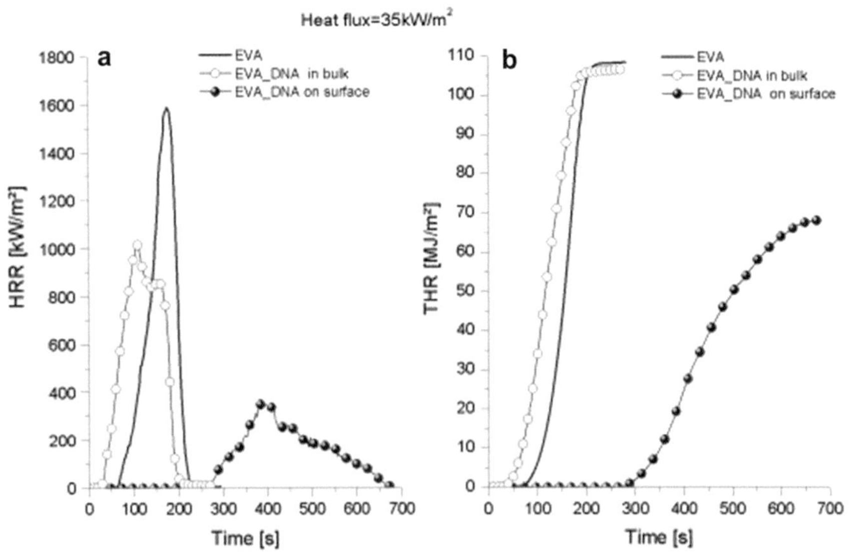

Fig. 9 HRR—Heat Release Rate (**a**) and Total Heat Release—THR (**b**) versus time curves of ethylene vinyl acetate (EVA), the copolymer embedding 15 wt.% of DNA (EVA_DNA in bulk) and the copolymer coated with the biomacromolecule (EVA_DNA on surface) (irradiative heat flux: 35 kW/m^2). Reprinted with permission from [20]. Copyright Elsevier, 2014

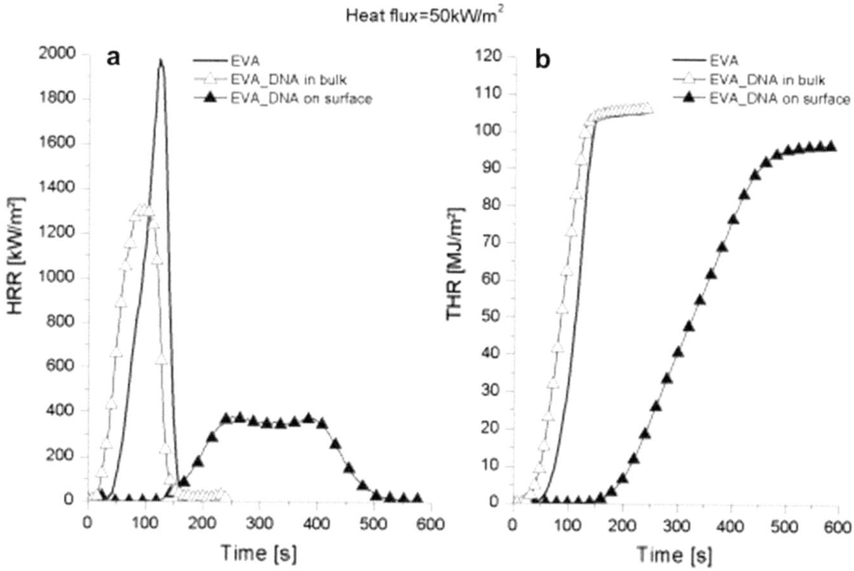

Fig. 10 HRR—Heat Release Rate (**a**) and Total Heat Release—THR (**b**) versus time curves of ethylene vinyl acetate (EVA), the copolymer embedding 15 wt.% of DNA (EVA_DNA in bulk) and the copolymer coated with the biomacromolecule (EVA_DNA on surface) (irradiative heat flux: 50 kW/m^2). Reprinted with permission from [20]. Copyright Elsevier, 2014

The effectiveness of the DNA coating approach was further extended to other thermoplastic matrices, namely polypropylene, acrylonitrile–butadiene–styrene, poly(ethylene terephthalate), and polyamide 6 [21].

In a further research effort, the flame-retardant performance of ethylene vinyl acetate copolymer containing DNA (at 10 or 20 wt.% loading) was compared with that of an analogous compound embedding ammonium polyphosphate at the same concentrations [22]. In particular, the char-forming character of both additives was demonstrated by means of thermogravimetric analyses and pyrolysis combustion flow calorimetry. Further, forced-combustion tests (35 kW/m^2) revealed a similar behavior of the two flame retardants, though the performance of ammonium polyphosphate was slightly better with respect to DNA (particularly referring to Total Heat Release values and the final residues), at constant loading (Table 1).

Then, DNA was melt compounded with low-density polyethylene (LDPE) at different loadings (namely, 5, 10, 20, 30, and 40 wt.%): its fire behavior was compared with that provided by melamine polyphosphate [23]. Enhanced compatibility between DNA and the polyolefin matrix was observed, despite the occurrence of some denaturation, fragmentation, and oxidation phenomena at the processing temperature (i.e., 145 °C), which did not significantly affect the fireproof features of the biomacromolecule. In particular, as assessed by horizontal flame spread tests, the presence of DNA accounted for the achievement of self-extinction at loading levels beyond 5 wt.%, unlike melamine polyphosphate. Besides, pyrolysis combustion flow

Table 1 Results from cone calorimetry tests (irradiative heat flux: 35 kW/m^2) for ethylene vinyl acetate (EVA) and its compounds with DNA (EVA_XXDNA) and ammonium polyphosphate (EVA_XXAPP), where XX represents the wt.% loading of the additives

Sample	TTI (s)	pkHRR (kW/m^2)	ΔpkHRR (%)	THR (MJ/m^2)	ΔTHR (%)	Residue (%)
EVA	65	1588	–	108	–	0
EVA_10DNA	28	974	−39	108	0	3
EVA_10APP	28	1030	−35	93	−14	6
EVA_20DNA	28	963	−39	100	−7	6.5
EVA_20APP	32	820	−48	88	−18	13

calorimetry tests highlighted a similar behavior provided by each flame-retardant additive, with a decrease in net THR of about 40% with respect to unfilled LDPE (700 vs. 1123 W/g, respectively), which trended linearly with increasing the loading levels of both DNA and melamine polyphosphate. In general, the residue yields of the compounds containing DNA were higher with respect to the counterparts embedding melamine polyphosphate.

Some attempts were also performed to investigate the combination of DNA with other co-additives, trying to achieve the best flame retardance with the lowest DNA loading level: indeed, DNA is currently very expensive, and any possible reduction in its amount for flame retardant formulations is very important.

In this view, Alongi and co-workers [24] melt-compounded DNA with ethylene vinyl acetate at three different loading levels, namely, 10, 15, and 20 wt.%, aiming at (i) decreasing the DNA amount required for reaching satisfying flame-retardant features, (ii) providing a carbon source, and (iii) assessing any synergism occurring between the additives. Both α-cellulose and β-cyclodextrins accounted for a decrease in pkHRR with respect to the unfilled EVA, as assessed by forced-combustion tests (Table 2) performed at 35 kW/m^2. Finally, the calculation of the synergistic effectiveness parameter [3] demonstrated a joint effect taking place between DNA and each carbon source additive.

Albite-Ortega and co-workers [25] aimed at reducing the loading levels of metal hydroxides in the formulation of flame-retardant thermoplastics. For this purpose, in the first set of compounds, magnesium hydroxide (20 wt.%), keratin fibers (10 wt.%), and DNA (10 wt.%) were melt-compounded with LDPE/EVA (70/30 in weight), in the presence of maleic anhydride-modified polyethylene (6 wt.%) as compatibilizer. Then, the same compounds were prepared without DNA that was coated (at 10 wt.%) onto the LDPE/EVA surface by means of hot pressing. All these formulations were compared with a standard one, i.e., incorporating 55 wt.% of magnesium hydroxide (and without compatibilizer). As assessed by forced-combustion tests (35 kW/m^2), the combination of $Mg(OH)_2$, DNA, and keratin fibers either in bulk or with the

Table 2 Results from cone calorimetry tests (35 kW/m^2) for ethylene vinyl acetate (EVA) and its compounds containing 20 wt.% DNA (EVA_20DNA), 10 wt.% DNA and 10 wt.% α-cellulose (EVA_10DNA_10CEL), and 10 wt.% DNA and 10 wt.% β-cyclodextrins (EVA_10DNA_10CD) [24]

Sample	TTI (s)	pkHRR (kW/m^2)	THR (MJ/m^2)	Residue (%)
EVA	65	1588	108	0
EVA_20DNA	28	963	100	6.5
EVA_10DNA_10CEL	28	872	108	5.0
EVA_10DNA_10CD	28	1074	115	5.1

DNA coated onto the surface of the compound turned out to provide flame retardant performances close to the standard formulation (Table 3), without a significant worsening in the ductility of the compound (as assessed for the standard flame retardant formulation).

Further, DNA has been employed in combination with nanofillers for flame retardant purposes. In this context, Rajczak and co-workers [26] intercalated low molecular weight DNA from herring sperm into two commercially available nanoclays (i.e., sodium Montmorillonite and Sepiolite); the resulting clays, at 10 wt.% loading level, were melt-compounded with EVA (vinyl acetate content: 19%). As revealed by thermogravimetric analyses, the presence of each modified nanoclay accounted for a significant increase in the thermal and thermo-oxidative stability of the copolymer. Besides, EVA flammability was not affected by the incorporation of the DNA-modified nanoclays. All the nanocomposites, irrespective of the type of nanofiller, were V-2 rated in vertical flame spread tests. The presence of these additives remarkably affected the combustion behavior of EVA, with a decrease in the HRR by ~7% and ~8% for DNA-modified sodium Montmorillonite and Sepiolite, respectively, pkHRR by ~24 and ~33%, respectively, TSP by ~ 17% for both the modified nanoclays and SEA by ~8 and 11%, respectively, in respect to unfilled copolymer.

Quite recently, Zabihi and co-workers [27] exploited ball milling for obtaining exfoliated DNA-functionalized graphene nanoplatelets from graphite. Then, the

Table 3 Results from cone calorimetry tests (35 kW/m^2) for LDPE/EVA and its compounds

Sample	TTI (s)	pkHRR (kW/m^2)	THR (MJ/m^2)
LDPE/EVA/55 $Mg(OH)_2$*	80	546	75
LDPE/EVA/20 $Mg(OH)_2$/10DNA	73	630	89
LDPE/EVA/20 $Mg(OH)_2$/10KF	72	1291	96
LDPE/EVA/20 $Mg(OH)_2$/10DNA/10KF	73	499	81
LDPE/EVA/20 $Mg(OH)_2$/10DNAsurf/10KF**	75	470	79

* Compound without compatibilizer (reference formulation)
**NA coated on the surface of the sample by hot compression
Legend LDPE/EVA/XXMg(OH)$_2$/YYDNA/ZZKF = compound containing XX wt.% of magnesium hydroxide, YY wt.% of DNA, and ZZ wt.% of keratin fibers [25]

functionalized nanofiller was dispersed in three different polymers, namely: epoxy resin, polyvinyl alcohol, and polystyrene, at 10 wt.% loading level. The resulting nanocomposites were V-0 rated for flammability tests (Fig. 11) and their LOI values were as high as 39% (for the epoxy system; neat resin: 21%), 32.8% (for polyvinyl alcohol; neat polymer: 18.2%), and 28.4% (for polystyrene, neat polymer: 17.6%).

Fig. 11 Snapshots of flammability tests carried out in vertical configuration on the three unfilled polymers and their nanocomposites containing 10 wt.% of exfoliated DNA-modified graphene (D-Gr). Reprinted with permission from [27]. Copyright Elsevier, 2020

5 Applications of DNA as Flame Retardant for Textiles

The first inventive work dates to 2013, when the excellent fireproof features of DNA were demonstrated for cotton fabrics [28]. In particular, the cellulosic fabrics were dipped into an acidified aqueous DNA suspension (2.5 wt.%) for 1 min in a climatic chamber (operating at 30% relative humidity and 30 °C) and dried to a constant weight (19 wt.% final dry add-on). The treated fabrics exhibited self-extinguishing behaviour in horizontal flammability tests and achieved LOI values of 28% (vs. 18% for the untreated fabric); besides, it was not possible to ignite them upon exposure to 35 kW/m^2 irradiative heat flux during cone calorimetry. All these findings were attributed to the intumescence of DNA, which accounted for the formation of a stable aromatic char upon decomposition under a direct flame or an irradiative heat flow.

Then, the same group optimized the minimum add-on of the biomacromolecule, which allowed for cotton self-extinction in flammability tests and for excellent fireproof features [29]. In doing so, different dry add-ons were investigated, namely 5, 10, and 19 wt.%. As revealed by horizontal flammability tests (Fig. 12), the minimum DNA amount required for achieving self-extinction was 10 wt.%, notwithstanding that, even at 5 wt.% add-on, a coherent residue was obtained.

Besides, forced-combustion tests (35 kW/m^2) highlighted that unlike cotton DNA-treated with 19 wt.% add-on, lower biomacromolecule add-ons were not able to

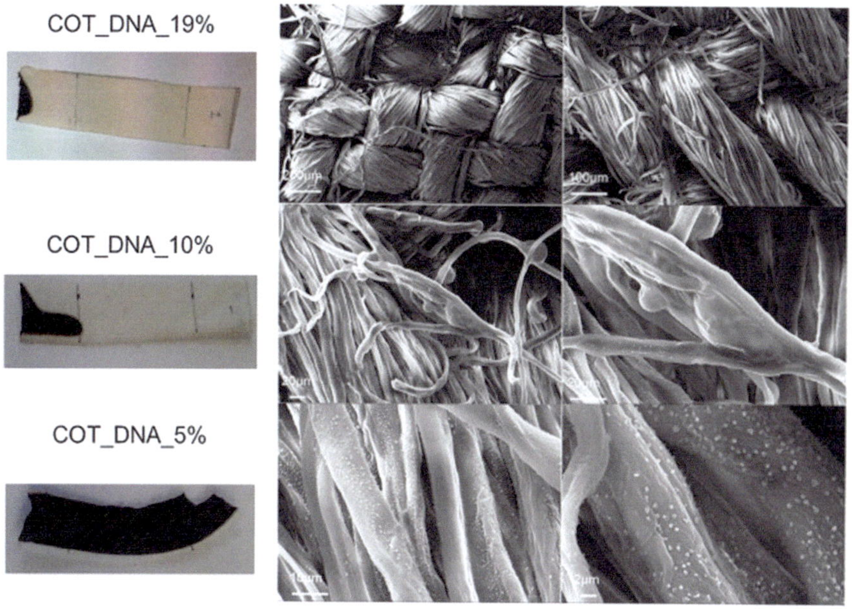

Fig. 12 Residues of the DNA-treated cotton fabrics after horizontal flammability tests and SEM micrographs of the residue of cotton treated with DNA (add-on: 19 wt.%) after flame spread tests. Reprinted with permission from [29]. Copyright Elsevier, 2013

block the fabric's ignition, although the pkHRR values were lower compared with the untreated fabric (Table 4). Raising the irradiative heat flux to 50 kW/m^2 accounted for the ignition of the DNA-treated fabric with 19 wt.% add-on, which still exhibited a significantly decreased pkHRR (-60% with respect to untreated cotton).

Then, instead of using an impregnation/exhaustion method, DNA was combined with chitosan (CHI) in bi-layered assemblies (namely, 5, 10, and 20 bi-layers), or alternatively, dipping the fabrics into 0.5 wt.% aqueous solutions of the two biomacromolecules [30]. As assessed by horizontal flammability tests, the fabrics treated with 20 bi-layers of DNA and chitosan achieved self-extinction (Fig. 13); besides, their LOI achieved 24% (vs. 18% for untreated cotton). Besides, the presence of bubbles in the burnt treated fabric clearly indicates the intumescent behavior of the deposited LBL architecture.

Table 5 collects the results of cone calorimetry tests (35 kW/m^2) [30]. It is worth noticing that TTI decreased with increasing the deposited bi-layers. In addition, the higher this number, the lower the pkHRR and THR values: thus, the best performances were achieved by the fabrics treated with 20 bi-layers.

Then, the effects of such parameters as pH of the DNA aqueous solutions (used for impregnating cotton fabrics), DNA molecular size, and the number of impregnation steps on flame retardance were thoroughly investigated [31]. To this aim, three DNA differing as far as their molecular size is considered (i.e., low, medium, and high molecular size DNA) were chosen as model biomacromolecules and applied to cotton fabrics at two different pH values (namely 4 and 8, respectively below and beyond the isoelectric point of the biomacromolecule); both single and multiple impregnation steps were employed, reaching a constant dry add-on (i.e., 8 wt.%).

Table 4 Data from cone calorimetry tests (at two irradiative heat fluxes: 35 and 50 kW/m^2) for cotton and the fabric treated with various DNA amounts

Sample	Heat flux: 35 kW/m^2				Heat flux: 50 kW/m^2			
	Time to ignition (s)	Flame out (s)	pkHRR (kW/m^2)	Residue (%)	Time to ignition (s)	Flame out (s)	pkHRR (kW/m^2)	Residue (%)
Untreated cotton	45	75	125	<3	16	40	128	<3
Cotton + 5 wt.% DNA	24	39	68	15				
Cotton + 10 wt.% DNA*	19	29	62	15				
Cotton + 19 wt.% DNA	No ignition				10	19	51	17

*No ignition for 2 specimens on 5

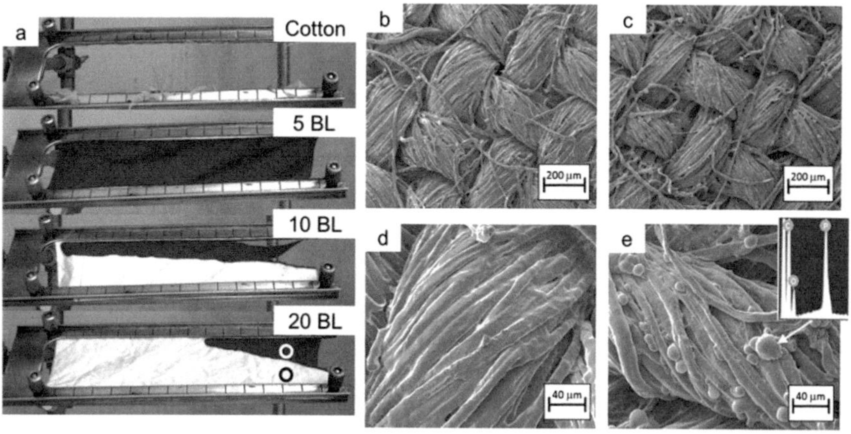

Fig. 13 Residues of untreated and layer-by-layer-treated (XXBL, where XX represents the number of deposited bi-layers) cotton fabrics after horizontal flammability tests (**a**); SEM micrographs of the unburned (**b**, **d**) and burned portion (**c**, **e**) of cotton coated with 20 bi-layered assemblies. Reprinted with permission from [30]. Copyright Elsevier, 2013

Table 5 Results from cone calorimetry tests (irradiative heat flux: 35 kW/m^2) for cotton and the layer-by-layer-treated fabrics

Sample	TTI (s)	pkHRR (kW/m^2)	ΔpkHRR (%)	THR (MJ/m^2)	ΔTHR (%)	Residue (%)
Cotton	39	97	–	1.9	–	2
Cotton + 5BL	17	73	−25	1.7	−10	11
Cotton + 10BL	20	60	−38	1.5	−21	12
Cotton + 20BL	23	57	−41	1.3	−32	13

Legend Cotton + XXBL: fabric treated with XX bi-layers of chitosan and DNA [30]

As assessed by horizontal flammability tests, the fabrics treated with the DNA with the lowest molecular size achieved self-extinction for around the 74% and 86% of the tested samples, for pH = 8 and 4, respectively. These findings were attributed both to the short DNA strands that were able to better penetrate the fibers' surface and to their higher thermo-oxidative stability. Besides, comparing the single and multiple impregnation techniques (with the same dry add-on), multiple impregnations with the low molecular size DNA at pH = 4 were found to be more effective than the single impregnation, leading to about 25 and 45% decrease in pkHRR and THR values with respect to neat cotton, as observed during forced combustion tests carried out at 35 kW/m^2 irradiative heat flux.

Two main issues concerning the usage of DNA as an effective flame retardant for cotton have been progressively addressed. The former refers to the durability of the treatments with DNA, as the biomacromolecule is deposited on the cellulosic fabric from aqueous solutions: therefore, it can easily be washed out during laundry cycles. The second issue refers to the current very high cost of DNA, which allows for its use at the lab scale only.

To fix the durability issues, Casale and co-workers [32] proposed depositing chitosan and DNA, either by impregnation or by layer-by-layer method, exploiting the reactivity of chitosan toward UV radiation (through a photografting reaction). The UV-cured layer-by-layer DNA-chitosan coatings exhibited the best flame-retardant properties (i.e., self-extinction in horizontal flammability tests, as well as a significant decrease in the pkHRR (about −38%) and THR (about −45%), when compared to the neat cotton). Besides, the fabrics treated with these coatings exhibited antibacterial activity against *S. Aureus*.

Another approach for conferring washing fastness to DNA-treated cotton fabrics was proposed by Ortelli and co-workers [33], who mixed two oppositely charged TiO_2 nanoparticle suspensions with a DNA solution, to obtain a hybrid flame retardant coating. The adopted strategy encompassed the inherent formation of electrostatic forces or H-bonds derived from the combination of the two phases in water, tuning the pH. Consequently, titania nanoparticles acted as a cross-linker between DNA and cotton, hence remarkably increasing the washing fastness of the latter. Besides, the concurrent presence of an intumescent biomacromolecule (i.e., DNA) and a ceramic phase (i.e., titania) accounted for the achievement of self-extinction in horizontal flammability tests and a strong decrease of the thermal parameters (Table 6) as assessed in cone calorimetry tests (35 kW/m^2).

Recently, Luo and co-workers proposed another strategy for enhancing the washing fastness of DNA coated onto cotton fabrics [34]: more specifically, they modified the fabric surface with citric acid, hence increasing the free carboxyl groups, the formation of esters, and the subsequent DNA chemical absorption. In particular, the latter followed a quasi-second-order-kinetic model. As assessed by vertical flammability tests, the fabrics modified with citric acid and then with DNA were self-extinguishing; besides, self-extinction was maintained even after 4 laundry cycles performed according to AATCC standard (Fig. 14).

At present, the issue related to the high cost of DNA for flame retardant purposes has not been solved yet, though Bosco and co-workers proposed to recover DNA from either vegetable scraps or spent brewer's yeast [35]. It was demonstrated the possibility of recovering DNA by means of a low environmental impact process, using an aqueous extraction buffer (containing 10% v./v. of liquid detergent, and

Table 6 Cone calorimetry data (35 kW/m^2) for cotton and the fabric treated with TiO_2 and DNA [33]

Sample	TTI (s)	Add-on (%)	pkHRR (kW/m^2)	ΔpkHRR (%)	THR (kJ/m^2)	ΔTHR (%)	Residue (%)
Cotton	30	–	47.0	–	1.0	–	1
Cotton + TiO_2/DNA	37	16.5	14.4	−69	0.35	−65	12

	Washing time					
	Original sample	1	2	3	4	5
Damaged length (mm)	31	38	98	187	211	350
Burning pictures						

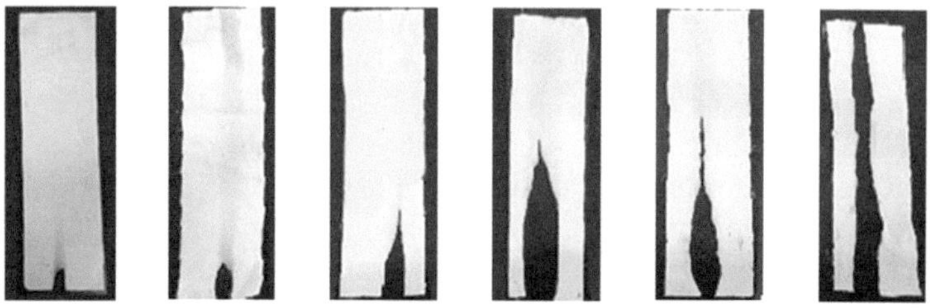

Fig. 14 Snapshots from vertical flammability tests carried out on cotton fabrics treated with citric acid and DNA: assessment of the washing fastness. Reprinted from [34] under Creative Commons CC-BY license

10% w./v. of NaCl), followed by precipitation in cold ethanol (either at +4 or − 20 °C). The recovered deoxyribonucleic acids were resuspended in water and applied to cotton fabrics, reaching a dry add-on of 8 wt.% (DNA from spent brewer's yeast) and 5 wt.% (DNA from vegetable scraps). 8 wt.% loading was enough for attaining self-extinction in horizontal flammability tests.

Quite recently, DNA was employed not only for providing cotton with fireproof characteristics but also for achieving superhydrophobicity [36]. More specifically, the cellulosic fabric was dipped into an aqueous suspension containing DNA (3 wt.% concentration), then into a 1 mM silver nitrate solution, and finally into an ethanolic solution of octadecyltriethoxysilane. The as-prepared and treated fabric showed a water contact angle as high as 157°, hence exhibiting a superhydrophobic surface that was maintained also after laundry cycles (water contact angle beyond 150° even after 6 washing cycles). Besides, as assessed by vertical flammability tests, the superhydrophobic cotton reached self-extinction (burnt char length: 0.7 cm).

As summarized in this paragraph, most of the applications of DNA as flame retardant for textiles refer to cotton. Only one paper describes the application of this biomacromolecule to other fibers, namely polyester, and nylon 6 [37]. In particular, the authors first introduced carboxyl groups for the successive coupling of DNA through enzymatic surface hydrolysis with *Humicola insolens cutinase*. Then, the DNA immobilization on the modified fabrics, carried out by means of three different crosslinking agents (namely: 1-Ethyl-3-(3-dimethylaminopropyl)carbodiimide/*N*-Hydroxysuccinimide, tyrosine, or dopamine) was confirmed by FTIR spectroscopy. As assessed by vertical flammability tests, the treated fabrics were self-extinguishing, irrespective of the type of crosslinking agent employed; besides, dopamine and tyrosine accounted for a good washing fastness, respectively retaining around 50 and 60% of DNA on both the fabrics, after repeated laundry cycles performed at 40 °C for 5 min, in the presence of a commercial detergent.

6 Applications of DNA as Flame Retardant for Foams

Compared to bulky polymers and textiles, the research on the potential use of DNA as an effective flame retardant for foamed materials is still at the beginning: in fact, only a few papers investigate this peculiar application.

More specifically, Li and co-workers [38] exploited a layer-by-layer approach [39] for the buildup of DNA-containing multi-layered architectures suitable for imparting flame-retardant features to a flexible polyurethane foam. DNA was combined with chitosan (giving rise to 10 bi-layered assemblies consisting of chitosan and DNA + sodium montmorillonite), or with chitosan and sodium montmorillonite (obtaining either 10 tri-layered assemblies of chitosan, montmorillonite, and DNA, or 5 quad-layered assemblies of chitosan, DNA, chitosan, and montmorillonite). In cone calorimetry tests (35 kW/m^2), all the designed assemblies were very effective in decreasing the average HRR and the pkHRR, promoting, at the same time, an increase in the final residues (Table 7).

Then, the effectiveness of DNA in replacing ammonium polyphosphate in quad-layered assemblies (made of poly(acrylic acid)/ poly(diallyldimethylammoniumchloride)/DNA or ammonium polyphosphate/ poly(diallyldimethylammoniumchloride)), suitable for imparting flame retardant features to closed cell poly(ethylene terephthalate) foams, was assessed [40].

The superior performance of ammonium polyphosphate-based coatings with respect to DNA-based counterparts was demonstrated through horizontal flammability tests and cone calorimetry measurements. In particular, the coating made of 4 quad-layered assemblies containing ammonium polyphosphate accounted for the suppression of melt dripping and the lowering of the pkHRR by about 25% (Table 8).

Table 7 Results from cone calorimetry tests (35 kW/m^2) performed on untreated and layer-by-layer treated polyurethane foams [38]

Sample	Average HRR (kW/m^2)	pkHRR (kW/m^2)	ΔpkHRR (%)	THR (MJ/m^2)	ΔTHR (%)	Residue (%)
Untreated PU	286	488	–	30.3	–	0
PU treated with 10 bilayers of chitosan and DNA	204	339	−30.5	26.6	−12.2	9
PU treated with 10 bilayers of chitosan and DNA + Montmorillonite	53	239	−51.0	26.5	−12.5	18
PU treated with 10 trilayers of chitosan, montmorillonite, and DNA	77	260	−46.7	25.4	−16.2	16
PU treated with 5 bilayers of chitosan, DNA, chitosan, and montmorillonite	69	251	−48.6	25.5	−15.8	11

Table 8 Results from horizontal flame spread tests and cone calorimetry measurements (35 kW/m^2) carried out on untreated and layer-by-layer treated poly(ethylene terephthalate) closed cell foams [40]

Sample	Cone calorimetry tests			Flame spread tests	
	pkHRR (kW/m^2)	THR (MJ/m^2)	Residue (%)	Burning time (s)	Dripping (%)
Untreated foam	650	20.0	9.5	193	Yes
Foam treated with 2 quadlayers containing ammonium polyphosphate	656	20.1	8.5	41	Yes
Foam treated with 4 quadlayers containing ammonium polyphosphate	487	17.4	11.0	50	No
Foam treated with 2 quadlayers containing DNA	634	20.0	9.0	195	195
Foam treated with 4 quadlayers containing DNA	607	19.5	8.0	44	44

Quite recently, Maddalena and co-workers [41] proposed a water-based single-step deposition of graphene oxide (GO) and DNA on open-cell polyurethane foams. These latter were previously activated by dipping in 1 wt.% poly(acrylic acid) solution, washing in ultrapure water, and subsequently dipping in 1 wt.% poly(diallyldimethyldiammonium chloride) solution. The subsequent deposition of GO and DNA accounted for the formation of a uniform coating on the foam, where DNA behaved as a ligand for graphene oxide nanoplatelets. Horizontal flammability tests revealed that the deposited DNA-GO coating was capable to prevent the flame ignition and spread, hence making the polyurethane foam self-extinguishing. Further, forced-combustion tests (35 kW/m^2) demonstrated that the deposited coating significantly lowered pkHRR (by about 30%), THR (by about 36%), and TSR (by about 30%), remarkably increasing the final residue with respect to the neat foam (Table 9).

Table 9 Results from cone calorimetry tests (35 kW/m^2) carried out on untreated and DNA + GO coated open-cell polyurethane foams [41]

Sample	pkHRR (kW/m^2)	THR (MJ/m^2)	TSR (m^2/m^2)	Residue (%)
Untreated foam	304	7.4	89	7
Foam treated with DNA + GO coating	76	4.7	62	20

7 Conclusions and Future Perspective

Despite the main use of DNA as a carrier of genetic information and gene expression, this biomacromolecule has shown interesting potentialities in the setup of efficient flame-retardant systems for bulk polymers, textiles, and even foamed materials. About 20 years ago, thinking about the use of DNA as a "green" alternative to traditional chemical flame retardants was nearly unreasonable. Conversely, the discovery of the high flame-retardant features provided by DNA has pushed academics toward a deep investigation of its possible uses as a fireproof component, alone or in combination with other additives, also considering the need for replacing the toxic, hazardous, and, in some cases, carcinogenic/mutagenic conventional flame retardants, as clearly indicated by both European and American directives and laws.

The scientific papers that have been summarized in this chapter clearly elucidate the high potential offered by DNA in flame retardance, notwithstanding that its fire performances are strictly related to many experimental parameters, such as molecular size, concentration, processing temperature, pH and isoelectric points of DNA aqueous suspensions, flame retardant mechanism behind the biomacromolecule, among a few to mention. In this view, it has been demonstrated that the tuning and optimization of the experimental condition/parameters allow for designing useful flame retardant "recipes", suitable for different polymers.

However, there are still some "open questions" that have to be solved, hence widening the use of DNA in flame retardant applications.

First, the DNA exhibits quite a low thermal stability (it starts degrading at about 200 °C), which does not allow its use when high-performing engineering or specialty polymers are processed at high temperatures (i.e., by melt compounding, extrusion, injection molding, among others). As a matter of fact, up to now, all the flame-retardant applications of DNA in bulk have involved low-melting polymers (such as polyolefins and EVA copolymers): the possibility of using DNA for high-performing engineering or specialty polymers is limited to surface approaches (i.e., coating these polymers with a DNA-rich coating).

Another important current issue refers to the low cost-effectiveness of DNA: indeed, being a very expensive biomacromolecule, its use is limited to small-scale applications (i.e., for scientific purposes). Therefore, up to now, it is hard to foresee the DNA potential at a larger scale (like a pre-industrial/pilot scale), although, specifically referring to flame retardant purposes, high purity is not needed. However, it is probable that any technological improvement in DNA extraction/recovery will allow for achieving higher yields and proper purity levels, particularly appropriate for flame retardant purposes.

In addition, although it has been demonstrated at the lab scale only, the possibility of recovering DNA from wastes and by-products derived from the food/fish industry paves the way toward the implementation of the circular economy concept, limiting the conventional landfill confinement of these wastes and by-products and providing them with a new life.

Specifically referring to textiles and foamed materials, for which DNA has been employed as a surface coating only, despite the quite successful attempts described in the present chapter, washing fastness is still a challenging issue that needs to be definitively solved, using, if possible, low environmental impact products and technologies. Besides, when DNA is applied as a surface coating on fabrics or foams, very often it alters the mechanical behavior of the treated substrates: this is a very detrimental effect, as the so-called "soft touch" of textiles or foams is very important and needs to be preserved in several application sectors.

To conclude, it is likely to foresee that further progress in the extraction, application, and exploitation of DNA will widen its applications in the flame-retardant field, paving the way toward improved sustainability and decreasing the current limitations of the biomacromolecule.

References

1. Wypych, G.: Handbook of Flame Retardants. ChemTech Publishing, Toronto (Canada) (2021)
2. Lewin, M.: Synergistic and catalytic effects in flame retardancy of polymeric materials—an overview. J. Fire Sci. **17**(1), 3 (1999)
3. Lewin, M.: Synergism and catalysis in flame retardancy of polymers. Polym. Adv. Technol. **12**, 215 (2001)
4. Weijs, L., Dirtu, A.C., Malarvannan, G., Covaci, A.: Bioaccumulation and biotransformation of brominated flame retardants. Compre. Anal. Chem. **67**, 433 (2015)
5. Howard, P.H., Muir, D.C.G.: Identifying new persistent and bioaccumulative organics among chemicals in commerce. Environ. Sci. Technol. **44**, 2277 (2010)
6. Howard, G.J.: Chemical alternatives assessment: the case of flame retardants. Chemosphere **116**, 112 (2014)
7. de Boer, J., Ballesteros-Gomez, A., Leslie, H.A., Brandsma, S.H., Leonards, P.E.G.: Flame retardants: dust—and not food—might be the risk. Chemosphere **150**, 461 (2016)
8. Hull, T.R., Law, R.J.: Environmental drivers for replacement of halogenated flame retardants. In: Papaspyrides, C.D., Kiliaris, P. (eds.) Polymer Green Flame Retardants, pp. 119–179. Elsevier, Amsterdam (2014)
9. Arruda, E.H., Brancalhão Melatto, R.A.P., Levy, W., de Melo Conti, D.: Circular economy: a brief literature review (2015–2020). Sustain. Oper. Comput. **2**, 79 (2021)
10. Bosco, F., Carletto, R.A., Alongi, J., Marmo, L., Di Blasio, A., Malucelli, G.: Thermal stability and flame resistance of cotton fabrics treated with whey proteins. Carbohydr. Polym. **94**, 372 (2013)
11. Carosio, F., Di Blasio, A., Cuttica, F., Alongi, J., Malucelli, G.: Flame retardancy of polyester and polyester-cotton blends treated with caseins. Ind. Eng. Chem. Res. **53**, 3917 (2014)
12. Alongi, J., Carletto, R.A., Bosco, F., Carosio, F., Di Blasio, A., Cuttica, F., Antonucci, V., Giordano, M., Malucelli, G.: Caseins and hydrophobins as novel green flame retardants for cotton fabrics. Polym. Degrad. Stab. **99**, 111 (2014)
13. Costes, L., Laoutid, F., Brohez, S., Dubois, P.: Bio-based flame retardants: when nature meets fire protection. Mater. Sci. Eng. R. Rep. **117**, 1 (2017)
14. Fitzgerald-Hayes, M., Reichsman, F.: DNA and Biotechnology. Academic Press, Burlington, USA (2010)
15. Sonmezoglu, S., Sonmezoglu, O.A.: Optical and dielectric properties of double helix DNA thin films. Mater. Sci. Eng. C **31**, 1619 (2011)
16. Kress, W.J., Erickson, D.L.: DNA barcodes: genes, genomics, and bioinformatics. Proc. Natl. Acad. Sci. U.S.A. **105**, 2761 (2008)

17. Teles, F.R.R., Fonseca, L.P.: Trends in DNA biosensors. Talanta **77**, 606 (2008)
18. Alongi, J., Di Blasio, A., Milnes, J., Malucelli, G., Bourbigot, S., Kandola, B., Camino, G.: Thermal degradation of DNA, an all-in-one natural intumescent flame retardant. Polym. Degrad. Stab. **113**, 110 (2015)
19. Alongi, J., Cuttica, F., Di Blasio, A., Carosio, F., Malucelli, G.: Intumescent features of nucleic acids and proteins. Thermochim. Acta **591**, 31 (2014)
20. Alongi, J., Di Blasio, A., Cuttica, F., Carosio, F., Malucelli, G.: Bulk or surface treatments of ethylene vinyl acetate copolymers with DNA: investigation on the flame retardant properties. Eur. Polym. J. **51**, 112 (2014)
21. Alongi, J., Cuttica, F., Carosio, F.: DNA coatings from byproducts: a panacea for the flame retardancy of EVA, PP, ABS, PET, and PA6? ACS Sustain. Chem. Eng. **4**, 3544 (2016)
22. Alongi, J., Cuttica, F., Bourbigot, S., Malucelli, G.: Thermal and flame retardant properties of ethylene vinyl acetate copolymers containing deoxyribose nucleic acid or ammonium polyphosphate. J. Therm. Anal. Calorim. **122**, 705 (2015)
23. Isarov, S.A., Lee, P.W., Towslee, J.H., Hoffman, K.M., Davis, R.D., Maia, J.M., Pokorski, J.K.: DNA as a flame retardant additive for low-density polyethylene. Polymer **97**, 504 (2016)
24. Alongi, J., Di Blasio, A., Cuttica, F., Carosio, F., Malucelli, G.: Flame retardant properties of ethylene vinyl acetate copolymers melt-compounded with deoxyribonucleic acid in the presence of α-cellulose or β-cyclodextrins. Curr. Org. Chem. **18**, 1651 (2014)
25. Albite-Ortega, J., Sanchez-Valdes, S., Ramirez-Vargas, E., Nunez-Figueredo, Y., Ramos deValle, L.F., Martínez-Colunga, J.G., Graciano-Verdugo, A.Z., Sanchez-Martínez, Z.V., Espinoza-Martínez, A.B., Rodriguez-Gonzalez, J.A., Castaneda-Flores, M.E.: Influence of keratin and DNA coating on fire retardant magnesium hydroxide dispersion and flammability characteristics of PE/EVA blends. Polym. Degrad. Stab. **165**, 1 (2019)
26. Rajczak, E., Arrigo, R., Malucelli, G.: Thermal stability and flame retardance of EVA containing DNA-modified clays. Thermochim. Acta **686**, 178546 (2020)
27. Zabihi, O., Ahmadi, M., Li, Q., Ghandehari Ferdowsi, M.R., Mahmoodi, R., Kalali, E.N., Wang, D.-Y., Naebe, M.: A sustainable approach to scalable production of a graphene based flame retardant using waste fish deoxyribonucleic acid. J. Clean. Prod. **247**, 119150 (2020)
28. Alongi, J., Carletto, R.A., Di Blasio, A., Carosio, F., Bosco, F., Malucelli, G.: DNA: a novel, green, natural flame retardant and suppressant for cotton. J. Mater. Chem. A **1**, 4779 (2013)
29. Alongi, J., Carletto, R.A., Di Blasio, A., Cuttica, F., Carosio, F., Bosco, F., Malucelli, G.: Intrinsic intumescent-like flame retardant properties of DNA-treated cotton fabrics. Carbohydr. Polym. **96**, 296 (2013)
30. Carosio, F., Di Blasio, A., Alongi, J., Malucelli, G.: Green DNA-based flame retardant coatings assembled through layer by layer. Polymer **54**, 5148 (2013)
31. Bosco, F., Casale, A., Mollea, C., Terlizzi, M.E., Gribaudo, G., Alongi, J., Malucelli, G.: DNA coatings on cotton fabrics: effect of molecular size and pH on flame retardancy. Surf. Coat. Technol. **272**, 86 (2015)
32. Casale, A., Bosco, F., Malucelli, G., Mollea, C., Periolatto, M.: DNA-chitosan cross-linking and photografting to cotton fabrics to improve washing fastness of the fire-resistant finishing. Cellulose **23**, 3963 (2016)
33. Ortelli, S., Malucelli, G., Blosi, M., Zanoni, I., Costa, A.L.: NanoTiO_2@DNA complex: a novel eco, durable, fire retardant design strategy for cotton textiles. J. Colloid Interface Sci. **546**, 174 (2019)
34. Luo, Q., Gao, P., Zhou, J., Zhang, J., Wu, W., Cao, J., Reddy, N., Ma, H.: Imparting flame resistance to citric acid–modified cotton fabrics using DNA. J. Engin. Fibers Fabr. **15**, 1 (2020)
35. Bosco, F., Casale, A., Gribaudo, G., Mollea, C., Malucelli, G.: Nucleic acids from agro-industrial wastes: a green recovery method for fire retardant applications. Ind. Crops Prod. **108**, 208 (2017)
36. Suryaprabha, T., Sethuraman, M.G.: Fabrication of a superhydrophobic and flame-retardant cotton fabric using a DNA-based coating. J. Mater. Sci. **55**, 11959 (2020)
37. Quartinello, F., Kremser, K., Vecchiato, S., Schoen, H., Vielnascher, R., Ploszczanski, L., Pellis, A., Guebitz, G.M.: Increased flame retardancy of enzymatic functionalized PET and nylon fabrics via DNA immobilization. Front. Chem. **7**, 685 (2019)

38. Li, Y.C., Yang, Y.H., Kim, Y.S., Shields, J.S., Davis, R.D.: DNA-based nanocomposite biocoatings for fire-retarding polyurethane foam. Green Mater. **2**, 144 (2014)
39. Richardson, J.J., Cui, J., Björnmalm, M., Braunger, J.A., Ejima, H., Caruso, F.: Innovation in layer-by-layer assembly. Chem. Rev. **116**, 14828 (2016)
40. Carosio, F., Cuttica, F., Di Blasio, A., Alongi, J., Malucelli, G.: Layer by layer assembly of flame retardant thin films on closed cell PET foams: efficiency of ammonium polyphosphate versus DNA. Polym. Degrad. Stab. **113**, 189 (2015)
41. Maddalena, L., Carosio, F., Fina, A.: In situ assembly of DNA/graphene oxide nanoplates to reduce the fire threat of flexible foams. Adv. Mater. Interfaces **8**, 2101083 (2021)

Recent Developments on the Use of Eco-Friendlier Phosphazenes as Flame-Retardants for Different Polymeric Materials

R. Bocibo, M. J. Mochane, M. T. Motloung, T. C. Mokhena, and K. Selatile

Abstract The study of phosphazenes as halogen free flame-retardants has gained attention in recent years to replace halogen flame retardants because of their negative impact on the environment. Most phosphazenes have a very distinctive structures which play a significant role on their excellent flame-retardant properties. They can be used as reactive flame-retardants (FRs), additives FRs or even modifiers in polymers. Researchers have synthesized and investigated their use in polymers, to observe the effect they have on the flammability properties of polymers. These investigations have successfully allowed researchers to outline the mechanisms by which phosphazenes work as effective flame-retardants, including the formation of protective char layers and restraint of volatile combustible gases. Most studies involving the use of phosphazene as a FR require curing agents on thermosets such as epoxy resin. The cone calorimeter tests, thermogravimetric analysis (TGA), vertical burning test (UL 94) and the limiting oxygen index test (LOI) have been used in most studies to evaluate the thermal stability and the flame retardancy properties of polymers. Research studies show a great improvement of flame retardancy in polymers where phosphazenes were used as FRs and this presents a promising venture that can be used to design advanced flame-retardant polymeric materials with enhanced fire safety and reduced flammability across an extensive number of industries and applications.

Keywords Phosphazenes · Flame retardancy · Flammability properties

---

R. Bocibo · M. J. Mochane (✉) · M. T. Motloung
Department of Life Sciences, Central University of Technology, Bloemfontein, Free State, South Africa
e-mail: mochanejonas@gmail.com

T. C. Mokhena
DSI/Mintek-Nanotechnology Innovation Centre, Advanced Materials, Mintek, Randburg 2125, South Africa

K. Selatile
Division of Polymer Technology, Department of Chemical, Metallurgical and Materials Engineering and Institute of Nano Engineering Research, Tshwane University of Technology, Pretoria, South Africa

T. C. Mokhena et al. (eds.), *Biomaterials as Green Flame Retardants*, Engineering Materials, https://doi.org/10.1007/978-981-97-6871-4_14

1 Introduction

The use of polymers has become an integral part of uses in everyday life, for instance its use in furniture, motor vehicle parts, food and non-food packagings, construction, and many other applications. The demand for the use of polymers grows with the increase in industrialisation, and new ways in which they are being used in numerous industries due to their excellent characteristics. Most of these polymers are highly flammable and can easily ignite when exposed to heat or a flame which poses danger to human health and loss of property. Polymers have a high content of combustible elements, such as carbon and hydrogen, in their molecular structure, and examples include polyethylene (PE) and polypropylene (PP) and others that can be seen in Table 1 including their applications. There are different levels of flammability and thermal stability in different polymers due to their chemical composition, molecular structure, and other factors. Due to polymers being highly flammable, the use of flame retardants (FRs) has gained motion throughout the years [1]. Halogenated flame retardants were effectively used widely at the beginning for applications in different industries. However, despite their effectiveness in enhancing fire resistance, halogenated flame retardants raise concerns about their impact on the environment and human health [2]. Some of these compounds have been found in the environment, including water bodies and wildlife, prompting concerns about their potential to persist, accumulate in organisms, and exhibit toxicity. Consequently, regulatory measures have been introduced in some countries to restrict the use of halogenated flame retardants [2]. This led to the study and adoption of halogen free flame retardants that are more ecologically friendly and less prone to bioaccumulation such as zinc borates, phosphorus containing FRs such as phytic acid, alginates and phosphazenes. Usually there are two ways in which flame retardancy can be incorporated in polymeric materials. The first one is introducing reactive type FR into polymer chains by means of copolymerization and the second one includes the integration of additive FR into materials through blending or coating [3]. Phosphazenes were originally known as phosphonitrillic polymers, they were later named phosphazenes to emphasize its main components which are phosphorus and nitrogen [4]. They are a family of hybrid organic–inorganic materials characterized by their alternating phosphorus and nitrogen atoms which can be cyclic or in linear chains [5]. They are known in various structural forms such as cyclic trimers, tetramers, pentamers, hexamers, or higher polymers [6–8]. Phophazenes can exist as microspheres, nanotubes, nanofibers, and nano chains. This is due to the existence of multiple binding sites (e.g., hydroxyl, amine, chlorine) that allow for tuning of the molecule through covalent or noncovalent methods. Thus, small molecular compounds can be incorporated into this class of materials to produce other materials They are used in many applications including their use in medicine for drug delivery due to their biocompatibility, they can used to manufacture artificial blood vessels, they can be used as catalysts in chemical reactions and most importantly they can be used as flame retardants in polymers. They can be used solely or in synergy with other flame retardants to enhance the flame-retardant properties of polymers. Studies have shown that phosphazene compounds

Table 1 Examples of polymers and their common uses in different industries [10]

Type of polymer	Common applications
Polypropylene (PP)	Furniture, bumpers, lining and external board of cars
Polystyrene (PS)	Toys, hard packaging, refrigerator trays and CD cases
High density polyethylene (HDPE)	Washing shampoos bottles, bleach bottles, water pipes and laundry detergents bottles
Low density polyethylene (LDPE)	Shopping bags
Polyethylene terephthalate (PET)	Cooking oil bottles, soft drink and soft drink bottles and water bottles
Polyvinyl chloride (PVC)	Doors, windows, construction materials

and their derivatives can be introduced to thermosets primarily through chemical reaction, and this has restricted their applications in many cases as flame retardants [9]. Over the past years, there has been significant attention charged regarding the use of phosphazene-based family of material due to their advantages of high efficiency and biodegradability with less toxic byproducts, diverse thermal and chemical stabilities. Moreover, they offer to enhance the thermal and flame-retardant characteristics of epoxy resins and their composites. In this chapter we review the recent use of phosphazenes as flame retradants in different polymeric materials.

2 History of Phosphazenes

In 1834, Wohler and Rose conducted an experiment on the reaction of ammonia (NH_3) and phosphorus pentachloride (PCl_5), which produced a white crystalline product that could be collected without decomposition as a pure compound [11–13]. However, the empirical formula for this compound was proposed as $NPCl_2$ in 1844. Further, Gladstone and Holmes verified this formula using the vapour density method and suggested the procedure $(NPCl_2)_3$ [11–13]. In the 1980s Gladstone, Besson, Couldridge and stokes conducted research on the earliest insights of hydrolytic, ammonolytic and thermal behaviors of chlorophosphazenes [14].

In the context of polyphosphazene chemistry, Stokes was the first to propose the cyclic structure of polyphosphazene and reported "inorganic rubber" by heating the chlorophosphazenes. After Stokes' contribution, little progress was made in phosphazene polymer and material chemistry until the mid-1960s. Allcock, Valan, and Kugel reported a series of experiments on polyphosphazene and revealed the basic features of phosphazene. These findings led to the current state of polyphosphazene chemistry [11–13].

Polyphosphazenes are different from other polymers in both structure and method of synthesis. The backbone consists of alternating phosphorus and nitrogen atoms with two organic side groups attached to each phosphorus [6, 15, 16]. This backbone has some unique and valuable properties that include combinations of molecular and materials flexibility, fire resistance, near-ultraviolet and visible transparency, and radiation stability [15, 16].

3 Structure and Properties

Polyphosphazenes have become a burning topic in the world of research of phosphorus flame retardants [17–19]. They are suitable for ground vehicles and aeronautical components because of properties like hydrophobicity and oil resistance that enable glass transition temperatures as low as −60 °C. They have also been used in biostable biomedical equipment. Glassy elastomers and brittle glasses are examples of phosphazene polymers. Aryloxy-side groups or hydrogen bonds between amino-side groups raise the glass transition temperature. When rubber-like materials come into contact with water, hydrolysis causes them to lose their beneficial characteristics. The *P–N* bond distances in these ring structures are often equal or very near to being equal. The single-bond lengths that were anticipated are less than these. These findings suggest that the atoms of *P* and *N* are connected by a large number of links and resonance structures. The most complex polymers have a linear structure.

As mentioned in the introduction, the form of organic–inorganic hybrid high molecular polymer known as polyphosphazene has an organic side chain in place of the side chain and a main chain that is alternately joined by single and double bonds made up of phosphorus and nitrogen atoms. Through nucleophilic substitution reactions between polydichlorophosphazene (PDCP) or hexa- chlorocyclotriphosphazene (HCCP) molecules and various nucleophiles, the –P–N– repeating unit can form high molecular weight linear chain, branched, dendritic architecture, and cross-linked architecture. Among polyphosphazenes, the linear and crosslinked polyphosphazenes shown in Fig. 1 are the ones that have been investigated the most. [17, 19, 20].

More specifically, linear polyphosphazenes possess high temperature resistance, high efficiency flame retardant effect, and mechanical enhancement effect due to the molecular structure and flexibility of the main chain, which has great potential application prospects in the fields of high temperature resistant materials, biomaterials, coating materials, etc. The crosslinked polyphosphazenes can be used as flame retardant additives for some compounds, such as phenolic resin, polyethylene glycol terephthalate (PET) epoxy resin (EP) and many more others. These have been extensively researched due to the synergistic flame-retardant effect of phosphorus-nitrogen elements and high thermal stability. Additionally, the properties of polyphosphazenes have changed because of the various side group types, including elasticity, thermal

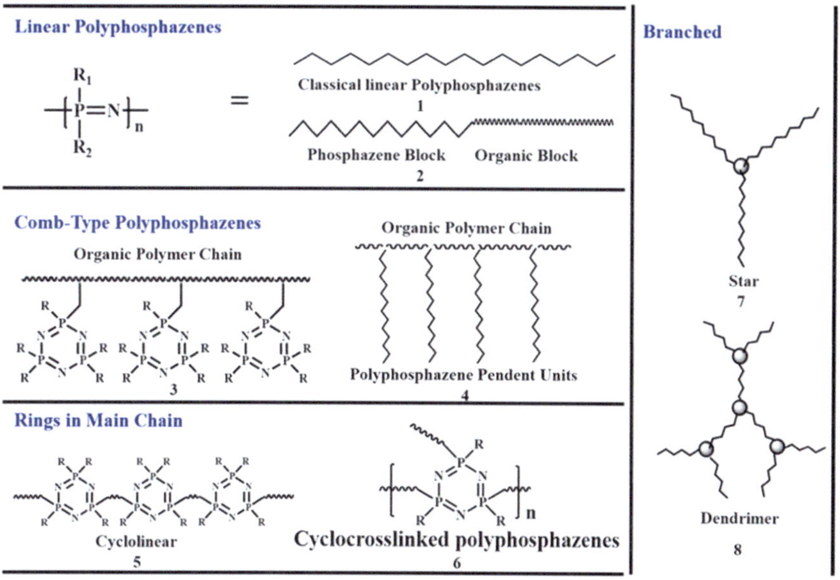

Fig. 1 An arrangement of different types of polyphosphazenes [17]

stability, solvent resistance, biological utility, electro-optical properties, and others. This shows that polyphosphazenes can be adjusted and controlled depending on the materials to which they are applied. Therefore, current studies on flame retardants based on polyphosphazenes are reviewed, and a flame-retardant mechanism for polyphosphazenes is also suggested. It is thought that by thoroughly comprehending the relationship between material structure, mechanism, and performance, the desired polymer materials may be developed. A benefit in the study of advanced functional polyphosphazene materials with multifunctional properties like electrical/electronic properties, mechanical properties, flame retardant properties, and thermal stability is the diversity of polyphosphazene, which endows itself with a variety of physical and chemical properties [21, 22].

3.1 Linear Phosphazenes

Linear polyphosphazenes, which are produced by nucleophilic substitutions between PDCP and nucleophiles, are an organic–inorganic hybrid linear polymer made up of phosphorus and nitrogen single and double bonds arranged alternately as flexible main chains and various functional substituents modified as side chains. Polymerization followed by substitution and substitution followed by polymerization are the two primary steps in the synthesis of linear polyphosphazenes. As shown in Fig. 2, due to the variety of linear polyphosphazenes structure, these compounds have a variety of physical and chemical properties, making them useful in a variety of applications, including conductive materials, special rubber and elastomers, optical materials, and film preparation [17, 23].

Particularly, the synergistic system made up of linear polyphosphazenes with high phosphorus/nitrogen ratios exhibit:

- outstanding flame retardancy
- excellent flexibility
- high temperature resistance
- high limiting oxygen index,
- and low smoke production.

Due to their fire resistance and self-extinguishing properties, linear polyphosphazenes are frequently employed in construction. Linear polyphosphazenes have excellent growth potential and may be employed as flame retardants in different polymeric materials. Another benefit for enhancing the overall performance of polymer

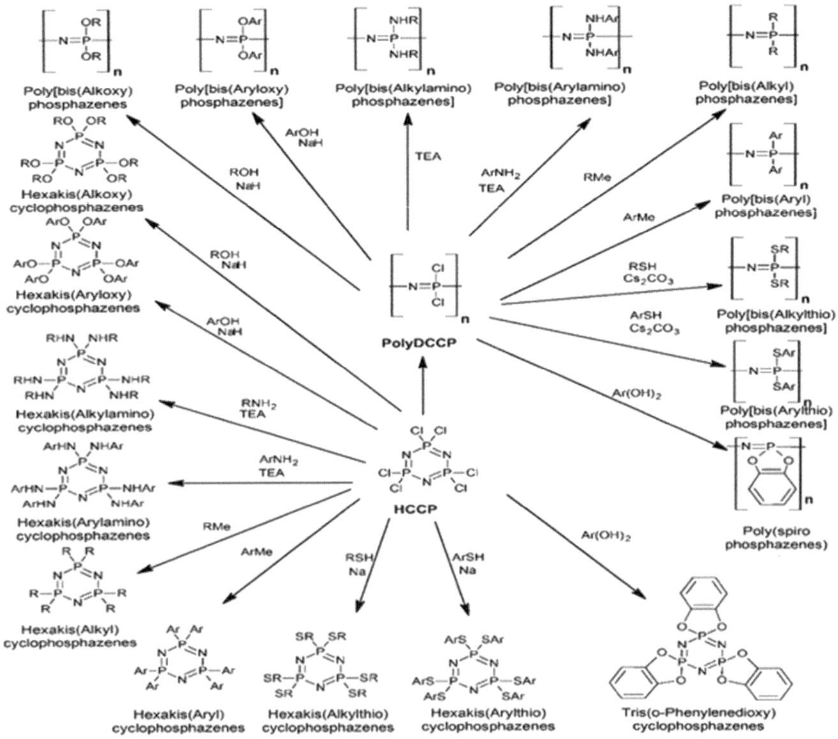

Fig. 2 HCCP and polydichlorocyclophosphazene (PolyDCCP) to a wide variety of derivatives by nucleophilic substitution [17]

materials is the high compatibility between linear poly-phosphazenes and polymer materials. Linear polyphosphazenes produce less hazardous gases like hydrogen halide during burning when compared to halogen-based flame-retardant materials, which is an important research direction of innovative ecologically friendly non-toxic flame-retardant materials [17, 23–26].

There have only been a few instances of linear polyphosphazenes being used as a flame retardant, which may be because of the demanding synthesis conditions and poor yield of the critical intermediate PDCP. Additionally, some linear polyphosphazene elastomers, like poly(aryloxyphosphazene)s and fluorinated linear polyphosphazenes, have excellent elasticity and flame retardancy, a low glass transition temperature, and high thermal stability. These materials are used as thermal protection materials and sealing materials in aerospace materials. Linear polyphospazenes are additionally created and intended to improve the flame-retardant qualities of polymers like fibers and plastics [17, 23, 24, 26, 27].

The safety of flame-retardant fiber is another crucial need for its application, and it is well recognized that flammability is a concern that must be overcome in the commercial application of fiber. In the present study, a functional linear polyphosphazene called poly [bis(methox- yethoxy)phosphazene] (PMEP) (Fig. 3) was devised and produced to improve the flame retardant qualities of viscose fiber by the use of wet spinning. It has been noted that pure viscose fiber has a far smoother surface than viscose fiber that has undergone PMEP modification. An investigation was conducted on the viscose fiber's thermal stability and combustion capabilities when combined with various PMEP concentrations. It appears that the presence of PMEP enhanced the viscose fiber's char residue, which was crucial in reducing the transmission of mass and heat. The long-term industrial applications of durable flame-retardant fibers are currently constrained. As seen in Fig. 3, a UV-induced grafting procedure was developed and used to create a noncombustible and highly foam-forming polyphosphazene derivative that can be fixed durably on cotton and other cotton/polyester blended textiles in high additions up to 40% [17, 23–26].

3.2 Cyclocrosslinked Polyphosphazene

Another organic–inorganic hybrid polymer created using HCCP as the nucleus and nucleophilic replacements of active chlorine with organic molecules is called cyclocrosslinked polyphosphazenes. Cyclocrosslinked polyphosphazenes have a higher yield up to 90% than linear polyphosphazenes, which attracts more and more interest from researchers. On the one hand, the phosphazene rings of cyclocrosslinked polyphosphazenes contain six N and P bonds that are of the same length, and the P atoms are evenly spaced above and below the rings.

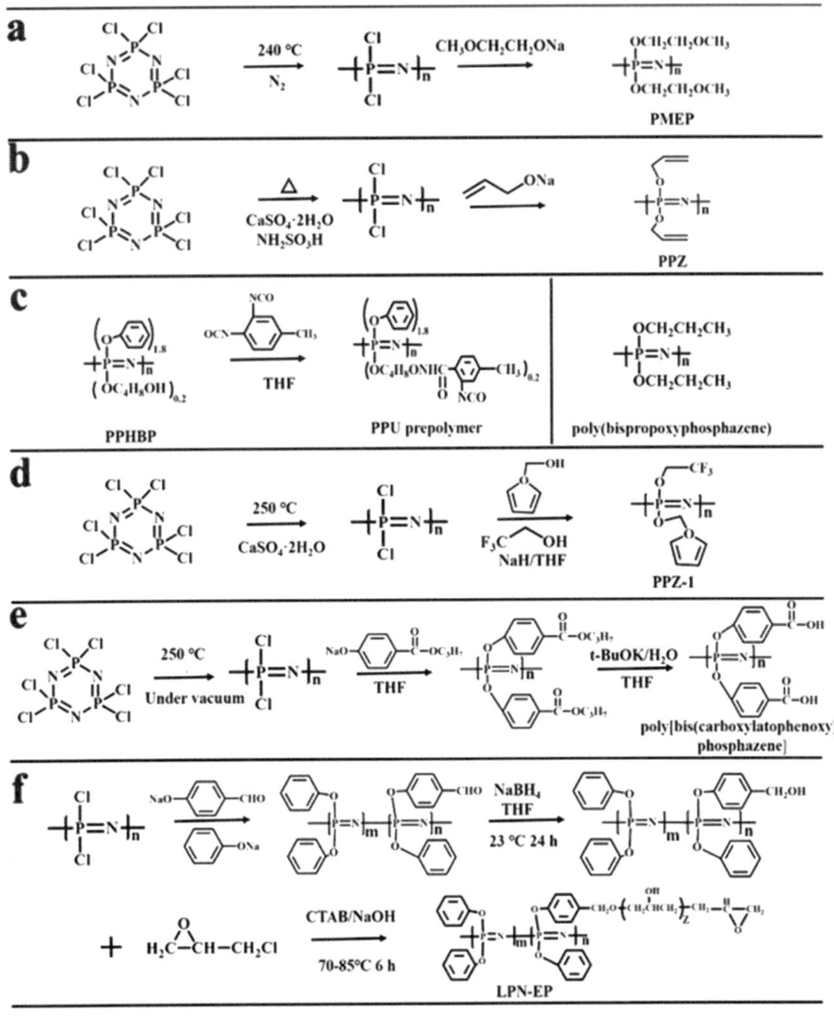

Fig. 3 Synthesis of linear phosphazenes [17]

As a result, the *d–p* hybrid orbital also forms in the phosphazene ring of cyclocrosslinked polyphosphazenes, indicating that these compounds have many properties with linear polyphosphazenes, such as good thermal stability and heat resistance. The high degree of crosslinking and hexatomic ring structure of the cyclocrosslinked polyphosphazenes, on the other hand, contribute to their stiffness. Additionally, unlike linear polyphosphazenes, crosslinked polyphosphazenes may be formed into a variety of micro-nano structures such nanotubes, microspheres, and nano-sheets [17, 25, 27, 28].

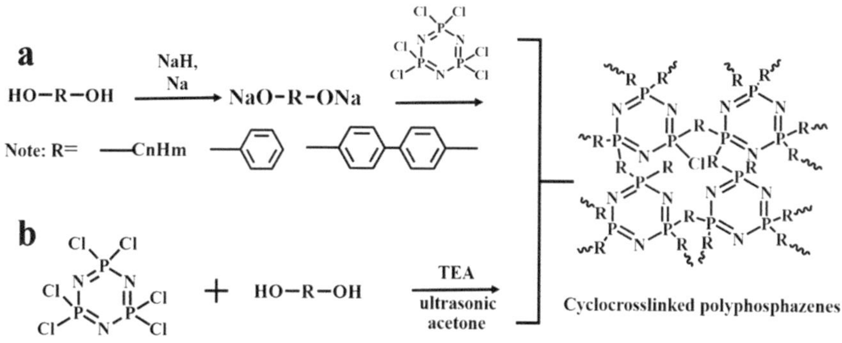

Fig. 4 Synthesis of cyclophosphazenes [17]

Cyclocrosslinked polyphosphazenes have a wide range of applications in flame retardant materials, special materials for the military industry, filling materials, medicine loading, adsorption materials, etc. because of their excellent thermal stability, heat and corrosion resistance, flame retardant, and micro-nano size effect. The fact that the phosphazene rings of cyclocrosslinked polyphosphazenes can also be altered by various functional groups and that the physicochemical properties can be directly tailored by varying the type and quantity of functional groups demonstrates their superiority to other organic polymers and their strong molecular designability. In conclusion, because of their distinct structure and properties, cyclocrosslinked polyphosphazenes have attracted increasing attention as flame retardant additives in recent years. They are thought to represent a significant branch of phosphorus flame-retardants and to be a potent alternative to halogen flame-retardants [17, 25, 28, 29] (Fig. 4).

3.3 Cyclocrosslinked Polyphosphazenes Micro-Nanostructure Flame Retardant Materials

Due to their unique effects, such as the quantum size effect, volume effect (or small size effect), surface and interface effect, macroscopic quantum tunneling effect, and dielectric confinement effect, nanomaterials have earned the title of "the most promising materials in the twenty-first century." Additionally, due to the small size impact and high specific surface area of the micro-nano morphology, flame retardant additives with micro-nanostructures are provided with good enhancing effect in both mechanical and flame retardant capabilities of polymers [17, 25, 28, 29] (Fig. 5).

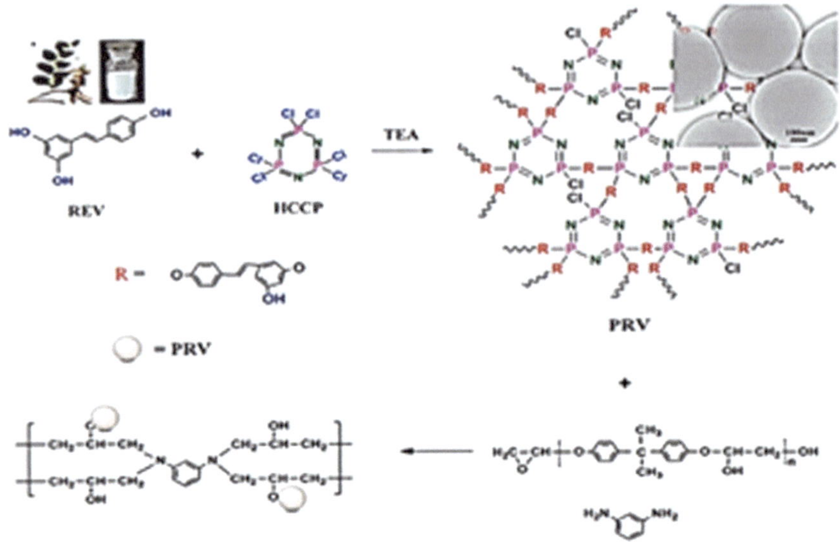

Fig. 5 Synthesis of cyclophosphazene flame retardants [17]

4 Phoshazenes (Linear and Cyclo) and Polymers: Flame Retardancy

The term flammability refers to how easily a material can catch fire when exposed to an open flame or any heat source and its ability to continue burning. Figure 6 clearly demonstrates factors affecting flammability of polymeric materials.

There are a number of tests that need to be conducted in order to test for flammability of materials and these include the limiting oxygen test (LOI), vertical burning test (UL-94), thermogravimetric analysis (TGA) and the cone calorimeter tests. These tests are further explained in Table 2. Polymeric materials are made through the polymerisation of organic molecules and are therefore highly flammable, which increases the risk of fire during their lifetime. To ensure the safety of both individuals and property, fire retardants are routinely applied to most polymer materials during the preparation cycle. The addition of non-combustible fillers decreases flammability by reducing fuel and slowing down the diffusion of oxygen and fuel within the polymer. Bio-based flame-retardants are commonly used in a variety of materials, including plastics and fabrics. Phosphazenes are emerging as the most promising flame-retardant candidates for polymer composites. Resultantly, they have been added to polymer matrices like epoxy and produced higher flame restriction properties which have been attributed to the blocking effect of phosphorous in the condensed phase, the quenching effect of phosphorous free radicals and the dilution effect of non-combustible gas that is released by nitrogen-containing flame retardants [5, 30, 31]

Fig. 6 Factors affecting flammability of polymeric materials

Table 2 Definitions of the types of flammability tests conducted on polymers

Type of flammability test	Explanation	References
Limiting oxygen index	This test measures the amount of oxygen that is needed to start and sustain the burning of a material. It takes place in a controlled atmospheric condition. Since air contains about 20.95% volume of oxygen, any material with an LOI value less than this will burn easily in air. A lot of researchers proposed that for a material to be regarded as "self-extinguishing" it needs to have an LOI value of 28% and above	[32–34]
Vertical burning test (UL-94)	Established by the Underwriters Laboratories, this testing tool measures the material's ability to start flames and self-extinguish, and the way in which it drips when exposed to a heat source. It has ratings that determines whether a material has passed or not, and it goes as follows: • V-0: flames self-extinguish after 10 s with no dripping • V-1: flames self-extinguish after 30 with no dripping • V-2: flames self-extinguish after 30 s with dripping For a material to effectively passed the test it needs to obtain V-0	[34, 35]
Thermogravimetric analysis (TGA) test	Measures the material's mass as it changes when exposed to different temperatures. The data is recorded down to form a thermal curve or thermogram which is defined as a graphical representation of the sample mass change vs time or temperature	[34, 36]
Cone calorimeter tests	Is a tool for evaluating the heat release rate of a material. It is implemented by the International Organization for Standardization (ISO 5660-1). It also measures time to ignition, mass loss, smoke released, and gas analyses and other factors related to combustion	[34, 37, 38]

5 Phosphazene/Polymer Composites

Phosphazenes represent a class of versatile hybrid materials at the interface between organic and inorganic that possess outstanding thermal stability [11]. Improved retardancy with phosphazenes has been reported in several thermosetting materials [11, 30, 39]. The combination of phosphorous and nitrogen in flame-retardant fillers results in a synergistic effect that considerably increases the flammability of the finished product. Numerous additives or reactive flame-retardants have been synthesized via side-group substitution, which serves to increase the thermal stability and flame retardancy of the compounds. Also, enhanced mechanical performance with polymer matrices can be achieved by increasing interfacial adhesion through active groups on their surface [5, 40, 41].

Phosphorous and nitrogen-containing compounds display enhanced flame retardancy in comparison to similar compounds containing phosphorous alone [42, 43]. By properly substituting phosphazene molecules membranes, solid ionic conductors, elastomers, flame-resistant materials, and inert biomaterials can be produced. This makes phosphazenes especially good candidates for the development of advanced fire-resistant materials for various applications. Flame-retardants in polymer materials primarily complete or obstruct the combustion chain reaction in a variety of ways. Two basic flame-retardant mechanisms exist, i.e., gas and condensed phase, and phosphazenes acts on both to extinguish polymer flammability. When phosphazenes are incorporated into polymer matrices, they act as intumescent, releasing non-combustible gases and producing a protective char layer when exposed to heat or flame. This layer functions as a barrier, lowering the availability of combustible gases and preventing fire spread [17, 31, 40]. The typical mechanism is depicted in Fig. 7.

The effect of phosphazenes as flame retardant fillers in various polymer matrices can differ depending on several factors, including the specific type of phosphazene compound, its concentration in the polymer composite, the nature of the polymer matrix, and the processing conditions. However, phosphazenes are known to be effective flame-retardants in general and can provide several advantages when mixed with diverse polymer matrices. In some cases, they improve not only fire protection but also compatibility, resulting in improved mechanical properties among others (see Table 3) [17, 39]. Synergies for better properties at low filler loading are also feasible. For instance, Ferry et al. used only 1.0 wt% of DM-Si (a phosphasene deriviate) in PLA to achieve a UL-94 V-0 rating and a high LOI value of 27.9%. Furthermore, the tensile toughness of PLA was improved from 2.8 to 4.7 MJ/m^3, while the mechanical strength is well-preserved. Finally, the resultant PLA/DM-Si materials showed comparable transparency to the PLA matrix. Hexaglycidyl cyclotriphosphazene (HGCP) was synthesized and incorporated in the epoxy resin diglycidyl ether of bisphenol A (DGEB) using 4,40-methylene-dianiline (MDA) as a curing agent for DGEB, with the aim to improve its flame retardancy. Epoxy resin offers numerous desirable qualities, including high strength, excellent linkage to various materials, good electrical insulating properties, resistance to chemicals, heat, and moisture. A

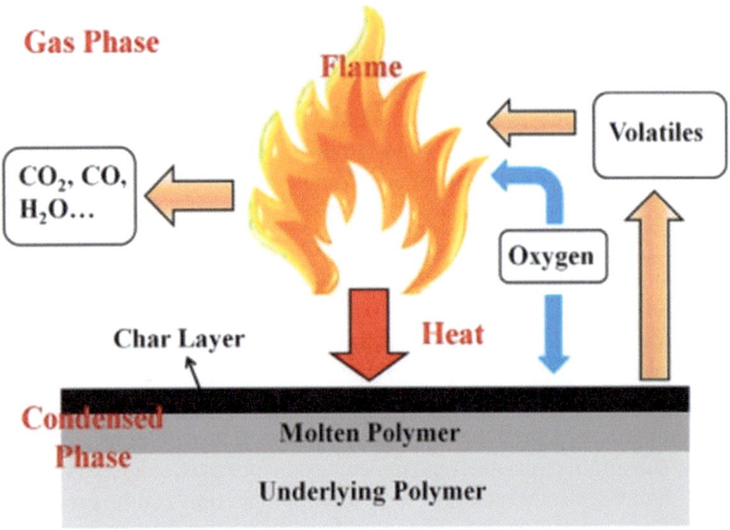

Fig. 7 The mode of action of phosphazene-based flame-retardants on polymer composites [31]

curing agent is used to enable the process of changing a liquid or semi-liquid material into a solid and durable form. The results indicated that DGEB/20%HGCP/MDA achieved a V-0 rating with light smoke [1].

6 Phosphazenes in Synergy with Other Flame Retardants in Polymers

In a recent study phosphazene-based epoxy resin (PN-EP) was effectively synthesized through the reaction between diglycidyl ether of bisphenol-A (DGEBA) and hexakis(4-hydroxyphenoxy)-cyclotriphosphazene. Cyclotriphosphazenes are compounds that are made up of rotating phosphorus and nitrogen atoms which aid in awarding this compound with thermal properties such as the ability to self-extinguish, enhancing its flame retardant properties [27]. Cyclotriphosphazene offers numerous benefits when employed as a dynamic reactive oligomer for flame retardancy. Cyclotriphosphazene has a number of benefits in its application as a reactive flame-retardant functional oligomer. First, preparation of cyclotriphosphazene-based copolymer with various substituents has a flexible synthetic approach which allows researchers to obtain multifunctional initiators or terminators with simplicity due to its adaptable synthetic methodology. Secondly thermal and non-flammable qualities of cyclotriphosphazene units can be conferred to the resultant polymers, particularly those possessing low molecular weights [1, 27]. The effectiveness of the flame retardancy was tested on for thermosets that had undergone curing. This type of

Table 3 Possible property improvements when all conditions are met

Property	Detail	References
Flame retardancy	Phosphorus and nitrogen containing compounds can act as effective flame-retardants in polymer composites, reducing flammability and delaying the spread of flames	[17, 31, 40]
Smoke suppression	Polymer composites with phosphazene can reduce smoke production during fires, minimizing visibility impairment and smoke inhalation risks	[17, 31, 40]
Reduced heat release rate	Phosphazenes can reduce the heat release rate during combustion, which is necessary for fire safety because it slows fire propagation and gives more time for evacuation or fire suppression	[17, 31, 40]
Mechanical properties	Sometimes incorporating additives such as phosphazenes can improve polymer composite properties such as tensile strength or impact resistance	[17, 31, 40]
Increased thermal stability	Adding phosphazenes to the polymer matrix can improve its thermal stability, allowing it to withstand higher temperatures before degrading	[41]
Compatibility	Phosphazenes are tunable and can be modified to suit a wide range of polymer matrices, making them versatile and applicable to various materials	[17, 44]
Less toxic by-products	Leads to the production of polymer blends with less toxic combustion products, which is an important factor to consider for applications where human exposure to combustion gases is a concern	[30]

phosphazene based epoxy resin passed the UL 94 rating with V-0 and LOI values of more than 36%.

Jiang et al. (2014) synthesized four unique intumescent flame-retardants containing phosphaphenanthrene and phosphazene groups by using the classic Atherton–Todd reaction and Kabachnik–Fields reactions. Firstly 9,10-dihydro-9-oxa-10-phosphaphenanthrene-10 oxide (DOPO) was reacted with hexa-aminophenoxy-cyclotriphosphazene (compound 2) in the presence of carbon tetrachloride and triethylamine (Atherton-Todd reaction) to get hexa-(phosphaphenanthrene-anmino-phenoxyl)-cyclotriphosphazene (compound 3). Compound 4 was synthesized by the condensation of hexa-aminophenoxy-cyclotriphosphazene with formaldehyde,salicylaldehyde, 4-hydroxybenzaldehyde respectively, followed by the addition of (Kabachnik-Fields reaction) to the imine linkage to get hexa-(phosphaphenanthrene-methyl-amino-phenoxyl)-cyclotriphosphazene (compound 4), hexa-(phosphaphenanthrene-methyl-(o-hydroxybenzal)-amino-phenoxyl)-cyclotriphosphazene (compound 5) and hexa-(phosphaphenanthrene-methyl-(p-hydrobenzal)-amino-phenoxylcyclotriphosphazene (compound 6). Amongst these compounds, compound 5 (hexa-(phosphaphenanthrene-methyl-(o-hydroxybenzal)-amino-phenoxyl)-cyclotriphosphazene) was used to enhance the flame retardancy

Table 4 Flame retardancy results of PLA applied with HTTCP and APP (sourced from [38])

Samples	PLA	HTTCP	APP	LOI	UL94	Dripping	Char residues at 700
Pure PLA	100	0	0	20.2	NR	Y	0.1
PLA-25%APP	75	0	25	28.7	V-2	Y	7.5
PLA-25%HTTCP	75	25	0	25.2	NR	Y	14.6
PLA-HTTCP-APP 1/1	75	12.5	12.5	41.3	V-0	N	13.9
PLA-HTTCP-APP1/2	75	8.3	16.6	40.8	V-0	N	14.9
PLA-HTTCP-APP 2/1	75	16.6	8.6	30.8	V-1	Y	16.3

of polylactic acid (PLA). The results show that they obtained a UL-94 rating of V-0 and an LOI value of 30% was higher than that of pure PLA which was only 21.5% [6].

A new flame retardant hexa-[N,N0,N00-tris-(2-amino-ethyl)-[1, 3, 5] triazine-2,4,6-triamine] cyclotriphosphazene (HTTCP) containing ammonium polyhosphate (APP) and triazine groups was synthesized and applied on the polylactic acid (PLA) polymer matrix with the goal to achieve an improvement in terms of flame retardancy and mechanical properties. Table 4 displays results obtained; pure PLA has the lowest LOI of 20.2% compared to when it is blended with either HTTCP or APP. It achieved the highest LOI value of 41.3% when the content compositions of HTTCP and APP are equal at PLA-HTTCP-PLA 1/1. Pure PLA was not rated (NR) in the UL-94 test, which means that the material did not meet the standards for any of the flame classifications specified by the test. PLA/25%/APP solely failed the UL-94 test with V-2 rating, PLA/25%HTTCP solely was also not rated, however when HTTCP and APP are combined at ratio 1/1 and ½ they get a good rating of V-0.

Yang et al. (2015) synthesized the hexa-(phosphite-hydroxyl-methyl-phenoxyl)-cyclotriphosphazene (HPHPCP) flame retardant from hexachlorocyclotriphosphazene, diethyl phosphite and p-hydroxybenzaldehyd as can be seen in Fig. 8. The HPHPCP was then applied to rigid polyurethane to observe if it would improve its flame retardancy. The HPHPCP content was 0, 5, 10, 15, 20 and 25% respectively, and the LOI values were 19, 21, 23, 25, 26 and 26%, meaning that the LOI value increases with an increase the HPHPCP content improving the flame retardancy of rigid polyurethane. The char residues formed in this study also increased with an increase in HPHPCP content, and this is observed as a strong evidence of flame retardancy [3]. They performed a horizontal burning test, this test is widely used to evaluate the flammability performance (ignition resistance, flame propagation and accompanied heat release) of foamed materials, it records the burning time, afterglow time of each specimen, dripping of burning specimen (ignition of cotton batting) and the length of the damaged specimen [45]. Pure rigid polyurethane foam and the one with 5%HPHPCP content failed the test whereas the other contents greater than them got a rating of HF-1.

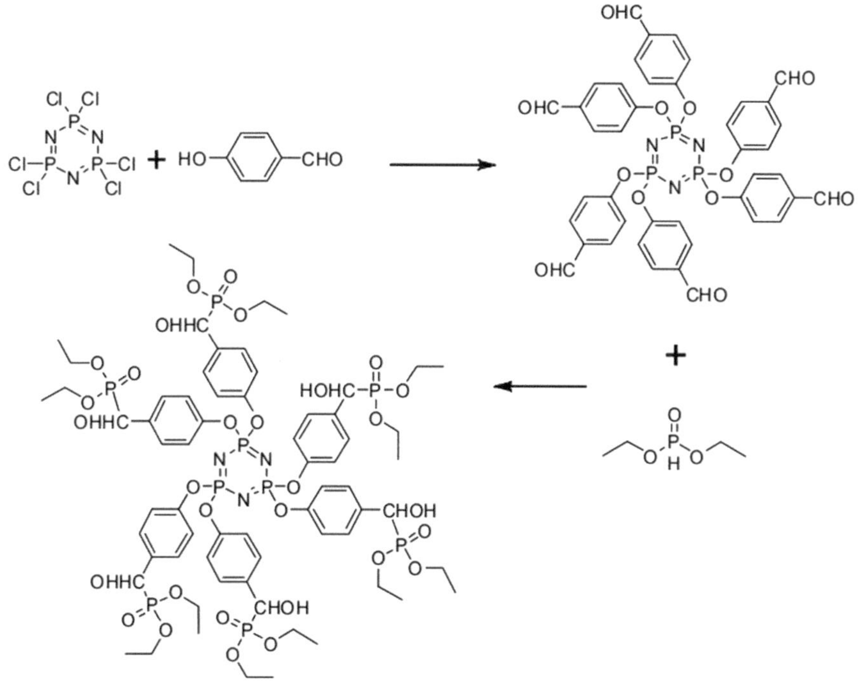

Fig. 8 Synthesis of HPHPCP [3]

Spirocyclic phosphazene-based epoxy resin was designed by a three-step synthetic pathway and was then cured using 4,4′-Diamino-diphenylmethane (DDM), 4,4′-Diamino-diphenyl sulfone (DDS), and Novolac as hardeners [46]. The hardener was added with a ratio that is equivalent to the epoxy resin of 1:1. The flame retardancy results differ with the kind of curing agent that was used, meaning that the curing agent does affect the flame retardancy of the phosphazene-based epoxy resin used. All three composites achieved a V-0 vertical burning test results, SP-PN epoxy/DDM got ~31%, while SP-PN epoxy/DDS obtained ~33% and SPN epoxy/novolac obtained a ~31%. This suggests that introducing the spiro-cyclotriphosphazene unit into the backbone of epoxy resin as a pendant group significantly enhances its flame retardancy. Therefore, the epoxy resin develops into a polymer with nonflammable properties [46]. The TGA results were obtained under air atmosphere and nitrogen atmosphere, and they obtained good thermal stabilities with high char yields (Table 5).

Table 5 Selective studies of phosphazenes as flame-retardants on polymers

Type of phosphazene + other flame retardant/polymer	Method of preparation	Highlight of results	Refs
Hexa-(4-aldehyde-phenoxy)-cyclotriphosphazene + 5,5-dimethyl-1,3,2-dioxaphosphinane 2-oxide to form hexa-(5,5-dimethyl-1,3,2-dioxaphosphinane-hydroxyl-methyl-phenoxyl)-cyclotriphosphazene (HDPCP)/rigid polyurethane (RPUF)	P–H addition reaction	Neat RPUF has a lower LOI value of 19% compared to RPUF/HDPCP15, 20 and 25, whose values were 23, 24 and 25% respectively. The LOI value increased with an increase in the content of HDPCP, showing an improvement in flame retardancy for the polyurethane	[47]
Amine-functionalized phosphazene nanotubes (APZ) + polypropylene-graft-maleic anhydride (PP-g-MA)/PP	Melt blending	• PP/PP-gMA/1%APZ got completely burned during the vertical burning test. Whereas, with an increase in the content of APZ like 5 and 10% got a V-0 rating • The LOI value of pure PP was 17.5%. the highest value attained was that of 10%APZ which got 23.3%, definitely showing an improvement in flame retardancy of PP	[48]
Phosphazene-based flame retardant (PBFA) + pentaerythritol derivatives (PEPA)/ PP	PBFA and PEPA were mixed together in a crusher to prepare evenly fame-retardant	The peak heat release rate for pure PP was 870 kW m^{-2}. This value decreased significantly with the introduction of PBFA/PEPA ratios • PP-PBFA got 403 kW/m^{-2} • PP-PEPA got 411 kW/m^{-2} • PP-PBFA/PEPA (1:1) got 244 kW/m^{-2}, • PP-PBFA/PEPA (2:1) got 214 kW/m^{-2} • PP-PBFA/PEPA (4:1) got 210 kW/m^{-2} They decreased by 53.58%, 52.69%, 71.95%, 75.38% and 75.76%, respectively	[49]
Poly(cyclotriphosphazene-co-pentaerythritol) (PCPP)/polylactide (PLA)	Melt blending	• The PHRR of pure PLA was 271 kW/m^2, compared to PLA/10PCPP and PLA/20PCPP which obtained 229 kW/m^2 and 122 kW/m^2, indicating a reduction of 16% and 55% from pure PLA • Pure PLA failed the UL 94 test, whereas an addition of PCPP got a rating of V-0 with no drippings • Pure PLA got an LOI value of 21%. An increase in the content of PCPP increased the LOI of PLA, with the highest LOI being 28% with 10% of PCPP on the PLA matrix	[50]

7 Conclusions

The research activity of the synthesis of novel phosphazenes as flame-retardants has increased over the years and it has shown to be a success. The mechanism of char formation on the polymers works well in enhancing the flammability properties of polymeric materials. They can function solely or in synergy with curing agents for thermosets or other flame-retardants. Various tests have been conducted to prove that they are good replacements of traditional halogen FRs because of their eco-friendliness, high efficiency, and biodegradability with less toxic byproducts during combustion aligning with sustainability objectives.

Although there are limited papers on linear phosphazenes, however their use and application as FRs have proven to be a success. More research is being done on cyclo-crosslinked polyphosphazenes which have proved to have a higher yield up to 90% than linear polyphosphazenes. Their compatibility with various polymers and the accessibility of different testing methods have helped their integration into a broad range of materials. The advantages of phosphazenes including their eco-friendliness, distinctive structures, adaptability, and the ability to enhance flame retardancy, places them as a promising and successful solution for establishing safer and more fire-resistant materials in different applications in the future.

References

1. El Gouri, M., El Bachiri, A., Hegazi, S.E., Rafik, M., El Harfi, A.: Thermal degradation of a reactive flame retardant based on cyclotriphosphazene and its blend with DGEBA epoxy resin. Polym. Degrad. Stab. **94**, 2101–2106 (2009). https://doi.org/10.1016/j.polymdegradstab.2009.08.009
2. Idumah, C.I., Hassan, A., Affam, A.C.: A review of recent developments in flammability of polymer nanocomposites. Rev. Chem. Eng. **31**, 149–177 (2015). https://doi.org/10.1515/revce-2014-0038
3. Yang, R., Hu, W., Xu, L., Song, Y., Li, J.: Synthesis, mechanical properties and fire behaviors of rigid polyurethane foam with a reactive flame retardant containing phosphazene and phosphate. Polym. Degrad. Stab. **122**, 102–109 (2015). https://doi.org/10.1016/j.polymdegradstab.2015.10.007
4. Atkins P.W., Shriver, D.F.: Inorg. Chem. 1–12 (2006)
5. Yang, G., Wu, W.H., Wang, Y.H., Jiao, Y.H., Lu, L.Y., Qu, H.Q., Qin, X.Y.: Synthesis of a novel phosphazene-based flame retardant with active amine groups and its application in reducing the fire hazard of Epoxy Resin. J. Hazard. Mater. **366**, 78–87 (2019). https://doi.org/10.1016/j.jhazmat.2018.11.093
6. Casella, G., Carlotto, S., Lanero, F., Mozzon, M., Sgarbossa, P., Bertani, R.: Cyclo- and polyphosphazenes for biomedical applications. Molecules **27** (2022). https://doi.org/10.3390/molecules27238117
7. Tian, Z., Hess, A., Fellin, C.R., Nulwala, H., Allcock, H.R.: Phosphazene high polymers and models with cyclic aliphatic side groups: new structure-property relationships. Macromolecules **48**, 4301–4311 (2015). https://doi.org/10.1021/acs.macromol.5b00946
8. Allcock, H.R., Desorcie, J.L., Riding, G.H.: The organometallic chemistry of phosphazenes. Polyhedron **6**, 119–157 (1987). https://doi.org/10.1016/S0277-5387(00)80783-4

9. Jiang, P., Gu, X., Zhang, S., Sun, J., Wu, S., Zhao, Q.: Syntheses and characterization of four phosphaphenanthrene and phosphazene-based flame retardants. Phosphorus, Sulfur Silicon Relat. Elem. **189**, 1811–1822 (2014). https://doi.org/10.1080/10426507.2014.902828
10. Gilbert, M.: Plastics materials: introduction and historical development. Brydson's Plast. Mater. Eighth Ed. 2–18 (2016). https://doi.org/10.1016/B978-0-323-35824-8.00001-3
11. Ahmad, M., Nawaz, T., Hussain, I., Chen, X., Imran, M., Hussain, R., Assiri, M.A., Ali, S., Wu, Z.: Phosphazene cyclomatrix network-based polymer: chemistry, synthesis, and applications. ACS Omega **7**, 28694–28707 (2022). https://doi.org/10.1021/acsomega.2c01573
12. Shaw, S.D., Blum, A., Weber, R., Kannan, K., Rich, D., Lucas, D., Koshland, C.P., Dobraca, D., Hanson, S., Birnbaum, L.S.: Phosphate esters. Rev. Environ. Health **25**, 261–305 (2010). https://doi.org/10.1515/REVEH.2010.25.4.261
13. Shaw, R.A.: The Phosphazenes (phosphonitrilic compounds) (1961)
14. Allcock, H.R.: Recent advances in phosphazene (phosphonitrilic) chemistry. Chem. Rev. **72**, 315–356 (1971)
15. Rothemund, S., Teasdale, I.: Preparation of polyphosphazenes: a tutorial review. Chem. Soc. Rev. **45**, 5200–5215 (2016). https://doi.org/10.1039/c6cs00340k
16. Weitkamp, R.F., Neumann, B., Stammler, H.G., Hoge, B.: Phosphorus-containing superbases: recent progress in the chemistry of electron-abundant phosphines and phosphazenes. Chem. A Eur. J. **27**, 10807–10825 (2021). https://doi.org/10.1002/chem.202101065
17. Zhou, X., Qiu, S., Mu, X., Zhou, M., Cai, W., Song, L., Xing, W., Hu, Y.: Polyphosphazenes-based flame retardants: a review. Compos. Part B Eng. **202**, 108397 (2020). https://doi.org/10.1016/j.compositesb.2020.108397
18. Gleria, M., De Jaeger, R.: Aspects of phosphazene research. J. Inorg. Organomet. Polym. **11**, 1–45 (2001). https://doi.org/10.1023/A:1013276518701
19. Bao, D.M., Wang, J.H., Hou, Z.M., Xu, Z.Y., Ye, X.L., Qi, Y.Z., Xu, S.J., Long, L.J., Wu, Z.L., Wen, Z.: Synthesis of a novel flame retardant with phosphaphenanthrene and phosphazene double functional groups and flame retardancy of poly (lactic acid) composites. Front. Mater. **9**, 1–9 (2022). https://doi.org/10.3389/fmats.2022.951515
20. He, Y., Chen, Y., Zheng, Q., Zheng, J., Chen, S.: Preparation and properties of flame-retardant viscose fiber modified with poly[bis(methoxyethoxy)phosphazene]. Fibers Polym. **16**, 1005–1011 (2015). https://doi.org/10.1007/s12221-015-1005-x
21. Seraji, S.M., Gan, H., Swan, S.R., Varley, R.J.: Phosphazene as an effective flame retardant for rapid curing epoxy resins. React. Funct. Polym. **164**, 104910 (2021). https://doi.org/10.1016/j.reactfunctpolym.2021.104910
22. Qian, L., Ye, L., Qiu, Y., Qu, S.: Thermal degradation behavior of the compound containing phosphaphenanthrene and phosphazene groups and its flame retardant mechanism on epoxy resin. Polymer (Guildf). **52**, 5486–5493 (2011). https://doi.org/10.1016/j.polymer.2011.09.053
23. Mu, X., Yuan, B., Hu, W., Qiu, S., Song, L., Hu, Y.: Flame retardant and anti-dripping properties of polylactic acid/poly(bis(phenoxy)phosphazene)/expandable graphite composite and its flame retardant mechanism. RSC Adv. **5**, 76068–76077 (2015). https://doi.org/10.1039/c5ra12701g
24. Chen, Y., Wang, W., Qiu, Y., Li, L., Qian, L., Xin, F.: Terminal group effects of phosphazene-triazine bi-group flame retardant additives in flame retardant polylactic acid composites. Polym. Degrad. Stab. **140**, 166–175 (2017). https://doi.org/10.1016/j.polymdegradstab.2017.04.024
25. Wang, L., Yang, B., Guo, Y., Zhang, Y., Wang, N., Li, F., Yu, H., Cui, J., Guo, J., Mu, B., Tian, L.: Synthesis of multielement phosphazene derivative and the study on flame-retardant properties of epoxy resin. High Perform. Polym. **32**, 1169–1180 (2020). https://doi.org/10.1177/0954008320937392
26. Mathew, D., Reghunadhan Nair, C.P., Ninan, K.N.: Phosphazene-triazine cyclomatrix network polymers: some aspects of synthesis, thermal- and flame-retardant characteristics. Polym. Int. **49**, 48–56 (2000). https://doi.org/10.1002/(SICI)1097-0126(200001)49:1%3c48::AID-PI309%3e3.0.CO;2-M
27. Liu, R., Wang, X.: Synthesis, characterization, thermal properties and flame retardancy of a novel nonflammable phosphazene-based epoxy resin. Polym. Degrad. Stab. **94**, 617–624 (2009). https://doi.org/10.1016/j.polymdegradstab.2009.01.008

28. Szczotok, A.M., Madsen, D., Serrano, A., Carmona, M., Van Hees, P., Rodriguez, J.F., Kjøniksen, A.L.: Flame retardancy of rigid polyurethane foams containing thermoregulating microcapsules with phosphazene-based monomers. J. Mater. Sci. **56**, 1172–1188 (2021). https://doi.org/10.1007/s10853-020-05389-6
29. Yusuf, A., Sai Avvaru, V., De la Vega, J., Zhang, M., Garcia Molleja, J., Wang, D.Y.: Unveiling the structure, chemistry, and formation mechanism of an in-situ phosphazene flame retardant-derived interphase layer in $LiFePO_4$ cathode. Chem. Eng. J. **455**, 140678 (2023). https://doi.org/10.1016/j.cej.2022.140678
30. Krishnadevi, K., Selvaraj, V.: Development of halogen-free flame retardant phosphazene and rice husk ash incorporated benzoxazine blended epoxy composites for microelectronic applications. New J. Chem. **39**, 6555–6567 (2015). https://doi.org/10.1039/c5nj00364d
31. Yang, Y., Díaz Palencia, J.L., Wang, N., Jiang, Y., Wang, D.Y.: Nanocarbon-based flame retardant polymer nanocomposites. Molecules **26**, (2021). https://doi.org/10.3390/molecules26154670
32. Nelson, M.I., Sidhu, H.S., Weber, R.O., Mercer, G.N.: A dynamical systems model of the limiting oxygen index test. Anziam J. **43**, 105–117 (2001). https://doi.org/10.1017/S1446181100011457
33. Suzanne, M., Delichatsios, M.A., Zhang, J.P.: Flame extinction properties of solids obtained from limiting oxygen index tests. Combust. Flame **161**, 288–294 (2014). https://doi.org/10.1016/j.combustflame.2013.08.026
34. Hodgson, D.M., Gras, E.: Recent developments in the chemistry of lithiated epoxides. Synthesis (Stuttg). **27**, 1625–1642 (2002). https://doi.org/10.1055/s-2002-33635
35. Wang, Y., Zhang, J.: Thermal stabilities of drops of burning thermoplastics under the UL 94 vertical test conditions. J. Hazard. Mater. **246–247**, 103–109 (2013). https://doi.org/10.1016/j.jhazmat.2012.12.020
36. Saadatkhah, N., Carillo Garcia, A., Ackermann, S., Leclerc, P., Latifi, M., Samih, S., Patience, G.S., Chaouki, J.: Experimental methods in chemical engineering: thermogravimetric analysis—TGA. Can. J. Chem. Eng. **98**, 34–43 (2020). https://doi.org/10.1002/cjce.23673
37. Schartel, B., Hull, T.R.: Cone calorimeter—a tool for measuring heat release rate. Fire Mater. **31**, 327–354 (2007)
38. Chen, Y., Wang, W., Liu, Z., Yao, Y., Qian, L.: Synthesis of a novel flame retardant containing phosphazene and triazine groups and its enhanced charring effect in poly(lactic acid) resin. J. Appl. Polym. Sci. **134** (2017). https://doi.org/10.1002/app.44660
39. Usri, S.N.K., Jamain, Z., Makmud, M.Z.H.: A review on synthesis, structural, flame retardancy and dielectric properties of hexasubstituted cyclotriphosphazene. Polymers (Basel) **13** (2021). https://doi.org/10.3390/polym13172916
40. Uslu, A., Tümay, S.O., Yeşilot, S.: Fluorescent materials based on phosphazene derivatives and their applications: sensors and optoelectronic devices. J. Photochem. Photobiol. C Photochem. Rev. **53** (2022). https://doi.org/10.1016/j.jphotochemrev.2022.100553
41. Velencoso, M.M., Battig, A., Markwart, J.C., Schartel, B., Wurm, F.R.: Molecular firefighting—how modern phosphorus chemistry can help solve the challenge of flame retardancy. Angew. Chemie - Int. Ed. **57**, 10450–10467 (2018). https://doi.org/10.1002/anie.201711735
42. Gao, F., Tong, L., Fang, Z.: Effect of a novel phosphorous-nitrogen containing intumescent flame retardant on the fire retardancy and the thermal behaviour of poly(butylene terephthalate). Polym. Degrad. Stab. **91**, 1295–1299 (2006). https://doi.org/10.1016/j.polymdegradstab.2005.08.013
43. Wang, D., Liu, Q., Peng, X., Liu, C., Li, Z., Li, Z., Wang, R., Zheng, P., Zhang, H.: High-efficiency phosphorus/nitrogen-containing flame retardant on epoxy resin. Polym. Degrad. Stab. **187**, 109544 (2021). https://doi.org/10.1016/j.polymdegradstab.2021.109544
44. Ogueri, K.S., Ogueri, K.S., Allcock, H.R., Laurencin, C.T.: Polyphosphazene polymers: the next generation of biomaterials for regenerative engineering and therapeutic drug delivery. J. Vac. Sci. Technol. B **38**, 030801 (2020). https://doi.org/10.1116/6.0000055
45. Vanspeybroeck, R.S.L., Sewell, R.A., Thoen, J.A.: Horizontal burn test at different oxygen levels: a small scale screening test for flexible polyurethane foam with excellent correlation to

the BS 5852 part 2 source 5. J. Fire Sci. **8**, 421–454 (1990). https://doi.org/10.1177/073490419000800603
46. Sun, J., Wang, X., Wu, D.: Novel spirocyclic phosphazene-based epoxy resin for halogen-free fire resistance: synthesis, curing behaviors, and flammability characteristics. ACS Appl. Mater. Interfaces **4**, 4047–4061 (2012). https://doi.org/10.1021/am300843c
47. Yang, R., Wang, B., Han, X., Ma, B., Li, J.: Synthesis and characterization of flame retardant rigid polyurethane foam based on a reactive flame retardant containing phosphazene and cyclophosphonate. Polym. Degrad. Stab. **144**, 62–69 (2017). https://doi.org/10.1016/j.polymdegradstab.2017.08.008
48. Rangaraj, V.M., Singh, S., Devaraju, S., Wadi, V.S., Alhassan, S., Anjum, D.H., Mittal, V.: Polypropylene/phosphazene nanotube nanocomposites: thermal, mechanical, and flame retardation studies. J. Appl. Polym. Sci. **137**, 1–14 (2020). https://doi.org/10.1002/app.49525
49. Zhang, K., Wu, H., Wang, T., Yao, M., Xie, J., Jiao, Y.: Flame-retardant effect of cross-linked phosphazene derivatives and pentaerythritol derivatives on polypropylene. J. Therm. Anal. Calorim. **145**, 3067–3075 (2021). https://doi.org/10.1007/s10973-020-09898-z
50. Tao, K., Li, J., Xu, L., Zhao, X., Xue, L., Fan, X., Yan, Q.: A novel phosphazene cyclomatrix network polymer: design, synthesis and application in flame retardant polylactide. Polym. Degrad. Stab. **96**, 1248–1254 (2011). https://doi.org/10.1016/j.polymdegradstab.2011.04.011

Future Trends for Bio-Based Flame-Retardant Materials

T. C. Mokhena, K. Shingange, S. Mapukata, M. J. Mochane, E. R. Sadiku, S. S. Ray, L. Twala, B. Ntsendwana, M. Motaung, and N. Miya

Abstract Biomaterials offer a promising solution for reducing the fire hazards of different materials. These materials are derived from natural sources, which makes them renewable, biodegradable, and environmentally friendly. They also have unique properties that can enhance the fire resistance of polymers, such as surface modification, thermal stability, flame retardancy, and smoke suppression. However, there are still many challenges and opportunities for improving the performance and application of biomaterials as fire retardants for various materials. This chapter provides an overview of the current research, issues, and future directions of biomaterials as fire retardants for different materials.

Keywords Biomaterials · Flame retardants · Future trends

T. C. Mokhena (✉) · K. Shingange · S. Mapukata · L. Twala · B. Ntsendwana · M. Motaung · N. Miya
DSI/National Innovation Centre, Advanced Materials Division, Mintek, 200 Malibongwe Drive, Private Bag X 3015, Johannesburg, Gauteng, South Africa
e-mail: TebohoM@mintek.co.za

M. J. Mochane
Department of Life Sciences, Central University of Technology, Bloemfontein, Free State, South Africa

E. R. Sadiku
Division of Polymer Technology, Department of Chemical, Metallurgical and Materials Engineering & Institute of NanoEngineering Research, Tshwane University of Technology, Pretoria, South Africa

S. S. Ray
Centre for Nanostructures and Advanced Materials, DSI-CSIR Nanotechnology Innovation Centre, Council for Scientific and Industrial Research, Pretoria 0001, South Africa

T. C. Mokhena et al. (eds.), *Biomaterials as Green Flame Retardants*, Engineering Materials, https://doi.org/10.1007/978-981-97-6871-4_15

1 Introduction

The globe is now dealing with a number of issues associated to resources availability, exponential population growth, and the effects of climate change [1]. Rapid technological advancement, urbanization (which creates bigger, denser cities), and population aging all occur together. All nations in the world, regardless of wealth, face these difficulties, and their ability to overcome them varies. One important topic that is frequently disregarded or ignored in discussions and legislation related to today's societal problems is the effect of fire on human welfare, the environment, and the global economy [1].

However, such discussions are critical because the impact of fire is significant [2]. According to data from the World Health Organization (WHO), 180 000 people worldwide die from burns each year. The majority of this figure originates from countries with low to middle incomes, many of which have limited laws or policies pertaining to fire safety [3]. This is as a result of urbanization and population growth that increase the exposure an vulnerability of these people to fire, especially in informal settlements and slums. There is a huge possibility that the fire-related deaths can increase in the future, unless effective measures are taken in these areas to mitigate fire risk and improve fire safety by applying strict policies and regulations. The developed nations of the globe has the same problems [2, 3]. For example, based on a five-year average, 3500 people die from fire per year in Europe alone [3]. Furthermore, there are about 70,000 injuries annually, which is 20 times more than the number of fatalities.

Approximately 1% of the GDP of developed nations is estimated to be lost due to fire [4]. An estimated 329 billion USD, or 1.9% of the US GDP, was the cost of the fire in the US in 2014 [4, 5]. According to recent reports, there is currently a greater global loss due to fire hazards than in the past. According to Holzheu et al. [5], insurance claims related to wildfires worldwide totalled USD 14 billion in 2017. Fires and explosions that cause billions of US dollars' worth of damage account for about 59% of the 1807 business interruption claims filed worldwide (https://commercial.allianz.com/news-and-insights/reports/claims-in-focus.html (accessed on 12 December 2023)).

There is a need for current and future solution for the growing fire problem that negatively affect our well-being. Promoting dialogues within all associations and goverments to work together in bringing cheaper and realible ways of improving fire safety for the whole world. Despite the use of flame-retardants as another suitable solution to address the issues related to fire safety hazard, there are few issues that need to be addressed. For instance the synthetic FRs are associated with few issues that led to their use being banned in some countries. These materials are highly persistent with potential for bioaccumulation and thus hazardous to humans and animals. The commonly used synthetic FRs are halogen-based. The synthetic Brominated-FRs are most effective FRs often used in electronics, furniture and construction, but are associated with endocrine disruption and thyroid dysfunction and the release of toxic chemicals during combustion. There was shift towards organophosphate FRs as alternative to halogenated FRs but they were also found to be hazardous. These FRs are reported to be a threat to brain and bone health. Therefore, FRs that are eco-friendlier and pose no threat to human and animal health are of essence.

Thus, there has been widely embraced as a paradigm change to green FRs, especially from natural resources such as plants, animals and organisms. In comparison to traditional FRs, these materials are thought to be less hazardous and more environmentally friendly. These materials fall into one of three categories: bio-based coatings, bio-based additives, or bio-based polymers. Bio-based FRs were estimated to have a 1.2 billion USD worldwide market in 2021, and growth of 9.7% was projected between 2022 and 2030 (https://www.globenewswire.com/news-release/2022/07/27/2486572/28124/en/Global-Flame-Retardants-Markets-2021–2022-2030-Opportunities-in-the-Development-Innovation-and-Launch-of-Fire-Retardant-Products.html (accessed on 12 December 2023)). This expansion is fuelled by the development of the building and construction industry, the consumption of electrical and electronic products that require FRs, and the enforcement of fire and safety regulations. This chapter explores the book's perspective on bio-based FRs, which are alternatives to synthetic FRs that may pose environmental and health risks. It also examines the factors that drive the demand for bio-based FRs. It then discusses the challenges and opportunities of developing and implementing bio-based FRs.

2 Fire Safety and Perspective

Fire is a major factor that affects the well-being of people, the environment and the economy worldwide [1]. Fire can cause devastating losses of life and property, as well as contribute to greenhouse gas emissions and air pollution. Fire can also disrupt the functioning of critical infrastructure, social services and economic activities. Fire risk is influenced by many factors, such as access to resources, population growth, climate change, urbanization and aging [1]. These factors vary across different regions and countries, and pose different challenges for fire prevention, protection and response. Therefore, it is important to have a holistic and integrated approach to fire management that considers the diverse needs and capacities of different stakeholders and communities. There has been unprecedented interest in the control or managing fire hazards which contribute to loss of lives and possessions.

Fire and explosion are major threats to human lives and properties, as they account for 21% of the global insurance claims, according to a report by Allianz (https://commercial.allianz.com/news-and-insights/reports/claims-in-focus.html). Despite the fire prevention measures that have been implemented over the years, the cost of fire remains high and affects the economic, social and ecological aspects of society. Therefore, a holistic approach that considers all these dimensions is needed to address this issue (Fig. 1). One possible solution is to use natural materials for fire safety, as they have several advantages. They are cheaper because

Fig. 1 Holistic approach for fire cost

they are widely available, they are environmentally friendly and they can create new opportunities for businesses, especially in low and medium income countries.

2.1 Alginate

Alginate is a natural occurring polymer isolated from cell wall of seaweed. Its properties are governed by degree of polymerization, monomer composition, and sequence of the alginate molecules [6–9]. It has good chelation behaviour towards divalent metal ions [10]. The presence of these metals significantly improves the carbonization of alginate making it potential green FR. The most used technique to treat material with alginate is layer-by-layer self-assembly [6, 11]. This relies on the electrostatic complexation of alginate with other materials (e.g., chitosan). For instance, Nabipour et al.[6] introduced hydroxyapatite (HAP) into a mixture of chitosan and alginate and applied them onto polyurethane (PU). It was pointed out that the as-prepared complex improved flame retardant and smoke suppression properties of PU. After coating with 9 bilayers of HAP-alginate/HAP-chitosan the peak heat release decreased by ~ 78%, total heat release by ~ 57% and smoke production release by ~ 54%.

2.2 Cellulose

Cellulose is the most abundant polymer that is capable of reinforcing the commonly used polymeric materials [12, 13]. It has been reported to have excellent flame resistance performance resulting from production of stable compact char that inhibit the diffusion of flammable products and shielding the underlying materials from heat and air [12]. The FR property of cellulose can further be enhanced through modification of combining with other materials including phytic acid, gaunazole, inorganic nanomaterials, and other many more [13, 14]. Besides unique properties that makes cellulose ideal FR, the main challenges include the complex extraction processes, and the FR system often need the optimization in order to achieve the desire fire resistance performance [14]. This calls for researchers to come up with novel strategies to explore cellulose-based materials as a green FR and to overcome challenges associated with its application to improve flame retardancy of different materials (https://www.businessresearchinsights.com/market-reports/flame-retardant-regenerated-cellulose-fibre-market-108685).

2.3 Lignin

Lignin is natural polyhydroxy aromatic polymer that is often isolated from plant cell walls [15–17]. It is amorphous natural polymer second to cellulose when comes to availability. Its aromatic structure makes it suitable carbon source in IFR systems. The resultant carbon content is ~ 60% after combustion without any traceable toxic gases. A lot of researchers have exploited lignin as FR alone, or in the presence of other substances [15, 16, 18–20]. Some have also introduced other flame retardant elements such as nitrogen, phosphorus, and metal ions to further enhanced the fire resistance performance [20–24]. The use of lignin nanoparticles have been also reported in literature [24–27]. The complexity of lignin structure is one of the issues that makes its flame resistance mechanism difficult to explain. This need more research in order to understand the driving mechanism behind this for reassurance so that it reach commercialization. The interaction between lignin with other materials and its dispersion are some of the issues when come to using lignin in advanced applications. The colour of lignin also may limit its application where aesthetics are important.

2.4 Chitin

Chitin is the second most available natural polymer with huge potential to be used for flame resistance of different polymeric materials [28]. It can be isolated from marine waste products including shrimp and crab shells. It contains large amount of nitrogen that makes it release incombustible gases such as NH_3 during combustion [28]. Chitin can further be modified by introducing phosphorus groups that can enhance charring and generation of incombustible products in the gas phase [29, 30]. The combination of chitin with other biomaterials (phytic acid, lignin, alginate etc.) to form polyelectrolyte complexes serve as alternative to further improve its flame retardance property.

Yet the combination or coating nanomaterials onto chitin can also be used to improve the overall fire resistance performance. Nanomaterials such as metal oxides, halloysite nanotubes, metal–organic frameworks can be play a huge role as synergists to provide chitin with fire retardant property in both gaseous and condensed phase.

The future of chitin as a flame retardant appears bright because it's a biodegradable, renewable, and eco-friendly material that may offer various polymers other functions in addition to excellent flame retardant performance. The low solubility of chitin, the difficult synthesis of certain chitin derivatives, and the development of flame retardant systems that combine chitin with other materials are some of the obstacles that still need to be overcome. Because of this, further investigation and advancement are required to fully explore chitin's potential as a flame retardant and get over the current obstacles.

2.5 Chitosan

Chitosan is obtained from chitin and is often used various fields such as pharmaceutical, wastewater treatment, biomedical, food packaging and cosmetics [31, 32]. The worldwide chitosan market is anticipated to reach a valuation of USD 10.88 billion in 2022 and increase at a compound annual growth rate (CAGR) of 20.1% between 2023 and 2030 (https://www.futuremarketinsights.com/reports/chitosan-market (accessed on 12 December 2023); https://www.reportsanddata.com/report-detail/chitosan-market (accessed on 12 December 2023)). The expansion is ascribed to the rising demand for goods produced from natural sources and their extensive use in medicinal, cosmetic, and wastewater treatment applications. The demand for chitosan is fuelled by a number of characteristics, including its capacity to produce films and gels, biodegradability, biocompatibility, and antibacterial activity. However, a few obstacles that can prevent the chitosan industry from expanding are the cost and accessibility of raw materials, legal concerns, and competition from artificial substitutes (https://www.reportsanddata.com/report-detail/chitosan-market (accessed on 12 December 2023)).

The use of chitosan as green flame retardants for different polymeric materials is emerging fields. It has been used as FR for thremoplastics, thermosets, fabrics and foams [29, 32]. It is known to be a good carbon source in the condensed phase. This promotes charring during combustion to afford compact char for protection of the underlying from heat and air as well as by inhibiting diffusion of gaseous products. On the other hand, the presence of nitrogen elements along this polymer plays a crucial role in the gaseous phase by releasing some of gaseous products that consumes flammable radicals during combustion. Nonetheless, the modification of chitosan with inorganic oxyacids salts (eg stannate, molybdate, and phosphate) were found to significantly improve the flame retardant property of chitosan-based materials [29, 31]. It was reported that these inorganic oxyacids salts are essential in enhancing the smoke suppression properties and fire resistance performance. The use of chitosan as flame retardant is expected to grow because of its unique features such as biocompatibility, biodegradability, antimicrobial activity, and ease processability to form different products such as films, gels etc. the challenges that can limit the use of chitosan as FR in the future include availability, and prices associated with raw materials, regulatory issues and competition from synthetic alternatives [29, 31].

2.6 Protein

Protein have been explored as FR for various products due to its made-up composition [33–35]. It is capable of charring under combustion which can protect the underlying materials. Modification of protein include the inclusion of other groups such as nitrogen and phosphorus to promote char formation and release of inflammable products. It has large amount of hydroxyl and amino groups that are easily modified [33]. The compatibility with hydrophobic materials is one of the main challenges of using protein-based FRs. The modification of these FRs with functionalities that improve compatibility without affect their flame retardant property is essential to achieve the overall desired properties. The inclusion of other materials such as, halloysite nanotubes [35], ammonium polyphosphate [35], titanium oxide [36], magnesium hydroxide [34] were reported to improve the flame retardant property of proteins. Therefore, the main challenge of protein as FRs is their hydrophilic character and the need for grafting modification, mixing with other molecules and other modification to achieve the desired flame retardant property.

2.7 Phytic Acid (PA)

PA is biomaterial that contains large amount of phosphorus due to the presence of negatively charged phosphate linked to twelve hydroxyl groups [7, 37–40]. It has chelating capability that is often exploited to attach other constituents that improve the overall flame retardant property. For instance, the grafting PA onto kaolinite to afford intumescent flame retardant (IFR) resulted in better fire resistance composite materials [37]. PA generated phosphoric acid during combustion that serve as catalyst to form cross-linked polymeric material and compact carbon char that protected the underlying material. The LOI value reached 30.8% indicating that resultant material had better fire resistance performance [37]. In general, the high P/C ratio plays a critical role on the flame retardant property of PA [38]. The acidic nature of PA can degrade some of materials that are susceptible to acid and thus affect the overall performance of the resultant materials [38]. The high hydrophilicity of PA can affect its interaction with hydrophobic materials as well as increasing moisture absorption of those materials that are susceptible to moisture absorption (e.g. wood, plant fibres etc.) [38].

2.8 Deoxyribonucleic Acid (DNA)

DNA is composed of phosphoric acid, five carbon sugar, and nitrogenous base connections that can act as acid source, carbon source and incombustible products to play synergistic role in the gaseous and condensed phase [38, 41]. During combustion the nitrogenous base generate inflammable products such as NH_3 that can enhance fire resistance property in two ways, viz. diluting flammable products during combustion, and the expansion of carbon layer resulting from the interlocked gaseous products. The resulting carbon layer then serve as barrier to protect the underlying material. Up to date, the main issues of DNA is the high cost [38]. Researchers have to look into different ways that the cost associated with DNA because it has good properties such as excellent compatibility and reactivity with a balance composition (carbon, nitrogen, and phosphorus) that does not need further modification when compared to other biomaterials [41, 42]. Thus, it is often called 'all-in-one' natural intumescent flame retardant (IFR). The main source DNA is fishery waste, i.e. discarded from salmon milt and roe sac [43]. It is often used in cosmetics and electronics industries, but with alternative sources such as biomass can reduce the price and eliminate the competition with these industries towards DNA [43].

2.9 Tannic Acid

TA is a polyphenol often isolated from different plants such as taro pod, gallnut, pomegranate, and many others [33, 38]. It has been reported to form electrostatic complexation with metal ions and often used for adsorption of phosphorus. Its interaction with these substances has been exploited to fabricate FRs that can preserve wood [44]. Therefore, the fire resistance performance of TA is often dependent on its modification or the inclusion of other materials. For instance, Gao et al. [45] investigated the effect of ferric tannin as shell layer of microencapsulated IFR with ammonium polyphosphate as core. The latter served as cis source while ferric tannin, obtained from reaction between ferric chloride and tannin acid, as a carbonizing agent and smoke suppression agent and surface modifier. It was reported that the presence of the microencapsulated IFR improved the flame retardant property and smoke suppression of melamine-urea formaldehyde (MUF) resin. This was ascribed to large number of –OH groups from tannin facilitating acting as good carbonizing agents while ferric tannin serve as smoke inhibitor. The presence of ferric tannin in IFR system facilitated charring resulting in expanded carbon layer that effectively shielded the underlying materials from further combustion. The LOI value increase from ~ 20% to ~ 37% with UL-94 rating of V-0 indicating that resultant product is less flammable.

The main issue with the use of tannin is that it has to be modified or mixed with other materials to produce multicomponent flame retardant system in order to achieve the desired flame resistance performance. In addition, the acidic nature of tannic acid may be issues when treating materials that are susceptible to acid and thus affect their strength.

2.9.1 Phosphazenes

A family of flame retardants known as 'biobased phosphazenes' comes from naturally occurring materials such sugars, amino acids, and plant oils [46, 47]. Their typical structure is P = N-R, where an organic group with a bio-based basis is R. Typical stuctures of phosphazene are shown in Fig. 2.

Polyphosphazenes have been used as FRs for a wide variety of materials [46]. These materials have significant levels of nitrogen and phosphorus content, which helps to provide great flame retardancy. As a result, condensed phase phosphazenes facilitate the formation of dense and compact materials as a result of strong graphitization and aromatic crosslinking. The resultant char necessary suppresses the flow of heat, combustible compounds, and oxygen into and out of the sample while shielding the underlying components. Phosphazene breaks down in the gas phase to produce

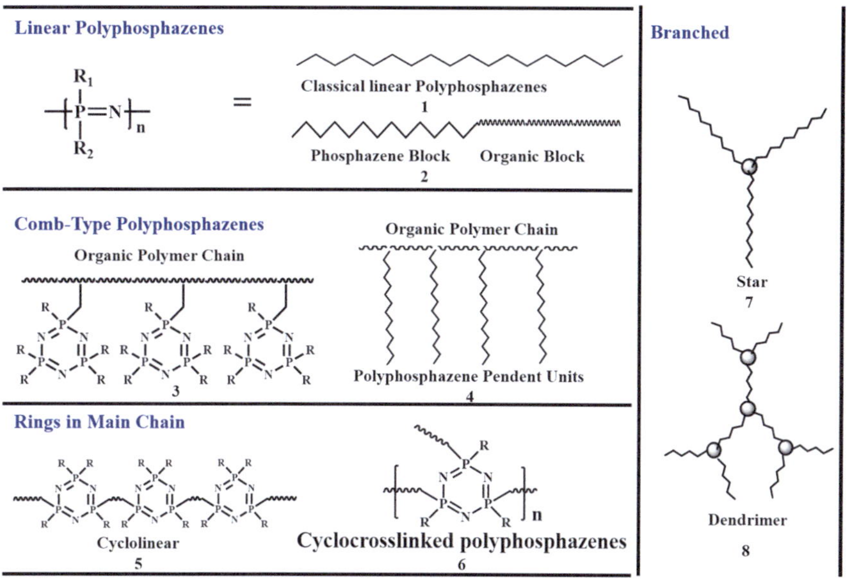

Fig. 2 Different polyphosphazenes structures [47]

products that react with the material to release inflammable gases, which lowers the concentration of flammable gases and oxygen and removes heat. Furthermore, the phosphorus oxygen radicals that are produced serve as scavengers, inhibiting the flame's chain thermal oxidation reaction and extinguishing it. Bio-based phosphazenes have a promising future as flame-retardants since they have a number of benefits over traditional ones like any other biobased materials. The complicated synthesis and purification of bio-based phosphazenes, the high cost and limited availability of bio-based precursors, and the optimization of flame retardant systems based on bio-based phosphazenes and their combinations with other materials are some of the issues that need to be addressed [46, 47].

2.9.2 Wood and Lignocellulosic Materials

Wood and lignocellulosic materials are widely used in construction, furniture, energy storage, and packaging applications [13, 48–52]. However, they are also highly flammable and need to be protected by flame-retardants [48–50]. There is a growing interest in developing eco-friendly and sustainable flame-retardants from wood and lignocellulosic materials [51]. These materials have inherent advantages such as renewability, biodegradability, low cost, and availability [50]. Moreover, they can provide synergistic effects with other flame retardants by enhancing the formation of char layers, which can act as physical barriers to heat and mass transfer during combustion. Some examples of wood and lignocellulosic materials that can be used as flame retardants are cellulose [13, 53], hemicellulose [13, 53, 54], lignin [13, 53], tannins [13], and nanocellulose [51, 52, 55, 56]. These materials can be applied to various substrates by different methods such as impregnation, coating, grafting, or blending [13]. Ghanadpour et al. [55] prepared phosphorylated cellulose nanofibrils using ammonium salt of phosphoric acid ($(NH_4)_2HPO_4$) in the presence of urea. The resulting paper sheet exhibited self-extinguishing properties despite being exposed to flame for 3 s (see Fig. 3) and did not ignite under heat flux of 35 kW m^{-2}. This was ascribed to the as-prepared paper sheet being able to form a thermally stable char that inhibit the production of flammable products and protecting underlying material from direct flame.

3 Market for Biomaterials as FRs

With increasing fire and safety restrictions from governments throughout the world enforcing stringent laws and standards on industry to lower fire risks and accidents flame retardant materials received attention as solution high demand. Since many petroleum-based or halogenated flame-retardants have been linked to toxicity, pollution, and health issues, more environmentally friendly and biodegradable bio-based alternatives are in high demand. With distinct molecular architectures, functions, and processes, scientists and industry participants are creating innovative and efficient

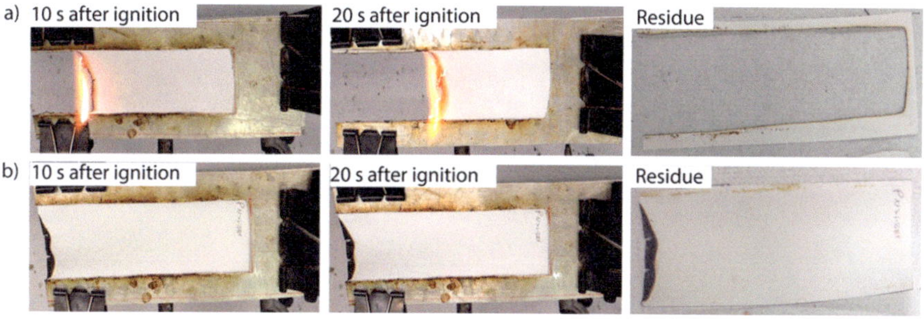

Fig. 3 Image of **a** neat paper and **b** phosphorylated paper sheet at fixed times of flammability test [55]

bio-based flame-retardants. Additionally, when combined with other materials or additives, several bio-based flame-retardants have synergistic effects that improve their efficacy and flame retardancy.

The market for flame-retardants was estimated to be worth USD 7.09 billion in 2019 and is expected to grow at a compound annual growth rate (CAGR) of 5.9% to reach USD 11.17 billion by 2027. It is anticipated that bio-based materials would garner a substantial portion of this market because of their benefits and potential. For instance, bio-based FRs were estimated to have a 1.2 billion USD worldwide market in 2021, and growth of 9.7% was projected between 2022 and 2030 (https://www.globenewswire.com/news-release/2022/07/27/2486572/28124/en/Global-Flame-Retardants-Markets-2021–2022-2030-Opportunities-in-the-Development-Innovation-and-Launch-of-Fire-Retardant-Products.html). This indicates that in future bio-based FRs will be centre of FR industry due to the benefits associated with these materials as well as their strict regulations for fire safety.

4 Application of Bio-Based Flame-Retardant Materials

Flame-retardants play a critical role on the to improve the fire resistance of various materials that are employed in different industries, such as furniture, textiles, packaging, construction, and automotive as illustrated in Fig. 4. Bio-based flame-retardants are a subject of ongoing research and development that promotes continuous innovation. Industrial manufacturers are able to use bio-based flame retardants without having to implement major changes to their production processes due to these materials can be integrated into already-existing manufacturing processes [1–4]. Though their price may be a barrier, producers may find that bio-based flame-retardant materials eventually become more affordable as a result of increasing demand and technological improvements. Manufacturers can meet regulatory standards, customer demands, and environmental sustainability objectives by

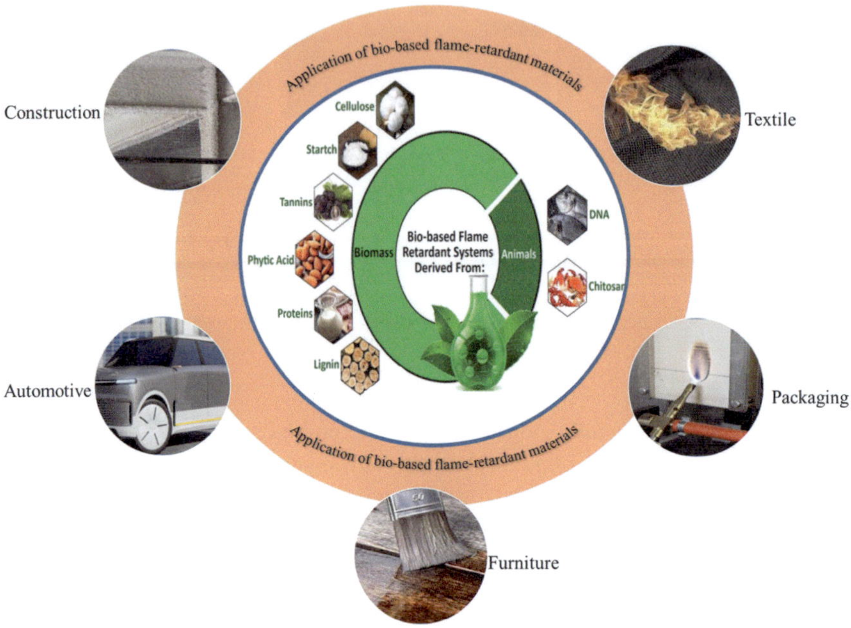

Fig. 4 Application of bio-based flame-retardant materials

implementing bio-based flame-retardant ingredients into their production processes [5–7].

The use of bio-based flame-retardant materials in industries is currently gaining attention as part of a broader movement toward sustainable and ecologically friendly techniques across a wide range of industries. Materials can have flame-retardants applied to them to lessen their flammability and delay the spread of fire. Strict laws are in place in several nations concerning the flammability of food packaging and furnishings. Manufacturers can comply with these rules and maintain environmental responsibility by using bio-based flame-retardants. When compared to conventional options, bio-based flame-retardants can have fire-retardant qualities that are on comparable levels with or even better. This guarantees that the furniture satisfies safety requirements without sacrificing functionality. Furniture and textile products can integrate bio-based additives, such as those made from natural fibers. Urbanization and industrialization have a significant impact on human activities, and the construction industry, which deals with infrastructure (streets, bridges, railroads, buildings, etc.), plays a major part in their development [1, 8]. However, as the global population grows annually, the strain from the environment, the economy, and society is also growing. According to research, construction materials is an essential component of the construction industry, require a huge amount of energy owing to the extraction of raw materials, manufacturing processing, and disposal of construction debris, as well as contributing the majority of greenhouse gas emissions [9]. Certain rules have been put in place in other countries to encourage the use of environmentally friendly building materials, with the aim of reducing the impact on the environment and optimizing the use of resources. Thus, the most effective strategy to achieve environmental sustainability is to use bio-based products in order to meet the restricted standards of low environmental impact, efficient resource usage, and economic efficiency. In order to meet the requirement for construction material waste disposal, biopolymers are among the materials that are meant to be employed as environmentally acceptable building materials. Sustainable hybrid materials are created by combining wood and natural fibers with other materials. In addition, the crop wastes could potentially used as a substitute for thermal insulation in building projects. Because these biobased materials such as polymeric materials, wood-based hybrids, and crop-based composites have a natural combustibility, they should be regarded as essential indicators when it comes to construction materials [1, 8, 10].

5 Outlook

Biomaterials are often used in various applications such as medicine, engineering, and biotechnology. One of the emerging uses of biomaterials is as flame-retardants, which are substances that can prevent or slow down the combustion of flammable materials. Flame-retardants are widely used in textiles, furniture, electronics, and other products to reduce the risk of fire hazards. However, many conventional flame-retardants are synthetic chemicals that can pose environmental and health problems. Therefore, there is a growing interest in developing biomaterials that can act as effective and eco-friendly flame-retardants. Some examples of biomaterials that have shown potential as flame-retardants are cellulose, chitin, lignin, proteins, and DNA. These biomaterials can either be used alone or in combination with other additives to enhance their flame-retardant properties. Biomaterials can offer several advantages over synthetic flame-retardants, such as biodegradability, biocompatibility, renewability, and low toxicity. However, there are also some challenges and limitations that need to be addressed before biomaterials can be widely adopted as viable flame-retardants. Some of these challenges include the availability and cost of raw materials, the scalability and stability of production processes, the compatibility and durability of biomaterials with different substrates, and the regulation and standardization of biomaterial-based flame-retardants. Therefore, more research and development are needed to explore the potential and feasibility of biomaterials as flame-retardants for various applications.

6 Conclusions

Most of biomaterials are capable to serve as FRs due to their capability of forming stable and compact char during combustion. This inhibits the diffusion of gaseous products from or into the material during combustion and protect the underlying material from heat and air and thus improve flame retardant property of the resultant material. In addition, the availability of accessible surface functionalities makes it easy for modification with other materials as well as exchange with other functionalities that are known to be highly inflammable. On the other hand, some of these materials (e.g. chitin, DNA and chitosan) have already equipped with functionalities that are inflammable which makes them the materials of interest when come to flame retardancy.

The biomaterials that contains phosphorus and carbon often generate phosphoric acid during combustion that facilitate the formation of char residue. The obtained char often protect the underlying materials from pyrolysis. The main issue is their acidic nature that usually decrease thermal stability of the resultant materials and may induce colour change. Research on the weakening acidic nature to avoid reduction of thermal stability of resultant materials without affecting the performance of the flame retardant is crucial for future application of such materials as alternative FRs of synthentic based FRs.

References

1. McNamee, M., Meacham, B., van Hees, P., Bisby, L., Chow, W.K., Coppalle, A., et al.: IAFSS agenda 2030 for a fire safe world. Fire Saf. J. **110**, 102889 (2019). https://doi.org/10.1016/j.firesaf.2019.102889
2. Evarts, B.: Fire loss in the United States during 2017. National Fire Protection Association 20 (2019)
3. Brushlinsky, N., Ahrens, M., Sokolov, S., Wagner, P.: World fire statistics. Center of fire statistics-2006-Report 10 (2016)
4. Zhuang, J., Payyappalli, V.M., Behrendt, A., Lukasiewicz, K.: Total cost of fire in the United States. Fire Protection Research Foundation Buffalo, NY, USA (2017)
5. Holzheu, T., Tamm, K., Lechner, R., Fan, I.: Profitability in non-life insurance: Mind the gap. Zurich: Swiss Reinsurance Company. Retrieved from (2018)
6. Nabipour, H., Wang, X., Song, L., Hu, Y.: A fully bio-based coating made from alginate, chitosan and hydroxyapatite for protecting flexible polyurethane foam from fire. Carbohyd. Polym. **246**, 116641 (2020). https://doi.org/10.1016/j.carbpol.2020.116641
7. Mokhena, T.C., Jacobs, N.V., Luyt, A.S.: Electrospun alginate nanofibres as potential biosorption agent of heavy metals in water treatment. Express Polym Lett **11**, 652–663 (2017). https://doi.org/10.3144/expresspolymlett.2017.63
8. Mokhena, T.C., Luyt, A.S.: Development of multifunctional nano/ultrafiltration membrane based on a chitosan thin film on alginate electrospun nanofibres. J. Clean. Prod. **156**, 470–479 (2017). https://doi.org/10.1016/j.jclepro.2017.04.073
9. Mokhena, T.C., Luyt, A.S.: Electrospun alginate nanofibres impregnated with silver nanoparticles: Preparation, morphology and antibacterial properties. Carbohyd. Polym. **165**, 304–312 (2017). https://doi.org/10.1016/j.carbpol.2017.02.068
10. Mokhena, T.C., Jacobs, N.V., Luyt, A.S.: Nanofibrous alginate membrane coated with cellulose nanowhiskers for water purification. Cellulose **25**(1), 417–427 (2018). https://doi.org/10.1007/s10570-017-1541-1
11. Pan, H., Wang, W., Shen, Q., Pan, Y., Song, L., Hu, Y., et al.: Fabrication of flame retardant coating on cotton fabric by alternate assembly of exfoliated layered double hydroxides and alginate. RSC Adv. **6**(113), 111950–111958 (2016). https://doi.org/10.1039/C6RA21804K

12. Mohamed, A.L., Hassabo, A.G.: Flame Retardant of Cellulosic Materials and Their Composites. In: Visakh, P.M., Arao, Y. (eds.) Flame Retardants: Polymer Blends, Composites and Nanocomposites, pp. 247–314. Springer International Publishing, Cham (2015)
13. Liang, Y., Jian, H., Deng, C., Xu, J., Liu, Y., Park, H., et al.: Research and application of biomass-based wood flame retardants: A review. Polymers (2023). https://doi.org/10.3390/polym15040950
14. Ding, H., Qiu, S., Wang, X., Song, L., Hu, Y.: Highly flame retardant, low thermally conducting, and hydrophobic phytic acid-guanazole-cellulose nanofiber composite foams. Cellulose **28**(15), 9769–9783 (2021). https://doi.org/10.1007/s10570-021-04159-0
15. Canetti, M., Bertini, F., De Chirico, A., Audisio, G.: Thermal degradation behaviour of isotactic polypropylene blended with lignin. Polym. Degrad. Stab. **91**(3), 494–498 (2006). https://doi.org/10.1016/j.polymdegradstab.2005.01.052
16. Gallina, G., Bravin, E., Badalucco, C., Audisio, G., Armanini, M., De Chirico, A., et al.: Application of cone calorimeter for the assessment of class of flame retardants for polypropylene. Fire Mater. **22**(1), 15–18 (1998). https://doi.org/10.1002/(SICI)1099-1018(199801/02)22:1%3c15::AID-FAM626%3e3.0.CO;2-3
17. Solihat, N.N., Hidayat, A.F., Taib, M.N.A.M., Hussin, M.H., Lee, S.H., Ghani, M.A.A., et al.: Recent developments in flame-retardant lignin-based biocomposite: Manufacturing, and characterization. J. Polym. Environ. **30**(11), 4517–4537 (2022). https://doi.org/10.1007/s10924-022-02494-2
18. Song, K., Ganguly, I., Eastin, I., Dichiara, A.B.: Lignin-modified carbon nanotube/graphene hybrid coating as efficient flame retardant. Int. J. Mol. Sci. (2017). https://doi.org/10.3390/ijms18112368
19. Song, K., Ganguly, I., Eastin, I., Dichiara, A.: High temperature and fire behavior of hydrothermally modified wood impregnated with carbon nanomaterials. J. Hazard. Mater. **384**, 121283 (2020). https://doi.org/10.1016/j.jhazmat.2019.121283
20. Lu, X., Lee, A.F., Gu, X.: Improving the flame retardancy of sustainable lignin-based epoxy resins using phosphorus/nitrogen treated cobalt metal-organic frameworks. Mater. Today Chem. **26**, 101184 (2022). https://doi.org/10.1016/j.mtchem.2022.101184
21. Fu, R.L., Cheng, X.S.: Synthesis and flame retardant properties of melamine modified EH lignin. Adv. Mater. Res. **236–238**, 482–485 (2011). https://doi.org/10.4028/www.scientific.net/AMR.236-238.482
22. Zhou, S., Tao, R., Dai, P., Luo, Z., He, M.: Two-step fabrication of lignin-based flame retardant for enhancing the thermal and fire retardancy properties of epoxy resin composites. Polym. Compos. **41**(5), 2025–2035 (2020). https://doi.org/10.1002/pc.25517
23. Liu, L., Qian, M., Song, Pa., Huang, G., Yu, Y., Fu, S.: Fabrication of Green Lignin-based Flame Retardants for Enhancing the Thermal and Fire Retardancy Properties of Polypropylene/Wood Composites. ACS Sustain. Chem. & Eng. **4**(4):2422–31 (2016). https://doi.org/10.1021/acssuschemeng.6b00112.
24. Yu, H., Zhan, W., Liu, Y.: Engineering lignin nanoparticles deposition on melamine sponge skeleton for absorbent and flame retardant materials. Waste Biomass Valorization **11**(8), 4561–4569 (2020). https://doi.org/10.1007/s12649-019-00763-1
25. He, T., Chen, F., Zhu, W., Yan, N.: Functionalized lignin nanoparticles for producing mechanically strong and tough flame-retardant polyurethane elastomers. Int. J. Biol. Macromol. **209**, 1339–1351 (2022). https://doi.org/10.1016/j.ijbiomac.2022.04.089
26. Chollet, B., Lopez-Cuesta, J.-M., Laoutid, F., Ferry, L.: Lignin nanoparticles as a promising way for enhancing lignin flame retardant effect in polylactide. Materials. (2019). https://doi.org/10.3390/ma12132132
27. Wang, X., Ji, S.-L., Wang, X.-Q., Bian, H.-Y., Lin, L.-R., Dai, H.-Q., et al.: Thermally conductive, super flexible and flame-retardant BN-OH/PVA composite film reinforced by lignin nanoparticles. J. Mater. Chem. C **7**(45), 14159–14169 (2019). https://doi.org/10.1039/C9TC04961D
28. Chen, F., Wang, C., Liu, X., Wang, Z., Fang, L., Zhang, Y.: Flame-retardant properties of chitin liquefaction-based polyurethane foam. J. Polym. Res. **30**(4), 143 (2023). https://doi.org/10.1007/s10965-023-03521-z

29. Malucelli, G.: Flame-retardant systems based on chitosan and its derivatives: State of the art and perspectives. Molecules (2020). https://doi.org/10.3390/molecules25184046
30. Riehle, F., Hoenders, D., Guo, J., Eckert, A., Ifuku, S., Walther, A.: Sustainable Chitin nanofibrils provide outstanding flame-retardant nanopapers. Biomacromol **20**(2), 1098–1108 (2019). https://doi.org/10.1021/acs.biomac.8b01766
31. Song, Q., Wu, H., Liu, H., Wang, T., Meng, W., Qu, H.: Chitosan-regulated inorganic oxyacid salt flame retardants: preparation and application in PVC composites. J. Therm. Anal. Calorim. **146**(4), 1629–1639 (2021). https://doi.org/10.1007/s10973-020-10170-7
32. Wang, G., Li, H., Shao, Y., Wu, X., He, Q., Niu, K.: Preparation method of biogenic phytic acid and chitosan-related epoxy resin flame retardant: A review. J. Polym. Res. **30**(7), 269 (2023). https://doi.org/10.1007/s10965-023-03580-2
33. Chen, X., Li, J., Pizzi, A., Fredon, E., Gerardin, C., Zhou, X., et al.: Tannin-furanic foams modified by soybean protein isolate (SPI) and industrial lignin substituting formaldehyde addition. Ind. Crops Prod. **168**, 113607 (2021). https://doi.org/10.1016/j.indcrop.2021.113607
34. Uddin, M., Kiviranta, K., Suvanto, S., Alvila, L., Leskinen, J., Lappalainen, R., et al.: Casein-magnesium composite as an intumescent fire retardant coating for wood. Fire Saf. J. **112**, 102943 (2020). https://doi.org/10.1016/j.firesaf.2019.102943
35. Wang, Y., Liu, C., Lai, J., Lu, C., Wu, X., Cai, Y., et al.: Soy protein and halloysite nanotubes-assisted preparation of environmentally friendly intumescent flame retardant for poly(butylene succinate). Polym. Testing **81**, 106174 (2020). https://doi.org/10.1016/j.polymertesting.2019.106174
36. Ortelli, S., Malucelli, G., Cuttica, F., Blosi, M., Zanoni, I., Costa, A.L.: Coatings made of proteins adsorbed on TiO2 nanoparticles: a new flame retardant approach for cotton fabrics. Cellulose **25**(4), 2755–2765 (2018). https://doi.org/10.1007/s10570-018-1745-z
37. Zhang, H., Meng, D., Wang, W., Li, H., Gu, X., Zhang, S., et al.: Fabrication of phytic acid embellished kaolinite and its effect on the flame retardancy and thermal stability of ethylene vinyl acetate composites. J. Appl. Polym. Sci. **138**(46), 51364 (2021). https://doi.org/10.1002/app.51364
38. Xinsheng, H., Yang, G., Zhang, Q.: Application status and development prospects of bio-based flame retardants in packaging materials. Eur. J. Wood Wood Prod. **81**(6), 1337–1357 (2023). https://doi.org/10.1007/s00107-023-01977-w
39. Mokhena, T.C., Sadiku, E.R., Ray, S.S., Mochane, M.J., Matabola, K.P., Motloung, M.: Flame retardancy efficacy of phytic acid: An overview. J. Appl. Polym. Sci. **139**(27), e52495 (2022). https://doi.org/10.1002/app.52495
40. Mokhena, T.C., Jacobs, V., Luyt, A.: A review on electrospun bio-based polymers for water treatment (2015)
41. Alongi, J., Carletto, R.A., Di Blasio, A., Carosio, F., Bosco, F., Malucelli, G.: DNA: a novel, green, natural flame retardant and suppressant for cotton. J. Mater. Chem. A **1**(15), 4779–4785 (2013). https://doi.org/10.1039/C3TA00107E
42. Alongi, J., Di Blasio, A., Milnes, J., Malucelli, G., Bourbigot, S., Kandola, B., et al.: Thermal degradation of DNA, an all-in-one natural intumescent flame retardant. Polym. Degrad. Stab. **113**, 110–118 (2015). https://doi.org/10.1016/j.polymdegradstab.2014.11.001
43. Wang, L., Yoshida, J., Ogata, N., Sasaki, S., Kajiyama, T.: Self-assembled supramolecular films derived from marine deoxyribonucleic acid (DNA)−cationic surfactant complexes: Large-scale preparation and optical and thermal properties. Chem. Mater. **13**(4), 1273–1281 (2001). https://doi.org/10.1021/cm000869g
44. Tondi, G., Wieland, S., Wimmer, T., Thevenon, M.F., Pizzi, A., Petutschnigg, A.: Tannin-boron preservatives for wood buildings: mechanical and fire properties. Eur. J. Wood Wood Prod. **70**(5), 689–696 (2012). https://doi.org/10.1007/s00107-012-0603-1
45. Gao, M., Chen, S., Wang, H., Chai, Z.: Design, preparation, and application of a novel, microencapsulated, intumescent flame-retard.-Based mimicking mussel. ACS Omega **3**(6), 6888–6894 (2018). https://doi.org/10.1021/acsomega.8b00364
46. Ma, H.-X., Li, J.-J., Qiu, J.-J., Liu, Y., Liu, C.-M.: Renewable cardanol-based star-shaped prepolymer containing a phosphazene core as a potential biobased green fire-retardant coating.

ACS Sustain. Chem. & Eng. **5**(1), 350–359 (2017). https://doi.org/10.1021/acssuschemeng.6b01714
47. Deniz, A., Zaytoun, N., Hetjens, L., Pich, A.: Polyphosphazene-tannic acid colloids as building blocks for bio-based flame-retardant coatings. ACS Appl. Polym. Mater. **2**(12), 5345–5351 (2020). https://doi.org/10.1021/acsapm.0c00574
48. Renner, J.S., Mensah, R.A., Jiang, L., Xu, Q., Das, O., Berto, F.: Fire behavior of wood-based composite materials. Polymers (2021). https://doi.org/10.3390/polym13244352
49. Zhang, Z.X., Zhang, J., Lu, B.-X., Xin, Z.X., Kang, C.K., Kim, J.K.: Effect of flame retardants on mechanical properties, flammability and foamability of PP/wood–fiber composites. Compos. B Eng. **43**(2), 150–158 (2012). https://doi.org/10.1016/j.compositesb.2011.06.020
50. Mokhena, T.C., Sadiku, E.R., Mochane, M.J., Ray, S.S.: Mechanical properties of fire retardant wood-plastic composites: A review. Express Polym Lett **15**, 744–780 (2021). https://doi.org/10.3144/expresspolymlett.2021.61
51. Aziam, H., Ouarga, A., Ettalibi, O., Shanmukaraj, D., Noukrati, H., Sehaqui, H., et al.: Phosphorylated cellulose nanofiber as sustainable organic filler and potential flame-retardant for all-solid-state lithium batteries. J. Energy Storage **62**, 106838 (2023). https://doi.org/10.1016/j.est.2023.106838
52. Gou, J., Liu, W., Tang, A., Xie, H.: A phosphorylated nanocellulose/hydroxypropyl methylcellulose composite matrix: A biodegradable, flame-retardant and self-standing gel polymer electrolyte towards eco-friendly and high safety lithium ion batteries. Eur. Polymer J. **158**, 110703 (2021). https://doi.org/10.1016/j.eurpolymj.2021.110703
53. Nishita, R., Kuroda, K., Ota, S., Endo, T., Suzuki, S., Ninomiya, K., et al.: Flame-retardant thermoplastics derived from plant cell wall polymers by single ionic liquid substitution. New J. Chem. **43**(5), 2057–2064 (2019). https://doi.org/10.1039/C8NJ04797A
54. Huang, C., Fang, G., Tao, Y., Meng, X., Lin, Y., Bhagia, S., et al.: Nacre-inspired hemicelluloses paper with fire retardant and gas barrier properties by self-assembly with bentonite nanosheets. Carbohyd. Polym. **225**, 115219 (2019). https://doi.org/10.1016/j.carbpol.2019.115219
55. Ghanadpour, M., Carosio, F., Larsson, P.T., Wågberg, L.: Phosphorylated cellulose nanofibrils: A renewable nanomaterial for the preparation of intrinsically flame-retardant materials. Biomacromol **16**(10), 3399–3410 (2015). https://doi.org/10.1021/acs.biomac.5b01117
56. Khakalo, A., Jaiswal, A.K., Kumar, V., Gestranius, M., Kangas, H., Tammelin, T.: Production of high-solid-content fire-retardant phosphorylated cellulose microfibrils. ACS Sustain. Chem. & Eng. **9**(36), 12365–12375 (2021). https://doi.org/10.1021/acssuschemeng.1c04403

MIX
Papier aus verantwortungsvollen Quellen
Paper from responsible sources
FSC® C105338

If you have any concerns about our products,
you can contact us on
ProductSafety@springernature.com

In case Publisher is established outside the EU,
the EU authorized representative is:
Springer Nature Customer Service Center GmbH
Europaplatz 3, 69115 Heidelberg, Germany

Printed by Libri Plureos GmbH
in Hamburg, Germany